FLORA OF BEDFORDSHIRE

GREAT EARTH-NUT (*Bunium bulbocastanum*)

FLORA

OF

BEDFORDSHIRE

BY

JOHN G. DONY

THE CORPORATION OF LUTON
MUSEUM AND ART GALLERY

1953

First published 1953

Sold at the Luton Museum
Wardown Park, Luton
Price £2 2s.

Printed by
HENRY BURT & SON LIMITED
MILL STREET   ·   BEDFORD

Made in Great Britain

To the Memory

of

CHARLES ABBOT

and

JAMES SAUNDERS

CONTENTS

ILLUSTRATIONS

9

FOREWORD

IT IS A pleasure to contribute a foreword to a Flora of Bedfordshire, since the youthful memory of the lack of a modern account of the wild flowers of that county is still vivid. It is true that Bedfordshire was one of the earliest areas to have a county flora, for Abbot's *Flora Bedfordiensis* had already been published by the end of the eighteenth century. Since that time, however, the contributions of various botanists and the activities of the local naturalists have greatly augmented our knowledge. James Saunders produced his Flora of the county in 1911, but this was meagre in respect of localities and much has been done since then, especially by Dr. Dony and others, to enrich our knowledge, not only of the flora as it exists today, but of its relations with that of the past. I can recall as a young man some forty-five years ago being shown by Saunders a specimen of the Club Moss (*Lycopodium clavatum*) that he believed to be the last specimen found growing in the south of the county, and I have myself witnessed the decline and disappearance of the Mountain Cat's-foot (*Antennaria dioica*) on Dunstable Downs. The diminutions and losses, especially owing to drainage and industrialization, have been very appreciable, whilst the plant introductions, whether by the direct activities of man or due to indirect human influence or other causes, are no less striking. Both losses and accretions are of interest equally to students of ecology and distribution.

Despite all the changes in modern times Bedfordshire is a county that retains many features of natural beauty. It is rich in beech woods on the chalk and in oak woods on the clay. The shady beech woods, with their saprophytic flowering plants such as the Yellow Bird's-nest and the Bird's-nest Orchid in the summer, and their wealth of fungi in the autumn, are as familiar here as elsewhere. The beechwood margin and the chalk scrub are amongst the naturalist's most cherished habitats, with a wealth of flowers only equalled by their rich and varied fauna, where such plants as the Wild Columbine, the Fly Orchid and the beautiful Crested Cow-wheat have their home.

Bedfordshire is a county characterized also by its rolling downs, gay in summer with their carpet of wild Thyme and Rock Roses, above which rise the delicate cream inflorescences of the Dropwort and the dainty blue flowers of the Small Scabious. Here the diligent observer may find the Pasque Flower's purple beauty in the spring and the drifts of Pyramidal and Fragrant Orchids in the summer; much more locally the Spotted Cat's-ear, the Great Earth-nut, *Seseli libanotis*, the Spring Gentian, the Greater Felwort and the Dwarf Orchid delight the knowledgeable botanist. The stream-sides are, like the watersides in other areas, gay with Purple Loosestrife and the heavily scented cream-coloured flowers of the Meadow-sweet, the brilliant yellow of the Yellow-flag with the bright blue of the Skull-cap, and in the shallow water we look for the charming pink umbels of the Flowering Rush standing erect amidst the triangular leaves and the handsome inflorescences of the Arrow-head with its tiers of three-petalled flowers. It is in the water-meadows that we find not only the display of blue-flowered Meadow Cranesbill but very locally the chequered bells of the Snake's-head Fritillary.

Although during the past century many of the flowers, especially those of marsh and bog, have sadly diminished or been lost, the county as a whole still presents a considerable diversity of habitats and a varied flora that embraces a large range of species, all the more interesting geographically since these include some members of the continental component that here find their western limits, and others, oceanic species, which have their eastern limits in this county. Bedfordshire furnishes us with few species that occur nowhere else in Britain, but it provides many which are local and thus of particular interest to those concerned with the relation of plant life to its environment.

If we may judge by the comparative paucity of early records for Bedfordshire plants, the county was not a favourite hunting-ground for the collector, not so much, we may well believe, because of any poverty, as because of its former relative inaccessibility. Nevertheless we find that in the sixteenth century Gerard recorded in his famous Herbal the occurrence of Elecampane 'in fieldes as you go from Dunstable to Puddle Hill', reminding us that Saxon herbal lore prized this plant for its supposed medicinal virtues. It was from the famous Barton Hills that the first British records were made by Gerard of the wild Sanfoin and the Purple Milk-vetch, also in the sixteenth century.

Probably few who consult county floras, apart from those directly or indirectly concerned in their production, are conscious of the vast amount of meticulous care that is involved in the assemblage and collation of numerous records, the checking of those that are doubtful, the elimination of the erroneous ones, and particularly in the field, in observations on occurrence and critical examination of specimens, all essential to a successful outcome of the final achievement.

Noteworthy features of this Flora are the records by Dr. Dony of the habitat conditions and species which occur in nearly a hundred selected localities. These help to provide a conspectus of the ecological conditions and floristic composition of typical environments which will enable the reader to form a picture of the communal plant relationships and give to the data concerning frequency and abundance an objective reality that is too often lacking.

We honour many for their study of the Bedfordshire flora, amongst whom the chief were Charles Abbot (1761–1817), W. Crouch (1817–46), J. McLaren (1815–88), W. Hillhouse (1850–1910), James Saunders (1839–1926) and J. E. Little (1861–1936). To this distinguished company must now be added the name of the present author of a Flora that takes a worthy place amongst its predecessors and contemporaries.   EDWARD J. SALISBURY.

PREFACE

JAMES SAUNDERS published his *Field Flowers of Bedfordshire* when I was eleven years old, and it was the first serious book I possessed. In later years I knew Saunders well and found that he was aware of the limitations of his small book. Saunders strove in his later years to get a museum established in Luton, as indeed one was only two years after he died. It is most appropriate that this fuller account of the Bedfordshire flora should be published by that museum, and it is my first duty to thank Luton Borough Council for undertaking it. Other local authorities might well follow its lead and publish scientific work dealing with their neighbourhoods.

I am grateful to the Royal Society for a generous grant of £300, which has enabled us to print a larger book than we had originally intended and to publish it at less than its cost of production.

Publication has thrown much additional work on the already overburdened curator of the museum, Mr. C. E. Freeman, which he has borne cheerfully, and his assistance to me throughout has been invaluable. Over a number of years the members of the museum staff have been very helpful, and I am indebted to Miss B. M. Vizard for her assistance with the maps.

In its general arrangement the Flora follows that adopted by convention in local floras in this country, as there seemed little to gain in a wide departure from custom. The section on the Mosses is more detailed than is usual in county floras and I have been most fortunate in having as competent a bryologist as Mr. T. Laflin to write it. Mr. Laflin must also be thanked for the section on Soils. The section on the *Fungi*, prepared by Mr. D. A. Reid, has been included only with the help of the Royal Society grant. The distribution of the species in this group is not fully known, as it has received the attention of few botanists, and the arrangement within this section differs from that adopted in the rest of the Flora. It takes its place somewhat appropriately as an appendix to the main work. Mr. P. Taylor, in addition to preparing the section on the *Hepaticae*, has revised my section on the *Pteridophyta*.

The relegation of the less permanent members of the *Spermatophyta* to a section at the end of the Flora was done with some reluctance when it was realized that the book would be cumbersome. It is hoped, however, that some useful purposes have been served by the division.

It has been my good fortune to be able to visit frequently the British Museum (Natural History) and the Royal Botanic Gardens, Kew. Mr. Laflin joins me in thanking the members of the scientific and library staffs of these and other institutions for their ready assistance. Mr. J. E. Dandy of the British Museum (Natural History) has given me considerable advice on the arrangement of the Flora and on problems of nomenclature in the *Spermatophyta*.

My revision of the flora has taken me into every parish in the county many times and I am grateful to the many landowners who have given me access to their properties. I must thank especially the Railway Executive, as much of the original work I have done has resulted from a study of railway property. It is my sincere hope that other botanists, wishing to visit places in which my co-workers and I have found interesting plants, will first seek permission from the owners.

An annotated copy of the Flora will be kept at Luton Museum, and in it will be incorporated additional notes and records of plants found in the county and the adjoining counties. I would welcome additional records, comments and corrections, and will assist local botanists at any time in naming their specimens or in furthering their studies.

J. G. Dony

Luton Museum

July 1953

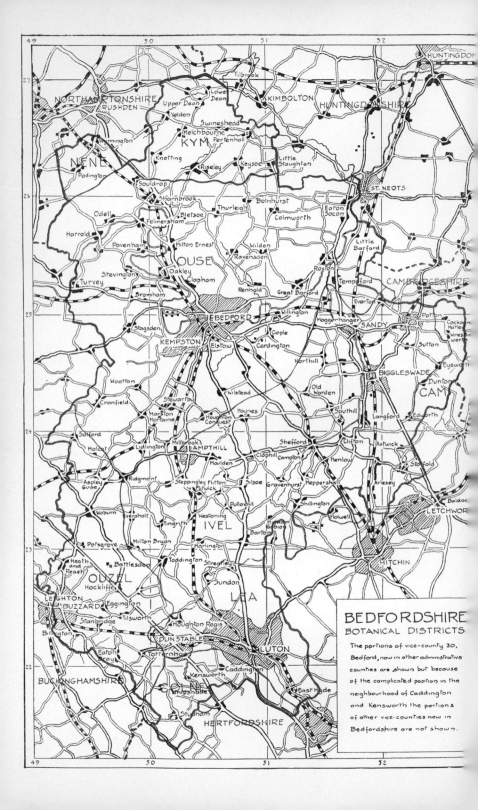

BEDFORDSHIRE
BOTANICAL DISTRICTS

The portions of vice-county 30,
Bedford, now in other administrative
counties are shown but because
of the complicated position in the
neighbourhood of Caddington
and Kensworth the portions
of other vice-counties now in
Bedfordshire are not shown.

THE STUDY OF BEDFORDSHIRE BOTANY

THE history of botany in Bedfordshire reflects in miniature the story of its study in the country as a whole. It is that of a few enthusiastic workers caught in the stream of periods of national enthusiasm and drawing around them bands of less active observers. There are many periods of comparative inactivity during which many records, now lost, must have been made, and there is little doubt that only part of the full story is known.

There are many references to plants in medieval documents, and place and field names are often based on plants; but, interesting as these may be, there is always some doubt as to the identity of the plants concerned. Most of the early records of plants for the county were made by travellers, and as is almost to be expected, many of these were made for the chalk hills, for not only do the more interesting plants grow there, but it was also there that passengers would have had to leave their coaches and walk. The very first records are in Gerard's *Herbal* (1597), which gives Purple Milk Vetch (*Astragalus danicus*), Sainfoin (*Onobrychis viciifolia*) and Elecampane (*Inula helenium*), all from the Chalk. Parkinson's *Theatrum Botanicum* (1640) records two other chalk-hill plants, Felwort (*Gentiana amarella*) and Grass of Parnassus (*Parnassia palustris*). At about the same time Francis Taverner noted Danewort (*Sambucus ebulus*) at Barton.

Notwithstanding the close proximity of Cambridge there is no evidence that John Ray, 'the father of British botany', ever visited Bedfordshire. He records, however, Maiden Pink (*Dianthus deltoides*) in *Synopsis Stirpium Britannicum* (1670), and Black Currant (*Ribes nigrum*), Greater Burnet Saxifrage (*Pimpinella major*) and Crested Cow Wheat (*Melampyrum cristatum*) in the second edition of the *Synopsis* (1696). The Small-leaved Lime was reported to him by John Aubrey from Chicksands, where it still grows, and Gibson's *Magna Britannia*, in a list compiled by Ray, adds Woad (*Isatis tinctoria*).

The *Synopsis* was re-edited by Dillenius in 1724 and included a number of records, mainly from the Markyate area, supplied by Thomas Knowlton (1692–1782). Knowlton was later gardener to James Sherard at Eltham and was a competent botanist: it would be helpful to know more of his early connection with Markyate. His only certain Bedfordshire record is Blue Mountain Anemone (*A. appenina*), probably a garden escape, from Luton Hoo. The same volume contains records also of Buckthorn (*Rhamnus catharticus*) and Marsh Violet (*Viola palustris*).

Bedfordshire was soon to have a distinguished but less fruitful botanical connection in John Stuart, 3rd Earl of Bute (1713–1792). After a chequered political career—he was probably the most unpopular Prime Minister the country ever had—he lived at Luton Hoo, where the gardens under his direction became famous for their variety of cultivated plants. He was well in the forefront of botanical controversy of his day and opposed the Linnean system of classification, which he countered with one of his own. His *Tabular Distribution of British Plants* (1780) and *Botanical Tables* (1785) are interesting works, the latter beautifully illustrated by Philip Miller, and limited to only twelve copies. The hope that some further mention might be made of Bedfordshire plants is not realized for only *Anemone appenina*, picked up from Dillenius, is recorded. In passing it is interesting to note that it was Bute's influence with the Princess of Wales, George III's mother, that was largely responsible for the founding of the Royal Botanic Gardens at Kew and the appointment of John Hill as its first director.

In the meantime a few odd records appear in miscellaneous works. *The Blecheley Diary of William Cole* (1765–7) mentions Lily of the Valley (*Convallaria majalis*) at Aspley. It is noted in most lists subsequently. Hudson's record of *Seseli libanotis* in *Flora Anglica* (1762), 'inter St. Albans et Stoney Stratford', is claimed for Hertfordshire, Buckinghamshire and Bedfordshire. If the record is a reliable one it would most likely have referred to the chalk hills near Dunstable in Bedfordshire. It is from there that Box (*Buxus sempervirens*) was recorded, on the authority of T. J. Woodward, in Stokes' edition of Withering's *Natural Arrangement* (1787).

The second edition of *Flora Anglica* (1778) adds Horseshoe Vetch (*Hippocrepis comosa*), Sulphur Clover (*Trifolium ochroleucon*), Yellow Bird's-nest (*Monotropa hypopitys*) and White Helleborine (*Cephalanthera damasonium*). Additions of a surprising nature come from a list provided by O. St. John Cooper, vicar of Podington, in notes on that parish in *Collections towards the history and antiquities of Bedfordshire* (John Nichols, 1783). It includes Vervain (*Verbena officinalis*), Shepherd's Purse (*Capsella bursa-pastoris*), Melilot (*Melilotus sp.*) and Wall Pellitory (*Parietaria diffusa*).

The study of Bedfordshire botany had made little progress by 1788; less than 30 species had been recorded and these for the most part were unusual plants or comparative rarities. It was, however, in this year that botanical study, both in Britain and in the county, gained fresh inspiration. J. E. Smith (1759–1828) founded the Linnean Society, and Charles Abbot (1761–1817) came to Bedfordshire. Abbot, an early member of the Linnean Society (1793), corresponded frequently with J. E. Smith and J. Sowerby, when the latter began in collaboration with Smith the incomparable *English Botany* in 1790. That Abbot was influenced by Smith there can be no doubt, but Smith gives cross-references to Abbot throughout *Flora Britannica* (1800).

Abbot, who was probably born at Blandford, was educated at Winchester and New College, Oxford. In addition to holding a post as usher or undermaster at Bedford Grammar School he was vicar of Oakley Reynes, vicar of Goldington, curate of St. Mary's and for a time of St. Paul's, Bedford. His greatest interest in life was field botany, but this did not diminish his other interests in entomology, numismatics and archaeology. As early as 1795 he submitted to the Linnean Society a manuscript, *Plantae Bedfordiensis*, containing a list of 200 less common plants of the county. This was followed three years later with a published work, *Flora Bedfordiensis*. The county thus became the first, apart from Cambridgeshire and Oxfordshire, to boast a published account of its flora. It was a model, and it was to be many years before any county had a better.

The flora lists 748 flowering plants and vascular cryptogams and 577 other species. In each case a brief description of the species is given and illustrations are cited. Habitats and notes on frequency follow, with stations for the rarer species. His own observations are added and still make interesting reading: for instance he writes of Smooth Cat's-ear (*Hypochoeris glabra*), 'cup smooth, purple at the point, flowers not bigger than a silver penny'. The thoroughness of the flora is continued in his own annotated copy, the most precious relic of the study of the county flora. This, with the Smith correspondence preserved at the Linnean Society Library, carries the story on to his death.

His herbarium—or should it more correctly be called Mrs. Abbot's?—is disappointing, for the beautifully mounted and still well-preserved specimens are not annotated in any way and it is impossible to say which, if any, were collected in the county. With his knowledge of the flora of the county his services were in great demand. A manuscript sent to Lysons for his *Magna Britannia* (1806) is still preserved, as is a useful account of the weeds of the county sent to Thomas Batchelor for his *General View of Agriculture in the County of Bedford* (1808).

A fuller account of Abbot's life and botanical work, of his part in the controversy regarding the Bedford willow, and of his entomological discoveries, has been told elsewhere. The story of other controversies, for instance his efforts to save the Harpur Trust for its original educational purpose, has yet to be told.

Abbot drew around him a circle of interested people; of some of these a certain amount is known. Thomas Martyn (1735–1825), having followed his father, John, in a distinguished professorship at Cambridge, became rector of Pertenhall in 1798. His many works contain no references to Bedfordshire plants and his correspondence with J. E. Smith is equally barren. He revised for Abbot the section of his flora dealing with the flowering plants, but still no records appear. Abbot's annotated copy does record Man Orchid (*Aceras anthropophora*) in Martyn's name from Pertenhall, but this was afterwards erased. T. Orlebar Marsh (1749–1831), who was born and died at Felmersham, was for a time vicar of Stevington and became a Fellow of the Linnean Society in 1797. He sent specimens to Sowerby, and Meadow Sage (*Salvia pratensis*) was recorded in his name in the flora. The best botanist among Abbot's Bedfordshire contemporaries was probably John Hemsted (1746–1824). He was born at Lynton, educated at Cambridge and eventually became vicar of St. Paul's, Bedford, and died at Bedford. He also corresponded with Smith, and in his name Spotted Medick (*Medicago arabica*) and *Stachys germanica* were recorded.

Of Abbot's other contemporaries less is known. Thomas Vaux of Bedford was of a Whipsnade family, and Abbot refers to him as 'my late regretted friend' and acknowledges the great use he made of his library. He apparently found Bladderwort (*Utricularia vulgaris*) and *Hypericum dubium*. Abbot also had the use of the library of the Countess of Ossory and to her should be attributed the discovery of the Grass Pea (*Lathyrus nissolia*). Joseph Sibley apparently lived in or near the south of the county and is referred to by Smith as 'a gentleman who has paid much attention to the orchis tribe'. He discovered Frog Orchid (*Coeloglossum viride*) and Ground Pine (*Ajuga chamaepitys*) in the county. D. Jenks, the recorder of *Tulipa sylvestris*, was a rector of Little Gaddesden.

Of the others nothing is known except the bare details given in the flora. James Payne, who added *Caucalis daucoides* from Oakley, was a herbalist of that village, and John Payne, who added Caraway (*Carum carvi*), was his brother. George Dixon of Eversholt was responsible for the record of Royal Fern (*Osmunda regalis*), J. Parker M.D. (d. 1809) for Golden Rod (*Solidago virgaurea*), Mr. Fowler for a station for Fly Orchid (*Ophrys insectifera*), P. Walker for Spider Orchid (*Ophrys sphegodes*) from Southill, no doubt in error, and Mr. Rugely, probably of Potton, for Bloody Cranesbill (*Geranium sanguineum*).

Abbot always hoped to add something to the British fauna or flora. He was successful in the former with the discovery of the Chequered Skipper butterfly, but less so in the latter. He came the nearest to success with the Bedford Willow (*Salix* × *russelliana*). His insistence that this was a new willow and should be named in honour of the Russell family is one of the more interesting things that emerge from the Smith correspondence.

That there was need to honour the Russell family there can be no doubt. When John Russell (1766–1839) succeeded his brother and became 6th Duke of Bedford in 1802, British botany gained one of its greatest patrons. The Woburn series of publications which he sponsored were devoted mainly to horticulture, but some had a distinct bearing on the native flora. Foremost was *Hortus Gramineus Woburnensis* (1816) by George Sinclair (1786–1834). Sinclair was born at Mellerstain, Berwick, became gardener to the Duke of Bedford, and died at Deptford. His work, which ran into many editions, was devoted to fodder plants generally and contains some records for the county. The only actual addition to the county flora was Upright Brome Grass (*Bromus erectus*). Another contributor to the series was James Forbes (1773–1861). His *Salictum Woburnense* (1829) recalls the Bedford Willow controversy but contains no new Bedfordshire records. In *Pinetum Woburnense* (1839) he describes the still unique Woburn pinetum which, while not concerned with the native flora, is of interest to all who have a regard for the plants of the county. Forbes died at Woburn Abbey.

Travellers still observed plants when passing through the county. The most distinguished of these recorders in this period was Samuel Goodenough (1743–1827), who added Field Fleabane (*Senecio integrifolius*)—another chalk-hill species. A few records for the county appeared also in the *Botanist's Guide* (1805), in which, on the authority of Sir T. G. Cullum, *Ornithogalum nutans* was added.

Interest was next shown in the flora of the county from outside. Bedfordshire has a long common boundary with Hertfordshire; Hitchin is almost on this boundary and it was here that Isaac Brown (1803–1895), a Quaker, kept a school from 1838 until it was burned down in 1845. In the days before inspectors and directors of education were known he was able to be in advance of his day, or ours, in his practice of field studies. It produced a well-repaid dividend in his scholars—the Ransoms, Joseph Pollard (*q.v.*), Joseph (Lord) Lister and Arthur Lister. The reunion of Brown's scholars in 1870, when they all went again to Lilley Hoo, must have been a memorable occasion. A number of Bedfordshire records stand to Brown's credit but his solitary addition to the county flora was Candytuft (*Iberis amara*).

One of Brown's pupils caught an early interest, retained but not carried so enthusiastically through life. This was Henry Brown (1824–1892) of Luton, later a successful timber merchant there. He was not related to Isaac Brown but was at the school from 1838 to 1842. He made a herbarium of about 250 sheets, 40 of which were of plants collected in the Luton area. Of these Fringed Water-lily (*Nymphoides peltata*) and Pale-blue Creeping-toadflax (*Linaria repens*) were new to the county.

Bedfordshire was soon to have another enthusiastic worker in William Crouch (1818–1846). He was born at Cainhoe near Clophill, and died there. His only contribution to the study of Bedfordshire botany was the compilation of two herbaria; the specimens in these, which are carefully annotated with dates and stations, are the first certain records of most of the species they represent. The specimens are arranged in alphabetical order of genera, but the numbers used in *Flora Bedfordiensis* were added. The larger herbarium was bound in four volumes, probably after his death. It contains about 440 specimens collected between 1841 and 1845, most of them from Bedfordshire. The smaller collection, bound in one volume, contains 176 specimens, and was supposed to have been compiled by Crouch for his cousin.[1] It was probably bound during his lifetime. Neither collection contains any sedges or grasses.

Most of Crouch's specimens were collected between Cainhoe and Lidlington where he held a curacy. This part of Bedfordshire has changed more than any other during the past century; enclosures and drainage have diminished the extensive bog and moorland which then skirted Maulden and Ampthill. Many of the species

[1] Possibly the Miss C. Crouch who discovered *Erica cinerea* on a heath near Markyate.

he collected are now extinct and the specimens of Petty Whin (*Genista anglica*) and Marsh Helleborine (*Epipactis palustris*) are the only ones known to exist from the county. Crouch added Alternate-leaved Golden Saxifrage (*Chrysosplenium alternifolium*)—the only record—to the county list. His *Veronica agrestis* is represented by *V. persica*, *Chenopodium murale* by *C. polyspermum* and *Potamogeton perfoliatus* by *P. alpinus*, in each case the first known occurrence of the species.

Had Crouch lived longer the story of the study of Bedfordshire botany would undoubtedly have been more complete. After his premature death the larger herbarium went to his eldest brother, James Frederick (1809–1889), a keen botanist and prebendary of Hereford Cathedral. He returned to the county from time to time and added Gold of Pleasure (*Camelina sativa*) and Yellow Star-thistle (*Centaurea solstitialis*) to the county flora. Both herbaria eventually passed to his nephew, Charles Crouch (*q.v.*), to become his most treasured possessions. When he died in 1944 they passed at his request to Luton Museum.

In the meantime the work being carried on in Hertfordshire was brought to fruition in Webb and Coleman's *Flora Hertfordiensis* (1849). Some of what was then Hertfordshire is now in the administrative county of Bedfordshire and consequently many records for our county appear in that work. It was the first county flora to be based on a river-drainage system and by the standards of its time is a good flora. The better botanist of the two was W. H. Coleman (1816–1863). He added Great Earth Nut (*Bunium bulbocastanum*) to the Bedfordshire flora, having previously found it for the first time in Britain at Cherry Hinton in Cambridgeshire. R. H. Webb (1805–1880) also made a few independent records for the county.

The middle years of the century was a period when field botany became a respectable study for gentlewomen, and we have a few interesting survivals of their pastime. A set of beautiful illustrations of plants from Odell Wood, made by Emma M. and Caroline M. Alston about 1830, is preserved in the County Record Office, as is a *List of Indigenous plants found at Basmead Manor* about 1860, presumably compiled by Peter Squire.[1] A small herbarium made by (Mrs.) J. M. Vipian of plants collected near Potton in 1855 has been presented to Luton Museum by John L. Gilbert of Wansford, Peterborough. In 1855 a long list of plants appeared in the *History of Luton* by F. Davis; for them the author thanked a Miss Paybody, of whom I know nothing more. Some of the records are interesting enough but others are suspect. At about this period a small collection of mosses was made, mainly in the neighbourhood of Sharnbrook, by E. Cooper, who was probably related to O. St. John Cooper (*q.v.*). In this collection are some specimens collected in the same neighbourhood by A. O. Black, not a local man, and in his day a well-known bryologist.

[1] Founder of the firm of P. Squire, chemists, Oxford Street, London (information given by Mr. E. O. Squire, Basmead).

It was in the middle of the century that John McLaren (1815–1888) came to Bedfordshire. Born at Methven, Perthshire, he came to England at the age of about 23 with Lord Lynadoch and became gardener to Samuel Whitbread at Cardington in 1846. He immediately set up a meteorological station, and forty years later was presented with a purse of sovereigns from eighty friends for his pioneer work on weather recording. He was elected a Fellow of the Royal Meteorological Society in 1850. It is not possible to say when his interest in field botany began, but he entered the Royal Horticultural Society's competition for the best collection of plants made in a county in 1864. The competition raised a volume of criticism from botanists all over the country, who feared the extermination of rarities, and was consequently not repeated. The Society defended itself by saying that it expected only about 200 specimens from any one county, and the more common species at that. In all 26 silver medals were presented and a number of bronze ones: it is doubtful whether a similar competition now would be so well supported. McLaren's entry contained about 700 specimens and every rarity he could lay his hands on. The fears of the botanists would thus appear to have been justified, but one doubts if the competition did as much harm as they claimed it would. The herbarium, bound in four large volumes, with the silver medal he won, was deposited at Luton Museum in 1944 by his grandson, S. McLaren Dynes. It is remarkable for the representative gatherings made, and the fine state of preservation of the specimens. The same holds for another and similar collection he made, presumably for the Whitbread family, and presented in 1939 to the British Museum (Natural History). McLaren's was the first attempt to make a reasonably complete collection of the plants of the county and the first serious work following the Benthamite system of classification.

It is not surprising that McLaren's work had some drawbacks. The greatest of these is the inadequacy of his labels. He gave no dates[1] and the stations are not always clearly described, e.g. riverside, side of park, etc., while all common species are described as 'Co.' A study of the flora will show that about thirty new county records must be ascribed to him and in many cases it is a McLaren specimen which constitutes a first certain record of a species. He was the first Bedfordshire botanist to be interested in adventives, and his records of these are made all the more interesting by the fact that in every case we still have the actual specimens.

McLaren wrote nothing, and records appear in his name only towards the end of his life. It is difficult to know whether he remained a gardener until he died. His obituary describes him as 'a true and faithful servant of the Whitbread family'.

An early contemporary of McLaren's was Thomas Corder (1812–1873). His career was an interesting one and demands further study.

---

[1] Dates were given, e.g. 14 May, throughout in the earlier herbarium, and we may assume that all refer to 1864.

He was born at Widford Hall, near Chelmsford, of a Quaker family, was educated at Isaac Payne's school at Epping, and was apprenticed to a miller at Royston. In the meantime he had found *Bupleurum falcatum*, a species new to Britain, in Essex. This secured him election as an Associate of the Linnean Society in 1833 when he was described as living at Writtle. He went to Mount Barker, near Adelaide, South Australia, about 1839, and returned to England in 1845.[1] He came later to manage Kempston Mill for Joshua Ransom, another Quaker, and in 1851 he married Elizabeth Anstey by whom he had one daughter. In 1852 he made another British record with *Claytonia perfoliata* from Ampthill, where it is still especially abundant. This was followed in 1860 by the addition of the Earth-nut Pea (*Lathyrus tuberosus*), jointly with his brother Octavius, to the British flora. He also added Wall Rocket (*Diplotaxis muralis*) and Hoary Cress (*Cardaria draba*) to the Bedfordshire flora. Later in life he managed or owned another mill at Cranfield where he died. He was buried, somewhat appropriately, in the Friends' Burial Ground at Ampthill. Nothing more seems to be known of his botanical work, but it would seem incredible that three British and two county records should be its sole extent.

The impetus that Smith gave to the study of British field botany in the early years of the century soon diminished, but was renewed by the adoption of the natural system of classification. Of even greater consequence was the influence of H. C. Watson, who saw clearly the need to study the status and distribution of plant species. Records flowed in to him from a host of correspondents, often supported by specimens, now happily preserved in his herbarium at Kew. His Bedfordshire correspondents were few. Isaac Brown was helpful, as were Susan and Miss J. Foster, apparently also Quakers and presumably of Luton: to Susan must be credited the first record of *Symphytum tuberosum*. Others were visitors to the county: Septimus Warner of Hoddesdon, J. E. Leefe, and Edward Forster (1765–1849), an Essex botanist, who added *Lotus tenuis*. Elizabeth Twining was credited with Bastard Toadflax (*Thesium humifusum*) but there is some doubt as to the validity of the record. Of J. Anderson, who sent the bare record of Lesser Snapdragon (*Antirrhinum orontium*), I know nothing. Great assistance came from William W. Newbould (1819–1896), one of the most competent botanists of his day. He held curacies at Bluntisham in Huntingdonshire and Comberton in Cambridgeshire, and eventually settled at Kew. His Bedfordshire records are mainly of 'splits'—plants previously considered to be one species but later considered to comprise more than one—or of species easily confused with others.

[1] These dates, which were given to me by Philip Corder, his great-nephew and assistant Secretary of the Society of Antiquaries, do not agree with those given of his movements in the membership lists of the Linnean Society. Dr. Corder also informs me that Thomas's botanical collection was supposed to have passed to his brother Octavius (1829–1910). Octavius eventually became a chemist at Norwich and died at Brundall. His daughter, Miss Grace Corder, is still living but has no knowledge of the collection.

The publication in 1873–4 of Watson's *Topographical Botany* gave for the first time a reasonably complete distribution of British plants on a county basis. There were gaps in most counties and a number in Bedfordshire, not so much because of a lack of thoroughness on the part of Abbot and his successors, but because of the appearance of 'new' species. Record clubs entered a fresh lease of life, new ones appeared, and there followed an almost indecent haste to make new county records. Again much of the interest in Bedfordshire came from Hertfordshire workers. Foremost in the field was Thomas Bates Blow (1854–1941), certainly the strangest person ever to dabble in Bedfordshire botany. He was born at Welwyn and died there, but his best botanical work was done in Madagascar, British Guiana and Japan, on *Charophyta*. He married, rather unsuccessfully, a Japanese lady of noble birth, and was honoured by the Emperor of Japan. His most noteworthy addition to the county flora was Purple-stalked Timothy (*Phleum phleoides*) jointly with H. Groves (1855–1912).

It was at this period that another pupil of Isaac Brown found a renewed interest in field botany. Joseph Pollard (1825–1909) was born at Ware but lived at High Down, Pirton, on the boundary of the county and scarcely a mile from Knocking Hoe, from 1841. A Quaker to the end of his life, he was a great correspondent. His herbarium is now at Hitchin Museum.

The fresh interest in Hertfordshire was soon centred on a revision of the county flora begun by Alfred Reginald Pryor (1839–1881) in 1875. Pryor was born at Hatfield and educated at Tonbridge School, where he was head boy, and University College, Oxford. He was the only Catholic whom I know to have worked on the flora of our county. He died at Baldock at the early age of 42. Pryor realised the necessity of knowing something of the wider distribution of species, which led him to take an active interest in the flora of Bedfordshire. He paid a number of visits to the county, and made a number of additions to the flora.

In the meantime, there had been a great and unexpected awakening of interest in the Bedfordshire flora. It arose out of a controversy in August, 1874, in the *Bedfordshire Times*, on the status of Sweet Flag (*Acorus calamus*) in the county, which ended with a letter from Thomas Elger suggesting that a local natural history society be formed. The response was surprisingly great and on 16th April, 1875, the Bedfordshire Natural History Society began. The moving spirit was William Hillhouse (1850–1910). He was at the time an assistant master at Bedford Modern School where he had been educated, but in 1877 he left the county to take up a scholarship at Trinity College, Cambridge. His only useful work for Bedfordshire botany was done between 1875 and 1877.

Hillhouse planned a revision of the Bedfordshire flora, and the 9th Duke of Bedford was persuaded to donate £25 towards its preparation. He planned well, and had he remained in the county no doubt the flora would have been completed. He was greatly influenced by C. C. Babington, whose *Flora of Cambridgeshire* (1860), based on a division of natural regions, rejected the division by catchment areas devised by Webb and Coleman and followed in most county floras. Hillhouse adopted the regional basis for Bedfordshire and in *On the Surface Geology and Physical Geography of Bedfordshire* suggested a division into seven districts, each with seven subdivisions based on geological formations. Lists based on *Flora Bedfordiensis* were printed and recorders urged to complete the all too many gaps. He started well with *A Contribution towards a new flora of Bedfordshire* (abbreviated in this flora as *Hillhouse List, 1875,* or *W.H. 1875*), a list of 430 records he made in 1875. This was followed by the *Bedfordshire Plant List for 1876* (abbreviated to *Hillhouse List, 1876,* or *W.H. 1876*). The list grows to 700 species, and the assistance of fifteen correspondents in various parts of the county is acknowledged. Such progress in one year must have been very gratifying.

Hillhouse stayed as student, lecturer, and assistant curator of the herbarium at Cambridge, until 1882, when he became the first professor of botany at Birmingham University. He himself added few species to the county list, but was beaten by only a short head by Pryor and Blow with many of the Watsonian splits.

The greatest contribution of Hillhouse was in leading the Natural History Society to do some useful work; in fact the only useful work the society did was its field botany. It probably killed the society, but was the means of bringing John McLaren into contact with younger botanists such as James Saunders (*q.v.*) and John Hamson (*q.v.*), and began a phase in the study of the county flora that can be traced almost continuously from that day to this.

When Hillhouse left the county the direction of the flora soon passed to Arthur Ransom (1832–1912), not to be confused with the Quaker and related Ransoms of Kempston Mill and Hitchin, supporters of the society. For many years he was editor of the *Bedfordshire Times*. Other members of the society who had some interest in botany included S. Hoppus Adams, a Bedford doctor, who in 1878 read a paper on mosses, F. W. Crick (1820–1895), known as 'Dr. Crick' in Bedford but only a 'quack,' and A. Poulton of Leighton Buzzard, who added *Medicago denticulata*. It is also interesting to note that Dora Martyn of Pertenhall, obviously a descendant of Thomas (*q.v.*), sent in a few records. More active were two ladies, Ada Stimson and M. L. Berrill.

Ada Stimson (1842–1915) was the eldest daughter of William Stimson of Moat Farm, Marston Mortaine. She left the county in 1879 when she married Sir John Thomas, a prominent Congregationalist, of Wooburn, Bucks., and from then her life was involved in the causes of Liberalism and total abstinence. In her earlier days she was a competent botanist and sent in many records from the Marston area. Her most noteworthy discovery was the correctly identified *Luzula forsteri*, from Marston Thrift.

M. L. Berrill was the daughter of a farmer and lived at Chestnut Farm, Maulden, where she kept a private school. Apart from the fact that she maintained an invalid sister, Fanny, I have been unable to find anything more about her. She sent in numerous records from the Maulden area.

The Natural History Society came to an end about 1885 but in the meantime had secured a valuable member in James Saunders (1839–1925). Saunders was born at Salisbury and had little education. His father died young, and having tried his hand as a pupil teacher James came to Luton in 1859. He worked for a brief period for a bookseller and then entered the rapidly expanding and flourishing hat industry. Before he came to Bedfordshire, he had been interested in geology and natural history generally. A chain of tragic domestic disasters brought a period of mental strain, and about 1877 he sought the advice of an old friend and as a result made field botany his life interest. That he became a great botanist in the spheres in which he specialised cannot be denied, but his attitude throughout was that of a man with a great capacity for work, concerned only with what he himself was doing. From the moment he joined the society the task of revising the flora became his.

His first botanical work was on the flowering plants, but it was not long before the mosses were attracting most of his attention, and these soon gave way to the slime fungi which he found all-absorbing. He was always much more at home with cryptogams than with flowering plants, for he had no visual memory and found it necessary to trace each individual specimen in a flora. His herbarium consequently contains a number of wrongly identified specimens. In 1900 he was elected an associate of the Linnean Society.

Saunders wrote more than any other Bedfordshire naturalist and he wrote well. *The Bedfordshire Plant List for 1882* was nominally, no doubt, Arthur Ransom's work (it was unsigned), but most of the records were from Saunders. This was followed by the *South Bedfordshire Plant List* (1885), entirely Saunders's work, and in the meantime he had written a number of articles on flowering plants, mosses and charophytes in the *Journal of Botany*, *Midland Naturalist* and *Transactions of the Hertfordshire Natural History Society*. In 1897–8 he contributed a long series of articles to a local newspaper entitled *The Wild Flowers of Bedfordshire* (abbreviated in the text *W.F.B.*). It was the first attempt since the *Flora Bedfordiensis* to account for the flora of the county as a whole but the stress was on the flora of the south of the county as Saunders knew it.

In 1911 this was reprinted in book form with few alterations as *The Field Flowers of Bedfordshire*, the only book other than *Flora Bedfordiensis* dealing with the flora of the county. Apart from its main defect of dealing only with the south of the county in detail, it took no account at all of the useful work published in the meantime, such as the account by Druce in the *Victoria County History* (1904) and Hamson's *Flora of Bedfordshire* (1906). Its chief virtue was that it was interleaved with blank pages which has allowed a number of annotated copies to be made.

Saunders acknowledged the help of many of his contemporaries, of whom a little is known. S. A. Chambers was an insurance agent who afterwards lived in South London. His herbarium is incorporated in that of J. E. Lousley. R. C. Chambers was probably related to him. Miss R. M. Welch, who discovered *Herminium monorchis*, was a grand-daughter of Thomas Sworder of Luton. J. Catt was probably a gardener at Luton Hoo. C. L. Higgins lived at Turvey and W. Green was headmaster of a Luton school. Mrs. Tindall lived at Leighton Buzzard and it is possible that Mrs. Twidell, whose name enters frequently in the flora, lived in the same neighbourhood. Miss Carruthers, Miss Lye and G. H. Latchmore were members of well-known Luton families and E. R. Hunt still lives in Luton. Of Mr. Anstee, C. F. Boultbee, Mrs. Dickinson, Mr. Hedge, Miss Mercer, Mr. Piffard and F. Wiseman I know nothing.

John Hamson (1858–1930) resembled in many respects his more celebrated contemporary. He was born at Vale Farm, Naseby, and also had little education. He left school at the age of twelve and worked on the farm until he was sixteen. He then became a draper's assistant at Walsall, where he learned shorthand. In 1878 he became a reporter for the *Leicester Post*, of which Arthur Ransom was then editor. In 1883 he followed Ransom to Bedford, and later became assistant editor of the *Bedfordshire Times*. Defects in early education were remedied with the study of geology, chemistry, history, archaeology, Greek, French, German and Esperanto. He taught shorthand in evening classes at Bedford for 30 years and botany for 27 years. His weekly 'Spectator' column in the paper contained frequent notes on plants.

He joined the Bedfordshire Natural History Society soon after his arrival and was for a year its president. He continued the society's Bedfordshire herbarium, which was contributed to by McLaren and Saunders. What might have been a valuable collection was allowed to disintegrate and rot in a room at Bedford Library which it was nobody's responsibility to keep in order. Hamson himself was not a good botanist; he made far too many mistakes in identification, and failed to send specimens to reliable authorities when he was not sure of their identity, but whatever may have been his shortcomings as a botanist, he did two worthwhile and really valuable things for Bedfordshire botany. In 1906 he followed Saunders's lead and published in the *Bedfordshire Times* a series of lists of plants from the county, which were afterwards published as a booklet, *The Flora of Bedfordshire* (1906). It gave most attention to the north of the county, and again it is to be regretted that Saunders and he did not work more closely together. Few stations are given in the booklet, but this omission is more than made up in a manuscript flora he kept, which was later bound into five volumes. As far as possible each species was allotted a separate page. I have referred to this valuable work in the text of this flora as *Hamson's Notes*, or more simply *J.H. Notes*.

Hamson's enthusiasm was unbounded and he was genuinely interested in plants and the countryside. In fairness to him it should be said that he was as interested in fungi as he was in the flowering plants, and I am unable to judge the value of his work in that field. Between them Saunders and Hamson were covering the whole of the plant world, but again almost in isolation. A man in Hamson's position was well placed to collect records from a number of workers. Details of some are known but most of them are lost in obscurity. Muriel Hamson was his daughter; W. Davis was Librarian of Bedford General Library for many years and retired about 1920; Edward Laxton, who died in 1951 at the age of 82, was a member of the famous Bedford firm of rose growers and fruit-raisers; W. Gifford Nash (1852–1935) was a popular Bedford doctor who came to Bedford in the early nineties and was primarily an entomologist; C. Hemsley was a vicar of Thurleigh; and G. Squire was headmaster of Marston Mortaine school in the early years of the present century; Miss Harradine was probably connected with Beadlow near Clophill, and J. Ekins with Sundon; Mr. Biffin and Mr. Halahan, the latter interested mainly in orchids, were probably from Bedford—the names appear in local directories; J. Woods, L. Perry, Charles Clarke, W. Odell, Sidney Crowsley, Edgar Evans and Ivor Evans could have been schoolboys, but no guess can be made as to the identity of Messrs. Burgoyne, Day, Horley, Robinson and Wyatt and Mrs. Cotton. Their records, however, were in most cases of little consequence.

In the meantime some botanists of greater consequence had interested themselves in the county flora. One of the most note-worthy was Worthington George Smith (1835–1917)—the Smith of Fitch and Smith (*Illustrations of the British Flora*). Smith, who lived an exceedingly full life, was born in London and died in Dunstable. His botanical interest was primarily in fungi but he found time when at home in Bedfordshire to give some attention to flowering plants.

E. M. Langley (1851–1933) was senior mathematics master at Bedford Modern School. He was mainly interested in brambles, on which he was an authority. It is interesting to note that W. Moyle Rogers (1835–1920), an even greater authority on brambles, visited the Woburn area and provided a useful list, which was used by Saunders in *Field Flowers of Bedfordshire*; but I have been unable to find the original. E. F. Linton (1848–1928), who was also interested in brambles among other plants, also visited the Woburn area and published lists. His brother W. R. Linton also made visits to the county.

Another local botanist soon appeared in the person of Charles Crouch (1855–1944). He was a nephew of William Crouch and it is thanks to him that the William Crouch material was preserved. He was originally a farmer at Mead Hook, near Pulloxhill, but failed about 1880. He became interested in field botany in the late eighties and soon became a friend of James Saunders and Joseph Pollard. It is doubtful if anyone knew the middle of the county and

the Greensand flora better than he, and with greater encouragement
he would have been a competent botanist. He never married,
mainly because of the fear of carrying tuberculosis, 'the evil of the
Crouch family'. I saw him frequently in the last eight years of his
life and remember him chiefly as a man of definite views and
violent prejudices. Botanists were by no means immune from his
criticisms and the splitters came off badly; 'I tried to keep up to
date but when they split *Bursa* I gave up'. Crouch added few
plants to the flora but found a number which had been presumed
extinct, and many stations stand to his credit. His specimens are to
be found at the end of William Crouch's herbarium and in James
Saunders's and Joseph Pollard's herbaria. I have in my possession
an annotated copy of *Wild Flowers of Bedfordshire* made by Crouch
with many additional records.

Another local botanist of some interest was D. Martha Higgins
(? 1856–1920). She lived in Luton with her two sisters and was
friendly with Saunders; she was also interested in mycetozoa but
appears to have worked for the main part independently. In
1886 she joined the Watson Botanical Exchange Club, to which she
belonged until her death, sending specimens over a number of
years, most of them of little interest. She was able to get into the
field more often than Saunders and to visit some of the more distant
parts of the county. A number of new county records stand to her
credit. Field Cow-wheat (*Melampyrum arvense*) was her best,
although in some cases she did not know that she had collected
anything of more than passing interest. Her specimens are incorporated in British Museum (Natural History) Herbarium, specimens
collected by her sister, Emily Kate, being labelled 'Emily'; others
are labelled 'Bertha'.

In the meantime the flora of Hertfordshire begun by Pryor was
published (1887), but although Pryor's name appears as the author
it would be more correct to credit it to B. Daydon Jackson. Still
more Bedfordshire records appear for the area of overlap between
the two counties.

After the passing of the Bedfordshire Natural History Society a
Bedford Natural History and Archaeological Society was founded
in 1888 by F. W. Crick. Two years later it was joined by W. N.
Henman who has continued ever since to be its moving spirit. In
addition to Crick it has had other good naturalists—E. M. Langley,
Edward Laxton and W. Gifford Nash—among its presidents, but it
appears to have done little useful botanical work.

A herbarium recently presented to Kew contains a number of
specimens collected in 1902 in the neighbourhood of Bedford by
A. B. Sampson, about whom I have been unable to find anything.

Serious work was being done in this period on the flora of the
neighbouring counties of Buckinghamshire and Northamptonshire.
In each case it was directed by G. C. Druce (1850–1932), whose
name must rank with those of John Ray and H. C. Watson in the
history of British botany. He was born a few miles from the

Bedfordshire border, at Potterspury in Northamptonshire, and spent his early years in that neighbourhood. He moved to Oxford in 1879 and for fifty years the study of British botany revolved around him. He never lost interest in the scene of his childhood and his *Flora of Buckinghamshire* (1926) and *Flora of Northamptonshire* (1930), both good floras in the orthodox style, contain much information of use to the Bedfordshire botanist. He also compiled the list of plants given in the *Victoria County History of Huntingdonshire* (1926). In a wider field he raised the Botanical Exchange Club (now the Botanical Society of the British Isles) from an obscure society into a flourishing one. For it he compiled the *British Plant List* and the *Comital Flora*, the latter carrying on the work so well begun with *Topographical Botany*.

Druce made a number of visits to Bedfordshire, and many specimens and records from the county were passed on to him. He was chosen to edit the section on botany in the *Victoria County History of Bedfordshire* (1904). In some ways the choice was a good one, for he alone at the time was capable of viewing the county as a whole, and he had the wisdom to seek the assistance of Saunders and Hamson. In other respects it was a pity, as he knew too little of the county and failed, as both Saunders and Hamson were failing, to collect all the available material. Druce's account of the flora is interesting if only for the fact that for the first time the county is divided into river-drainage areas. These may be criticized, in that he lumped together the Colne and Lea catchment areas to make one division, and arbitrarily divided the large Ouse area into two, East and West Ouse. From one of his districts, the Cam, he had no plants to record, and from another, the Nene, he recorded only one, picked up from Newbould. His division of the county brought Bedfordshire into line with the work so far done in Hertfordshire and the work he was doing in Buckinghamshire and Northamptonshire; but he rejected the river district basis in Huntingdonshire.

His work had some serious defects. He had little respect for a county boundary, and his records were often incomplete. The *Flora of Buckinghamshire* and the *Comital Flora* give records for Bedfordshire with no details, and search has failed to reveal either specimens in his herbarium or further information in *B.E.C. Reports* to support the records. Druce was an incurable splitter—but perhaps it is as well to keep an open mind on the question of splitting. Druce added a large number of species to the county although it is not certain that he found them first.

In recent years the most useful work done on the flora was by J. E. Little (1861–1935). He was born at Tonbridge and was educated at Tonbridge and Lincoln College, Oxford. An excellent classical scholar he became headmaster of Hitchin Grammar School in 1885. He was not temperamentally suited for a life as a schoolmaster and resigned his post in 1897, after which he undertook temporary work at Rugby and Haileybury, but continued to make Hitchin his home until his death. He became interested in field botany shortly after he came to Hitchin, but it was not until after his retirement that his most serious work was done. He took charge of the botanical side of the Hitchin and District Regional Survey Association's work, and as a member, first of the Watson Botanical Exchange Club and later the Botanical Exchange Club, his work was soon known to botanists throughout the country. His area of greatest interest was the Ivel district, both in Hertfordshire and Bedfordshire, but I am especially grateful for the few visits he made to the Cam district. Little was one of the most competent botanists who has worked on the Bedfordshire flora. He was especially concerned with the identity and variability of plants rather than with problems of their life history and associations. Aliens probably interested him as much as native species. He was one of the best of the school of botanists whose names must be associated with Druce and who have now almost all passed away.

Little wrote two major papers on the Bedfordshire flora, but more useful than these are his carefully kept note-books and an annotated copy of *Field Flowers of Bedfordshire*. His large herbarium he presented to the University of Cambridge before his death.

He corresponded with many of the other leading botanists of the day. Nearer home there was no one to equal him, and although he met James Pollard and Charles Crouch he seems to have worked mainly alone,[1] except that Margaret Brown, daughter of Henry Brown, who in the later years of Little's life was most active in the Wild Flower Society, was able to take him in her car to places he might have found it difficult to visit. Miss Brown compiled a Bedfordshire Diary for the W.F.S., which contains some useful records. She left Bedfordshire in 1934 and now lives at Ilminster.

The most noteworthy addition made by Little to the flora was without doubt Spotted Cat's-ear (*Hypochoeris maculata*), but his total makes, for his day, an impressive list.

It was from Hitchin, too, that great interest came from the Phillips family. Hugh Phillips, still living, collected in the thirties very thoroughly in the then critical groups, and a number of county and British records and 'types' of *Taraxacum* and *Capsella* are based on his gatherings. It was on his many gatherings of roses that E. B. Bishop based his paper, *Notes on the Roses of Bedfordshire* (1939). To Mr. Phillips, who unfortunately has lost interest in field botany, should also be acknowledged a few other records. His sister, Enid MacAlister Hall (1885–1941), lived at Clifton House and was also a keen botanist.

---

[1] Little became the recipient of records from a small circle of keen workers in north Hertfordshire, some of whom provided Bedfordshire records. Ruth Bott was a pupil of Hitchin Grammar School—she is now married and lives in Brazil; Joshua Lamb (1856–1943) was a Quaker farmer of Sibford Lewis, Warwickshire; Robert Long (1859–1935) was a farmer and lived at Stondon Manor; H. Read was for many years headmaster of Biggleswade School; H. C. Littlebury is still alive and lives at Hitchin; Miss M. C. Williams taught botany at Hitchin Grammar School and now lives in Cornwall; records also came from Dame Margaret Tuke who then lived at Pegsdon.

A. Bruce Jackson (1876–1947) had a double contact with Bedfordshire; his cousin lived in Luton, and he was for many years adviser to the Duke of Bedford on forestry questions. He recorded a number of plants from Woburn, including *Carex divisa*. It was through Jackson that the 11th Duke of Bedford recorded *Xanthium spinosum* in 1937, probably the first wool alien to be observed in the county.

H. N. Dixon (1861–1944), who during his lifetime was Britain's leading bryologist and lived at Northampton, had relations in Bedfordshire. He collected mosses with Saunders and later independently in various parts of the county.

In 1932 H. J. Riddelsdell (1866–1941) paid a visit to Whipsnade Zoo and recorded some brambles. Isolated records also came from W. W. Mason (1853–1932) and A. Templeman (1887–1945), botanists of wider reputation. (Sir) Sidney Peel (1870–1938) of Sandy Lodge was responsible for some isolated records, and Mrs. Smith (*née* Stansfield) of Harlington is to be envied for her discovery of Lizard Orchid (*Himantoglossum hircinum*).

In the newly developed Letchworth, also, an interest was being shown in the Bedfordshire flora. Richard Morse, for many years editor of *Countryside*, made many records and added what is without doubt the most important discovery of this century to the flora, *Seseli libanotis*. Almost as important was the reward that Harry and Doris Meyer, keen field workers, had in eventually finding the genuine Oxlip (*Primula elatior*) in the county. S. Bowden is primarily an entomologist. These workers are still with us. T. A. Dymes (1866–1941), also of Letchworth, was an excellent botanist and made a few records.

My own interest in the county flora began in 1916 and continued until about 1926 when it became necessary for me to seek professional qualifications and my interest ceased with the feeling that I was working too much in isolation. It was renewed in 1935, when I determined that I would attempt a revision of the flora. Many of the botanists already mentioned were my contemporaries and a few still are. It has been my good fortune to have the active assistance of a vast body of local workers, and my greater fortune to be able to persuade almost every botanist of national repute to visit the county.

Some local botanists of my own period demand more than passing mention. V. H. Chambers of Luton was, in the thirties, very much my co-worker. It was he who persuaded me to undertake a revision of the flora, and much of the plan it follows is the result of our discussions. I had hoped at one time that it might be a joint authorship, for the clarity of his views is always refreshing. He felt, no doubt wisely, that entomological problems were more pressing, and they have in more recent years absorbed his whole attention. Equally helpful in the early days, and always ready to discuss problems since, has been A. W. Guppy of Bromham. He has a good knowledge of the flora, especially of the north of the county, and many records would stand to him had he collected specimens. Miss G. H. Day of Harrold, a niece by marriage of H. N. Dixon, was botanizing in the neighbourhood of Roxton before I was born, though I have often found it difficult to believe when I have panted to keep pace with her. She has sent me more records than any other botanist, and all from inaccessible and little worked parts of the county. L. W. Wilson (1887–1951) was for a number of years an inspector of taxes at Bedford and Luton. He gave me many records and I met him frequently after he left the county and became a member of the B.S.B.I. He had a great love of the countryside. F. Seymour Lloyd and his daughter Mrs. B. Garratt now live at Battle in Sussex. Dr. Lloyd was in practice in Luton for many years and knew James Saunders and D. M. Higgins. Miss Winifred Kitchin, daughter of a previous rector of Podington, came back to the county and recalled a happy childhood and memories of plants she had seen in the small Nene area. For the same area R. R. B. Orlebar (1862–1950) of Hinwick, whose knowledge of his own parish was profound, was able to fill many gaps. Miss M. Dalton, formerly of Dean, likewise remembered many plants she had seen in the poorly-worked Kym area.

T. Laflin came to Bedfordshire in 1947 to manage the C.W.S. apple orchards at Cockayne Hatley and left in 1950. In this brief period he was able to revise thoroughly the moss flora of the county and the great value of his work in this field may be judged from that section of the Flora. In addition he has provided lists of flowering plants from the Cam district.

Among more active botanists W. D. Coales, for many years second master of Dunstable Grammar School, has proved most observant. He did not become a field botanist until comparatively late in life, but has made many records, and added *Valerianella rimosa* to the flora. C. C. Foss (1885–1953), retired from a distinguished military career and, living at Bedford, was a keen field botanist. My most constant companion in excursions in recent years has been H. B. Souster of Luton. Many of the records credited to me were made jointly with him.

Younger botanists have been exceptionally helpful. I met Bernard Verdcourt for the first time when he was a pupil at Luton Grammar School. He was torn even then between the rival interests of conchology and field botany. Studies and employment in other parts of the country, and now as a botanist at Nairobi in East Africa, have meant that all too little time was spent in his native county, but he made many records.

Younger is Peter Taylor, also of Luton and now at Kew. An exceedingly observant botanist he has made ferns and horsetails his special study. He has also done some useful work on bryophytes and collected for the Flora all available material on the *Hepatics*. In a short period of field work he added *Carex binervis*, *Ceterach officinarum*, *Dryopteris borreri* and *Thelypteris oreopteris* to the flora. Also at Kew is another young Bedfordshire botanist, Derek A. Reid of Leighton Buzzard. He is interested mainly in fungi and dragon-flies, but has provided some useful flowering plant records.

Of botanists of a national reputation who have taken an interest in the flora priority must be given to E. Milne-Redhead of Kew. Early in 1944 he was stationed with the Royal Artillery at Dunstable, remained there until the war ended, and has since made many visits to the county. In many respects his coming to the county brought a change in the study of its flora, as he, even more than I, has been instrumental in interesting other botanists in its flora.

Most of the professional botanists now at Kew have visited the county. Those whose names appear in the Flora are J. P. M. Brenan, R. B. Drummond, C. E. Hubbard, R. D. Meikle, R. Melville, N. Y. Sandwith, V. S. Summerhayes and H. K. Airy Shaw. Acknowledgment of the great assistance of these and other members of the staff at Kew is made elsewhere.

During the war years A. J. Wilmott (1888–1950) of the British Museum (Natural History) made a number of visits. He was the leading British botanist of his day and was very interested in the revision of the Bedfordshire flora. J. E. Dandy, also of the British Museum, has made visits to study the distribution of pondweeds. He has given great assistance in many directions and his sound views on the arrangement of the Flora and nomenclature have been followed throughout.

Other visitors have done valuable work in various critical and other groups of plants. E. C. Wallace in a few visits added much to our knowledge of the *Bryophytes*. W. C. R. Watson, with great experience of brambles, did good work in one long and two shorter visits. The section on *Rubus* in the Flora would have made poor reading but for his work. In the same cause W. H. Mills (1873–1951) of Cambridge made a few fruitful visits. Also from Cambridge S. M. Walters made one visit to add *Eleocharis uniglumis* to the flora and P. D. Sell and C. West came to study *Hieracia*. G. M. Ash made one memorable visit to study *Epilobia* and L. J. Tremayne, *London Catalogue* in hand, over a number of years made visits and noted all he saw.

In recent years R. A. Graham and J. E. Lousley, often together but sometimes independently, have made many visits. Lousley is, without doubt, the best living non-professional British botanist. His greatest interest in the Bedfordshire flora has been its wool adventives and some of the records credited to us jointly could, more strictly, be credited to him, although Graham and I had a large haul on one never-to-be-forgotten day when Lousley could not join us.

Among other visitors the names of the following appear: R. W. Butcher, well known to all British botanists, R. C. L. Burges, J. F. G. Chapple, J. Codrington, A. E. and E. M. Ellis, H. Gilbert-Carter, Miss C. M. Goodman, J. F. Hope-Simpson, Miss C. W. Muirhead, J. Ounsted, F. Rose, S. P. Rowlands, Mrs. J. Russell, N. D. Simpson and D. P. Young, an authority on *Epipactis*.

Since I began the revision of the flora I have received records from many local workers, some of whom would not claim to be botanists. They are listed below with the towns or villages in which they lived. Those shown (D) have since deceased and those shown (L) have now left the county.

Ampthill: J. K. Horne, Misses M. and H. G. Oldfield (L).
Aspley Guise: R. Palmer (L).
Barton: I. J. O'Dell (L).
Battlesden: H. E. Pickering (D).
Bedford: Rev. R. L. R. Bearman, Mrs. W. J. Brown (*née* Dalton), J. W. Cardew, W. Durant, H. A. S. Key, R. Lucas, Miss E. Proctor, F. G. R. and Mrs. Soper, R. Townsend, B. B. West (useful lists), K. E. West, Miss A. Wooding, J. Wooding.
Caddington: L. J. Margetts.
Carlton: G. A. Battcock.
Clifton: Miss A. K. Smith.
Cranfield: J. Burrell.
Dunstable: A. S. Johnston, H. B. Sargent (L), Miss S. Tearle.
Felmersham: Lady (Mrs. S. R.) Wells.
Harpenden: Miss A. M. Buck, Mrs. C. Swain.
Hitchin: Miss B. M. Vizard.
Flitwick: C. M. Crisp, M. Holdsworth (L).
Houghton Regis: E. T. Blundell.
Kempston: Mrs. B. F. Haylock (D).
Kensworth: Mrs. C. Dunham.
Luton: K. G. Bull, R. L. Chambers, J. Clifford; H. Salmon, D. B. Slope (pupils at Luton Grammar School), F. L. Chesham, H. Cole, H. Cook, Miss M. W. Cornish (L), Rev. J. C. Culshaw (L), Miss K. Goodwin (L), Mrs. G. K. Long (L), H. A. J. Martin (L), Miss Tabb (L), W. N. Thorpe.
Marston Mortaine: G. E. Brown, Rev. R. H. Goode.
Northill: Miss Mary Nisbet (D).
Potton: E. A. Baxter, Mrs. C. Reynolds.
Sandy: Miss I. J. Allison, Miss I. M. Allison, G. Cope, J. Lawrie, Miss J. Lawrie.
Streatley: L. C. Chambers.
Tempsford: Miss M. Dawes.
Toddington: Mrs. M. E. Boutwood, Sir F. Mander.
Westoning: G. D. Nicholls (L).
Whipsnade: G. M. Vevers.

My own work on the Bedfordshire flora has been carried out almost entirely through Luton Museum where I became honorary keeper of botany shortly after serious work was begun on the revision. I have built up a large Bedfordshire herbarium which takes its place at the Museum with Abbot's (on long loan), William Crouch's, McLaren's earlier one, and Saunders's. A good working library has been built up and an attempt made to include all works dealing with the flora of the county and neighbouring counties. I soon joined the Botanical Exchange Club and was elected to its council in 1946, taking an active part in its emergence into the Botanical Society of the British Isles, and became its Field Secretary in 1949 and Meetings Secretary in 1953. In 1946 I helped to form a new Bedfordshire Natural History Society, and became a member of its council, its recorder for botany and a contributor to its journal.

THE BASIS OF THE FLORA

THE ADMINISTRATIVE COUNTY

Bedfordshire is a small inland county in the South Midlands. Its length from north to south is 35 miles and its greatest breadth is 23 miles. The lines of communication are mostly from north to south or radiate from Bedford, the county town, which is approximately in the centre of the county. Its area compared with the neighbouring counties is:

Bedfordshire	302,942 acres
Huntingdonshire	233,985 ,,
Hertfordshire	404,520 ,,
Buckinghamshire	479,360 ,,
Cambridgeshire (with the Isle of Ely) .	563,241 ,,
Northamptonshire (with the Soke of Peterborough)	635,143 ,,

THE WATSONIAN VICE-COUNTY

With a few exceptions this coincides with the administrative county, as H. C. Watson in *Topographical Botany* (1873) took the then existing counties as the basis of his vice-counties[1] By the Watsonian system Bedfordshire is designated v.c. 30, Bedford, and the neighbouring counties become v.c. 20, Herts, v.c. 24, Bucks, v.c. 29, Cambridge, v.c. 31, Hunts, and v.c. 32, Northampton. Watson did not use a full stop after his abbreviated names and his rule, regrettably dropped by subsequent botanists, will be followed in this Flora. When the full stop is used, *e.g.* Beds., Herts., Bucks., Cambs., Hunts. and Northants.[2], it refers to the administrative counties.

County boundaries have changed considerably since 1873, but to ensure continuity of records the vice-counties must remain. Parts of the vice-county are now in other administrative counties, and parts of the administrative county lie in other vice-counties.

[1] For a closer study of the Watsonian system and its application to Bedfordshire see J. G. Dony, What Bedfordshire Is, *Journ. Beds. Nat. Hist. Soc.*, 1 (1927), 8–12.
[2] For convenience the Isle of Ely is considered as part of Cambridgeshire, and the Soke of Peterborough as part of Northamptonshire.

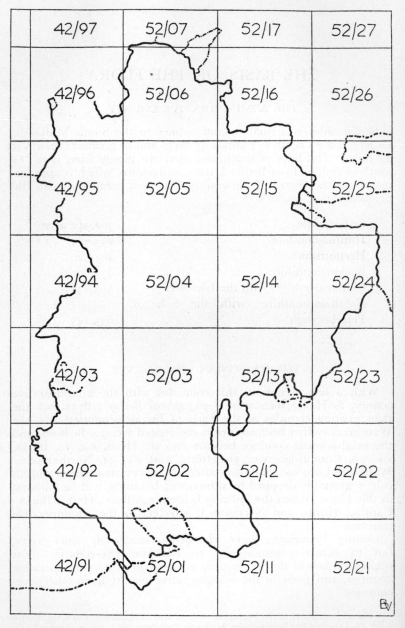

FIG. I   THE NATIONAL GRID

This is shown on all maps of the county in the text.

The changes[1] as they affect Bedfordshire are:

A. Four parts of v.c. 30, Bedford are now in other administrative counties:

1. The parish of Tilbrook, 1,342 acres, v.c. 30 [Hunts.][2].
2. A detached portion of the parish of Tetworth, about 900 acres, v.c. 30 [Hunts.][3].
3. The parish of Holwell (in two detached portions), about 900 acres, v.c. 30 [Herts.].
4. Part of the parish of Markyate (formerly the hamlet of Humbershoe), about 370 acres, v.c. 30 [Herts.].

There is no record of any plant species for any of these areas which is not also recorded for the rest of the county.

B. Four parts of other vice-counties are included in the present administrative county:

1. A large area, including the whole of the parish of Kensworth, a part of the parish of Caddington and a small part of the borough of Dunstable, area about 4,200 acres, v.c. 20 [Beds.].
2. The south-eastern corner of the parish of Studham, area about 1,000 acres, v.c. 20 [Beds.].
3. Most of the parish of Aspley Heath, area about 510 acres, v.c. 24 [Beds.].
4. The hamlet of Farndish in the parish of Podington, area about 150 acres, v.c. 32 [Beds.].

These areas include some country interesting to the botanist and the following species not recorded for the rest of the county have been recorded from them: *Galium pumilum* on an extension of Blow's Downs in the parish of Caddington, and *Sphagnum subsecundum*, *Lycopodium inundatum*, *Thelypteris oreopteris* and *Juncus squarrosus* at New Wavendon Heath in the parish of Aspley Heath.

The revisions of boundaries outlined above reduce the area of the vice-county to about 300,600 acres.

This Flora has as its field the administrative county with those parts of the vice-county now in other administrative counties. This has an area of about 306,500 acres. When there is no indication to the contrary it may be understood that reference is made to areas and stations which are in both the administrative county and the vice-county.

[1] I have ignored minor changes of boundary from one side of the road to the other.
[2] This form of abbreviation should be read as 'a part of Watsonian vice-county 30, Bedford, now in the administrative county of Huntingdonshire'.
[3] This was at no time in Bedfordshire but Watson included detached portions of counties in the vice-counties with which the greater part of their boundaries coincided.

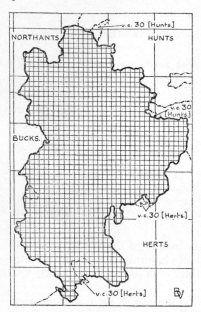

FIG. 2   THE ADMINISTRATIVE COUNTY

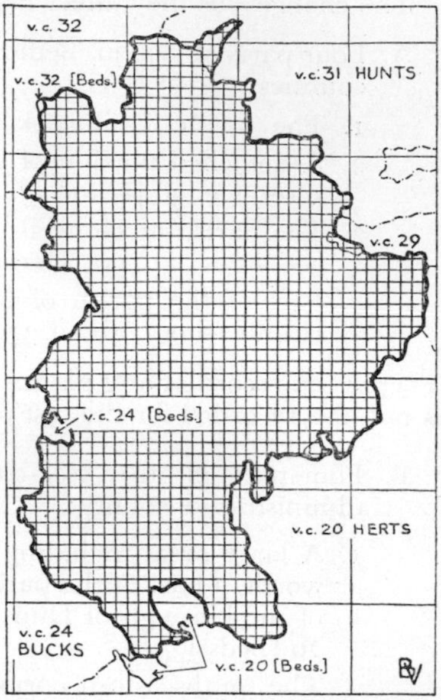

FIG. 3   THE WATSONIAN VICE-COUNTY

ALTITUDE

Three upland ridges cross the county in a south-westerly to north-easterly direction: a northern upland ridge, the greensand ridge and the chalk hills. Between these are two areas of comparatively low land: the Ouse and Ivel valley, and the gault plain. The county generally is higher on its western side, and slopes gradually to the eastern border.

The northern upland ridge is little more than a gradually rising cap of Boulder Clay marking the northern limit of the Ouse valley. Sharnbrook Summit is 337 ft., and 344 ft. is reached on the site of the disused airfield on the Forty Foot Lane. To the north-west the land drops sharply in the Nene valley to 175 ft. at the county boundary near Irchester Grange.

The Ouse and Ivel valleys must be considered as one. Above Bedford the Ouse, which enters the county at 146 ft. at Turvey, pursues a winding course. Some stretches of the river are here scenically attractive, but below Bedford, where the river follows a straighter course and is joined by the closely similar Ivel, the landscape at times is almost monotonous. The combined rivers leave the county at Eaton Socon, 45 miles downstream from Turvey, at only 48 ft. above sea level. The river valley, narrow above Bedford, widens considerably in its lower parts. Beyond the valley slightly undulating hills reach 239 ft. at Oakley Hill, 310 ft. at Pict's Hill between Bromham and Turvey, 270 ft. at Keysoe Row, 287 ft. at Galsey Wood, and 275 ft. at Twin Wood.

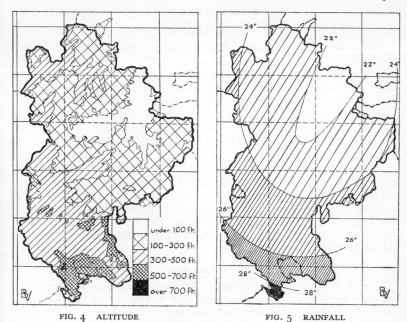

FIG. 4   ALTITUDE   FIG. 5   RAINFALL

The greensand ridge rises sharply on its northern face, giving pleasantly open and wooded hill country. It is broken by the Ivel valley, and to the east of this there is a smaller and less elevated greensand ridge on which lies Potton. The highest point on the Lower Greensand is Hill Farm at Potsgrove, 517 ft., a part of Aspley Wood is 443 ft., Brogborough Hill 346 ft., Jackdaw Hill 408 ft., Deadman's Hill 370 ft. and Hammer Hill 290 ft. In the eastern portion Castle Hill at Sandy, more prominent in its surroundings than any of the preceding, is only 219 ft. The Greensand is drained by the Ouzel, 259 ft. at Heath and Reach, and the Ivel, 89 ft. at Biggleswade. The lowest parts of Flitwick Moor are 207 ft.

The gault plain follows closely the southern limit of the Greensand and is in no place more than five miles wide. The otherwise flat and intensively cultivated country is relieved by a few hills, usually of glacial origin, as at Toddington, 485 ft., and Topler's Hill, a well-known landmark, 224 ft.

The chalk hills, like the Greensand, rise sharply on their northern face. For the greater part they are open country and scenically the Chilterns are, by common consent, the most attractive part of the county. The highest point in Bedfordshire, a little to the south of Dunstable Downs, is 796 ft. Galley Hill is 614 ft., Deacon Hill 566 ft., and Sharpenhoe Clappers 524 ft. To the south the chalk hills are drained by the Gade, 416 ft. at Milebarn, and the Lea, 304 ft. at East Hyde.

CLIMATIC FACTORS
Rainfall
Probably only Huntingdonshire and Essex have less rainfall than Bedfordshire. Within a limited area rainfall varies directly with altitude, and the eastern parts of the county are consequently drier than the western parts. In the lower parts of the Ouse valley (altitude less than 100 ft.) the average annual rainfall is about 20 ins. but in the higher parts of the Chilterns (altitude over 700 ft.) the annual rainfall is above 28 ins. Rainfall records are too few to allow an accurate rainfall map to be made but the one given (Figure 5) is approximately correct.

The more reliable statistics show that the months of greatest rainfall are October, November and December, which account for 30 per cent of the total annual rainfall. February, March and April are the months of least rainfall, accounting for only 20 per cent of the yearly total. It should be noted that the driest months are those of the growing period, and this must have a direct bearing on the plant life of the county.

Temperature
There are even fewer statistics of temperatures, but the more reliable records show a wide range. In January and February the average temperature (corrected to sea-level conditions) is less than 40°F., about the same as it is in places of comparable altitude in the north of Scotland. In July the average temperature is between 61°F. and 62°F. The annual range of temperature of 22°F. is higher than in any other part of the country except East Anglia.

Humidity
Plant distribution depends a great deal upon humidity, records of which are lacking for the county. Records for a wider field show that Bedfordshire lies to the east of the area within Britain with the least humidity. It can, however, be assumed that within the county, other things being equal, greater humidity prevails in the more wooded parts and in the regions of marshes, rivers and ponds. It also varies greatly during the day and at different times of the year.

The physical features outlined above show that the county, at least in its eastern parts, shares the relatively continental conditions of East Anglia. The comparatively large range of temperature with low rainfall of the eastern border gives way to milder conditions in the western parts of the county. These changes are too small to affect greatly plant distribution within the county, but a few species, limited presumably by climatic factors to Eastern England, have their western limit in the county, and other species which are common in Western Britain but rare or absent in East Anglia are frequent in the west of the county but absent in the east. These species, which help to give individuality to the Bedfordshire flora, are considered in some detail on pp. 134–139.

GEOLOGICAL FACTORS

A study of the geology is of prime importance as it determines very largely the distribution of plants within the county. The solid geology is based on five main formations, the relation of which to each other and to the minor formations is seen in the table below:

Formation	Estimated thickness in feet
Upper Chalk	100
Middle Chalk (including Chalk Rock and Melbourn Rock) . . .	200–220
Lower Chalk (including Totternhoe Stone and Chalk Marl) . .	150–180
Gault	170–300
Lower Greensand	220–280
Kimmeridge Clay	10
Ampthill Clay	40–60
Oxford Clay	300–400
Kellaways Rock	10–50
Cornbrash	2–15
Great Oolite Clay	5–10
Great Oolite Limestone . . .	25–30
Upper Estuarine Series . . .	15–20

The oldest rocks (the last six in the above table) are exposed only in the north-west of the county, in the Upper Ouse Valley, in the small Nene valley and in the deep cuttings of the railway; otherwise they are covered by a thick cap of Boulder Clay. In earlier times the oolitic limestone was quarried for building-stone (many of the houses in this area, *e.g.* at Pavenham, Felmersham and Harrold, are built of local stone) and for making lime for agricultural use.

Apart from the older rocks the whole of the northern half of the county rests on the Oxford Clay and the almost indistinguishable Ampthill Clay. In the north of the county the Oxford Clay is shallow, but further south in the neighbourhood of Stewartby and Marston Mortaine, where it is quarried extensively for brick making, it reaches a depth of 400 ft. To the north of the Ouse it is overlain by a large cap of Boulder Clay, which is here as much as 100 ft. thick. In the Lower Ouse and Ivel valleys there are considerable stretches of alluvium and river gravels. The former provide some of the richest agricultural land in the county and the latter are becoming extensively worked for all the many purposes to which sand and gravel can be put.

To the south of the Oxford Clay is the Lower Greensand, which stretches in a narrow belt never more than five miles wide in a north-easterly direction from Leighton Buzzard to the Ivel valley,

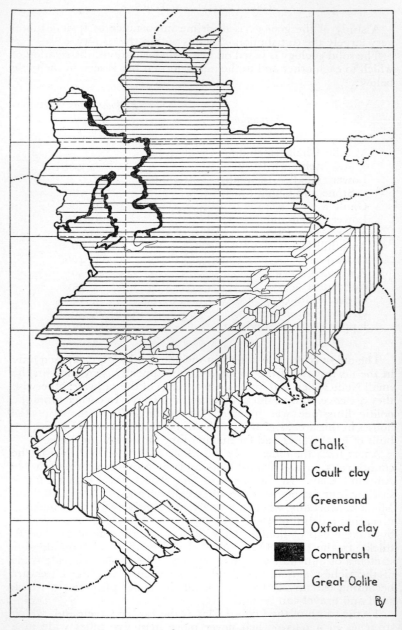

FIG. 6   SOLID GEOLOGY

Chalk
Gault clay
Greensand
Oxford clay
Cornbrash
Great Oolite

where it is broken, and reappears to the east of the valley in the neighbourhood of Sandy and Potton. The county possesses more greensand than any other county north of the Thames. The sand itself is very variable and includes strata useful for glass making, moulding and building purposes, and in consequence is extensively quarried, especially in the western part of the county, where the overburden is slight and the sand is at its greatest thickness. Elsewhere it is overlain by drift deposits, mainly Boulder Clay, and there are considerable alluvial deposits in most of the river valleys. In the 18th century much of the alluvium was undrained moor and bog land, isolated survivals of which may still be seen at Flitwick Moor, Westoning Moor, Cooper's Hill and Sutton Fen. The drift deposits in the neighbourhood of Biggleswade are most complicated, for in addition to river gravels and alluvial deposits there are large deposits of Glacial Gravel and some Brick Earth, the origin of which is still to be explained.

A belt of Gault, also at no place more than five miles wide, lies to the south of the Lower Greensand. It is overlain by Boulder Clay, especially in the east of the county, and there are caps of Glacial Gravel, the most prominent of which is the hill upon which Toddington stands.

The Chalk is the underlying rock of the south of the county. Its northern face forms a fine escarpment, below which the Lower Chalk has washed down to the Gault to form the Chalk Marl. Until recent drainage this was marshy and of considerable botanical interest. Dividing the two formations is the Totternhoe Stone, at one time much used for building. At Totternhoe itself it is about 20 ft. thick. Among the hills of the Lower Chalk are Totternhoe Knolls, Sharpenhoe Clappers, Knocking Hoe and the lower parts of Barton Hills. Both the Chalk Marl and the Lower Chalk are extensively quarried for cement manufacture.

Above the Lower Chalk and dividing it from the Middle Chalk is the Melbourn Rock, which is scarcely visible in the county. The Middle Chalk forms fine hills in Deacon Hill, the higher parts of Barton Hills and Galley Hill. Capping the Middle Chalk is the Upper Chalk, the two divided by the almost negligible Chalk Rock. The Upper Chalk gives the highest hills of the county with Whipsnade Downs, Dunstable Downs, Blow's Downs and Warden Hill. There are small deposits of Glacial Gravel on the Middle Chalk, which is also the southern limit of the Boulder Clay.

Much of the Upper Chalk is covered with a cap of Clay-with-Flints, the oldest drift deposit in the county. This is up to 30 ft. thick and was until comparatively recently quarried for brick making at Stopsley and Caddington. Of the streams draining the Chalk to the south, the Lea is the only one of importance. In association with this there are river gravel deposits at Leagrave and alluvial deposits at East Hyde.

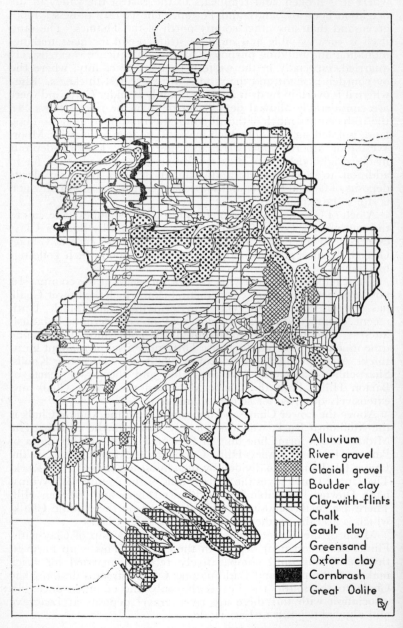

Alluvium
River gravel
Glacial gravel
Boulder clay
Clay-with-flints
Chalk
Gault clay
Greensand
Oxford clay
Cornbrash
Great Oolite

FIG. 7   SURFACE GEOLOGY

SOILS

by T. Laflin

As no soil survey has been carried out in Bedfordshire accurate knowledge of the distribution of the soil types is lacking. Classification under Series headings would, no doubt, correlate many of our soils with Series already described from other parts of England. Some of these are, in fact, recognisable. The soil map given here is compiled from personal knowledge and observation, and attempts to give no more than an outline of the general distribution of the main soil types as an aid to the study of the flora.

In a highly cultivated agricultural county the brown-earth type of soil predominates over wide areas, but because much of the farm land is derived from heavy, calcareous clays, naturally poorly drained, the gley horizon often extends practically to the surface, and such soils have been segregated as calcareous surface-water gleys. The flood plains of the great rivers, too, tend towards the brown-earth type, but the relatively high, though fluctuating, water table, with impeded drainage, places them in the ground-water gley or meadow-soil group. Reasonably well-drained river valley soils with a moderately high water table have been separated as alluvia, and peats have been mapped separately. Finally, the brick-earths have been shown separately, as representing a fine-grained, loamy type, possibly of loessic origin. It must be remembered that agricultural practice may constantly modify the differences between the gleys and the brown earths, drainage and cultivation transforming the first into the second, whilst with the cessation of cultivation the reverse happens.

Over small areas of the brown earths of the Lower Greensand and the Clay-with-Flints (too small to map) paucity of basic salts and rapid leaching have caused podsols to develop. Still more rarely, waterlogging of the surface of the podsolized soil has built up a surface humus to a depth almost justifying their classification as raised bogs. Mermaid's Pond and the silted-up lower pool on Wavendon Heath are examples. All phases are found, from degenerate brown earth through ordinary podsol to the extreme type with a deep, wet surface humus layer, although they are not common. Again, the condition is not always static, drainage and cultivation changing the podsol types back to brown earths and *viceversa*.

Rendzinas derive their character mainly from the nature of the parent rock rather than from climate, which is the major factor in the formation of brown earths and, generally, though not in Bedfordshire, of podsols. Rendzinas derived from the Chalk are termed white rendzinas and those from the Jurassic Limestones red-brown rendzinas, because of their colour. The rare and tiny areas of alkaline soil on coprolitic rock on the Lower Greensand, as exposed in a quarry at Houghton Conquest, are also red-brown rendzinas.

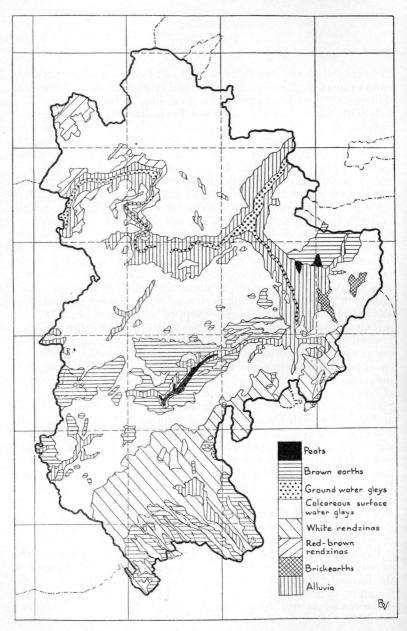

FIG. 8   SOILS

PLATE 2   SHARNBROOK SUMMIT

This shows the cutting to the south of the Forty-Foot Lane. Habitat Studies 1a and
1b were made a short distance to the north of the lane

PLATE 3   WEST WOOD (Habitat Study 2a)

PLATE 5   OAKLEY BRIDGE (Habitat Study 7)

PLATE 4   JUDGE'S SPINNEY (Habitat Study 5)

On gentle slopes, with cultivation and the incorporation of humus, the rendzinas incline towards the brown-earth type.

Soil Series have been described from similar geological formations in other parts of south-east England. Soils derived from the Oolite and Cornbrash do not entirely agree with the descriptions of any of the Series of the Cotswold group described by Kay (1934) or Robinson (1948), probably because in Bedfordshire these formations are limited to comparatively narrow bands on slopes bordering the Upper Ouse and the Nene. Close proximity of both solid and drift formations above and below, with erosion on the slopes mixing the surface material, make the soils difficult to classify. Robinson describes a Newport Pagnell Series as 'fluviatile drift of flinty and oolitic sand and sandy clay over oolitic limestone or clay', a light to medium, acid to neutral loam with imperfect drainage, which approximates to some of our valley soils along the Upper Ouse.

The Oxford and Ampthill Clays are stiff, generally calcareous clays which, with the chalky Boulder Clay which is usually similar in texture, give a rather monotonous vegetation. The Ampthill Clay, often rather dark in colour, may be non-alkaline in places. Oxford Clay soils in Bedfordshire are fairly homogeneous, the main modifications being those caused by cultural practices. I can find no Series descriptions which fit our soils of these formations. The Lower Greensand in Bedfordshire is, with the exception of the tiny alkaline areas mentioned, an acid sand, highly ferruginous over most of the county and with ironstone sometimes near the surface. Many of the soils are similar to the Bearsted and Lowlands Series described by Bane and Jones (1934) from the Folkestone Beds of the Lower Greensand in Kent. The first is a typical shallow, sandy, sedentary soil overlying sand. The second is a colluvial sand-loam of valley bottoms. Podsolized soils approximate to the Hoth-field Series of the same authors. The peaty soils of the flood valley of the Flit, of Sutton Fen, and of other small areas, have no parallel with described soils, nor have the few red-brown rendzina soils. The soils of the Gault are a fairly uniform heavy, calcareous clay. Kay's Series from the Vale of the White Horse may possibly be applicable to them.

Along the foot of the Chalk are colluvial marls, recognisable as the Wantage and Blewbury Series of Kay and the Gore Series of Lee (1931). These three soils are rather similar but vary in the depth of the downwash material. The chalk soils, too, fit well into Kay's Series described from Berkshire: Icknield, a chalk-upland soil under pasture, and Upton, similar, but under arable conditions. Much of the Chalk is capped by deposits of Clay-with-Flints. Soils corresponding to Lee's Rattle Series are widespread; Batcombe (Robinson), Wallop (Kay), and Charity (Kay), probably also exist here. In addition there are areas of Eocene sands and gravels, usually acid.

The Boulder Clay covers large tracts and, as with most drift deposits, shows considerable variation, even over limited areas.

D

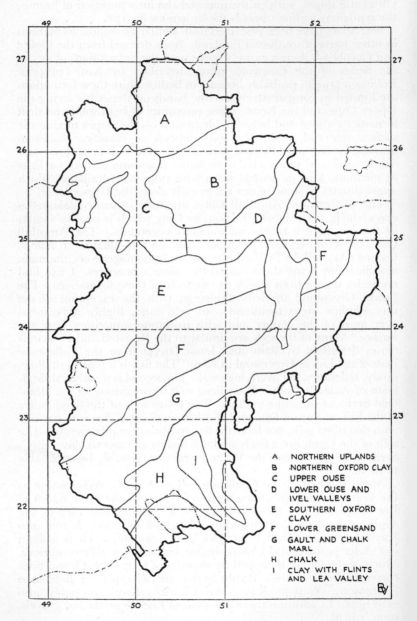

FIG. 9   NATURAL REGIONS

A   NORTHERN UPLANDS
B   NORTHERN OXFORD CLAY
C   UPPER OUSE
D   LOWER OUSE AND
    IVEL VALLEYS
E   SOUTHERN OXFORD
    CLAY
F   LOWER GREENSAND
G   GAULT AND CHALK
    MARL
H   CHALK
I   CLAY WITH FLINTS
    AND LEA VALLEY

Generally, however, it is calcareous and quite stiff in texture, although even sandy loams and beds of flints with chalky material between are found. The alluvial soils are also exceedingly variable in composition and drainage, and would be very complex to classify.

Bibliography

Bane, W. A., and Jones, G. H. G. (1934), Fruit Growing Areas on the Lower Greensand in Kent, *Bull. Min. Agric. and Fish.*, 80 (London).

Kay, F. F. (1934), *A Soil Survey of the Eastern Portion of the Vale of the White Horse* (Reading).

Lee, Linwood L. (1931), The Possibilities of an International System for the Classification of Soils, *Journ. S.E. Agric. Coll.*, 28, pp. 65–114.

Robinson, K. L. (1948), in Good, R., *Geographical Handbook of the Dorset Flora* (Dorchester).

A STUDY OF THE FLORA

NATURAL REGIONS

The physical and geological factors outlined above allow a division of the county into well-marked natural regions. There were a number of earlier attempts to do this. Batchelor (1808) made five regions based on soils, but his soil classification would not be acceptable today. A more reliable basis was suggested by Hillhouse (1877). He was obviously influenced by Babington, whose *Flora of Cambridgeshire* (1860) was based on eight natural districts. Hillhouse found no difficulty with the southern half of the county, which he conveniently divided into three approximately equal districts based on the Lower Greensand, Gault and Chalk. The northern half gave him more difficulty and for no apparent reason he divided it into four more or less equal districts, each roughly equal to the southern three. He did not name his districts, but they may be described as

1. Upper Ouse and Oolite.
2. North Western.
3. Lower Ouse.
4. Eastern Oxford Clay.
5. Lower Greensand
6. Gault.
7. Chalk.

Each of his seven districts he further divided into seven sub-districts 'to induce thorough and systematic working', for he intended his system to be a basis for the study of plant distribution. His boundaries were 'either river, road or rail'.

Attractive as is Hillhouse's system I have, with some regret, rejected it as it groups together the very different floras of the Chalk and Clay-with-Flints, separates the very similar Lower Ouse and Ivel valleys and introduces artificial divisions into the uniform area north of the Ouse. It can legitimately be abandoned, for no useful field work has been based on it.

The natural regions adopted in this Flora have some of the drawbacks of Hillhouse's system, since the large area of the county north of the Ouse, which is almost entirely capped with Boulder Clay, defies satisfactory division. For reasons given below the regions are not used as a basis for the study of plant distribution, so no attempt has been made to define them precisely.

The regions are:

A. *Northern Uplands*
B. *Northern Oxford Clay*
C. *Upper Ouse*
D. *Lower Ouse and Ivel Valleys*
E. *Southern Oxford Clay*
F. *Lower Greensand*, in two portions
G. *Gault and Chalk Marl*
H. *Chalk*
I. *Clay-with-Flints and Lea Valley*

It would have been useful to make the regions the basis for the study of plant distribution but I have not done so for five reasons:

a. Drift deposits overlap the underlying rocks and arbitrary lines would have to be drawn through areas having a uniform flora.

b. The shape of some regions, *e.g.* the *Gault and Chalk Marl*, would have made field work based on them difficult. To have divided them further would have been a reversion to Hillhouse and serious work would have been almost impossible.

c. The detached portions of the *Lower Greensand* would have required separate study; otherwise a false impression would have been given of plant distribution.

d. As Hillhouse realized, boundaries must be related to something visible in the field. It was found impossible to devise such dividing lines.

e. No serious field work had been done in the county on a basis of natural regions, and in Cambridgeshire alone among neighbouring counties had such a system at any time been used. On the other hand J. E. Little had done some useful work on the Ivel District based on the traditional botanical district system of catchment areas, and this system had also been used in published accounts of the floras of Hertfordshire, Buckinghamshire and Northamptonshire.

Natural regions are used in this work as a basis for the study of the various types of flora which are to be found in the county. This study has been built largely on a close examination of a few small areas. In choosing this method I have been influenced by R. Good's *Geographical Handbook of the Dorset Flora* (1948). My 'habitat studies' are much less numerous than his 'stands' but are probably more detailed.

It is hoped that the habitat studies will serve some useful purposes. The most important is to describe in detail a number of plant communities typical of the flora of the county. With two exceptions the sites were previously determined. Had they been chosen in a purely mechanical manner, *e.g.* points of intersection of grid lines on the map, it would have given 60 per cent of the studies on arable land and 10 per cent on urban sites. Few would have appeared in natural habitats.

The studies may give more information regarding the frequency and distribution of the very common plant species than is usual in a local flora. The cross-references in the body of the Flora give as many as fifty stations for some of the more common species and within these stations details of local frequency, plant associates and types of soils on which the species grows.

They may enable field botanists of a future generation to determine with some degree of accuracy changes which may occur in the flora. For this purpose the centre of the station has been clearly defined by map references and field directions. At least two stations have changed materially since the studies were made but a number will no doubt undergo little change as time passes.

Each station was visited at least three times, in spring, summer and autumn. Except where indicated the study was limited to an area having a radius of five yards from the centre of the station. The frequencies of the species were calculated by visual impression—the work could not have been done on any other basis. In cases where a difference in frequency was noted at different seasons that which showed the greatest abundance was finally adopted.

The list of bryophytes appended to each study should be taken to be complete only when it was made in company with T. Laflin.

No attempt was made to estimate frequencies of weeds when arable fields were the subject of studies. The cleanness of the crop determines the frequency of the weeds but the lists of weeds given may be helpful to future botanists.

Abbreviations used in the lists are those in common use by ecologists:

d	dominant
a	abundant
la	locally abundant (*i.e.* in one part of the area of survey only)
f	frequent
o	occasional
r	rare
1	one plant only
L	a species found, but not abundantly, in one part of the area only
M	a species found growing on the margin of the area
sp.	an unidentified species of a genus
subsp.	subspecies
agg.	the name of the species is used in the broad and not in the restricted sense

The pH values of the soil were made with a B.D.H. Capillator from samples taken at a depth of 6 ins. A pH value of 7 indicates that the soil is neutral, a pH value of more than 7 marks an alkaline soil and a pH value of less than 7 an acid soil.

A.　The Northern Uplands Region

This small region has a mixed geological basis. The underlying rock of the area to the north of the Forty Foot Lane, and including the parishes of Wymington and Podington, is the Great Oolite Series. The rest of the region lies on the Oxford Clay. The higher land is overlain with large caps of Boulder Clay, the sedentary rocks being exposed only in the valleys of the streams flowing into the Nene and Kym, or in railway or road cuttings.

When the oolite comes to the surface or is exposed it is made evident by a flora in which either *Brachypodium pinnatum* or *Bromus erectus* is dominant. In rough pastures there are usually to be found *Viola hirta, Silaum silaus, Pimpinella saxifraga* and *Primula veris*, and there may appear *Genista tinctoria, Lotus tenuis, Sanguisorba officinalis, Pimpinella major, Erigeron acris, Lithospermum officinale, Orchis morio* and *Anacamptis pyramidalis*.

Habitat Study 1a   Sharnbrook Summit (west aspect)
Parish: Wymington   District: Nene   Grid Ref. 42/968624
Alt. 330 ft.   Soil: Great Oolite Limestone   pH 7.4
Surveyed: 10 July 1949 with H. B. Souster; 28 Aug. 1949 with P.T. and W. Durant; 16 April 1950 with T.L. and H. B. Souster.
The large railway cutting known as Sharnbrook Summit and the baulk covering the tunnel adjacent to it provide plant associations not to be found elsewhere in the county. In spring the colonies of *Cerastium* spp. are of especial interest, and later in the year the *Hieracia* well repay study. *Arabis hirsuta* is in greater quantity here than elsewhere in the county. The tunnel baulk is dominated by patches of *Trifolium medium* and *Melilotus officinalis*. The study was made 100 yards north of the bridge crossing the Forty-Foot Lane.

Abundant
Erophila verna   Lotus corniculatus   Bromus erectus
Linum catharticum   Trifolium campestre

Frequent
Achillea millefolium   Leontodon hispidus   Poa pratensis
Brachypodium pinnatum   Medicago lupulina   subsp. angustifolia
Chrysanthemum leucan-   Melilotus officinalis   Potentilla reptans
  themum

Occasional
Bellis perennis   Cirsium acaulon   Filipendula vulgaris
Clinopodium vulgare   Daucus carota   Vicia cracca
Centaurea nigra

Local
Genista tinctoria (M)   Lathyrus pratensis
Geranium robertianum (on stonework)   Lithospermum officinale (M)
Hieracium exotericum   (do.)   Thymus pulegioides
H. pilosella   (do.)   Trifolium medium

Rare
Briza media   Plantago lanceolata   Stellaria media
Cirsium arvense (1)   Polygala vulgaris   Taraxacum officinale
Crepis capillaris   Ranunculus bulbosus   Tragopogon minor
Dactylis glomerata   Rubus caesius × ? (1)   Trifolium pratense
Galium mollugo   Rumex acetosa   Veronica polita
G. verum   Senecio erucifolius   Viola hirta
Picris hieracioides

Bryophytes (confined mainly to the stone-filled drainage channels on the margin of the survey): Amblystegium serpens (r), Barbula convoluta (a), Brachythecium rutabulum (a), B. velutinum (a), Bryum caespiticium (a), B. capillare (a), Ceratodon purpureus (a), Eurhynchium swartzii (o), Fissidens taxifolius (o), Funaria hygrometrica (r) Mnium longirostrum (o), Tortula ruralis (o).

Habitat Study 1b   Sharnbrook Summit (east aspect)
This study was made on the bank immediately opposite the previous study. The
plant growth here was noticeably thicker and the species common to each study
were more robust on the east aspect. The details are the same as for Habitat
Study 1a except that the pH value was 7.5.

Abundant

Arrhenatherum elatius	Melilotus officinalis

Frequent

Brachypodium pinnatum	Ononis spinosa	Potentilla reptans
Chrysanthemum leucan-	Poa pratensis	Poterium sanguisorba
themum	subsp. angustifolia	Vicia cracca
Medicago lupulina		

Occasional

Achillea millefolium	Daucus carota	Senecio erucifolius
Briza media	Lathyrus pratensis	Silaum silaus
Centaurea nigra	Leontodon hispidus	Trifolium medium
Cirsium acaulon	Pimpinella saxifraga	T. campestre
Clinopodium vulgare	Plantago lanceolata	Viola hirta
Convolvulus arvensis	Rubus caesius × ?	

Local

Linaria vulgaris	Valeriana officinalis

Rare

Acer campestre 3 ft. (1)	Galium verum	Primula veris
Crepis capillaris	Geranium robertianum	Rosa 'canina' 3 ft. (1)
Dactylis glomerata	Heracleum sphondylium	Silene vulgaris
Dactylorchis fuchsii	Holcus lanatus	Stachys sylvatica
Equisetum arvense (1)	Plantago media	Tamus communis
Fragaria × ananassa		

Bryophytes: Amblystegium serpens (o), Barbula convoluta (o), Brachythecium
rutabulum (a), B. velutinum (a), Bryum caespiticium (a), B. capillare (a), Ceratodon purpureus (a), Dicranella varia (o), Eurhynchium confertum (o), Eurhynchium swartzii (o), Fissidens taxifolius (o), Mnium longirostrum (o), Pohlia nutans
(1a), Rhacomitrium lanuginosum (1a), Tortula ruralis (o).

Open areas in this region include Yelden Wold and Newton
Gorse, once rough grassland but now mainly under the plough.
They contain little of botanical interest. Gorse itself (*Ulex europaeus*)
is rare and a few bushes in the Forty-Foot Lane account for the
total population. The Boulder Clay soon develops a scrub flora if
left untended and there are few examples of good well-established
pasture. A transitional stage is seen in the wide rides of West Wood.

Habitat Study 2a   West Wood (open ride)
Parish: Knotting and Souldrop   District: Kym   Grid Ref. 42/989625
Alt. 310 ft.   Soil: Boulder Clay overlying Oxford Clay   pH 7.4
Surveyed: 10 July 1949 with H. B. Souster; 16 April 1950 with H. B. Souster and
T.L.; 7 Sept. 1950 with W. Durant. This study was made on the wide main ride
of the wood at a point 200 yards from the Bedford-Rushden (A.6) road.

Abundant

Festuca arundinacea

Frequent

Cynosurus cristatus	Hypericum hirsutum	Trisetum flavescens
Dactylis glomerata	Lotus corniculatus	Valeriana officinalis
Filipendula ulmaria	Plantago lanceolata	Vicia cracca
Galium verum	Trifolium medium	

Occasional

Agrimonia eupatoria	Hypericum perforatum	Prunella vulgaris
Arrhenatherum elatius	Heracleum sphondylium	Succisa pratensis
Clinopodium vulgare	Holcus lanatus	Trifolium pratense
Fragaria vesca	Lotus uliginosus	Viola hirta
Galium mollugo	Odontites verna	

Local

Agrostis tenuis	Festuca longifolia	P. sterilis
Angelica sylvestris	Potentilla anserina	

Rare

Achillea millefolium	Euphrasia nemorosa	Prunus spinosa 3 ft. (M)
Bromus ramosus	Festuca rubra	Quercus robur 4 ft. (M)
Centaurea nigra	Filipendula vulgaris	Rosa 'canina' 3 ft.
Cerastium holosteoides	Helictotrichon pratense	Rubus sp. 2 ft.
Chrysanthemum leucan-	Leontodon autumnalis	Sanguisorba officinalis
themum	L. hispidus	Senecio jacobaea
Cirsium arvense	Lolium perenne	Stachys sylvatica
Dactylorchis fuchsii	Medicago lupulina	Thelycrania sanguinea
Deschampsia cespitosa	Phleum nodosum	Torilis japonica
Dipsacus fullonum	Poa pratensis	Trifolium repens

Bryophytes: Brachythecium rutabulum, Eurhynchium praelongum, Fissidens taxifolius, Pseudoscleropodium purum.

A fine belt of natural woodland stretches across the ridge of Boulder Clay. This includes Harrold Park Wood, Dungee Wood Odell Great Wood, Great Hayes Wood, Halsey Wood, West Wood, Galsey Wood, Worley's Wood and Swineshead Wood. Most of the woods suffered badly by becoming ammunition stores during the war and have not yet fully recovered. They are usually oak-ash woods with coppiced hazel. The spring carpet includes *Anemone nemorosa, Viola riviniana, V. reichenbachiana, Primula vulgaris* and *Orchis mascula*. In summer the woods contain various species of *Epilobium, Arctium minus, Carex sylvatica, Dactylorchis fuchsii, Dryopteris filix-mas, Milium effusum, Neottia nidus-avis, Paris quadrifolia* and *Platanthera chlorantha*. In autumn most woods add *Euphrasia nemorosa, Odontites rubra, Centaurium minus, C. pulchellum* and *Epipactis latifolia*. In open places in many woods *Daphne laureola* may be found.

Habitat Study 2b   West Wood (woodland)
Details as for Habitat Study 2a.   pH 7.4
This study was made about 20 yards along a narrow path to the north of site H.S. 2a. The wood here had been cleared a few years previously and the undergrowth allowed to grow freely. By 1949 it had grown to a height of about 12 feet. *Shrub layer*: Acer campestre (r), Crataegus monogyna (f), Euonymus europaeus (r), Ligustrum vulgare (f), Lonicera periclymenum (o), Prunus spinosa (f), Rosa arvensis (f), Salix caprea (r).

Frequent

Anemone nemorosa	Endymion non-scriptus	Ranunculus ficaria
Bromus ramosus	Filipendula ulmaria	Vicia cracca
Deschampsia cespitosa	Primula vulgaris	

Occasional

Agrimonia eupatoria	Geranium robertianum	Listera ovata
Ajuga reptans	Heracleum sphondylium	Mercurialis perennis
Brachypodium sylvaticum	Juncus effusus	

Rare

Aegopodium podagraria   Fragaria vesca   Primula veris × vulgaris
  (1)   Glechoma hederacea   Valeriana officinalis
Angelica sylvestris (1)   Luzula pilosa   Vicia sepium
Arum maculatum   Platanthera chlorantha   Viola reichenbachiana
Dactylorchis fuchsii   Potentilla sterilis

Bryophytes: Acrocladium cuspidatum (r), Brachythecium rutabulum (f), Cirriphyllum piliferum (o), Eurhynchium praelongum (f), Fissidens taxifolius (o),
Mnium undulatum (o), Pseudoscleropodium purum (o), Thuidium tamariscinum
(o).

Habitat Study 3a   Great Hayes Wood (ride)
Parish: Podington   District: Nene   Grid Ref. 42/96576185
Alt. 330 ft.   Soil: Boulder Clay overlying Great Oolite Limestone   pH 7.0
Surveyed: 10 July 1949 with H. B. Souster; 28 Aug. 1949 with W. Durant and
P.T.; 16 April 1950 with H. B. Souster.
This wood has an unusual flora; it is the only natural wood of any size in the county
in which Primrose (*Primula vulgaris*) is absent. In one corner there is a good fern
colony containing *Athyrium filix-femina*, *Dryopteris borreri* and *D. dilatata*; this is of
especial interest as ferns are exceedingly rare in this part of the county. The study
was made 63 yds. to the east of the first crossing of paths on the south-east side of
the wood. The vegetation here is very rich and marsh plants intrude as they
often do in damp rides and paths in woods on the Boulder Clay.

Frequent

Carex sylvatica   Juncus effusus   Poa pratensis
Glechoma hederacea   J. conglomeratus   Potentilla reptans
Hypericum hirsutum   Lysimachia nummularia   Sagina procumbens

Occasional

Brachypodium sylvati-   Filipendula ulmaria   Mercurialis perennis (L)
  cum   Fragaria vesca   Myosotis arvensis
Briza media   Holcus lanatus   Prunella vulgaris
Carex flacca   Juncus inflexus   Rumex sanguineus
Centaurium minus   Medicago lupulina   Valeriana officinalis
Dactylorchis fuchsii   Mentha aquatica   Viola hirta

Rare

Agrostis stolonifera   Epilobium montanum   Lolium perenne
Arctium minus   Galium mollugo   Picris echioides
Bellis perennis   G. palustre   Plantago major
Carex otrubae   Geranium robertianum   Potentilla anserina
Cirsium palustre   Geum urbanum   Ranunculus repens
Cynosurus cristatus   Hypericum tetrapterum   Senecio erucifolius
Deschampsia cespitosa   Juncus articulatus   Trifolium repens

Habitat Study 3b   Great Hayes Wood (woodland)
Details as for Habitat Study 3a   pH 5.2
This study was made in the wood itself, immediately to the north of Habitat Study
3a. The trees were mainly *Betula pendula* growing to a height of about 50 ft., with
boles of a girth of 27 ins. There were occasional trees of *Acer pseudoplatanus* and
Fraxinus excelsior, and an undergrowth mainly of *Corylus avellana*. The wood was
close-growing and few herbaceous plants were evident after the spring.
Shrubs (limited mainly to the margin of the path): Crataegus monogyna, Ligustrum
vulgare, Lonicera periclymenum, Prunus spinosa, Rosa arvensis, Rubus ? vestita,
Salix ? atrocinerea, Thelycrania sanguinea, Viburnum lantana.

Frequent

Endymion non-scriptus   Mercurialis perennis (L)

Occasional

Ajuga reptans	Carex sylvatica	Valeriana officinalis (M)
Arum maculatum	Fragaria vesca	Veronica officinalis
Circaea lutetiana	Geum urbanum	Viola riviniana

Rare

Epipactis helleborine   Orchis mascula

Bryophytes (incomplete): Eurhynchium striatum, Hypnum cupressiforme, do. var. resupinatum, Thuidium tamariscinum.

A large area of scrub known as Tilbrook Bushes, v.c. 30 [Hunts.], contains a large quantity of *Agrimonia odorata* and merits closer study. At the east end of Forty-Foot Lane is a small wood, Wymington Scrub, which is damp in parts and has an unusual mixture of marsh and woodland plants. Elms are especially fine in this part of the county, the more frequent species being *Ulmus carpinifolia, U. coritana* and *U. plotii,* but hybrids are more frequent than the species themselves.

There are no marshes, but some small marshy areas to the north of Wymington contain *Carex disticha* and *Juncus compressus.* Ponds are few, but the lakes at Melchbourne Park and Hinwick House repay study. Both are colonized in parts by *Hippuris vulgaris,* and their marshy fringes add other species which would otherwise be absent from this region. The streams provide little of interest except for one ditch near Pertenhall which has some quantity of *Samolus valerandi.*

What the region lacks in its native flora is compensated by the variety of the weeds of its arable land. When the Oolite is near to the surface, as at Wymington and Podington, most fields contain *Melandrium noctiflorum, Torilis arvensis, Galium tricorne, Kickxia spuria, K. elatine* and *Euphorbia platyphyllos.* The common mosses of these fields are *Barbula unguiculata, Phascum cuspidatum, Pottia intermedia* and *P. truncata.* Further east where the Boulder Clay overlies the Oxford Clay the weeds are of less interest and it should be recorded that *Knautia arvensis* seems to be absent.

Habitat Study 4   Arable field, Wymington
Parish: Wymington   District: Nene   Grid Ref. 42/96406275
Alt. 330 ft.   Soil: Boulder Clay overlying Great Oolite Limestone   pH 7.5
Surveyed: 10 July 1949 with H. B. Souster; 16 April 1950.

Aethusa cynapium	E. platyphyllos	Poa trivialis
Alopecurus myosuroides	Galium aparine	Polygonum aviculare
Anagallis arvensis	Geranium dissectum	P. convolvulus
Arrhenatherum elatius	Kickxia spuria	Prunella vulgaris
Bromus commutatus	Lapsana communis	Ranunculus repens
Carduus crispus	Lithospermum arvense	Rumex crispus
Cerastium holosteoides	Lolium multiflorum	Scandix pecten-veneris
Cirsium arvense	Matricaria recutita	Trifolium pratense
Convolvulus arvensis	Medicago lupulina	T. repens
Daucus carota	Odontites rubra	Vicia hirsuta
Euphorbia exigua	Phleum nodosum	

The walls in this part of the county are usually built of oolitic limestone and add *Sagina ciliata, Sedum album, Saxifraga tridactylites* and

Poa compressa to the flora. Their moss flora is also of interest, with *Tortula muralis*, *T. intermedia*, *Grimmia pulvinata*, *Barbula unguiculata*, *B. fallax*, *Pterygoneurum ovatum*, *Camptothecium sericeum*, *Bryum capillare* and *B. argenteum*. The railway walls add *Asplenium adiantum-nigrum* and *Phyllitis scolopendrium*.

B. The Northern Oxford Clay Region

There is a dull uniformity about this part of the county that, botanically at least, makes it uninteresting. The Oxford Clay is overlain with large caps of calcareous Boulder Clay, and the undulating country is intensively cultivated. Grain crops predominate, but permanent pasture is maintained in some fields. To the south-west of Colmworth there was until about 1941 an area of rough pasture known as the Wilderness; but the only relic of its once relatively interesting flora is *Ranunculus parviflorus*, which still survives in fields on the northern side. The pastures usually have *Primula veris*, *Orchis morio* and *Carex caryophyllea*, and when the calcareous content of the Boulder Clay is high[1] *Linum catharticum*, *Blackstonia perfoliata*, *Thymus pulegioides*, *Ophrys apifera* and *Anacamptis pyramidalis* may appear.

The roadsides have two species which relieve the otherwise monotonous nature of this region: *Ornithogalum pyrenaicum* is abundant in the eastern parts, appearing also in less quantity in small copses, and in the same area *Trifolium ochroleucon* frequently dominates considerable stretches of the roadside.

This is the least wooded part of the county. Twin Wood, which was found profitable by Abbot and other Bedford naturalists, suffered badly in the late war, and it is possible that *Geum rivale*, *Melampyrum cristatum* and *Elymus europaeus* are lost here. *Melampyrum cristatum* is still to be seen near Colmworth, as is *Helleborus viridis*. Among other woods the inaccessible Clapham Park Wood would probably repay more study. The many copses in the region usually contain *Daphne laureola* and *Iris foetidissima*.

The wood which has interested me most is Judge's Spinney.

Habitat Study 5   Judge's Spinney
Parish: Oakley   District: Ouse   Grid Ref. 52/01755414
Alt. 225 ft.   Soil: Boulder Clay overlying Oxford Clay   pH 7.3
Surveyed: 17 July 1949 with W. Durant; 16 April 1950 with H. B. Souster and T.L.; 7 Sept. 1950 with W. Durant.
This is the only example of a beech wood not on the Chalk in Bedfordshire. It is of some antiquity, and notwithstanding its small size (about 5 acres) has some interesting species, including *Monotropa hypopitys*. The study was made 50 yards along the main path on the north-west side of the wood. The trees (*Fagus sylvatica*) were about 60 ft. high and about 15 ft. apart.
Shrubs: Acer campestre 6 ft. (r), A. pseudoplatanus 6 ins. (1), Carpinus betulus 12 ft. (1), Corylus avellana 10 ft. (r), Crataegus monogyna 4ft. (r), Fraxinus excelsior 6 ins. (1), Hedera helix (1a), Ligustrum vulgare 6 ft. (M), Quercus robur 6 ft. (r), Ribes uva-crispa 2 ft. (1), Rosa canina 4 ft. (r), Rubus ? vestita 3 ft. (r), Thelycrania sanguinea 6 ft. (r), Ulmus sp. 4 ft. (1), Viburnum lantana 4 ft. (r).

---

[1] At Cleat Hill, the best example, in addition to the species listed, *Inula helenium* is also found.

Abundant
Sanicula europaea

Frequent
Bromus ramosus

Occasional

Arum maculatum	Daphne laureola	Rumex sanguineus
Brachypodium sylvati-	Festuca gigantea	Tamus communis
cum	Fragaria vesca	Viola reichenbachiana

Rare

Arctium minus	Galium aparine (1)	Iris foetidissima
Cephalanthera dama-	Geranium robertianum	Ranunculus acris (1)
sonium (1)	Geum urbanum	Taraxacum officinale
Cirsium vulgare (1)		

Bryophytes: Amblystegium serpens (r), Barbula unguiculata (o), Brachythecium velutinum (o), Bryum capillare (o), Fissidens bryoides (r), F. taxifolius (r), Hypnum cupressiforme (f).

There are no marshes or ponds of more than passing interest and the streams have little worthy of attention.

C.   The Upper Ouse Valley Region

I have given considerable thought to the limits of this region and have finally decided to include only the area directly influenced by the river. This comprises the river itself, the alluvium and river gravels formed by it, and the oolitic and other sedentary rocks exposed on either side of the river valley. Included also is the lower part of the valley of the Crawley Brook, which extends from Bromham to Stagsden. The eastern limit of the region is the town of Bedford. The geology of the region is exceedingly complex and it is almost impossible to follow in the field the pattern so neatly portrayed on the map. Scenically and botanically it is one of the more interesting parts of the county.

The aquatic flora demands some attention. *Nymphaea alba* is the common water-lily of the river, but *Nuphar lutea* is by no means uncommon. In mid-stream the pondweeds are usually *Potamogeton lucens*, *P. perfoliatus* and *P. pectinatus*, and sheltered places by the side of the river usually add *P. natans*, *Zannichellia palustris* and *Polygonum amphibium*. There are fine stands of *Scirpus lacustris* in many stretches of the river, and *Oenanthe fluviatilis*, *Ranunculus fluitans* and *Sparganium erectum* manage to compete very well with the pondweeds and water-lilies. At the river's edge, except where the banks are steep, there is usually to be seen *Achillea ptarmica*, *Acorus calamus*, *Sium latifolium*, *Butomus umbellatus*, *Sagittaria sagittifolia*, *Carex riparia*, *Phalaris arundinacea* and *Glyceria maxima*. A newcomer, *Impatiens capensis*, appears in places in some quantity, especially where the river overflows in winter. Here there are usually some interesting plant communities which include *Arctium lappa*, *Petasites hybridus*, *Lythrum salicaria* and *Stachys palustris*. It is here also that *Cuscuta europaea* often appears.

Habitat Study 6   Riverside, Felmersham
Parish: Felmersham   District: Ouse   Grid Ref. 42/99055785
Alt. 125 ft.   Soil: Alluvium   pH 7.4
Surveyed: 18 June 1949 with W. M. Cornish; 12 April 1950; 7 Sept. 1950 with
W. Durant. This study was made immediately to the south-east of Felmersham
Bridge.

In river

Myriophyllum spicatum   Nymphaea alba (r)   Scirpus lacustris (la)
  (f)

At water's edge

Glyceria maxima (a)   Phalaris arundinacea (a)   Sparganium erectum (f)
Lythrum salicaria (r)

Marshy area beyond water's edge

Epilobium hirsutum (f)   Myosotis scorpioides (o)   Scrophularia aquatica (f)
Filipendula ulmaria (f)   Myosoton aquaticum (f)   Stachys palustris (r)
Galium palustre (o)   Rorippa amphibia (r)   Stellaria media (f)
Impatiens capensis (f)   R. microphylla (o)   Veronica beccabunga (r)
Mentha aquatica (r)

Drier area nearer road

Atriplex patula (r)   Poa annua (r)   Rumex crispus (f)
Alliaria petiolata (o)   P. trivialis (r)   R. conglomeratus (r)
Eupatorium cannabinum   Polygonum hydropiper   R. obtusifolius (r)
  (r)   (r)   Salix fragilis (r)
Glechoma hederacea (o)   P. persicaria (o)   Solanum dulcamara (f)
Petasites hybridus (la)   Potentilla anserina (r)   Sonchus oleraceus (r)
Phleum pratense (r)   Ranunculus ficaria (o)   Urtica dioica (f)
Plantago major (r)

Bryophytes: Fontinalis antipyretica (aquatic), Pohlia delicatula (on river bank).

The following is a similar study, with the zones equally well
marked.

Habitat Study 7   Oakley Bottom
Parish: Oakley   District: Ouse   Grid Ref. 52/00855292
Alt. 115 ft.   Soil: Alluvium   pH 7.3
Surveyed: 10 Aug. 1949 with C. C. Foss; 1 July 1950. The study was made at
the village end of Oakley Bridge. There is a backwater to the river here which adds
to the aquatic flora.

In water

Callitriche intermedia   L. minor   Polygonum amphibium
Ceratophyllum demer-   L. trisulca   Potamogeton pectinatus
  sum   Nymphaea alba   Ranunculus circinatus
Elodea canadensis   Oenanthe fluviatilis   Scirpus lacustris
Lemna gibba

Muddy shore of water

Callitriche stagnalis (l)   Phragmites communis   Scirpus lacustris (ld)
Glyceria maxima (la)   (la)   Sparganium erectum (f)
Myriophyllum verticilla-
  tum (l)

Marshy area

Species listed as local were restricted to the part nearer the muddy shore, those
listed as marginal to the drying-out portion; the rest showed no restricted distribution.

Achillea ptarmica (lf)
Agropyron repens (Mr)
Agrostis stolonifera (Ma)
Alisma plantago-aqua-
    tica (r)
Alliaria petiolata (Mr)
Alopecurus pratensis
    (Mf)
Angelica sylvestris (r)
Apium nodiflorum (lr)
Atriplex hastata (Mr)
Calystegia sepium (o)
Carex otrubae (o)
Cirsium arvense (Mr)
C. vulgare (Mr)
Dipsacus fullonum (r)

Epilobium hirsutum (o)
Filipendula ulmaria (lo)
Galium aparine (Mr)
Geranium dissectum
    (M1)
Hypochoeris radicata
    (Mr)
Juncus acutiflorus (f)
J. articulatus (o)
J. compressus (f)
J. inflexus (o)
Lactuca serriola (M1)
Lychnis flos-cuculi (r)
Lycopus europaeus (f)
Lysimachia vulgaris (r)
Lythrum salicaria (lo)

Malva sylvestris (Mr)
Mentha aquatica (la)
Myosotis scorpioides (la)
Phleum nodosum (Mr)
Poa trivialis (Mr)
Prunella vulgaris (Mo)
Pulicaria dysenterica (r)
Rorippa amphibia (lr)
R. islandica (Mr)
Rumex sanguineus (Mr)
R. crispus (Mr)
R. hydrolapathum (r)
Scrophularia aquatica (r)
Sinapis arvensis (Mr)
Trifolium repens (Mr)
Urtica dioica (Mr)

The marshy shores of the river often merge almost imperceptibly into the water meadows beyond them. Such is frequently the case between Bromham and Bedford where the valley widens.

Habitat Study 8   Riverside, Biddenham
Parish: Biddenham   District: Ouse   Grid Ref. 52/01254965
Alt. 95 ft.   Soil: Alluvium   pH 7.0
Surveyed: 17 June 1950; 5 Sept. 1950. This study was made on the riverside below Biddenham Church at the last marshy depression before the meadow is crossed by a big ditch.

Marshy area

Acorus calamus (la)
Alisma plantago-aqua-
    tica (o)
Berula erecta (r)
Callitriche stagnalis, on
    mud (r)
Caltha palustris (f)
Cardamine pratensis (r)
Glyceria fluitans (r)

G. maxima (f)
Iris pseudacorus (r)
Lythrum salicaria (r)
Mentha aquatica (f)
Myosotis scorpioides (f)
Phalaris arundinacea (1)
Plantago major (r)
Ranunculus repens (o)
Rorippa amphibia (o)

R. microphylla (f)
Rumex hydrolapathum
    (la)
Scirpus lacustris (la)
Senecio aquaticus (o)
Sparganium erectum (r)
Stachys palustris (r)
Veronica beccabunga (r)
V. catenata (o)

Meadow adjoining above

Alopecurus pratensis (a)
Bellis perennis (a)
Cardamine pratensis (r)
Carex hirta (r)
Cirsium arvense (o)
Cynosurus cristatus (f)
Festuca pratensis (f)
Hordeum secalinum (r)

Juncus inflexus (f)
Leontodon autumnalis
    (o)
Lolium perenne (f)
Phleum nodosum (f)
Plantago lanceolata (o)
P. major (o)
P. media (r)

Potentilla anserina (o)
Ranunculus acris (f)
R. ficaria (o)
Rumex crispus (r)
R. sanguineus (f)
Taraxacum officinale (o)
Trifolium pratense (f)
T. repens (a)

Above Bromham the river bends considerably and water meadows are restricted to the land between the bends. They are often inaccessible and much remains to be fully explored. Below Stevington Church the meadows are dominated by *Petasites hybridus* and *Equisetum telmateja*. At Moor End, Felmersham, there are marshy meadows with some quantity of *Rumex maritimus*, otherwise the usual species of the water meadows in this area are *Caltha palustris*, *Geranium pratense*, *Senecio aquaticus*, *Juncus compressus* and *Carex disticha*.

The meadows are often drained by shallow ditches which add little to the flora.

Habitat Study 9   Water meadow, Felmersham
Parish: Sharnbrook   District: Ouse   Grid Ref. 42/98955815
Alt. 130 ft.   .   Soil: Alluvium   pH 7.5
Surveyed: 18 June 1949 with W. M. Cornish; 12 April 1950; 7 Sept. 1950 with
W. Durant. This study was made in the meadow opposite Felmersham Church
at a point 25 yards from the road by a ditch parallel to the river.

Ditch

Dominant
Glyceria fluitans

Abundant
Callitriche stagnalis	Glyceria maxima	Ranunculus trichophyllus

Frequent
Apium nodiflorum	Myosotis scorpioides	Rumex conglomeratus
Carex hirta	Sium latifolium	

Rare
Alisma plantago-aquatica	Epilobium parviflorum	Oenanthe fistulosa
Alopecurus geniculatus	Galium palustre	Ranunculus repens
Carex disticha	Juncus articulatus	Sparganium simplex
C. otrubae	Lycopus europaeus	Vicia cracca

Meadow adjoining ditch

Abundant
Hordeum secalinum	Poa pratensis	Ranunculus repens
Lolium perenne		

Frequent
Bromus racemosus	Deschampsia cespitosa	Trifolium pratense
Cirsium arvense		

Occasional
Bellis perennis	F. rubra	Taraxacum officinale
Dactylis glomerata	Holcus lanatus	Trifolium dubium
Geranium dissectum	Ranunculus ficaria	Urtica dioica
Festuca pratensis	Rumex crispus	

Rare
Alopecurus pratensis	Cirsium vulgare	Plantago lanceolata
Bromus mollis (1)	Juncus inflexus	Ranunculus acris
Capsella bursa-pastoris	Phleum nodosum	Sonchus asper
Cerastium holosteoides		

The Oolitic Series is often exposed on the banks of the river;
between Pavenham and Stevington these give rise to springs which
form small interesting marshes, which are probably the Stevington
Bogs from which Abbot recorded so many plants. The area needs
further exploration and similar marshy places may yet be found.

Habitat Study 10a   Marsh, Stevington
Parish: Stevington   District: Ouse   Grid Ref. 42/98255470
Alt. 125 ft.   Soil: Alluvium overlying Great Oolite Limestone   pH 7.2
Surveyed: 10 Aug. 1949 with C. C. Foss; 12 April 1950; 9 July 1950 with H. B.
Souster. This marsh is about 20 yards in diameter and drains into the river about
30 yards below. A detailed study was not made of the adjoining pasture, which it
was noted contained *Cirsium acaulon* and *Campanula rotundifolia*.

PLATE 6 FELMERSHAM GRAVEL PITS (Habitat Study 11)

PLATE 7 WATER MEADOWS, EATON SOCON (Habitat Study 14b)

PLATE 8   THE HEATH, HEATH AND REACH (Habitat Study 22)

PLATE 9   RIDE, KING'S WOOD, HEATH AND REACH (Habitat Study 23)

Dominant
Juncus subnodulosus

Abundant

Anagallis tenella	Juncus inflexus	Valeriana dioica

Frequent

Agrostis stolonifera	Juncus articulatus	Ranunculus acris
Galium uliginosum		

Occasional

Carex hirta	Holcus lanatus	Triglochin palustris
Cerastium holosteoides	Lotus uliginosus	Veronica beccabunga
Cirsium palustre	Mentha aquatica	
Equisetum palustre	Ranunculus repens	

Rare

Bellis perennis (M)	Festuca arundinacea	P. reptans
Cardamine pratensis	Filipendula ulmaria	Ranunculus ficaria
Carex flacca	Lathyrus pratensis	Succisa pratensis
C. panicea	Plantago lanceolata	Taraxacum officinale
Cirsium arvense	Poa trivialis	Trifolium pratense
Epilobium parviflorum	Potentilla anserina	T. repens

Bryophytes: Acrocladium cuspidatum, Brachythecium rutabulum forma, Cratoneuron filicinum.

Habitat Study 10b   Marsh, Stevington
Parish: Stevington   District: Ouse   Grid Ref. 42/98255471
Alt. 125 ft.   Soil: Alluvium overlying Great Oolite Limestone   pH 7.0
Surveyed: 12 April 1950; 9 July 1950 with H. B. Souster; 23 Aug. 1951 with C. C. Foss. This marsh is comparable in size with the one described above, which is about 50 yds. away.

Abundant
Apium nodiflorum

Frequent

Juncus articulatus	J. inflexus	Veronica beccabunga

Occasional

Carex distans	Mentha aquatica	Prunella vulgaris
Cirsium palustre	Potentilla anserina	Ranunculus acris
Holcus lanatus	P. reptans	Rumex sanguineus
Juncus subnodulosus	Plantago lanceolata	Scrophularia aquatica

Rare

Anacamptis pyramidalis (M)	Centaurea nigra	Pimpinella saxifraga (M)
Angelica sylvestris	Cirsium arvense	Ranunculus ficaria
Bellis perennis	Epilobium parviflorum	Taraxacum officinale
Brachypodium sylvaticum	Lysimachia nummularia (L)	Trifolium repens
	Myosotis scorpioides (L)	Scabiosa columbaria (M)
		Succisa pratensis (M)

Bryophytes: Acrocladium cuspidatum, Brachythecium rutabulum forma, Cratoneuron filicinum.

Habitat Study 10c   Marsh, Pavenham
Parish: Pavenham   District: Ouse   Grid Ref. 42/98255472
Alt. 125 ft.   Soil: Alluvium overlying Great Oolite Limestone   pH 7.2
Surveyed: 10 Aug. 1949 with C. C. Foss; 12 April 1950; 9 July 1950 with H. B. Souster; 23 Aug. 1951 with C. C. Foss. This marsh is nearer to Pavenham and is slightly larger than the other two (10a and 10b). It is drained by a stream which flows into the river.

Dominant
Equisetum telmateja

Frequent
Cirsium palustre   Juncus inflexus   J. subnodulosus
Epilobium hirsutum

Occasional
Apium nodiflorum   Epilobium parviflorum   Hypericum tetrapterum
Angelica sylvestris   Festuca arundinacea   Lathyrus pratensis
Carex hirta   Filipendula ulmaria (M   Lychnis flos-cuculi
Centaurea nigra   stream)   Ranunculus ficaria
Cirsium arvense   Holcus lanatus (M)   Scrophularia aquatica
Eupatorium cannabinum

Rare
Agrimonia eupatoria   Juncus acutiflorus   Pulicaria dysenterica
Agrostis tenuis (M)   Lysimachia nummularia   Ranunculus acris
Arrhenatherum elatius   (1)   R. repens
   (M)   Lythrum salicaria   Rumex sanguineus
Calystegia sepium   Potentilla reptans   Taraxacum officinale
Deschampsia cespitosa   Prunella vulgaris (1)   Veronica beccabunga
Equisetum palustre

Bryophytes: Acrocladium cuspidatum, Brachythecium rutabulum forma, Cratoneuron filicinum.

There are a large number of disused gravel pits near the river, and these have their own distinctive floras depending very much on the extent to which they have flooded. At Clapham some similar pits have been used as rubbish dumps and alien species and garden escapes have appeared.

To the east of Bromham there are similar pits where *Chrysanthemum segetum* makes a fine show. Newer and more extensive pits on the opposite side of the river to Felmersham are undisturbed and, having flooded, are already developing an interesting aquatic flora. *Utricularia vulgaris* and *Scirpus setaceus* are, with other species, recorded here.

Habitat Study 11   Felmersham Gravel Pits
Parish: Sharnbrook   District: Ouse   Grid Ref. 42/98805815
Alt. 130 ft.   Soil: River gravel   pH 7.4
Surveyed: 18 June 1949 with W. M. Cornish; 12 April 1950; 7 Sept. 1950 with W. Durant. This study was made in the pit on the western side of the Felmersham-Sharnbrook road. Its centre was a small peninsula about 30 ft. long on the north side of the large pool formed here.

In water

Alisma plantago-aquatica   Elodea canadensis (o)   Potamogeton natans (la)
   (r)   Polygonum amphibium   Utricularia vulgaris (a)
Callitriche stagnalis (L)   (o)   Zannichellia palustris (r)

At water's edge

Epilobium hirsutum (f)   Juncus articulatus (o)   J. inflexus (a)

On the peninsula

Abundant
Holcus lanatus
   Frequent
Equisetum arvense (L)   Poa pratensis   Tussilago farfara

Occasional

Centaurium minus
Cerastium holosteoides
Cirsium vulgare

Crepis capillaris
Dactylis glomerata
Picris echioides

Plantago lanceolata
Rumex conglomeratus

Rare

Anagallis arvensis (1)
Anthemis cotula
Carex hirta
Cirsium arvense
Hypochoeris radicata
Leontodon hispidus

Lolium perenne
Lotus corniculatus
Phleum pratense
Prunella vulgaris
Rumex crispus
Salix alba 3 ft.

Sonchus asper
Stachys palustris
Trifolium dubium
T. repens (1)
Veronica serpyllifolia

Bryophytes: Barbula convoluta, B. unguiculata, Brachythecium rutabulum, Ceratodon purpureus.

Woods are few in this region. Woodcraft, near Stevington, has more *Milium effusum* and *Pteridium aquilinum* than is usual in woods in this part of the county. There are a few osier beds, in which *Salix triandra* and *S. viminalis* appear to be most frequently planted. Arable fields are also few and their weeds lack interest.

Habitat Study 12   Arable field, Sharnbrook
Parish: Sharnbrook   District: Ouse   Grid Ref. 42/989582
Alt. 130 ft.   Soil: ? River gravel   pH 7.5
Surveyed: 18 June 1949 with W. M. Cornish; 12 April 1950; 7 Sept. 1950 with W. Durant. This field adjoins the gravel pit (H.S. 11). The crop, wheat, was reasonably clean and few weeds were evident.

Alopecurus myosuroides
Bromus sterilis
Capsella bursa-pastoris
Chenopodium album
Cirsium arvense
Convolvulus arvensis
Equisetum palustre
Euphorbia peplus
Galium aparine

Lamium purpureum
Lithospermum arvense
Myosotis arvensis
Papaver rhoeas
  var. pryorii
Poa pratensis
Polygonum aviculare
P. convolvulus
P. persicaria

Ranunculus repens
Senecio vulgaris
Sinapis alba
S. arvensis
Sonchus asper
S. oleraceus
Stellaria media
Veronica polita
V. hederifolia

Melampyrum arvense has been known for many years in fields near Stagsden, and in a field near Harrold *Falcaria vulgaris* is equally well established.

The walls in this region are invariably of limestone and have a flora similar to that already described (p. 59). Around Felmersham *Chenopodium bonus-henricus*, usually found at the base of walls, is more abundant than elsewhere in the county.

The adventive flora is limited to gravel pits and the railway. At Bedford there are extensive railway sidings with the usual mixture of British colonists and species of alien origin.

D.   The Lower Ouse and Upper Ivel Region

This is in many respects a sharp contrast with the previous region. The Ouse below Bedford becomes sluggish and follows an almost straight course and the valley becomes much wider. At Tempsford it is joined by the Ivel which has a volume probably equal to that of the main river and an even wider valley. This, the flattest part of the county, has the most monotonous natural scenery, a dullness reflected in its flora which, both in its flowering plants and bryophytes, is less varied than that of the Upper Ouse valley. As the land is intensively cultivated it is to the river and water meadows, and to the ever-increasing number of gravel pits, that one must look for plant life.

In the river itself *Potamogeton praelongus* replaces *P. lucens* between Bedford and Cardington, but downstream from Cardington the course tends to be choked by *Ceratophyllum demersum*. *Nymphaea alba* is still the common water-lily of the river; it is apparently absent in the Ivel, and *Ranunculus fluitans* is more plentiful here than it is in the river above Bedford. The Ivel Navigation, the only artificial waterway in the county, has long fallen into disuse and its flora may be considered to have become static; but so far it has produced none of the interesting species found elsewhere in disused canals.

On the river banks *Acorus calamus* and *Impatiens capensis* still appear, and near Blunham *I. glandulifera* has become well established. In the shallow margins of the rivers *Butomus umbellatus*, *Sagittaria sagittifolia*, *Typha latifolia*, *Scirpus lacustris*, *Carex riparia* and *Glyceria maxima* usually appear.

Habitat Study 13   Riverside, Willington
Parish: Willington   District: Ouse   Grid Ref. 52/11905045
Alt. 65 ft.   Soil: Alluvium   pH 7.2
Surveyed: 12 July 1949; 2 Sept. 1949 with C. C. Foss; 12 April 1950; 22 Aug. 1951 with C. C. Foss. This study was made by the side of a backwater of the river near Willington Mill.

In water

Glyceria maxima (la)   Polygonum amphibium   Scirpus lacustris (r)
Nymphaea alba (f)   (r)   Sparganium erectum (r)
Phragmites communis   Potamogeton natans (la)   Typha latifolia (la)
   (lo)   Rorippa microphylla (la)

On wet margin

Abundant
Myosotis scorpioides (L)   Ranunculus ficaria (L)

Frequent
Agropyron repens   Phalaris arundinacea   Scrophularia aquatica
Galium palustre   Rumex conglomeratus   Stachys palustris
Mentha aquatica   R. crispus

Occasional
Apium nodiflorum (L)   Epilobium hirsutum   Polygonum hydropiper
Carex otrubae (L)   Filipendula ulmaria   Urtica dioica
C. riparia (L)   Juncus bufonius

Rare
Agrostis stolonifera   Iris pseudacorus   Polygonum persicaria
Angelica sylvestris (1)   Juncus articulatus   Ranunculus sceleratus
Arctium lappa   J. inflexus   Senecio aquaticus
Bidens tripartita   Lycopus europaeus   Solanum dulcamara
Equisetum arvense   Lythrum salicaria   Sonchus asper
Galium aparine   Myosoton aquaticum   Vicia cracca
Gnaphalium uliginosum

On dry margin

Achillea millefolium (o)
Alliaria petiolata (r)
Alopecurus pratensis (o)
Anthriscus sylvestris (o)
Arctium lappa (r)
A. minus (r)
Arrhenatherum elatius (r)
Artemisia vulgaris (o)
Atriplex hastata (r)
Calystegia sepium (r)
Centaurea nigra (o)
Cirsium arvense (o)

C. vulgare (o)
Deschampsia cespitosa (r)
Dipsacus fullonum (r)
Erigeron canadensis (r)
Glechoma hederacea (f)
Hordeum secalinum (o)
Lolium perenne (f)
Medicago lupulina (o)
Phleum pratense (o)
Plantago lanceolata (f)
P. media (o)

Potentilla anserina (o)
P. reptans (r)
Polygonum convolvulus (r)
Rumex conglomeratus (r)
Salix purpurea 8 ft. (1)
Solanum nigrum (o)
Taraxacum sp. (r)
Trifolium hybridum (r)
Valeriana officinalis (r)
Viola arvensis (o)

A study in less disturbed surroundings was made at Eaton Socon.

Habitat Study 14a   Riverside, Eaton Socon
Parish: Eaton Socon   District: Ouse   Grid Ref. 52/17956048
Alt. 46 ft.   Soil: Alluvium   pH 7.4
Surveyed: 19 June 1949 with W. Durant; 6 Aug. 1949; 18 May 1951. This study
was made on the edge of the river about half-way along the meadow below St.
Neots Bridge.

In river

Ceratophyllum demersum (a)   Nuphar lutea (a)

On riverside

Abundant

Glyceria maxima   Rumex hydrolapathum

Frequent

Apium nodiflorum   Lycopus europaeus   Scrophularia aquatica
Berula erecta   Ranunculus sceleratus

Occasional

Mentha aquatica   Stellaria alsine

Rare

Acorus calamus   Galium palustre   Lythrum salicaria
Butomus umbellatus   Epilobium parviflorum

The water meadows in the Lower Ouse valley well repay study.

Habitat Study 14b   Water meadow, Eaton Socon
Details as for H.S. 14a
This study was made in the meadow immediately adjoining that of the previous
study. It is the lowest point in the county, the level of the meadow being low compared with the river and therefore frequently flooded.

Abundant

Alopecurus geniculatus   E. uniglumis   Trifolium repens
Eleocharis palustris   Glyceria plicata   Lotus uliginosus
   subsp. microcarpa   Rorippa microphylla   var. glaber

Frequent

Bromus racemosus   Glyceria fluitans   Oenanthe fistulosa
Callitriche stagnalis   Holcus lanatus   Poa pratensis
Caltha palustris   Juncus articulatus   Veronica beccabunga
Equisetum palustre   Mentha aquatica

Occasional

Cardamine pratensis   Ranunculus acris   Trifolium fragiferum (M)
Dactylorchis incarnata   R. trichophyllus   Valeriana dioica

Rare

Apium nodiflorum	Festuca pratensis	Ranunculus repens
Bellis perennis (1)	F. rubra	Rumex conglomeratus
Carex disticha	Glyceria × pedicellata	R. obtusifolius
C. hirta	Hypochoeris radicata	Senecio aquaticus
Catabrosa aquatica	Juncus compressus	Taraxacum officinale (1)
Cerastium holosteides	J. inflexus	Trifolium pratense (M)
Eleocharis palustris subsp.	Lychnis flos-cuculi	Veronica catenata
microcarpa × uniglumis	Myosotis scorpioides	

Bryophyte: Drepanocladus aduncus (a).

Near Tempsford is a ditch with *Apium inundatum*, *Hottonia palustris* and *Baldellia ranunculoides*, all extremely rare species in the county. From meadows near Fenlake *Polygonum minus* has been recorded. The meadows by the Ivel are on alluvium washed down from the Lower Greensand and tend to be more acid. Biggleswade Common, which until comparatively recently had an interesting flora, has now been drained and is mainly cultivated.

Habitat Study 15   Biggleswade Common
Parish: Sandy   District: Ivel   Grid Ref. 52/19404755
Alt. 95 ft.   Soil: Alluvium overlying Lower Greensand   pH 6.0
Surveyed: 16 July 1949 with T.L.; 23 April 1950 with W. Durant; 6 Sept. 1950.
This study was made in an isolated part of the common to the north-east side of the footbridge over the railway.

Abundant

Agrostis tenuis	Filipendula ulmaria	Juncus effusus

Frequent

Arrhenatherum elatius	Galeopsis tetrahit	Lotus uliginosus
Cirsium palustre	Holcus lanatus	Lythrum salicaria

Occasional

Angelica sylvestris	Equisetum palustre	R. repens
Carex disticha	Heracleum sphondylium	Rumex acetosa
Chamaenerion angusti-folium	Linaria vulgaris	Scutellaria galericulata
	Ranunculus flammula	

Rare

Cirsium arvense	Galium palustre	Stachys palustris
Dactylis glomerata	Hydrocotyle vulgaris	Taraxacum officinale
Epilobium hirsutum	Quercus robur 1 ft. (1)	Vicia tetrasperma (1)

Bryophytes: Aulacomnium androgynum, Brachythecium rutabulum, on mud, Ceratodon purpureus, on wood in marsh, Physcomitrium pyriforme.

Gravel pits occur at intervals along the Ouse, mainly on the south side from the outskirts of Bedford to Willington, in the neighbourhood of Eaton Socon, and around Henlow and Langford in the Ivel valley. They develop an interesting flora and colonists soon become established. When water settles *Myriophyllum spicatum* and *Potamogeton berchtoldii* usually appear. The only known stations at present for *P. alpinus* are gravel pits in this region.

The pits at Cople and Chawston were used as rubbish dumps and a number of adventives have appeared. Before they were filled the pits at Eaton Socon produced more alien species than any other single station in the county.

Habitat Study 16   Gravel pit, Willington
Parish: Willington   District: Ouse   Grid Ref. 52/01554965
Alt. 80 ft.   Soil: River gravel   pH 7.0
Surveyed: 12 July 1949; 12 April 1950; 22 Aug. 1951 with C. C. Foss. This survey
was made in the western end of the pits three years after working had ceased. The
pits are now (1953) being filled in.

Abundant

Stellaria media   Chenopodium album

Frequent

Capsella bursa-pastoris   Poa pratensis   Rumex crispus
Erysimum cheiranthoides   Polygonum aviculare   Senecio vulgaris
  (L)   P. convolvulus

Occasional

Agropyron repens   Gnaphalium uliginosum   Scrophularia aquatica
Cirsium arvense   (L)   (L)
Crepis capillaris   Juncus bufonius (L)   Sonchus asper
Epilobium adenocaulon   Lactuca serriola   Trifolium dubium
E. lamyi   Matricaria maritima   Typha latifolia (L)
E. parviflorum   Reseda luteola   Veronica persica
E. tetragonum   Rumex acetosa

Rare

Agrostis stolonifera   Lamium purpureum   Ranunculus bulbosus (1)
Anthemis cotula   Melandrium album   Senecio jacobaea (1)
Bellis perennis   Papaver dubium   Sisymbrium officinale
Cerastium holosteoides   P. rhoeas   Trifolium repens
Epilobium hirsutum   Plantago major   Tussilago farfara
Erigeron canadensis   Poterium sanguisorba   Vulpia myuros
Holcus lanatus

There are no woods in this region and few osier holts. By the
Ouse the common tree is *Populus nigra* var. *betulifolia*. The roadside
trees are oak, ash and the various species of elm. *Trifolium ochroleucon*
and *Ornithogalum pyrenaicum* appear frequently by roadsides in the
eastern part of the region, and *Rumex pulcher* is common on the
roadside verges near Cardington.

The weeds of the arable land depend much on the fertilizer used,
but *Solanum nigrum*, *Chenopodium album* and *Urtica urens* are common
in most fields. Shoddy is used a great deal on the market gardening
land. The fields in which it has been used are rich in wool
adventives.

Habitat Study 17   Arable field, Willington
Parish: Willington   District: Ouse   Grid Ref. 52/10504965
Alt. 75 ft.   Soil: River gravel   pH 7.0
Surveyed: 12 July 1949; 12 April 1950; 22 Aug. 1951 with C. C. Foss; 12 June
1952. This study was made in a field adjoining the gravel pit (H.S. 16).

Aphanes arvensis   Crepis vesicaria   Senecio jacobaea
Atriplex patula   Equisetum arvense   S. vulgaris
Bromus commutatus   Lamium amplexicaule   Solanum nigrum
B. sterilis   Matricaria maritima   Sonchus oleraceus
Capsella bursa-pastoris   Papaver rhoeas   Stellaria media
Chenopodium album   Poa annua   Veronica persica
Cirsium arvense   Polygonum aviculare   V. polita
Convolvulus arvensis   P. convolvulus   Urtica urens

E.   The Southern Oxford Clay Region

This region has a comparatively simple geological basis. When the Oxford Clay comes to the surface it is covered with only a shallow overburden; its exploitation is therefore easy, and there are many clay pits and brick works, especially in the neighbourhood of Stewartby and Marston Mortaine. Elsewhere the Oxford Clay is overlain with belts of Boulder Clay of varying depth. This is one of the best wooded parts of the county.

The northern part of the region lying to the north of the Crawley Brook is intensively cultivated and little wooded, but White's Wood is probably the best station in the county for *Helleborus foetidus*. To the south of the Crawley Brook are Astey Wood and Hanger Wood, two well-botanized woods. They are good oak-ash woods, differing little in their flora from the larger woods further north; but *Helleborus viridis* still survives in Astey Wood and *Platanthera bifolia* has been recorded from Hanger Wood.

Further south is a finer belt of wooded country containing Holcot Wood, Marston Thrift and Wootton Wood, which have a more varied flora and demand greater attention. Holcot Wood is delightfully situated and has *Campanula latifolia*, *Epipactis helleborine*, *Neottia nidus-avis* and *Paris quadrifolia*, all in some quantity.

Habitat Study 18   Holcot Wood
Parish: Holcot and Salford   District: Ouse   Grid Ref. 42/95904022
Alt. 285 ft.   Soil: Boulder Clay overlying Oxford Clay   pH 7.2
Surveyed: 9 July 1949 with H. B. Souster; 20 April 1950 with W. Durant; 5 Sept. 1950 with J. W. Cardew. This survey was made on a narrow path on the south side of the wood at a point where it is crossed by a ditch. The trees are mainly ash (*Fraxinus excelsior*) growing to a height of about 50 ft. A thick undergrowth gives intense shade. Species listed below as local were limited to the path.
Shrubs: Acer campestre 8 ft. (r), Corylus avellana 12 ft. (a), Crataegus monogyna 6 ft. (r), Rubus sp. (o), Sambucus nigra 1 ft. (r), Thelycrania sanguinea 2 ft. (r).

Abundant
Mercurialis perennis

Frequent

Circaea lutetiana	Poa trivialis (L)	Rumex sanguineus
Glechoma hederacea	Primula vulgaris	

Occasional

Ajuga reptans	Endymion non-scriptus	Potentilla reptans (L)
Anemone nemorosa	Fragaria vesca	Ranunculus ficaria
Arum maculatum	Galium aparine (L)	Veronica chamaedrys (L)
Carex sylvatica	Geum urbanum (L)	Viola reichenbachiana

Rare

Campanula latifolia	Lysimachia nummularia	Potentilla sterilis (L)
Dryopteris filix-mas (1)	(L)	Taraxacum officinale (1)
Epilobium montanum	Orchis mascula	Urtica dioica
Filipendula ulmaria	Poa annua (L)	Viola riviniana
Heracleum sphondylium		

Bryophytes: Atrichum undulatum, Thamnium alopecurum.

Marston Thrift is a mixed wood. The northern and higher part of the wood is very much like any other Bedfordshire wood on the

Boulder Clay, but the lower part is acid and, as water settles there, plant associations, worthy of study, are formed. This part of the wood is the only place in the county where *Luzula forsteri* is known.

Habitat Study 19   Marston Thrift
Parish: Marston Mortaine   District: Ouse   Grid Ref. 42/97504175
Alt. 185 ft.   Soil:? Boulder Clay overlying Oxford Clay   pH 5.2
Surveyed: 9 July 1949 with H. B. Souster; 20 April 1950 with W. Durant; 5 Sept. 1950 with J. W. Cardew. This study was made 60 yds. along the main ride from the south side of the wood.

(a) Path community

Abundant
Juncus effusus (L)

Frequent
Filipendula ulmaria (L)   Poa trivialis   Sagina procumbens
Holcus lanatus   Ranunculus repens

Occasional
Agrostis tenuis   Epilobium tetragonum   Lotus uliginosus (L)
Carex flacca   Galium palustre (L)   Odontites verna
C. otrubae   Glechoma hederacea   Prunella vulgaris
Cirsium palustre   Juncus articulatus (L)   Rumex sanguineus
Deschampsia cespitosa

Rare
Agrimonia eupatoria   Hypericum hirsutum   Poa annua
Agrostis stolonifera   H. tetrapterum   Potentilla anserina
Anagallis arvensis   Juncus bufonius   P. reptans
Bellis perennis   J. conglomeratus   Rumex crispus
Calamagrostis epigejos   J. inflexus   Scrophularia aquatica
Carex hirta   Lolium perenne   S. nodosa
C. pendula   Lychnis flos-cuculi   Succisa pratensis
C. remota   Myosotis arvensis   Taraxacum officinale
Dactylis glomerata   Phleum nodosum   Veronica officinalis
Epilobium parviflorum   Plantago major (M)

(b) Woodland community (to west side of path)

Trees and shrubs: Corylus avellana 10 ft. (f), Crataegus monogyna 6 ft. (r), Lonicera periclymenum (o), Populus tremula, with many suckers, 40 ft. (1), Quercus robur, close growing, 40 ft., Rosa canina 6 ft. (r), Rubus sp. (f), Salix caprea 6 ft. (r).

Frequent
Anemone nemorosa   Anthoxanthum odoratum   Luzula forsteri
Occasional
Carex sylvatica   Primula vulgaris   Ranunculus ficaria
Fragaria vesca   Prunella vulgaris   Veronica officinalis
Luzula pilosa

Rare
Ajuga reptans   Geum urbanum   Succisa pratensis
Circaea lutetiana   Poa nemoralis   Viola reichenbachiana
Epipactis purpurata   Scrophularia nodosa

Bryophytes: Atrichum undulatum, Dicranella heteromalla, Dicranum scoparium, Eurhynchium praelongum, Hypnum cupressiforme, Isothecium myosuroides, Mnium hornum, Polytrichum formosum, Thuidium tamariscinum.

Wootton Wood is less interesting but is worth study. In the higher parts of the wood *Cynoglossum officinale* is relatively abundant and towards the lower end *C. montanum* was found once.

Habitat Study 20   Wootton Wood
Parish: Wootton   District: Ouse   Grid Ref. 42/997450
Alt. 200 ft.   Soil: Boulder Clay overlying Oxford Clay   pH 7.6
Surveyed: 9 July 1949 with H. B. Souster; 20 April 1950 with W. Durant; 5 Sept.
1950 with J. W. Cardew. This study was made on the main ride through the wood
at a point 80 yards to the north of the main transverse ride.

(a) Path community

Abundant
Lysimachia nummularia

Frequent

Epilobium montanum	Potentilla reptans	Ranunculus repens
Glechoma hederacea	Prunella vulgaris	Veronica serpyllifolia

Occasional

Ajuga reptans	Epilobium tetragonum	Potentilla anserina
Circaea lutetiana (L)	Fragaria vesca	Senecio jacobaea
Dipsacus fullonum	Myosotis arvensis	

Rare

Anagallis arvensis	Geum urbanum	Poa annua
Arctium minus	Hypericum hirsutum	P. trivialis
Bellis perennis	H. tetrapterum	Rumex crispus
Centaurium minus	Juncus inflexus	R. sanguineus
Cirsium arvense	Leontodon taraxacoides	Sagina procumbens
Epilobium hirsutum	Plantago lanceolata	Scrophularia nodosa
Galium aparine	P. major	Vicia sepium
G. palustre		

(b) Woodland community (to west of the path)

The wood here is a mixture of oak (*Quercus robur*) and ash (*Fraxinus excelsior*),
close growing, to a height of 40 ft. There is one aspen (*Populus tremula*) with many
suckers.
Shrubs: Acer campestre 20 ft. (1), Corylus avellana 10 ft. (f), Crataegus monogyna
6 ft. (o), Ligustrum vulgare, on path, 5 ft. (L), Prunus spinosa 6 ft. (r), Rosa
canina (r), Rubus ? vestitus (o), Thelycrania sanguinea 6 ft. (r), Ulmus sp. 10 ft. (1).

Frequent

Circaea lutetiana	Primula vulgaris

Occasional

Anemone nemorosa	Viola sp.

Rare

Epilobium montanum	Glechoma hederacea	Ranunculus ficaria

Bryophytes: Cirriphyllum piliferum, Eurhynchium praelongum, E. striatum,
Isothecium myosuroides, Thamnium alopecurum, Thuidium tamariscinum.

There are few marshy places and almost no streams in this region.
Some of the meadows are damp and ponds are plentiful. The latter
are usually rich in water crowfoots but some are completely choked
with *Potamogeton natans*. At the sides of the ponds there is usually to
be found *Ranunculus sceleratus, Rorippa microphylla, Oenanthe fistulosa*
and various species of *Glyceria*. There is little to distinguish the
pastures on the Oxford Clay from those on the Boulder Clay except
the greater abundance of *Trifolium fragiferum* and *Ononis spinosa*
and the lack of *Linum catharticum* and *Filipendula vulgaris* in the
former.

The claypits soon develop a flora of their own. When water settles *Polygonum amphibium*, *Typha latifolia*, *Potamogeton natans* and *Charophytes* soon appear, followed at the margin of the pools by *Epilobium hirsutum*, *Sparganium erectum*, *Juncus bufonius*, *J. inflexus* and *Phragmites communis*. Willows follow and the pits take a changed appearance, the sides quickly becoming colonized by *Epilobia*, *Erigeron canadensis*, *Tussilago farfara*, *Crepis vesicaria*, *Lactuca serriola* and *Picris echioides*.

Habitat Study 21   Brogborough Pit
Parish: Lidlington   District: Ouse   Grid Ref. 42/969389
Alt. *c.* 200 ft.   Soil: Oxford Clay   pH 7.6
Surveyed: 9 July 1949 with H. B. Souster; 20 April 1940 with W. Durant; 5 Sept. 1950 with J. W. Cardew. This study was made on the northern shore of the pool formed at the bottom of the pit. Working ceased in the pit in 1937.

In and on margin of pool

Abundant
Lemna minor   Potamogeton natans   Typha latifolia
Myriophyllum spicatum

Frequent
Potamogeton crispus   Ranunculus trichophyllus

Occasional
Glyceria maxima (M)

Rare
Alisma plantago-aquatica   Epilobium hirsutum   Salix sp.

On side of pit

Abundant
Potentilla reptans   Senecio erucifolius   Tussilago farfara

Frequent
Agropyron repens   Lotus corniculatus   Poa pratensis
Daucus carota

Occasional
Agrostis stolonifera   Cirsium arvense   Prunella vulgaris
Arrhenatherum elatius   Juncus conglomeratus (L)   Rumex crispus
Carex flacca   J. inflexus (L)   Trifolium hybridum
Chrysanthemum leucan-   Lathyrus pratensis   T. pratense
  themum   Medicago lupulina

Rare
Agrimonia eupatoria   Hypochoeris radicata   Pulicaria dysenterica
Agrostis tenuis   Lathyrus nissolia   Ranunculus acris
Carex hirta   Leontodon hispidus   Rosa canina
Crataegus monogyna 6 ft.   Linum catharticum   Rubus caesius × ?
Cynosurus cristatus   Lolium perenne   Silaum silaus
Dactylis glomerata   Ophrys apifera   Trisetum flavescens
Deschampsia cespitosa   Phalaris arundinacea   Typha angustifolia
Epilobium sp.   Picris echioides   Vicia cracca
Galium verum   Plantago lanceolata

The arable fields of this region appear to have no weeds of special interest. Adventives often grow in the neighbourhood of brickworks and *Senecio squalidus* is especially plentiful at Stewartby.

F. The Lower Greensand Region

Of all the regions this is the most varied geologically and the most interesting botanically. The Greensand is in two disconnected areas. One, on the western side of the county, stretches from Leighton Buzzard and Heath and Reach, diminishing in altitude to the Ivel valley. The other, a smaller portion, is to the east of the Ivel in the neighbourhood of Sandy and Potton. They are sufficiently different in their topography and flora to merit the separate treatment given here.

The larger area is overlain, especially in the centre of the county, with Boulder Clay, and considerable stretches of alluvial deposits in the river valleys add to the geological complexity.

When exposed the Greensand forms natural heath that can be seen at Cooper's Hill, Heath and Reach, Rowney Warren, and elsewhere in isolated patches that have escaped cultivation or afforestation. Here *Calluna vulgaris* is usually dominant but where clearings have been made spring annuals are to be seen at their best. *Cerastium semidecandrum*, *Spergularia rubra*, *Ornithopus perpusillus*, *Aphanes microcarpa* and *Filago minima* are among the many species which occur. In summer such clearings are often dominated by *Senecio jacobaea*, but this flora is usually only a stage in the succession to *Pteridium aquilinum*, which has complete control over much of the heathy land. When the soil is kept permanently disturbed, as in the neighbourhood of rabbit-warrens, *Corydalis claviculata* is almost certain to appear.

A thin topsoil produces turf in which *Vulpia bromoides* and *Festuca tenuifolia* are co-dominant. When this is insufficiently grazed *Ulex europaeus* and *Sarothamnus scoparius* appear. The margins of heathy places and roadside verges are also rich in various species of *Hieracium*.

The Greensand is also rich in micro-species of bramble, especially on the higher land. W. C. R. Watson found numerous brambles in the neighbourhood of Heath and Reach, Cooper's Hill, Aspley Wood and Clophill; but with equally diligent search found only four at Rowney Warren. Sandy and Potton were almost as unproductive. This confirmed his view, based on the bramble populations elsewhere, that the higher ground, having escaped glaciation, is usually more productive of brambles.

Habitat Study 22   The Heath, Heath and Reach
Parish: Heath and Reach   District: Ouzel   Grid Ref. 42/91952835
Alt. 360 ft.   Soil: Lower Greensand   pH 5.8
Surveyed: 8 June 1949; 2 Aug. 1949; 10 April 1950.

(a) open heathland, to east of the main path at
a point 100 yards from the village allotments

Dominant

Calluna vulgaris (in patches up to 50 ft. across)
Scattered trees and saplings of Betula pubescens, and intermediates with B. pendula.

In small clearing

Locally abundant
Deschampsia flexuosa

Frequent

Festuca ovina	Pteridium aquilinum	Sarothamnus scoparius
F. tenuifolia	Rumex acetosella	Stellaria media
Galium saxatile		

Rare

Campanula rotundifolia	Luzula multiflora	Ulex europaeus 3 ins. (1)
Capsella bursa-pastoris	Spergularia rubra (1)	Veronica arvensis
Filago minima	Stellaria graminea	

Bryophytes: Ceratodon purpureus, Dicranum scoparium, Hypnum cupressiforme, Polytrichum juniperinum.

(b) Previously cultivated plot about 30 yds. to the west of the above study. This was incorporated into a timber yard in 1952.

Abundant

Aira praecox	Senecio jacobaea	Vulpia bromoides

Frequent

Agrostis tenuis	Myosotis discolor	S. procumbens
Aphanes microcarpa	Ornithopus perpusillus	Stellaria media
Arabidopsis thaliana	Rumex tenuifolius	Trifolium dubium
Cerastium holosteoides	Sagina ciliata	Veronica arvensis

Occasional

Geranium molle	Luzula campestris	Myosotis hispida

Rare

Arenaria serpyllifolia	Filago minima	Taraxacum officinale
Chamaenerion angusti-	Galium saxatile (M)	Teesdalia nudicaulis
folium	Geranium pusillum	Urtica dioica (1)
Cirsium vulgare	Hieracium pilosella	Valerianella eriocarpa
Crepis capillaris	Holcus lanatus	Veronica chamaedrys
Erigeron canadensis	Hypochoeris radicata	Vicia angustifolia
Erodium cicutarium	Medicago lupulina (1)	Viola arvensis
Festuca tenuifolia (M)	Pteridium aquilinum	

Bryophytes: Ceratodon purpureus, Hypnum cupressiforme, Pohlia nutans, Polytrichum juniperinum, Pseudoscleropodium purum.

A contrast is to be seen in the well-grazed pastures on the Greensand.

Habitat Study 23   Path, King's Wood, Heath and Reach
Parish: Heath and Reach   District: Ouzel   Grid Ref. 42/92482985
Alt. 400 ft.   Soil: Lower Greensand   pH 5.4
Surveyed: 8 June 1949; 2 Aug. 1949; 10 April 1950; 26 Aug. 1950 with W. D. Coales and D. A. Reid. This study was made at the meeting of the main rides at the lower side of the wood. The wood here is birch (*Betula pubescens*), and *Convallaria majalis* is abundant. The rides are well grazed by rabbits; the herbage is no more than an inch high.

Abundant
Agrostis tenuis

Frequent

Aira praecox	Festuca rubra	Rumex acetosella
Ajuga reptans (M)	Galium saxatile	Sagina procumbens
Anthoxanthum odoratum	Glechoma hederacea (M)	Stellaria graminea
Aphanes microcarpa	Luzula campestris	Trifolium dubium
Calluna vulgaris	Potentilla erecta	Veronica officinalis
Cerastium holosteoides	Pteridium aquilinum	Vulpia bromoides

Occasional

Agrostis canina	Holcus mollis	Potentilla anserina (L)
Centaurium minus	Lotus uliginosus	Primula veris
Cerastium glomeratum	Lysimachia nemorum	Teucrium scorodonia
Cirsium arvense (L)	(M)	Trifolium micranthum
Crepis capillaris	Polygala serpyllifolia	T. repens
Endymion non-scriptus		

Rare

Betula pubescens 6 ins.	Hieracium pilosella	Poa annua
(M)	Hypericum sp. (1)	Potentilla anglica
Cirsium palustre	Lysimachia nummularia	Primula vulgaris (M)
Convallaria majalis	(L)	Ranunculus acris (1)
Galium palustre	Mentha arvensis	Rumex acetosa
Gnaphalium sylvaticum	Myosotis arvensis	Senecio jacobaea
(M)		

Bryophytes: Aulacomnium androgynum, Atrichum undulatum, Dicranella heteromalla, Dicranum scoparium, Eurhynchium praelongum, Pleurozium schreberi, Pohlia nutans, Polytrichum formosum, Pseudoscleropodium purum, Rhytidiadelphus triquetrus.

The limited flora of relatively undisturbed greensand heath is to be seen at Cooper's Hill.

Habitat Study 24   Cooper's Hill
Parish: Ampthill   District: Ivel   Grid Ref. 52/02753760
Alt. 355 ft.   Soil: Lower Greensand   pH 5.8
Surveyed: 3 July 1949 with W. M. Cornish and T.L.; 21 Aug. 1949 with P.T.; 11 April 1950.
This study was made at the crossing of the paths on the top of the heath. Here bracken (*Pteridium aquilinum*) and heather (*Calluna vulgaris*) are each dominant in considerable patches. The list below was made in a small clearing by the side of the path.

Abundant

Agrostis tenuis	Festuca tenuifolia	Rumex acetosella
Deschampsia flexuosa		

Occasional

Cerastium glomeratum	Spergularia rubra	Stellaria media
Galium saxatile		

Rare

Aira praecox	Quercus robur 3 ins.	Ulex europaeus
Hypochoeris radicata		

Bryophytes: Ceratodon purpureus, Dicranum scoparium, Funaria hygrometrica, Hypnum cupressiforme var. ericetorum, Pohlia nutans, Polytrichum piliferum.

Plant associations of a different nature are to be seen when the bare Greensand is exposed and left undisturbed. Such conditions are seen at their best in the many sand quarries and in road widenings. A good example of the latter is at Clophill, where road-widening plans were abandoned in 1939.

Habitat Study 25   Roadside, near Clophill
Parish: Silsoe   District: Ouse   Grid Ref. 52/08203714
Alt. 225 ft.   Soil: Lower Greensand   pH 6.4
Surveyed: 17 June 1949, 16 Aug. 1949, 11 April 1950, 4 Sept. 1950. This study was made on the western verge of the new road by the side of A.6 at a point about 50 yds. from the southern edge of Simpson's Hill Plantation. Some unexpected species appeared on these verges and in one part *Scabiosa columbaria* was apparently well established.

Abundant
Agrostis canina

Frequent

Agrostis tenuis	Rumex acetosella	Trifolium arvense
Arabidopsis thaliana	Stellaria media	Vulpia bromoides

Occasional

Agrostis gigantea	Filago germanica	Hypericum perforatum
Aira caryophyllea	F. minima	Lamium amplexicaule
A. praecox	Fragaria vesca	Ornithopus perpusillus
Chamaenerion angusti-	Geranium pusillum	Vicia angustifolia
folium	Holcus lanatus	

Local

Arrhenatherum elatius	Geranium robertianum	Sedum acre
Cerastium holosteoides	(hedge)	Trifolium striatum
Chaerophyllum temulum	Inula conyza	Veronica officinalis
(hedge)	Potentilla argentea	Vicia hirsuta
Dactylis glomerata (hedge)		

Rare

Achillea millefolium	Lathyrus nissolia (1)	S. sylvaticus (1)
Bromus mollis	Lolium perenne	S. vulgaris
Capsella bursa-pastoris	Lotus corniculatus	Sonchus asper (1)
Crepis capillaris	Medicago lupulina	Taraxacum laevigatum
Erodium cicutarium	Melandrium album	Tragopogon minor
Festuca rubra	Plantago lanceolata	Trifolium campestre
Geranium molle	Poa pratensis	T. dubium
Hypericum humifusum	Reseda luteola	T. repens
(1)	Scleropoa rigida	Urtica dioica
Hypochoeris radicata	Senecio erucifolius	Verbascum thapsus

Contrasted with this is the stable flora of an adjacent area.

Habitat Study 26   Simpson's Hill Plantation
Parish: Silsoe   District: Ivel   Grid Ref. 52/08163714
Alt. 225 ft.   Soil: Lower Greensand   pH 5.4
Surveyed: 17 June 1949; 4 Aug. 1949; 11 April 1950; 4 Sept. 1950. This study was
made at the meeting of the wide paths in the plantation. The turf is short and is
well grazed by rabbits. The plantation itself is entirely sweet chestnut (*Castanea
sativa*).

Abundant

Agrostis tenuis	Trifolium micranthum	Veronica serpyllifolia

Frequent

Aira praecox	Ornithopus perpusillus	Rumex acetosella
Aphanes microcarpa	Poa annua	Sagina procumbens
Holcus mollis (M)	P. pratensis (M)	Trifolium repens
Moehringia trinervia		

Occasional

Hypericum perforatum (L)	Luzula campestris	Stellaria media
Hypochoeris radicata (L)		

Rare

Agrostis stolonifera	Fragaria vesca	Senecio jacobaea
Cerastium holosteoides	Holcus lanatus	Taraxacum officinale
Chamaenerion angusti-	Juncus effusus	Trifolium dubium
folium	Lolium perenne	Vicia angustifolia
Dactylis glomerata	Plantago lanceolata	Vulpia bromoides

To the north of Shefford is Rowney Warren, at one time a fine piece of heathland but now largely planted with conifers. Broom (*Sarothamnus scoparius*) is abundant in the remnants which remain and parasitic upon it *Orobanche rapum-genistae*, one of the remaining treasures of the warren. An interesting flora may be studied in the wide rides, where *Trifolium subterraneum* remains in one of its few native stations.

Habitat Study 27   Rowney Warren
Parish: Southill   District: Ivel   Grid Ref. 52/12374040
Alt. 210 ft.   Soil: Lower Greensand   pH 6.6
Surveyed: 16 July 1949 with T.L.; 22 April 1950; 1 June 1950 with T.L.; 6 Sept. 1950. This study was made on the main ride opposite New Rowney Farm at the junction of the first ride on the left. The plantation here is almost entirely scots pine (*Pinus sylvestris*), 25 ft., with larch (*Larix europaea*) (o), 25 ft., and birch (*Betula pendula*) (r). At the edge of the plantation are *Pteridium aquilinum* (a), *Lonicera periclymenum* (o), *Bryonia dioica* (1) and *Sambucus nigra* (1), but the interior supports no vegetation. The wide verge is closely grazed by rabbits and disturbed by passing lorries.

Abundant

Agrostis tenuis	Filago minima	Trifolium dubium
Deschampsia flexuosa	Sagina procumbens	T. micranthum (L)

Frequent

Aira praecox	C. semidecandrum	Holcus mollis
Aphanes microcarpa	Erodium cicutarium	Myosotis sp.
Arrhenatherum elatius	Filago germanica	Spergularia rubra
Cerastium glomeratum	Geranium molle	Veronica arvensis

Occasional

Cerastium holosteoides	Geranium pusillum	Rumex acetosella
Chamaenerion angusti-	Galium saxatile (L)	Senecio jacobaea
folium	Poa annua	Stellaria media (L)
Crepis capillaris	Rubus sp. 2 ft.	Teucrium scorodonia

Rare

Achillea millefolium	Erigeron canadensis	Plantago lanceolata (1)
Arctium minus (M)	Festuca tenuifolia	P. major
Bellis perennis (1)	Glechoma hederacea	Potentilla reptans
Calluna vulgaris	Heracleum sphondylium	Sagina apetala
Chaerophyllum temulum	(M)	Senecio vulgaris (1)
(1)	Hypochoeris radicata	Rumex tenuifolius
Cirsium arvense	Matricaria matricarioides	Taraxacum officinale (1)
C. vulgare (1)	Moehringia trinervia	Thymus pulegioides
Cynosurus cristatus (1)	Myosotis arvensis (1)	Ulex europaeus 2 ft.
Dactylis glomerata		

Bryophytes: Aulacomnium androgynum, Barbula convoluta, B. unguiculata, Brachythecium albicans, B. rutabulum, Bryum capillare, Ceratodon purpureus, Dicranoweissia cirrata, Dicranum scoparium, Eurhynchium confertum, Hypnum cupressiforme, Orthotrichum affine, Pleurozium schreberi, Polytrichum juniperinum, Pseudoscleropodium purum (many of these were limited to some old stonework inside the plantation).

When turf develops with a certain amount of topsoil the Greensand flora changes. It is in these conditions that *Viola canina* appears, usually in the shelter of gorse (*Ulex europaeus*), always a feature of such situations.

PLATE 10   KING'S WOOD, HEATH AND REACH (Habitat Study 30)

PLATE II   FOLLY WOOD, FLITWICK (Habitat Study 40)

Habitat Study 28   Horsemoor Farm
Parish: Woburn   District: Ouzel   Grid Ref. 42/936332
Alt. 400 ft.   Soil: Lower Greensand   pH 6.2
Surveyed: 11 June 1949 with H. B. Souster; 4 Aug. 1949 with T.L.; 10 April 1950;
30 Aug. 1950 with D. A. Reid. This study was made on sloping ground (about
1 in 6) above the marsh to the west of the farm. It is well grazed by rabbits.
Ulex europaeus grows 10 ft. apart to a height of about 5 ft.

Abundant
Agrostis tenuis   Festuca ovina

Frequent
Achillea millefolium   Festuca rubra   Trifolium repens
Cerastium holosteoides   Rumex acetosella

Occasional
Campanula rotundifolia   Luzula campestris   Stellaria graminea
Cirsium acaulon   Rumex acetosa   Trifolium dubium
Hieracium pilosella   Sieglingia decumbens   Veronica chamaedrys (L)

Rare
Anthoxanthum odoratum   Dactylis glomerata   Poa annua
Arenaria serpyllifolia   Digitalis purpurea   Potentilla erecta (1)
Bellis perennis   Galium saxatile   Quercus robur 3 ins. (1)
Carex pilulifera   G. verum   Senecio jacobaea
Cirsium arvense   Holcus lanatus   Taraxacum officinale
C. vulgare (1)   Leontodon autumnalis   Veronica arvensis
Crepis capillaris   Lolium perenne   V. officinalis
Cynosurus cristatus   Lotus corniculatus   Viola canina

Bryophytes: Dicranella heteromalla, Pseudoscleropodium purum.

A study of a similar habitat was made on Jackdaw Hill, which
was even more disturbed by rabbits.

Habitat Study 29   Jackdaw Hill
Parish: Lidlington   District: Ouse   Grid Ref. 42/99673865
Alt. 300 ft.   Soil: Lower Greensand   pH 4.8
Surveyed: 9 July 1949 with H. B. Souster; 20 April 1950 with W. Durant; 5 Sept.
1950 with J. W. Cardew. This study was made about 100 yards west of the stream
at the foot of the hill and opposite to the crossing over the stream. The slope of the
hill is about 1 in 6 and is well grazed by rabbits. *Ulex europaeus* grows to 8 ft. apart
and about 5 ft. high. Brambles are frequent.

Abundant
Agrostis tenuis   F. rubra   Rumex acetosa
Festuca ovina

Frequent
Anthoxanthum odoratum   Lotus corniculatus

Occasional
Dactylis glomerata   Phleum nodosum

Rare
Achillea millefolium   Crepis vesicaria   R. crispus (M)
Arrhenatherum elatius   Hieracium pilosella   Senecio sylvaticus (L)
Bryonia dioica (1)   Holcus lanatus   Stellaria media (L)
Cerastium holosteoides   Hypochoeris radicata   Trifolium repens
  (M)   Melandrium album (1)   Trisetum flavescens
Cirsium arvense   Rumex acetosella   Urtica dioica (L)

The woods of this region are varied. On the Greensand itself
there is no native woodland, but large areas are planted with conifers
or sweet chestnut, neither of which support much ground vegetation.
Much of the Greensand is, however, overlain with Boulder Clay,

which is well wooded. Marshy areas support alder and birch woods, which add to the variety both of the scenery and plant life.

Of considerable interest are the boulder-clay woods, especially where the clay deposit is shallow. This is the case with Baker's Wood and King's Wood at Heath and Reach and parts of Aspley Wood. In King's Wood are three distinct zones of vegetation which merge into each other. On the southern side oak is dominant, with *Euphorbia amygdaloides* and *Luzula maxima* features of the shade. This gives way to aspen (*Populus tremula*) and *Tilia cordata* in the middle of the wood. On the north-west side the Greensand is close to the surface and birch (*Betula pubescens*) becomes dominant, with *Convallaria majalis* abundant (see H.S. 23).

Habitat Study 30   King's Wood
Parish: Heath and Reach   District: Ouzel   Grid Ref. 42/92952950
Alt. 475 ft.   Soil: Boulder Clay overlying Lower Greensand   pH 5.4
Surveyed: 8 June 1949; 2 Aug. 1949; 10 April 1950; 26 Aug. 1950 with W. D. Coales and D. A. Reid. This study was made on the south-east side of the crossing of the main rides in the south side of the wood.

(a) wood

Trees and shrubs: Quercus robur 30 ft. (a), Populus tremula 25 ft. (r), Betula pubescens 25 ft. (r), Corylus avellana 10 ft. (o), Fraxinus excelsior 4 ft. (1), Lonicera periclymenum (o), Rubus ? vestitus (o), Viburnum opulus 10 ft. (1).

Frequent

Anthoxanthum odoratum	Luzula multiflora	Veronica officinalis
Fragaria vesca	Succisa pratensis	

Occasional

Agrostis tenuis	Chamaenerion angusti-folium	Holcus mollis

Rare

Ajuga reptans	Luzula sylvatica	Serratula tinctoria
Dryopteris filix-mas	L. pilosa	Teucrium scorodonia
Euphorbia amygdaloides	Primula vulgaris	

(b) path adjoining above

Abundant
Succisa pratensis

Frequent

Ajuga reptans	Potentilla erecta	Rubus ? vestitus
Anthoxanthum odoratum	P. anglica	Viola riviniana

Occasional

Agrostis tenuis	Fragaria vesca	Luzula multiflora
Chamaenerion angusti-folium	Juncus conglomeratus	Teucrium scorodonia

Bryophytes: Dicranum scoparium, Eurhynchium praelongum, Polytrichum formosum, Thuidium tamariscinum.

Aspley Wood was a botanists' paradise but little of the original wood now remains. Mermaid's Pond (H.S. 36) is almost dry and birches grow there freely, but the rest of the wood is planted with conifers or sweet chestnut. The bramble flora is rich and clearings quickly become colonized with *Epilobia*, but most of the species recorded by earlier botanists are now gone—I myself saw the passing of *Frangula alnus* in 1940.

Habitat Study 31   Aspley Wood
Parish: Aspley Guise   District: Ouzel   Grid Ref. 42/938343
Alt. 375 ft.   Soil: Lower Greensand   pH 5.4
Surveyed: 11 June 1949 with H. B. Souster; 4 Aug. 1949 with T.L.; 10 April 1950;
30 Aug. 1950 with D. A. Reid. This study was made on the northern side of the
most southerly ride of the wood at a point 150 yds. from the road.
Trees: an equal mixture of sweet chestnut (*Castanea sativa*) and beech (*Fagus
sylvatica*), growing 5–20 feet apart and to a height of 70 ft. Apart from brambles no
shrubs were evident.

(a) verge of ride

Abundant
Agrostis tenuis   Poa nemoralis

Frequent
Agrostis stolonifera   Brachypodium sylvaticum   Stellaria holostea
Arrhenatherum elatius   Glechoma hederacea

Occasional
Alopecurus pratensis   Lapsana communis   Poa pratensis
Deschampsia flexuosa   Mercurialis perennis   Rubus sp.

Rare
Ajuga reptans   Dactylis glomerata   Scrophularia nodosa
Arctium minus   Myosotis arvensis   Urtica dioica
Bromus ramosus   Potentilla sterilis   Vicia sepium
Cerastium holosteoides

Bryophytes: Dicranella heteromalla, Hypnum cupressiforme, Mnium hornum.

(b) 8 yds. from verge

Abundant
Glechoma hederacea

Frequent
Brachypodium sylvaticum   Mercurialis perennis   Myosotis arvensis
Melandrium dioicum

Occasional
Rubus sp.   Teucrium scorodonia

Rare
Ajuga reptans   Poa pratensis   Urtica dioica
Lapsana communis (1)

(c) 16 yds. from verge

Abundant
Mercurialis perennis

Frequent
Ajuga reptans   Glechoma hederacea

Rare
Primula vulgaris   Sambucus nigra 6 ins. (1) Urtica dioica
Pteridium aquilinum

On the northern escarpment of the Greensand the woods are on
thicker deposits of Boulder Clay and have floras similar to the woods
in the north of the county. The lower parts of the woods are often
marshy but produce no species of more than passing interest.

King's Wood, Houghton Conquest, is a large oak-ash wood with
some fine beeches in the middle. In the higher parts of the wood long
branches of *Clematis vitalba* trail over the tree tops with sinister effect.
Along the western edge of the wood *Dipsacus pilosus* is plentiful and
under the beeches in the wood *Epipactis sessilifolia* grows sparingly.

Habitat Study 32   King's Wood
Parish: Houghton Conquest   District: Ouse   Grid Ref. 52/04504047
Alt. 225 ft.   Soil: Boulder Clay overlying Lower Greensand   pH 7.2
Surveyed: 7 July 1949 with R. H. Goode; 19 April 1950; 23 Aug. 1951. This
study was made at the north end of the main ride.
Trees and shrubs: Fraxinus excelsior 70 ft. and Quercus robur 70 ft. close-growing
and giving intense shade, Ulmus sp. 60 ft. (1), Acer campestre 10 ft. (f), Corylus
avellana 15 ft. (f), Crataegus oxyacantha 10 ft. (o), Ligustrum vulgare 4 ft. (r),
Lonicera periclymenum (o), Rosa arvensis 6 ft. (o), Rubus sp. 6 ft. (o)

Abundant

Bromus ramosus	Hedera helix	Mercurialis perennis

Frequent

Brachypodium sylvaticum	Carex sylvatica	Rumex sanguineus

Occasional

Alliaria petiolata	Endymion non-scriptus	Milium effusum
Arum maculatum	Geranium robertianum	Primula vulgaris
Dactylis glomerata (L)	Geum urbanum	Stellaria holostea
Deschampsia cespitosa		

Rare

Anemone nemorosa	Heracleum sphondylium	Stachys sylvatica
Anthriscus sylvestris	Holcus lanatus	Tamus communis
Arrhenatherum elatius	Lapsana communis	Taraxacum officinale (1)
Galium aparine	Poa trivialis	Viola riviniana
Glechoma hederacea	Ranunculus ficaria	

Habitat Study 33   King's Wood
Parish: Houghton Conquest   District: Ouse   Grid Ref. 52/04654015
Alt. 315 ft.   Soil: Boulder Clay overlying Lower Greensand   pH 6.0
Surveyed: 19 April 1950; 25 July 1950 with R. H. Goode. This study was made on
the south-east side of the crossing of the main paths of the wood. Here the trees are
well spaced, allowing freedom to the undergrowth.
Trees and shrubs: Acer campestre 40 ft. (1), Carpinus betulus 40 ft. (1), Corylus
avellana 20 ft. (o), Crataegus oxyacanthoides 3 ft. (r), Daphne laureola 18 ins. (1),
Fraxinus excelsior 30 ft. (o), Lonicera periclymenum (r), Quercus robur 40 ft. (o),
Rosa canina 6 ft. (o), Rubus sp. 4 ft. (f), Thelycrania sanguinea 6 ft. (1), Viburnum
lantana 10 ft. (o).

Abundant
Hedera helix

Frequent

Anemone nemorosa	Conopodium majus	Sanicula europaea

Occasional

Endymion non-scriptus ·	Mercurialis perennis	Viola reichenbachiana

Rare

Ajuga reptans	Dryopteris dilatata	Rumex sanguineus
Arum maculatum	Platanthera sp.	Taraxacum officinale (1)
Carex sylvatica	Ranunculus ficaria	Vicia sepium

Bryophytes: Atrichum undulatum, Eurhynchium praelongum; E. striatum, Isothecium myosuroides, Mnium hornum, M. longirostrum.

Wilstead Wood is similar but has been inadequately searched.

Habitat Study 34a   Wilstead Wood
Parish: Wilshamstead   District: Ouse   Grid Ref. 52/073426
Alt. 190 ft.   Soil: Boulder Clay overlying Lower Greensand   pH 6.0
Surveyed: 19 July 1949 with W. M. Cornish; 30 April 1950 with H. B. Souster;
4 Sept 1950. This study was made 85 yards along the first path on the west of the
main ride from the northern side of the wood. The trees are *Fraxinus excelsior* and
Quercus robur, 20–30 ft., growing some distance apart, the wood having obviously
been cleared in recent years.
Shrubs: very dense with little ground flora; Acer campestre 20 ft. (r), Corylus
avellana 10 ft. (o), Crataegus oxyacanthoides 5 ft. (o), Ligustrum vulgare 4 ft. (f),
Lonicera periclymenum (o), Populus tremula 10 ft. (o), Rosa canina (r), Rubus
caesius × ? (o), Salix caprea 10 ft. (r), Thelycrania sanguinea 4 ft. (r).
    The species listed were limited to the path, in which were deep water-filled
cart-ruts.

Frequent

Glechoma hederacea	J. effusus	Primula vulgaris
Juncus articulatus	Poa trivialis	

Occasional

Ajuga reptans	Glyceria plicata	Potentilla reptans
Carex otrubae	Hypericum tetrapterum	Prunella vulgaris (L)
C. pendula	Juncus conglomeratus	Ranunculus repens
Epilobium parviflorum	Lysimachia nummularia	Rumex sanguineus
Galium palustre	Polygonum aviculare	Sagina procumbens

Rare

Anemone nemorosa	Dipsacus fullonum	Plantago major
Angelica sylvestris (M)	Endymion non-scriptus	Poa annua
Arctium minus	Epilobium hirsutum	Polygonum persicaria
Arum maculatum	E. tetragonum	Potentilla sterilis
Callitriche stagnalis	Filipendula ulmaria	Rumex crispus
Capsella bursa-pastoris	Fragaria vesca	Scrophularia nodosa
Cardamine pratensis	Galium aparine	Senecio vulgaris
Carex sylvatica (M)	Hypericum hirsutum	Solanum nigrum
Centaurium minus (M)	Juncus bufonius	Sonchus oleraceus
C. pulchellum	Lapsana communis	Stellaria media
Chenopodium polysper-	Lychnis flos-cuculi	Taraxacum officinale
mum	Matricaria maritima	Urtica dioica
Cirsium arvense	Mentha aquatica	Veronica chamaedrys
Crepis capillaris	Myosotis arvensis	Viola reichenbachiana
Deschampsia cespitosa		

Bryophytes: Acrocladium cuspidatum, Atrichum undulatum, Brachythecium
rutabulum, Cirriphyllum piliferum, Eurhynchium praelongum, E. striatum,
E. swartzii, Fissidens taxifolius, Thuidium tamariscinum.

Habitat Study 34b   Wilstead Wood
Details as for H.S. 34a except pH = 6.4
This study was made 65 yds. along the first path on the north beyond the previous
study. The trees here are *Fraxinus excelsior* 30 ft., *Populus tremula* 20 ft., *Quercus
robur* 40 ft., and *Ulmus* sp. 40ft., close-growing and in about equal numbers.
Shrubs: Corylus avellana 15 ft. (o), Crataegus oxyacanthoides 6 ft. (o), Lonicera
periclymenum (r), Ligustrum vulgare 6 ft. (o), Prunus spinosa 10 ft. (1), Rosa
canina 6 ft. (o), Sambucus nigra 10 ft. (r), Thelycrania sanguinea 20 ft. (o).

Abundant

Glechoma hederacea (f on path)

Frequent

Ajuga reptans	Poa trivialis (path)	Viola reichenbachiana

Occasional

Anemone nemorosa	H. tetrapterum (path)	Veronica serpyllifolia
Arum maculatum	Mercurialis perennis	(path)
Galium palustre (path)	Primula vulgaris	Viola riviniana
Hypericum hirsutum	Rumex sanguineus (path)	
(path)		

Rare

Arctium minus	Epipactis helleborine	Prunella vulgaris (path)
Cardamine pratensis	Euphorbia amygdaloides	Ranunculus repens (path)
(path)	Filipendula ulmaria (M)	Rumex crispus (path)
Carex pendula (path)	Lysimachia nummularia	Sagina procumbens
Circaea lutetiana	(path)	(path)
Cirsium palustre (path)	Plantago major (1)	Tussilago farfara
Dactylorchis fuchsii	Potentilla sterilis (path)	Veronica chamaedrys
Endymion non-scriptus		

Bryophytes: Acrocladium cuspidatum, Atrichum undulatum, Ctenidium molluscum, Eurhynchium praelongum, E. striatum, E. swartzii, Fissidens taxifolius, Mnium affine, M. undulatum, Thuidium tamariscinum.

Two large woods, Exeter Wood and Sheerhatch Wood, lie further westward on the face of the escarpment. Both were found profitable by McLaren but I have been unable to make many visits. Sheerhatch Wood is now badly overgrown.

To the south of the escarpment there are a number of woods. Flitwick Wood and Maulden Wood stand almost isolated in an area given up to market-gardening. They are good oak-ash woods but produce little worthy of mention. Chicksands Wood has a fine show of *Campanula trachelium*, and *Tilia cordata* is here apparently just as abundant as it was in John Ray's time. In the same neighbourhood are Warden Great and Warden Little Woods, to which I have made only passing visits. Most of the remaining woods are planted with conifers and contain little of their virgin flora.

The areas of greatest interest in this region are the relics of fen and moorland. Some interesting bogs near Heath and Reach have long since been drained and it is only in isolated patches, as at Stockgrove, that a few of the more interesting species survive.

A series of ponds on the edge of New Wavendon Heath makes interesting study. The ponds are artificial and drain a boggy area in three stages.

Habitat Study 35   Pond, New Wavendon Heath
Parish: Aspley Guise, v.c. 24 [Beds.]   District: Ouzel   Grid Ref. 42/93203374
Alt. 395 ft.   Soil: Alluvium overlying Lower Greensand   pH 5.6
Surveyed: 11 June 1949 with H. B. Souster; 4 Aug. 1949 with T.L.; 10 April 1950; 30 Aug. 1950 with D. A. Reid. This study was made to the east of the island in the middle of the lowest pond. The dominant species is *Polytrichum commune*, and *Sphagnum cuspidatum* and *S. palustre* are abundant.

Frequent

Agrostis canina

Occasional

Calluna vulgaris (L)	Juncus effusus	Polygonum hydropiper
Carex echinata	J. squarrosus	Sagina procumbens
Hydrocotyle vulgaris		

Rare

Betula pendula 3 ft.
B. pubescens 4 ft.
Blechnum spicant
Calamagrostis epigejos
  (M)
Carex nigra

Dryopteris dilatata
Eleocharis palustris
Galium saxatile
Juncus bulbosus
J. conglomeratus
J. acutiflorus

Lotus uliginosus (1)
Luzula multiflora (1)
Molinia coerulea
Potentilla erecta
Quercus robur 2 ft.
Ranunculus flammula

Mermaid's Pond, in Aspley Wood, is now almost dried up, but a record of its present flora is of interest.

Habitat Study 36   Mermaid's Pond
Parish: Aspley Guise   District: Ouzel   Grid Ref. 42/93803475
Alt. 375 ft.   Soil: Lower Greensand   pH 4.2
Surveyed: 11 June 1949 with H. B. Souster; 4 Aug. 1949 with T.L.; 10 April 1950; 30 Aug. 1950 with D. A. Reid. This study was made on the western side of the pond. Around the pond is a ring of birch trees (*Betula pubescens*) about 40 ft. high.

10 ft. from edge of pond

Abundant
Juncus effusus

Frequent
Carex curta

Occasional
Chamaenerion angustifolium   Eriophorum angustifolium

5ft. from edge of pond

Abundant
Carex rostrata

Occasional
Juncus effusus

Rare
Carex curta   Eriophorum angustifolium   Pteridium aquilinum (1)
Epilobium adenocaulon (1) Juncus articulatus

Bryophytes: Ceratodon purpureus, Pohlia nutans, Sphagnum palustre.

In the same area is a marsh of a different nature.

Habitat Study 37   Marsh, Horsemoor Farm
Parish: Woburn   District: Ouzel   Grid Ref. 42/93633355
Alt. 375 ft.   Soil: Alluvium overlying Lower Greensand   pH 6.0
Surveyed: 11 June 1949 with H. B. Souster; 10 April 1950; 4 Aug. 1949 with T.L.; 30 Aug. 1950 with D. A. Reid. This study was made towards the upper end of the marsh, about 30 yds. from H.S. 28.

Abundant
Juncus effusus

Frequent

Agrostis stolonifera
Carex ovalis
Cirsium palustre
Deschampsia cespitosa

Glyceria fluitans
Lotus uliginosus
Montia fontana
  subsp. intermedia

Poa trivialis
Rumex acetosa

Occasional

Agrostis tenuis (M)
Anthoxanthum odoratum
Cardamine pratensis
Carex hirta
Cerastium holosteoides
Epilobium parviflorum

Festuca pratensis
Galium uliginosum
Hypochoeris radicata
Juncus sp.
Lychnis flos-cuculi
Myosotis cespitosa

M. scorpioides
Ranunculus flammula
R. repens
Sieglingia decumbens (M)
Stellaria alsine
S. graminea (M)

Rare

Ajuga reptans (1)	Equisetum fluviatile	Luzula campestris (M)
Alopecurus pratensis (M)	Galium palustre	Potentilla erecta (M)
Carex disticha	G. saxatile (M)	Rumex obtusifolius
C. nigra	Holcus lanatus	Trifolium pratense (M)
Cynosurus cristatus (M)	Lythrum salicaria	

Bryophytes: Acrocladium cuspidatum, Brachythecium rutabulum, Leptobryum pyriforme, Physcomitrium pyriforme.

Flitwick Moor is the largest marshy area left in the county. At one time it extended to Maulden and Clophill through an area now drained. The western side of the moor has largely dried out and has become acid. A large area here is dominated by *Phragmites communis*; and on the edge are usually fine specimens of *Dactylorchis maculata* × *praetermissa*.

Habitat Study 38   Flitwick Moor
Parish: Flitton   District: Ivel   Grid Ref. 52/04433457
Alt. 210 ft.   Soil: Alluvium overlying Lower Greensand   pH 4.4
Surveyed: 9 June 1949; 4 Aug. 1949 with T.L.; 15 April 1950. This study was made 30 yds. west of the Flitwick-Greenfield road and 100 yds. north of the footpath skirting the edge of the moor.
Trees: Alnus glutinosa 6 ft. (1), Crataegus monogyna 10 ft. (1), Salix aurita 4 ft. (1), S. atrocinerea 6 ft. (1), S. × russelliana 6 ft. (1).

Abundant

Agrostis tenuis	Anthoxanthum odoratum

Frequent

Carex ovalis	Hydrocotyle vulgaris	Poa pratensis
Conopodium majus	Juncus acutiflorus	Potentilla erecta
Dactylorchis maculata	J. effusus	Ranunculus acris
Festuca rubra	Linaria vulgaris	Stellaria graminea
Galium palustre	Lotus uliginosus	Valeriana dioica
Holcus lanatus	var. glaber	Vicia cracca

Occasional

Angelica sylvestris	Cirsium palustre	Luzula campestris
Cardamine pratensis	Equisetum palustre	Rumex acetosella
Carex paniculata (tussocks 1 ft.)	Galium uliginosum	Valeriana officinalis

Rare

Arrhenathcrum elatius	Deschampsia cespitosa	Juncus articulatus
Caltha palustris	Epilobium obscurum	Polygala serpyllifolia
Carex nigra	E. palustre	Ranunculus repens
Chamaenerion angusti-folium	Glechoma hederacea (L)	Stellaria palustris
Dactylis glomerata	Glyceria maxima (ditch)	Succisa pratensis
Dactylorchis praetermissa	Hypochoeris radicata	Taraxacum officinale
	Iris pseudacorus (ditch)	

Bryophyte: Brachythecium rutabulum.

To the east of the Flitwick-Greenfield road is the wettest part of the moor, but this is drying out in places and in these *Spiraea ulmaria* has a complete hold. In the wetter patches are fine stands of *Potentilla palustris*, *Menyanthes trifoliata*, *Scirpus sylvaticus* and *Carex curta*. Until comparatively recently *Pedicularis palustris* and *Potamogeton polygonifolius* appeared here but are now apparently extinct.

Habitat Study 39a   Flitwick Moor
Parish: Flitwick   District: Ivel   Grid Ref. 52/04453470
Alt. 205 ft.   Soil: Alluvium overlying Lower Greensand   pH —
Surveyed: 9 June 1949; 4 Aug. 1949 with T.L.; 15 April 1950; 24 Sept. 1950. This
study was made in the small area coloured blue on the 2½ inch O.S. map. The
water level is almost at the surface even in dry weather.
Trees: Alnus glutinosa 12 ft. (f), Salix atrocinerea 12 ft. (o), growing up to 20 ft.
apart.

Abundant

Agrostis stolonifera	Juncus subnodulosus	Potentilla palustris (L)

Frequent

Carex acutiformis	Galium palustre	Scutellaria galericulata
Equisetum fluviatile	Mentha aquatica	Vicia cracca

Occasional

Angelica sylvestris	C. rostrata	Lotus uliginosus
Caltha palustris	Equisetum palustre	Lythrum salicaria
Carex nigra	Eriophorum angusti-	Menyanthes trifoliata
C. ovalis	folium	Solanum dulcamara
C. paniculata (tussocks 4 ft.)	Juncus effusus	

Rare

Arrhenatherum elatius	Galium uliginosum	Phalaris arundinacea
(M)	Glyceria maxima	Poa trivialis
Cardamine pratensis	Holcus lanatus	Rubus idaeus 4 ft. (M)
Cirsium palustre	Juncus acutiflorus	Scirpus sylvaticus
Dactylorchis praetermissa	Lathyrus pratensis	Stellaria palustris
Epilobium hirsutum	Lychnis flos-cuculi	Valeriana dioica
E. palustre		

Habitat Study 39b
Details as above, except pH 6.0
This study was made about 20 yds. to the south of the above to show the rapid
transition to heath conditions so characteristic of the moor.

Abundant
Deschampsia flexuosa

Frequent

Anthoxanthum odoratum	Potentilla erecta	Stellaria graminea
Galium saxatile	Quercus robur 6 ins.	

Occasional
Rumex acetosa

To the eastern side of the moor there has been a considerable
amount of peat-digging. Some of the older beds are overgrown
with birch and the lowering of the water table has left neighbouring
areas dominated by *Chamaenerion angustifolium* and *Deschampsia
flexuosa*, attractive to view but of limited botanical interest. On the
extreme eastern side of the moor is Folly Wood, which has apparently
developed within the past sixty years on the site of previous peat-
digging. Saunders found *Drosera rotundifolia* here when the peat
beds were first abandoned, but the wood is now almost entirely
alder (*Alnus glutinosa*), growing to a height of 80 feet. In the southern
end of the wood *Ribes sylvestre* is abundant, and at the northern end,
where there are clearings, are small areas of bog in which *Viola*

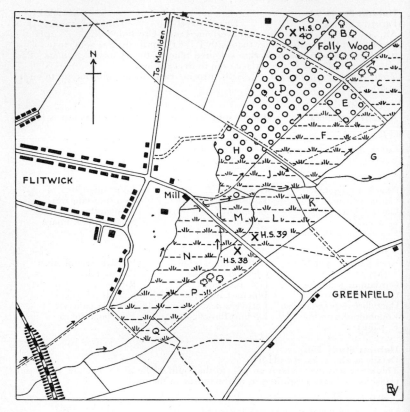

FIG. 10   FLITWICK MOOR

The area A is largely felled woodland now regenerating with birch and alder.
In open parts there is a considerable amount of sphagnum-bog. B (Folly
Wood) is largely alder. D, now overgrown with birch, was reputed, 50 years
ago, to be the wettest part of the Moor. E is the site of the more recent and
present peat-digging. F is rough meadow-land, and C and G meadow-land
now regularly grazed. H, dominated by birch at its northern end, is marshy in
the south. J to Q are rough meadow-land, N is regularly grazed and P very
acid. M contains the wettest parts of the Moor.

palustris grows. T. Laflin is of the opinion that the *Sphagnum* is
spreading, and calls attention to the fluctuating frequency of *Leptobryum pyriforme* in the recent clearings in the wood. In 1949, when
much bare peat was exposed, it was abundant, but in 1951, when the
peat was overgrown with other vegetation, the moss had disappeared
except from the vertical edges of bare peat in some of the ditches.
It behaved similarly at Sutton Fen, although there it was most
common in 1951, after the area had been flooded in the wet spring
of that year. Standing water killed other vegetation and, as the
water receded, the moss covered the drying peat.

The ferns of Folly Wood are of special interest, as there are fine specimens of *Dryopteris spinulosa* and *D. dilatata*, with apparent intermediates. On the edge of the wood *Carex curta* and *Molinia caerulea* appear in some quantity.

Habitat Study 40   Folly Wood
Parish: Flitwick   District: Ivel   Grid Ref. 52/04803538
Alt. 200 ft.   Soil: Alluvium overlying Lower Greensand   pH 6.0
Surveyed: 3 July 1949 with W. M. Cornish and T.L.; 15 April 1950; 24 Sept. 1950.
This study was made in the sphagnum bog area. The alder had been felled and was replenishing itself but there were few young trees within the area of survey. Other trees were Betula pubescens 4 ft. (r) and Quercus robur 1 ft. (1).

Abundant

Agrostis sp.	Hydrocotyle vulgaris	Molinia coerulea

Frequent

Carex acutiformis	C. rostrata

Occasional

Anthoxanthum odoratum	Equisetum fluviatile	Lotus uliginosus
Carex nigra	Juncus effusus	Luzula multiflora

Rare

Chamaenerion angusti-folium	Dryopteris spinulosa	Lonicera periclymenum
	Glyceria maxima	Solanum dulcamara
Cirsium palustre	Holcus lanatus	Viola palustris

Bryophytes: Aulacomnium palustre (o), Funaria hygrometrica (L), Eurhynchium praelongum (r), Hypnum cupressiforme (o), Marchantia polymorpha (L), Plagiothecium denticulatum var. aptychus (r), Pohlia nutans (f), Polytrichum commune (a), Sphagnum fimbriatum (a), S. palustre (a).

Westoning Moor resembles Flitwick Moor and has within its limited area all the same plant associations and most of the same species.

Habitat Study 41   Westoning Moor
Parish: Tingrith   District: Ivel   Grid Ref. 52/02203245
Alt. 240 ft.   Soil: Alluvium overlying Lower Greensand   pH 5.2
Surveyed: 3 July 1949 with W. M. Cornish and T.L.; 31 Aug. 1949 with R. H. Goode; 9 April 1950; 2 Sept. 1951 with W. D. Coales and H. B. Souster. This study was made about 30 yds. from the small alder wood to the south side of a small marshy area. A ditch was on the margin of the survey.

Frequent

Agrostis stolonifera	Festuca rubra	J. effusus
Carex paniculata (tus-socks 2 ft.)	Galium uliginosum	Menyanthes trifoliata
	Juncus acutiflorus	Stellaria graminea

Occasional

Alopecurus pratensis	Epilobium palustre	Poa pratensis
Angelica sylvestris	Equisetum palustre	Potentilla palustris
Caltha palustris	Holcus lanatus	Rhinanthus minor
Cirsium palustre	Lotus uliginosus	Rumex acetosa
Dactylorchis praetermissa	Lychnis flos-cuculi	

Rare

Agrostis tenuis (M)	Galium aparine (1)	Phleum pratense
Arrhenatherum elatius (M)	G. palustre	Ranunculus repens
Carex ovalis (1)	Glyceria maxima (ditch)	Senecio aquaticus (1)
Cerastium holosteoides	Helictotrichon pratense	Stellaria alsine
Deschampsia cespitosa	Hydrocotyle vulgare (M)	Urtica dioica
Filipendula ulmaria	Hypericum tetrapterum	

To the north of Westoning Moor is Moors Plantation, probably the roughest area in the county. Notwithstanding its name there is no evidence that it was ever planted. In some respects it resembles Folly Wood, except that it has fewer alders and no *Sphagnum*. It would repay closer study if one could find suitable clothing for such a task.

Habitat Study 42   Moors Plantation
Parish: Tingrith   District: Ivel   Grid Ref. 52/02103295
Alt. 240 ft.   Soil: Alluvium overlying Lower Greensand   pH 6.4
Surveyed: 9 April 1950; 25 June 1950 with H. B. Souster; 24 Sept. 1950. This study was made on the eastern side, about 90 yards to the south of the cattle trough in the stream.
Trees: Betula pubescens 40 ft., 10–20 ft. apart, Quercus robur 12 ft. (r), Salix cinerea 12 ft. (r), Sambucus nigra 10 ft. (1), Sorbus aucuparia 12 ft. (1).

Abundant

Chryosplenium oppositifolium	Scutellaria galericulata

Frequent

Carex paniculata (no tussocks)	Festuca gigantea	Rubus sp.
Circaea lutetiana	Mercurialis perennis	Scirpus sylvaticus

Occasional

Brachypodium sylvaticum	Galium aparine	Glechoma hederacea
Dryopteris spinulosa	Geranium robertianum	Rumex sanguineus
	Geum urbanum	Scrophularia aquatica

Rare

Arctium minus	Dactylis glomerata	Ribes sylvestre
Arrhenatherum elatius	Dryopteris dilatata	Rubus idaeus
Calystegia sepium	Filipendula ulmaria	Sparganium erectum (1)
Carex acutiformis	Holcus lanatus	Stellaria holostea
C. pendula	Juncus effusus	Tamus communis
C. remota	Lotus uliginosus	Urtica dioica
Cirsium palustre		

Bryophytes: Brachythecium rutabulum, Cirriphyllum piliferum, Pohlia nutans, Mnium hornum.

A contrast is seen in the marsh at the foot of Cooper's Hill. Beneath the trees *Equisetum telmateja* is abundant, and there is a plant association to be seen in few other places in the county.

Habitat Study 43   Marsh, Cooper's Hill
Parish: Ampthill   District: Ivel   Grid Ref. 52/02563740
Alt. 260 ft.   Soil: Alluvium overlying Lower Greensand   pH 6.2
Surveyed: 3 July 1949 with W. M. Cornish and T.L.; 21 Aug. 1949 with P.T.; 11 April 1950. This study was made about 20 yds. from the northern end of the marsh. *Betula pubescens* 60 ft., and *Fraxinus excelsior* 60 ft., are up to 30 ft. apart. Other trees are *Corylus avellana* 10 ft. (1), *Quercus robur* 40 ft. (1) and *Salix cinerea* 7 ft. (1).

Abundant

Equisetum telmateja

Frequent

Holcus lanatus	Mentha aquatica	Scutellaria galericulata
Lotus uliginosus		

Occasional

Ajuga reptans	Juncus acutiflorus	Rubus sp.
Carex acutiformis	J. effusus	Rumex sanguineus
Circaea lutetiana	Lonicera periclymenum	Solanum dulcamara
Cirsium palustre	Lychnis flos-cuculi	Valeriana officinalis
Epilobium obscurum	Poa pratensis	

Rare

Angelica sylvestris	Galeopsis tetrahit (1)	Pteridium aquilinum
Carex paniculata (M)	Geranium robertianum	Ranunculus repens
Epilobium hirsutum	Glechoma hederacea	Ribes sylvestre
Festuca gigantea	Myosotis arvensis	Scrophularia aquatica
Galium aparine	M. scorpioides	Urtica dioica

Bryophytes: Brachythecium rutabulum, Leptodictyum riparium, Mnium longirostrum, M. undulatum.

Mention must be made of one other marshy area, Beckeringspark Moor, now largely dried out. Here *Athyrium filix-femina* is abundant. There are numerous other isolated marshy areas that would be wearisome to enumerate.

Lakes in this region are mainly artificial. Southill Lake, the largest stretch of water in the county, has been well studied and J. E. Little found *Potamogeton trichoides*. On the muddy shores of Lower Drakelow Pond in Woburn Park *Limosella aquatica* is comparatively abundant and *Potamogeton compressus* is found. In a small pond near Warden Abbey *Rumex limosus* was found by J. E. Little and still survives there. But generally lakes and ponds here as elsewhere soon develop large patches of *Polygonum amphibium* and *Potamogeton natans* which exclude all other vegetation.

In the increasing number of sand quarries is a more interesting flora than is usually found in the brick pits and chalk quarries. Quarries near Leighton Buzzard have some quantity of *Draba muralis*, *Lotus tenuis*, *Euphorbia esula* and *Anchusa sempervirens*, and the moss *Bryum bicolor* is abundant in some pits.

Arable fields add considerably to the flora. Here the commonest weeds are *Chenopodium album* and *Urtica urens*, but in many fields *Chrysanthemum segetum* makes a fine show. Shoddy is used extensively in the neighbourhood of Flitwick and Maulden, and brings in numerous wool adventives. It would give a wrong impression to insert here a list of weeds from a field so treated and I give one from a more normal piece of arable land.

Habitat Study 44   Allotment, Heath and Reach
Parish: Heath and Reach   District: Ouzel   Grid Ref. 42/920282
Alt. 325 ft.   Soil: Lower Greensand   pH 6.2
Surveyed: 8 June 1949; 2 Aug. 1949; 10 April 1950. This study was made on a neglected patch on the northern side of the main path. It was not possible to take the same patch on each visit owing to very thorough weeding.

Aira caryophyllea
Antirrhinum orontium
Arenaria serpyllifolia
Arrhenatherum elatius
Bromus mollis
Capsella bursa-pastoris
Cerastium glomeratum
Chenopodium album
Cirsium arvense
Claytonia perfoliata
Convolvulus arvensis
Crepis capillaris
Dactylis glomerata
Epilobium adenocaulon
Equisetum arvense

Erigeron canadensis
Erodium cicutarium
Erophila verna
Erysimum cheiranthoides
Filago germanica
Geranium molle
G. pusillum
Holcus lanatus
Lamium amplexicaule
L. hybridum
Leontodon autumnalis
Melandrium album
Melilotus indica
Papaver argemone
P. dubium

Plantago lanceolata
Poa annua
Polygonum convolvulus
Rumex acetosella
R. obtusifolius
Sonchus asper
Spergula arvensis
Stellaria media
Trifolium campestre
T. dubium
T. repens
Urtica urens
Veronica polita
Vicia sepium
Viola arvensis

The main railway line crosses the region and its embankments are always worth study. On either side of Ampthill Tunnel is a large cutting in which the soft sand embankment has been covered with broken brick and reinforced with stonework. It was in the excavations here that Ampthill Clay was first distinguished from Oxford Clay. Possibly the lower part of the cutting is through Ampthill Clay, but it is now impossible to see any change in the strata.

Habitat Study 45a   Railway cutting, Ampthill (west aspect)
Parish: Ampthill   District: Ivel   Grid Ref. 52/02163745
Alt. 260 ft.   Soil: doubtful (see above)   pH 7.2
Surveyed: 3 July 1949 with W. M. Cornish and T.L.; 21 Aug. 1949 with P.T.; 11 April 1950. This study was made at a point opposite the fourth telegraph post to the north of the bridge at Ampthill Station.

Abundant

Erophila verna
Hieracium pilosella

Melilotus officinalis

Ulex europaeus (L)

Frequent

Arenaria serpyllifolia
Arrhenatherum elatius

Chrysanthemum leucan-
    themum
Filago germanica

Myosotis discolor
Tussilago farfara

Occasional

Crepis capillaris

Taraxacum laevigatum

Rare

Aphanes arvensis
Campanula medium
Cardamine hirsuta
Centaurea nigra
Chaenorhinum minus
Cirsium vulgare
Daucus carota

Erigeron acris
Holcus lanatus
Lamium amplexicaule
Leontodon hispidus
Plantago lanceolata
Rumex acetosa

Scleropoa rigida
Senecio erucifolius
S. vulgaris
Taraxacum officinale (1)
Trisetum flavescens
Vicia hirsuta

Bryophytes: Brachythecium velutinum, Bryum capillare, Ceratodon purpureus, Pseudoscleropodium purum, Tortula ruralis.

Habitat Study 45b   Railway cutting, Ampthill (east aspect)
Details as above. Vegetation more luxuriant than in the previous study.
Shrubs: Crataegus monogyna 2 ft. (r), Fraxinus excelsior 6 ins. (r), Pinus sylvestris 6 ft. (1), Populus nigra agg. 8 ft. (1), Ulex europaeus (1).

Abundant
Arrhenatherum elatius

Frequent

Festuca rubra	Leontodon hispidus	Myosotis arvensis
Galium verum	Linum catharticum	Silaum silaus
Hieracium pilosella	Medicago lupulina	

Occasional

Daucus carota	Potentilla reptans	T. dubium
Lathyrus pratensis	Senecio erucifolius	Trisetum flavenscens
Lotus corniculatus	Trifolium campestre	Tussilago farfara
Plantago lanceolata		

Rare

Achillea millefolium	Chrysanthemum leucan-	Erigeron acris
Bellis perennis	themum	Heracleum sphondylium
Cardamine hirsuta	Cirsium arvense	Pimpinella saxifraga
Centaurea nigra	C. vulgare	Ranunculus acris
	Dactylis glomerata	

In the detached portion of the Greensand Region it has been necessary to include some of the Oxford (probably more strictly Ampthill) Clay. This occupies a small intensively cultivated area and contains little worthy of a botanist's attention.

The Greensand comes to the surface at lower altitudes than in the west of the county and this, combined with greater extremes of climate and less rainfall, produces a slight change in flora. Galley Hill, Sutton, is typical of the open Greensand in this part of the county.

Habitat Study 46   Galley Hill
Parish: Sutton   District: Ivel   Grid Ref. 52/21904844
Alt. 140 ft.   Soil: Lower Greensand   pH 4.8
Surveyed: 19 June 1949 with W. Durant; 23 April 1950 with W. Durant and T.L.;
27 Aug. 1950 with W. Durant. This study was made about 50 yds. from the road at about the middle of the hill. The slope of the hill is here about 1 in 10. *Ulex europaeus* 18 ins. is scattered. In dry seasons the vegetation is sparse, *Polytrichum juniperinum* being dominant.

Abundant
Agrostis tenuis

Frequent

Aira praecox	Hypochoeris radicata	Rumex acetosella
Galium saxatile (L)	Ornithopus perpusillus	Vulpia bromoides
Holcus mollis (L)		

Occasional

Crepis capillaris	Filago minima

Rare

Chamaenerion angusti-	Holcus lanatus	Senecio jacobaea
folium	Poa pratensis	Trisetum flavescens
Dactylis glomerata	Quercus robur 3 ins.	

Bryophytes: Ceratodon purpureus (o), Polytrichum juniperinum (a).

A different plant association is seen to the south of Bunker's Hill, where there has been a certain amount of planting of conifers.

Habitat Study 47   Bunker's Hill
Parish: Sandy   District: Ivel   Grid Ref. 52/19204768
Alt. 125 ft.   Soil: Lower Greensand   pH 4.6
Surveyed: 16 July 1949 with T.L.; 23 April 1950 with W. Durant and T.L.;
6 Sept. 1950. This study was made on the left-hand side of the public footpath
through Sandy Lodge at a point about 25 yds. from its beginning. The slope is
gradual, the hill well grazed by rabbits. Species listed as marginal were growing
in a sheltered depression on the edge of the survey.
Trees, etc.: Pinus sylvestris 30 ft. (1), Quercus robur 30 ft. (1), Rubus ? cardiophyllus
8 ft. (f), Sambucus nigra 8 ft. (M), Sarothamnus scoparius 5 ft. (r).

Abundant

Calluna vulgaris   Deschampsia flexuosa

Frequent

Agrostis tenuis   Galium saxatile   Rumex acetosa

Occasional

Luzula campestris   Teesdalia nudicaulis

Rare

Aira praecox	Claytonia perfoliata (M)	Myosotis hispida
Bryonia dioica (M)	Conium maculatum (M)	Pteridium aquilinum
Cerastium holosteoides	Erophila sp.	Senecio sylvaticus (M)
(M)	Hieracium pilosella	S. vulgaris
Chamaenerion angusti-	Holcus mollis	Urtica dioica (M)
folium	Melandrium album (M)	U. urens (M)

Bryophytes: Aulacomnium androgynum (r), Bryum capillare (r), Ceratodon purpureus (f), Dicranum scoparium (a), Hypnum cupressiforme (o), Pleurozium
schreberi (o), Pohlia nutans (o), Polytrichum juniperinum (f), P. piliferum (o).

In Sandy itself, on the open ground marked 'Camp' on the O.S.
map, is a hill with a different plant association. It is a recreation
ground and as such has suffered from human interference.

Habitat Study 48   Hill, Sandy
Parish: Sandy   District: Ivel   Grid Ref. 52/17984892
Alt. 205 ft.   Soil: Lower Greensand   pH 6.0
Surveyed: 19 June 1949 with W. Durant: 23 April 1950 with W. Durant; 6 Sept.
1950. This study was made at the top of the hill about midway between two seats.
The hill is flat on the top but at the place of the study it falls steeply into rough
grassland.

Abundant

Agrostis tenuis   Plantago lanceolata   Rumex acetosella

Frequent

Aira caryophyllea   Ornithopus perpusillus   Vulpia bromoides (L)

Occasional

Achillea millefolium	Hieracium pilosella	Lycopsis arvensis (L)
Cerastium semidecan-	Holcus lanatus	Poa annua
drum	Hypochoeris radicata	Stellaria media
Crepis capillaris		

Rare

Bromus mollis	Filago minima	Sisymbrium officinale
Capsella bursa-pastoris	Geranium molle	Taraxacum laevigatum
Convolvulus arvensis	Melandrium album	Trifolium arvense
Dactylis glomerata	Sarothamnus scoparius	T. dubium
Erodium cicutarium	Senecio jacobaea	Vicia angustifolia
Festuca rubra		

PLATE 12   JACKDAW HILL. (Habitat Study 29)

PLATE 13   FLITWICK MOOR (Habitat Study 39a)

PLATE 14   PEAT WORKINGS, FLITWICK MOOR

These are but a few of the open spaces in this part of the county. Sandy Warren has still some wild patches, no doubt previously cultivated, as was rough ground between Sutton Fen and Sutton Cross Roads. Here there is a large area completely dominated by *Deschampsia flexuosa*, which presents an attractive scene. There has been little quarrying but the sides of the small disused pits are a habitat favoured by *Hypochoeris glabra*. In damp tracks in a disused pit near Sutton Fen is some quantity of *Montia fontana* subsp. *chrondosperma*.

The detached portion of Huntingdonshire is partly on the greensand and partly on clay. During World War II it was almost completely given over to a large airfield, now derelict. To the east of this is White Wood, part of which is in v.c. 30. In earlier days it was a profitable wood for Cambridgeshire naturalists.

There are few marshy places and I have been unable to locate Potton Marshes from which Abbot recorded many species, and which was still profitable in McLaren's day. Sutton Fen, still remaining, has largely dried out.

Habitat Study 49 Sutton Fen
Parish: Sutton District: Ivel Grid Ref. 52/20724751
Alt. 95 ft. Soil: Alluvium overlying Lower Greensand pH 4.4
Surveyed: 19 June 1949 with W. Durant; 23 April 1950 with W. Durant and T.L.; 6 Sept. 1950. This study was made approximately in the middle of the fen. Unfortunately there are no local landmarks to make it more precise. The trees are mainly *Betula pubescens* 60 ft., growing up to 20 ft. apart. Other trees were *Quercus robur* 60 ft. (1) and *Castanea sativa* 1 ft. (1).

Abundant
Deschampsia flexuosa

Frequent

Agrostis tenuis	Rubus sp.	Stellaria graminea
Pteridium aquilinum (L)		

Occasional

Dryopteris spinulosa	Luzula multiflora	Stellaria media (M)
Galium saxatile		

Rare

Chamaenerion angusti-	Holcus mollis	Lonicera periclymenum
folium	Juncus effusus	Moehringia trinervia
Dryopteris dilatata		

Bryophytes: Aulacomnium androgynum, Ceratodon purpureus, Dicranella heteromalla, Funaria hygrometrica, Pleurozium schreberi, Pseudoscleropodium purum, Rhytidiadelphus squarrosus.

Roadsides near Potton Church have *Salvia verbenaca* in some quantity and near Sandy *Ornithogalum pyrenaicum* and *Trifolium ochroleucon* appear though less abundantly than elsewhere in the county.

The weeds of arable land are little different from those in the rest of the Greensand Region. *Arnoseris minima* appears from time to time and the use of shoddy adds an adventive flora.

Habitat Study 50   Arable field, Sutton
Parish: Sutton   District: Ivel   Grid Ref. 52/20874880
Alt. 120 ft.   Soil: Lower Greensand   pH 7.2
Surveyed: 16 July 1949 with T.L.; 23 April 1950 with W. Durant; 6 Sept. 1950.
This study was made in the bottom corner of the first field on the east side of the
road from Deepdale to Sutton Cross-roads. In each year the crop was a poor one,
the soil having apparently become sour.

Achillea millefolium	Lamium amplexicaule	S. vulgaris
Agropyron repens	Matricaria maritima	Sisymbrium officinale
Artemisia vulgaris	Papaver dubium	Solanum nigrum
Capsella bursa-pastoris	P. rhoeas	Sonchus oleraceus
Chenopodium album	Poa annua	Stellaria media
Crepis capillaris	Polygonum aviculare	Urtica urens
Erodium cicutarium	Senecio jacobaea	Veronica persica
Galinsoga parviflora		

G.   The Gault and Chalk Marl Region

This awkwardly-shaped region is difficult to interpret. In the eastern part of the county the Gault is overlain to a considerable degree. As it emerges in the south of the county there are large caps of Glacial Gravel, and the Gault itself merges almost imperceptibly into the Chalk Marl which lies at the foot of the chalk escarpment. There is much to be gained by treating these broad divisions separately.

The eastern part stretches from the Great North Road to Cockayne Hatley and approximates closely to the Cam District. Apart from Potton Wood and Cockayne Hatley Wood it is intensively cultivated and there are extensive apple orchards at Cockayne Hatley.

Cockayne Hatley Wood is of some interest as it is one of the best stations in the county for *Melampyrum cristatum* and also contains some quantity of *Campanula latifolia*. Potton Wood is one of the largest in the county. Its chief claim to the botanist's attention is that it is on the fringe of the Oxlip (*Primula elatior*) country. In some parts of these woods *Orthodontium lineare*, rare elsewhere in the county, is common on rotting tree stumps.

Habitat Study 51   Potton Wood
Parish: Potton   District: Cam   Grid Ref. 52/25205009
Alt. 255 ft.   Soil: Boulder Clay overlying Gault   pH 6.2
Surveyed: 19 June 1949 with W. Durant; 23 April 1950 with W. Durant; 27 Aug.
1950. This study was made 100 yds. along the path in a north-easterly direction
from the crossing of main rides in the wood.
Trees, etc.: Acer campestre 10 ft. (1), Corylus avellana 12 ft. (a), Crataegus monogyna 6 ft. (f), Hedera helix (r), Fraxinus excelsior 30 ft. (o), Ligustrum vulgare
5 ft. (M), Lonicera periclymenum (r), Rosa canina (f), Rubus sp. (f), Sambucus
nigra 8 ft. (f), Tamus communis (r), Thelycrania sanguinea 8 ft. (r).

(a) woodland

Abundant
Bromus ramosus   Deschampsia cespitosa

Frequent
Ajuga reptans   Primula vulgaris

Occasional

Anemone nemorosa	Dactylorchis fuchsii	Mercurialis perennis
Arum maculatum	Endymion non-scriptus	Ranunculus ficaria
Brachypodium sylvaticum	Glechoma hederacea	Viola sp.
Carex sylvatica	Listera ovata	Vicia sepium

Rare

Angelica sylvestris	Epilobium parviflorum (1)	Neottia nidus-avis
Circaea lutetiana	Heracleum sphondylium	Orchis mascula
Cirsium palustre (1)	Hypericum hirsutum (M)	Potentilla sterilis (1)
Conopodium majus (1)	Luzula pilosa	Rumex obtusifolius (1)

Bryophytes: Acrocladium cuspidatum, Eurhynchium praelongum, Hypnum cupressiforme, Isothecium myosuroides, Rhytidiadelphus triquetrus, Thuidium tamariscinum.

(b) path

Between April and August 1950, the path was cleared for the erection of telegraph poles; this accounts for some of the weeds of arable land found.

Frequent

Arrhenatherum elatius	Centaurea nigra	Hypericum hirsutum

Occasional

Festuca gigantea	Sonchus asper	S. oleraceus
Plantago lanceolata		

Rare

Achillea millefolium	Galium aparine (1)	S. jacobaea
Agrimonia eupatoria	Potentilla anserina	S. vulgaris
Cirsium arvense (1)	Lotus corniculatus	Taraxacum officinale
C. vulgare (1)	Senecio erucifolius	Veronica chamaedrys

There are few ponds and little meadow-land in this part of the county. Near the stream on the county boundary at Brook Farm, Eyeworth, the rough meadows have some quantity of *Cynoglossum officinale*. The roadsides are rich with *Trifolium ochroleucon*, and *Smyrnium olusatrum* occurs by the roadside at Millow and again more plentifully at Topler's Hill.

The weeds of the arable land are of some interest. *Picris echioides* is more plentiful at Thistly-grounds Farm, Eyeworth, than elsewhere in the county, and *Bupleurum rotundifolium* appears regularly in fields at Cockayne Hatley, Wrestlingworth and Eyeworth.

To the south, where the Gault is in closer proximity to the Chalk, the soil conditions are more varied. Much of the Gault here is overlain by Boulder Clay and there is little on the surface to distinguish one from the other. The Boulder Clay is less deep than in the north of the county. The clays are well cultivated although there is still a considerable amount of woodland. Near Tilsworth, on the edge of the Gault and the Chalk Marl, there are two interesting woods. Stanbridge Wood is an elm wood and in it *Anthriscus sylvestris* is almost the only plant to be seen. Blackgrove Wood has an unusually large amount of *Epipactis helleborine*.

Habitat Study 52   Blackgrove Wood
Parish: Tilsworth   District: Ouzel   Grid Ref. 42/97902362
Alt. 320 ft.   Soil: Gault   pH 6.4
Surveyed: 24 June 1949; 19 April 1950; 27 Aug. 1950 with H. B. Souster. This study was made at a point 100 yds. along the main path of the wood which at this point is close-growing with intense shade.

Trees, etc.: Acer campestre 10 ft. (1), Bryonia dioica (r), Crataegus monogyna 6 ft.
(f), Quercus robur 30 ft. (1), Rhamnus catharticus 8 ft. (1), Rosa canina 6 ft. (1),
Rubus sp. 5 ft. (o), Ulmus glabra 30 ft. (f).

Abundant

Ajuga reptans	Poa trivialis	Ranunculus ficaria

Frequent

Circaea lutetiana	Glechoma hederacea	Rumex obtusifolius
Dactylorchis fuchsii	Myosotis arvensis	Urtica dioica (L)
Geum urbanum		

Occasional

Anemone nemorosa	Dipsacus fullonum	Geranium robertianum
Arum maculatum	Endymion non-scriptus	(L)
Bromus ramosus	Festuca gigantea	Lysimachia nummularia
Carex sylvatica	Filipendula ulmaria	Ranunculus repens
Cirsium palustre	Galium aparine	Rumex sanguineus
Deschampsia cespitosa		Viola reichenbachiana

Rare

Arctium nemorosum	Epipactis helleborine	Ranunculus auricomus
Brachypodium sylvaticum	Orchis mascula	Scrophularia nodosa
Carduus crispus (M)	Potentilla reptans	Taraxacum officinale (1)
Cirsium vulgare	Primula vulgaris	Veronica chamaedrys
Epilobium parviflorum	P. veris × vulgaris	Viola riviniana

Bryophytes: Brachythecium rutabulum, Eurhynchium praelongum, E. swartzii,
Fissidens taxifolius, Hypnum cupressiforme, Isothecium myosuroides, Thamnium
alopecurum.

Between these woods and Totternhoe was a place known as the
Litany. At one time waterlogged, attempts to cultivate it were
abandoned and it became the largest area of scrub in the county.
Hawthorn grew to a height of 20 ft., but it supported little ground
vegetation. It was cleared in 1952 and is now ploughed.

To the south of Eversholt are two woods partly on a cap of Glacial
Gravel. One, Washer's Wood, is on a hill slope, at the foot of which
the Lower Greensand is close to the surface. In the wood rises a
stream along which there were some fine alders, but they were felled
in indiscriminate clearing of the wood a few years ago. By the side
of the stream is some quantity of *Chrysosplenium oppositifolium* and
Equisetum telmateja. The other, Daintry Wood, is almost entirely on
Glacial Gravel.

Habitat Study 53a   Washer's Wood
Parish: Tingrith   District: Ivel   Grid Ref. 42/99323185
Alt. 350 ft.   Soil: Juncture of Boulder Clay and Lower Greensand   pH 5.6
Surveyed: 9 April 1950; 25 June 1950 with H. B. Souster and P.T.; 30 Aug. 1950
with D. A. Reid. This study was made to the east of the main ride at a point 60
yds. from the bottom of the wood. The dominant tree had been oak, but this was
cleared in 1949.
Trees, etc.: Acer campestre 6 ft. (1), Corylus avellana 6 ft. (r), Fraxinus excelsior
6 ft. (1), Lonicera periclymenum (1), Quercus robur, many suckers.

Abundant

Juncus conglomeratus (L)

Frequent

Galium aparine	Juncus effusus	Myosotis arvensis
Glechoma hederacea	Mercurialis perennis (L)	Veronica serpyllifolia

Occasional

Chamaenerion angusti-   Deschampsia cespitosa   Ranunculus repens
    folium   Holcus lanatus   Vicia sepium
Cirsium palustre   Prunella vulgaris

Rare

Ajuga reptans   E. obscurum   P. sterilis
Aphanes microcarpa   Filipendula ulmaria   Primula vulgaris
Arctium minus   Geum urbanum   Ranunculus ficaria
Carex sylvatica   Heracleum sphondylium   Rumex crispus
Cerastium holosteoides   Hypericum tetrapterum   Senecio jacobaea
Cirsium arvense   Lysimachia nemorum   Sonchus oleraceus
C. vulgare   L. nummularia   Stachys officinalis
Crepis capillaris   Lychnis flos-cuculi   Taraxacum officinale (1)
Epilobium adenocaulon   Plantago major   Trifolium pratense
E. adenocaulon × obscu-   Poa annua (M)   T. repens (1)
    rum   Potentilla anserina (M)   Urtica dioica
E. montanum   P. reptans   Veronica chamaedrys

Habitat Study 53b

This study was made in the marshy area in the lower part of the same wood.
Felling had removed all local landmarks but it should not be difficult to identify
the site.
Trees, etc.: Betula pendula 6 ft. (1), Fraxinus excelsior 20 ft. (1), Lonicera periclymenum (r), Viburnum lantana 10 ft. (1).

Abundant

Chamaenerion angusti-   Lysimachia nemorum   Hypericum perforatum
    folium

Frequent

Polygonum hydropiper (L)   Prunella vulgaris   Sagina procumbens
Potentilla reptans   Ranunculus repens   Veronica officinalis

Occasional

Centaurium minus   Geum urbanum   Potentilla sterilis
Cirsium palustre   Lychnis flos-cuculi   Pteridium aquilinum (L)
Fragaria vesca   Poa annua

Rare

Ajuga reptans   Cirsium arvense   Myosotis arvensis
Anagallis arvensis (M)   Dryopteris spinulosa   Potentilla erecta
Athyrium filix-femina   Filipendula ulmaria   Primula vulgaris
Callitriche stagnalis   Glechoma hederacea   Scrophularia nodosa
Carex sp.   Lysimachia nummularia   Valeriana officinalis
C. remota   Lythrum salicaria   Viola sp.
Cerastium glomeratum   Mercurialis perennis   Urtica dioica
Circaea lutetiana

Bryophytes: Acrocladium cuspidatum, Brachythecium rutabulum, Eurhynchium
praelongum, Physcomitrium pyriforme.

Habitat Study 54   Daintry Wood

Parish: Tingrith   District: Ivel   Grid Ref. 42/99753135
Alt. 400 ft.   Soil: Boulder Clay overlying Lower Greensand   pH 5.5
Surveyed with H.S. 53. This study was made 100 yds. along the main ride from
the public way between the wood and Washer's Wood. The wood was cleared in
1948 and most of the species listed were growing on the path.
Trees, etc.: Corylus avellana 6 ft. (f), Crataegus monogyna 6 ft. (r), Rubus sp. (f),
Sambucus nigra 6 ft. (1)

Abundant (in wooded part)

Mercurialis perennis   Pteridium aquilinum

Frequent

Anemone nemorosa
Chamaenerion angusti-
  folium

Endymion non-scriptus
Potentilla reptans

Veronica officinalis (L)

Occasional

Ajuga reptans
Epilobium montanum
Glechoma hederacea

Primula vulgaris
Prunella vulgaris
Ranunculus ficaria

Viola riviniana
Urtica dioica

Rare

Anagallis arvensis
Aphanes microcarpa
Arctium minus
Centaurium minus
Cerastium glomeratum
Cirsium arvense
C. palustre
Epilobium adenocaulon

Fragaria vesca
Galium aparine
Geum urbanum
Melandrium album
Moehringia trinervia
Myosotis arvensis
Plantago major

Poa annua (M)
Potentilla sterilis
Sagina procumbens
Scrophularia nodosa
Senecio jacobaea
Stellaria media
Veronica serpyllifolia

Bryophytes: Atrichum undulatum, Eurhynchium praelongum.

One of the best of the remaining woods in the region is Thrift
Wood, which lies to the west of Silsoe. It is well-kept, with the usual
flora of a clay wood.

Habitat Study 55   Thrift Wood
Parish: Silsoe   District: Ivel   Grid Ref. 52/070355
Alt. 275 ft.   Soil: Boulder Clay overlying Gault   pH 6.0
Surveyed: 11 April 1950; 25 June 1950 with H. B. Souster; 4 Sept. 1950. This
study was made at the fork in the rides on the east side of the wood.

(a) woodland

Close-growing with intense shade.
Trees, etc.: Crataegus monogyna 15 ft. (f), Fraxinus excelsior 30 ft. (f), Quercus
robur 40 ft. (f), Hedera helix (f).

Abundant

Endymion non-scriptus

Frequent

Anemone nemorosa
Arum maculatum

Circaea lutetiana

Ranunculus ficaria

Occasional

Urtica dioica

Viola riviniana

Rare

Ajuga reptans
Deschampsia cespitosa

Geum urbanum
Mercurialis perennis

Primula vulgaris

Bryophytes: Cirriphyllum piliferum, Eurhynchium praelongum, E. swartzii.

(b) path (adjoining above)

Shrubs: Rosa arvensis 2 ft. (o), Rubus sp. (f), Salix sp. 2 ft. (1).

Frequent

Holcus lanatus
Poa trivialis

Prunella vulgaris

Rumex sanguineus

Occasional

Brachypodium sylvati-   Epilobium montanum   Mercurialis perennis
  cum   Geranium dissectum   Myosotis arvensis
Cerastium holosteoides   G. robertianum   Potentilla reptans
Circaea lutetiana   Geum urbanum   Ranunculus repens
Cirsium palustre   Glechoma hederacea   Vicia sepium
Deschampsia cespitosa   Juncus effusus

Rare

Alopecurus pratensis   Dryopteris filix-mas (1)   Plantago major (1)
Angelica sylvestris   Epilobium parviflorum   Poa annua
Arctium minus   Galium aparine   Potentilla anserina
Bromus ramosus   Juncus conglomeratus   Taraxacum officinale
Carex sylvatica   Lapsana communis   Torilis japonica
Cirsium vulgare   Lathyrus pratensis   Urtica dioica
Dactylis glomerata (1)   Pimpinella major (1)   Veronica serpyllifolia
Dactylorchis fuchsii (1)

In an area so geologically varied there are a number of marshy places. Those already mentioned in Washer's Wood extend outside and on its western side, where *Equisetum telmateja* is again plentiful, *Pedicularis sylvatica* grows. Elsewhere interesting marshy places are to be found at the base of the Glacial Gravels at Chalgrave and Fancott.

Marshes formed by streams emerging from the Totternhoe Stone are always worth study. The finest examples are at Totternhoe itself where they have suffered from and are still affected by drainage. Totternhoe Mead has some quantity of *Blysmus compressus*, and various species of *Carex* and *Carum carvi*, no doubt native here. *Fritillaria meleagris* still grows in a similar meadow at Billington.

Habitat Study 56   Cow Common
Parish: Totternhoe   District: Ouzel   Grid Ref. 42/98002297
Alt. 305 ft.   Soil: Alluvium overlying Chalk Marl   pH 7.8
Surveyed: 24 June 1949; 27 Aug. 1949 with H. B. Souster; 19 April 1950. This study was made along the stream crossing the common 100 yds. from the point at which it leaves it. There was no evidence that the common had at any time been cultivated. It was drained in 1952 and ploughed in 1953.

(a) stream

Abundant

Eleocharis palustris   Glyceria fluitans

Frequent

Carex distans   Juncus inflexus   Ranunculus flammula
C. lepidocarpa

Occasional

Juncus articulatus   Poa trivialis   Veronica beccabunga

Rare

Alopecurus geniculatus   Epilobium hirsutum   Rumex conglomeratus
Blysmus compressus   Equisetum palustre   Sparganium erectum
Carex disticha   Galium palustre   Trifolium repens
C. hirta   Ranunculus repens   Triglochin palustre
Cirsium palustre   Rorippa microphylla

(b) meadow

This is to the south-east of the above. Species listed as local were limited to some large ant-hills, a feature of the common.

Abundant

Erophila sp. (L)

Frequent

Briza media	Festuca ovina (L)	Leontodon taraxacoides
Cynosurus cristatus	F. rubra	

Occasional

Bellis perennis	Dactylis glomerata	Prunella vulgaris
Carex disticha	Juncus inflexus	Succisa pratensis
C. flacca	Koeleria gracilis (L)	Thymus pulegioides
Cirsium acaulon	Phleum pratense	

Rare

Achillea millefolium	Hordeum secalinum	P. reptans
Arenaria serpyllifolia (L)	Linum catharticum	Primula veris
Centaurea nigra	Lolium perenne	Ranunculus acris
Cerastium holosteoides	Lotus corniculatus	Scleropoa rigida (L)
Chrysanthemum leucan-	Medicago lupulina	Sieglingia decumbens
themum	Mentha aquatica	Silaum silaus
Cirsium arvense (1)	Oenanthe lachenalii	Taraxacum officinale
C. palustre	Ononis spinosa	T. palustre
C. vulgare	Plantago lanceolata	Trifolium pratense
Deschampsia cespitosa	P. major	Trisetum flavescens
Festuca pratensis	Poa pratensis	Veronica arvensis (L)
Galium verum	Polygala oxyptera	Vicia cracca (1)
Holcus lanatus	Potentilla anserina	Viola hirta

At Well Head there is a marsh of a different nature.

Habitat Study 57   Well Head
Parish: District: Ouzel   Grid Ref. 42/999204
Alt. 400 ft.   Soil: Alluvium overlying Chalk Marl   pH 7.6
Surveyed: 28 June 1949 with W. M. Cornish; 6 April 1950; 1 Sept. 1950.

(a) stream

This study was made 20 yds. from the spring.

Abundant

Glyceria plicata

Frequent

Apium nodiflorum	Hypericum tetrapterum	Myosotis scorpioides
Epilobium hirsutum	Juncus inflexus	Scrophularia aquatica
E. parviflorum		

Occasional

Carduus crispus	Pulicaria dysenterica	Veronica beccabunga
Cirsium palustre	Rumex conglomeratus	

Rare

Achillea millefolium (M)	Daucus carota (M)	Prunella vulgaris (M)
Arrhenatherum elatius	Holcus lanatus (M)	Rumex obtusifolius
(M)	Glyceria × pedicellata	Urtica dioica
Dactylis glomerata (M)	Plantago lanceolata (M)	Verbena officinalis (M)

(b) marsh

This is small and has a uniform flora. The study was made in the centre of it.

Abundant

Juncus effusus

Frequent

Cynosurus cristatus (L)   Ranunculus acris

Occasional

Carduus crispus   Festuca arundinacea   Potentilla anserina
Cirsium palustre   Lathyrus pratensis   Pulicaria dysenterica (L)
Deschampsia cespitosa   Poa pratensis   Ranunculus ficaria

Rare

Arrhenatherum elatius   Festuca pratensis   Ranunculus repens
Carex hirta   Holcus lanatus   Rumex crispus (1)
C. spicata (L)   Myosotis scorpioides   R. obtusifolius (1)
Dactylis glomerata   Plantago lanceolata   Sonchus asper
Eleocharis palustris   Poa trivialis   Vicia cracca
Epilobium hirsutum

Bryophytes: Brachythecium rutabulum, Acrocladium cuspidatum, Fissidens taxifolius, Pohlia delicatula.

The Gault is quarried in the neighbourhood of Arlesey for making
bricks and tiles. Like the quarries on the Oxford Clay these soon
develop a flora worthy of study.

Habitat Study 58   Claypit, Arlesey
Parish: Arlesey   District: Ivel   Grid Ref. 52/187353
Alt. *c.* 100 ft.   Soil: Gault   pH 7.6
Surveyed: 16 July 1949; 14 May 1951; 1 Sept. 1951. This study was made on the
edge of a pool about 30 ft. in diameter on the north-eastern side of the main pit.

(a) species growing in and around pool

Abundant

Potamogeton natans   Typha latifolia (L)

Frequent

Juncus articulatus   Polygonum amphibium   Zannichellia palustris
J. inflexus   (L)

Occasional

Alisma plantago-aquatica   Carex otrubae

(b) species observed on side of pit

Abundant

Melilotus officinalis   Pastinaca sativa (L)

Frequent

Daucus carota   Leontodon autumnalis   Plantago lanceolata

Occasional

Cirsium arvense   Potentilla reptans   Senecio erucifolius
Medicago lupulina   Pulicaria dysenterica   Tussilago farfara
Poa pratensis

Rare

Achillea millefolium   Deschampsia cespitosa   Picris echioides
Agrostis stolonifera   Dipsacus fullonum   P. hieracioides
Arrhenatherum elatius   Equisetum sp. (1)   Ranunculus repens
Brachypodium sylvaticum   Erigeron acris   Rumex crispus
Bellis perennis (1)   E. canadensis   Salix ? caprea 1 ft.
Centaurea nigra   Holcus lanatus   Sinapis arvensis
Centaurium minus (1)   Hypericum perforatum   Sonchus arvensis
Cirsium vulgare   Inula conyza   Taraxacum officinale (1)
Crepis capillaris   Matricaria maritima (1)   Torilis japonica
C. vesicaria   Medicago sativa (1)   Trifolium pratense

The Chalk Marl is quarried usually with the Lower Chalk for cement manufacture and consideration of the flora of chalk pits is deferred until the next section. Apart from the disused brick pits and the railway there are few places which support an alien flora.

H. The Chalk Region

This closely rivals the Lower Greensand in its interest for the botanist. There are three zones of chalk in the county, the Lower Chalk and Middle Chalk, which account for the greater part, and the Upper Chalk, which was largely removed by glacial action and replaced by Clay-with-Flints.

The lower chalk escarpment produces some of the more attractive hills of the county in Totternhoe Knolls, Sharpenhoe Clappers (the county's best known landmark) and Knocking Hoe. Each hill has its own distinctive flora. Variations in the flora are not, however, the result of local differences in the texture of the soil and of changing climatic conditions, but have been caused by man's various usage of the downs over many centuries. These differences are strikingly shown in a comparison of the flora of Knocking Hoe (in the opinion of J. F. Hope-Simpson virgin chalk downland), Barton Hills (a good example of well established downland), Totternhoe Knolls (the site of a medieval castle), and Dunstable Cutting, where a chalk flora is developing.

A few species are limited to the lower-chalk downland and are put on record so that comparison may be made with their occurrence elsewhere. They include the three treasures of Knocking Hoe: *Seseli libanotis*, *Hypochoeris maculata* and *Spiranthes spiralis*, with *Herminium monorchis*, *Aceras anthropophora*, *Orchis ustulata* and *Anemone pulsatilla*—there are unconfirmed records of the two last-named species also from the Middle Chalk.

In addition to its rarities Knocking Hoe has sufficient to attract the botanist in the abundance of a trailing form of *Onobrychis viciifolia*, which retains its habit in cultivation and is assuredly native here. Another interesting species, now extinct here, is *Pinguicula vulgaris*, first found on the hill in 1875. It was in a similar category to other bog plants, *Parnassia palustris* and *Carex pulicaris*, found on a hillside at Streatley, also on the Lower Chalk. I kept *Parnassia* under observation for a number of years but unwisely showed it to an entomologist who uprooted all the plants under the mistaken notion that his action would add value to specimens already collected from this place. There can be little doubt that in both cases the plants grew in a small pocket made by the Totternhoe Stone, and that they were not, as Saunders imagined, relics of a vast prehistoric bog which had its southern limit at the base of the chalk escarpment. Laflin has drawn my attention to the fact that marshes on chalk slopes are not unusual and that they are often temporary in nature. Water in the Chalk will come to the surface when there is resistance to its free flow underground.

The rarities of Knocking Hoe are limited to a comparatively small area and do not 'invade' adjacent parts of the hill which have at some time been cultivated.

Habitat Study 59   Knocking Hoe
Parish: Shillington   District: Ivel   Grid Ref. 52/1331[1]
Alt. 325 ft.   Soil: Lower Chalk   pH 7.6
Surveyed: 30 June 1949 with W. M. Cornish; 19 Aug. 1949 with S. P. Rowlands; 19 April 1950.

Abundant

| Asperula cynanchica | Festuca ovina | Linum catharticum |

Frequent

Campanula rotundifolia	Hypochoeris maculata	Plantago lanceolata
Cirsium acaulon	Onobrychis viciifolia	Poterium sanguisorba
Euphrasia nemorosa	Pimpinella saxifraga	Primula veris
Filipendula vulgaris		

Occasional

Briza media	Helianthemum nummu-	Leontodon hispidus
Centaurea scabiosa	larium	Plantago media
Chrysanthemum leucan-	Hippocrepis comosa	Senecio integrifolius
themum	Koeleria gracilis	

Rare

Anacamptis pyramidalis	Crataegus monogyna	Lotus corniculatus
Anemone pulsatilla	3 ins.	Orchis ustulata
Bromus erectus	Euphrasia pseudokerneri	Polygala vulgaris
Campanula glomerata	Fagus sylvatica 6 ins. (1)	Scabiosa columbaria
Carex caryophyllea	Galium verum	Senecio erucifolius
C. flacca	Gentianella amarella	Seseli libanotis
Carlina vulgaris	Gymnadenia conopsea	Succisa pratensis
Centaurea nigra	Helictotrichon pratense	Thymus pulegioides

Bryophyte: Pseudoscleropodium purum (T.L. adds Ctenidium molluscum, Fissidens adianthoides, Weissia microstoma, as almost certainly there).

Parts of Knocking Hoe are now over-grazed by rabbits, which is allowing some of the rarer species to increase in numbers. At the same time it is checking the growth of *Bromus erectus*. On many of the hills the latter is spreading to the exclusion of most other vegetation, as on Barton Hills where, in common with most of the hills, sheep-grazing ceased in 1931. Barton Hills are scenically more attractive than Knocking Hoe, but less productive botanically. They have, however, some attractive plants: *Anemone pulsatilla*, scattered over the downland, *Antennaria dioica*, now possibly extinct, and *Orchis ustulata*.

Habitat Study 60   Barton Hills
Parish: Barton   District: Ivel   Grid Ref. 52/089302
Alt. 400 ft.   Soil: Lower Chalk[2]   pH 7.7
Surveyed: 2 July 1949 with H. B. Souster; 7 April 1950; 29 Aug. 1950. This study was made on the saddle of the hill to the south of the church.

Abundant
Bromus erectus

Frequent

| Cirsium acaulon | Koeleria gracilis | Poterium sanguisorba |

[1] For obvious reasons this is given in a misleading form.
[2] The higher parts of Barton Hills are on the Middle Chalk and it is possible that the site of the above study was indeed on the Middle Chalk.

Occasional

Asperula cynanchica
Briza media
Campanula rotundifolia
Festuca ovina

Filipendula vulgaris
Helianthemum nummu-
  larium
Linum catharticum

Primula veris
Senecio integrifolius
Thymus pulegioides

Rare

Anacamptis pyramidalis
Campanula glomerata
Carex flacca
Carlina vulgaris
Centaurea nigra
Chrysanthemum leucan-
  themum
Cirsium arvense
Euphrasia nemorosa (M)

Galium verum
Gentianella amarella
Gymnadenia conopsea
Hieracium pilosella
Hippocrepis comosa
Leontodon hispidus
Lotus corniculatus
Ophrys apifera

Pimpinella saxifraga
Polygala vulgaris
Prunella vulgaris
Reseda luteola
Scabiosa columbaria
Succisa pratensis
Vicia cracca
Viola hirta

Barton Hills include some varied habitats. To the west is Leete Wood (H.S. 69), at the foot of which is one of our best chalk streams. A considerable patch of Glacial Gravel overlies the eastern part of the hills. This has a flora of its own. Mingled with the normal chalk-hill species is some quantity of *Echium vulgare*, *Geranium columbinum* and *Filago germanica*. On this part of the hills appears also *Ajuga chamaepitys*.

Habitat Study 61   Near Ravensburgh Castle, Barton Hills
Parish: Barton   District: Ivel   Grid Ref. 52/09752960
Alt. 450 ft.   Soil: Glacial Gravel overlying Lower Chalk   pH 7.7
Surveyed: 2 July 1949 with H. B. Souster; 19 Aug. 1949; 7 April 1950; 29 Aug. 1950. This study was made at a point 100 yds. to the west of the gate to the castle site. The ground is very rough and is closely grazed by rabbits. It is probable that attempts have been made to cultivate this area; but I do not remember it having been ploughed.

Abundant

Hieracium pilosella

Frequent

Aphanes sp.
Arenaria serpyllifolia

Crepis capillaris
Erophila verna

Potentilla anserina
Trisetum flavescens

Occasional

Agrostis stolonifera
Arenaria leptoclados
Bromus erectus
Cirsium acaulon
Erodium cicutarium

Festuca ovina (L)
Lotus corniculatus
Medicago lupulina
Potentilla reptans

Poterium sanguisorba
Succisa pratensis
Thymus pulegioides
Trifolium dubium

Rare

Acinos arvensis
Agrimonia eupatoria
Brachypodium sylvati-
  cum
Carduus nutans
Carlina vulgaris
Centaurea scabiosa
Cerastium holosteoides
Cirsium arvense
C. vulgare
Convolvulus arvensis

Crataegus monogyna
  3 ins. (1)
Galium verum
Geranium molle
Helianthemum nummu-
  larium
Holcus lanatus
Knautia arvensis
Linum catharticum
Myosotis arvensis
Phleum nodosum

Plantago lanceolata
Prunella vulgaris
Reseda luteola
Scleropoa rigida
Senecio jacobaea
Sherardia arvensis (1)
Silene vulgaris
Veronica chamaedrys
V. serpyllifolia
Trifolium repens (1)

Bryophytes: Brachythecium rutabulum, Camptothecium lutescens, Pseudosclero-podium purum.

Between Barton Hills and Sharpenhoe Clappers are a few pieces of rough downland. On Smithcombe Hill is an abundance of *Anemone pulsatilla*; but, apart from the fact that *Antennaria dioica* has been found, the Clappers are botanically disappointing. The escarpment changes little between Sharpenhoe and Sundon. It was to the south-west of Streatley that *Parnassia palustris* and *Carex pulicaris* appeared, and *Ophrys insectifera* is still relatively plentiful. Nearer Sundon *Himantoglossum hircinum* and *Herminium monorchis* grew in scrub on the hills, and in disused chalk quarries is some quantity of *Gentianella germanica*.

There is little evidence of the Lower Chalk between Sundon and Dunstable. To the north-west of Dunstable between Sewell and Totternhoe the whole of the escarpment has been worn away by quarrying, and Totternhoe Knolls now stands as an island, retaining all its botanical treasures. Two species which arrest attention are *Anthyllis vulneraria*, growing in abundance as it does in all places on the Lower Chalk when the soil has at some time been disturbed, and *Bunium bulbocastanum*, another species of disturbed chalk soils, growing here in some quantity but rarely in association with *Anthyllis vulneraria*. Other species of the hill are *Aceras anthropophora*, *Herminium monorchis* and *Coeloglossum viride*.

Habitat Study 62   Totternhoe Knolls
Parish: Totternhoe   District: Ouzel   Grid Ref. 42/978222
Alt. 500 ft.   Soil: Lower Chalk   pH 7.6
Surveyed: 21 June 1949; 27 Aug. 1949 with H. B. Souster; 19 April 1950. This study was made at the closed end of the deep gully to the north-east side of the castle mound.

Abundant

Cynosurus cristatus	Koeleria gracilis	

Frequent

Asperula cynanchica	Cirsium acaulon	Hippocrepis comosa
Anthyllis vulneraria	Festuca ovina	Linum catharticum
Briza media	Helictotrichon pratense	Plantago media

Occasional

Carex flacca	Centaurea nigra	Plantago lanceolata
Carlina vulgaris	Leontodon hispidus	Polygala oxyptera

Rare

Bellis perennis (M)	Galium verum	Pimpinella saxifraga
Campanula glomerata (1)	Gentianella amarella	Primula veris
C. rotundifolia	Herminium monorchis (1)	Ranunculus repens
Centaurea scabiosa	Hieracium pilosella	Senecio erucifolius
Dactylis glomerata (M)	Lotus corniculatus	Taraxacum sp.
Daucus carota (1)	Onobrychis viciifolia (M)	Thymus pulegioides
Euphrasia nemorosa	Ophrys apifera (1)	Tragopogon minor (1)
Festuca rubra		

Bryophytes: Fissidens taxifolius, Pseudoscleropodium purum, Weissia crispa.

There are a number of places on the Lower Chalk where the development of a chalk flora can be studied. It may be seen at its best in the railway cuttings at Chalton and Sewell, and at Barton and Dunstable Cuttings where main roads cross the Chalk.

Habitat Study 63   Dunstable Cutting
Parish: Houghton Regis   District: Ouzel   Grid Ref. 52/004235
Alt. 450 ft.   Soil: Lower Chalk   pH 7.6
Surveyed: 8 June 1949; 2 Aug. 1949; 20 April 1950; 31 Aug. 1950. This study was
made about 50 yds. north of the turning to Sewell. The cutting was made in
1837 but has since been widened.

(a) East aspect—vegetation luxuriant

Shrubs: Corylus avellana 5 ft. (1), Crataegus monogyna 5 ft. (o), Rubus caesius ×
? (L).

Abundant
Festuca rubra   Hedera helix

Frequent
Chrysanthemum leucan-   Galium mollugo   Plantago lanceolata
  themum   Pimpinella saxifraga

Occasional
Achillea millefolium   Knautia arvensis   Pastinaca sativa
Anthyllis vulneraria   Leontodon hispidus   Scabiosa columbaria
Briza media   Linum catharticum   Senecio erucifolius
Centaurea nigra   Lolium perenne   Torilis japonica
Festuca ovina

Rare
Arrhenatherum elatius   Lapsana communis   Senecio vulgaris
Artemisia vulgaris   Lathyrus pratensis   Sonchus arvensis
Cerastium holosteoides   Lotus corniculatus   S. oleraceus
Cirsium arvense   Medicago lupulina   Taraxacum officinale
Crepis capillaris   Picris hieracioides   Trifolium pratense
Heracleum sphondylium   Polygonum aviculare   T. repens
Hippocrepis comosa   Ranunculus bulbosus   Veronica chamaedrys
Hypericum perforatum

Bryophyte: Ctenidium molluscum.

(b) West aspect
Vegetation less luxuriant, some bare patches supporting no growth.

Shrubs, etc.: Clematis vitalba, trailing (r), Crataegus monogyna 5 ft. (o), Prunus
spinosa 6 ft. (1), Rosa canina 3 ft. (o), Rubus caesius × ? (1).

Abundant
Hippocrepis comosa

Frequent
Arrhenatherum elatius   Festuca ovina   Helictotrichon pratense
Daucus carota   F. rubra   Scabiosa columbaria

Occasional
Anthyllis vulneraria   Pimpinella saxifraga   Thymus pulegioides
Galium mollugo   Reseda lutea   Tussilago farfara
Linum catharticum

Rare
Achillea millefolium   Hypericum perforatum   Picris hieracioides
Asperula cynanchica   Koeleria gracilis   Plantago lanceolata
Campanula rotundifolia   Linaria vulgaris   Sinapis alba
Carlina vulgaris   Leontodon hispidus   Taraxacum officinale
Centaurea nigra   Lotus corniculatus   Trifolium pratense
C. scabiosa with white-
  flowered form

Bryophytes: Acrocladium cuspidatum, Amblystegium serpens, Brachythecium
rutabulum, Bryum caespiticium, Camptothecium lutescens, Campylium chrysophyllum, Ctenidium molluscum.

An earlier stage of colonization is seen on the bare chalk in a quarry. Here the species are many but the number of individual plants is small. The pools, compared with those in sand and clay pits, are slow in developing a flora.

Habitat Study 64   Chalk quarry, Houghton Regis
Parish: Houghton Regis   District: Lea   Grid. Ref. 52/013232
Alt. *c*. 400 ft.   Soil: Lower Chalk   pH 7.8
Surveyed: 6 July 1949; 20 April 1950; 19 Sept. 1950. This study was made at the base of the main pit at a point almost opposite the offices of the Associated Portland Cement Company.

(a) in pool

Epilobium hirsutum (r)	J. inflexus (o)	Ranunculus repens (M)
Juncus articulatus (r)	Lemna minor (L)	Typha latifolia (r)

(b) on floor of pit

Acer pseudoplatanus 2 ft. (1)	Equisetum arvense (1)	P. pratensis (1)
Agropyron repens (1)	Erigeron acris (r)	Ranunculus repens (1)
Anthyllis vulneraria (r)	Festuca rubra (1)	Reseda luteola (1)
Arrhenatherum elatius (1)	Fumaria officinalis (1)	Rosa canina 3 ft. (1)
Carduus crispus (1)	Heracleum sphondylium (1)	Rumex crispus (r)
Centaurea scabiosa (1)	Holcus lanatus (1)	Salix caprea 5 ft. (1)
Chamaenerion angusti-folium (r)	Knautia arvensis (1)	S. viminalis 4 ft. (1)
Chrysanthemum leucan-themum (1)	Leontodon hispidus (o)	Scleropoa rigida (1)
	L. taraxacoides (1)	Senecio erucifolius (f)
Cirsium arvense (r)	Linum catharticum (1)	S. jacobaea (1)
C. vulgare (r)	Lolium perenne (1)	Solanum dulcamara (1)
Convolvulus arvensis (r)	Matricaria maritima (1)	Sonchus arvensis (1)
Crepis capillaris (r)	Medicago lupulina (f)	S. asper (1)
C. vesicaria (r)	Melilotus officinalis (1)	Sinapis alba (1)
Dactylis glomerata (r)	Papaver rhoeas (1)	Taraxacum officinale (r)
Daucus carota (o)	Phleum nodosum (r)	Trifolium hybridum (r)
Epilobium hirsutum (r)	Plantago lanceolata (r)	T. pratense (f)
	P. major (1)	T. repens (1)
	Poa compressa (r)	Trisetum flavescens (1)
		Tussilago farfara (o)

Bryophytes: Brachythecium rutabulum, Bryum affine, Funaria hygrometrica.

To the south of the Lower Chalk is a second and higher escarp-ment formed by the Middle Chalk, above which is a cap of Upper Chalk.

It begins in Bedfordshire with Whipsnade Downs, now occupied by the Zoo and much affected by human activity but with some plant-study problems. I have resisted the temptation to make a study of plant colonization on the Whipsnade Lion and of that part of the downs grazed by wallabies. East of Whipsnade Downs is the long sweep of Dunstable Downs, the largest stretch of chalk down-land in the county. It is botanically uninteresting and again this may be owing to human interference, as gliding there has made it popular. *Coeloglossum viride*, the only unusual species of the downs, appears at intervals. Here *Antennaria dioica* once grew, and there is a solitary record of *Himantoglossum hircinum*.

Habitat Study 65   Dunstable Downs
Parish: Dunstable   District: Ouzel   Grid Ref. 52/007210
Alt. 700 ft.   Soil: Upper Chalk   pH 7.7
Surveyed: 28 June 1949 with W. M. Cornish; 6 April 1950; 1 Sept. 1950. This study was made about 20 yds. to the west of the tumuli on the top of the hill.

Abundant
Festuca ovina   Helictotrichon pubescens

Frequent
Asperula cynanchica   Koeleria gracilis   Pimpinella saxifraga
Carex flacca   Linum catharticum   Poterium sanguisorba
Hippocrepis comosa

Occasional
Anthyllis vulneraria (L)   Campanula rotundifolia   Scabiosa columbaria
Briza media   Cirsium acaulon   Thymus pulegioides

Rare
Bromus erectus   Galium mollugo (1)   Plantago lanceolata
Campanula glomerata   G. verum   (2 forms)
Carlina vulgaris   Gentianella amarella   P. media
Centaurea nigra   Helianthemum nummu-   Primula veris
C. scabiosa   larium   Senecio integrifolius
Crataegus monogyna   Hieracium pilosella   Taraxacum officinale
  3 ins.   Leontodon hispidus   Tragopogon minor
Fagus sylvatica 6 ins. (1)   Lotus corniculatus   Trifolium pratense
Filipendula vulgaris

Bryophyte: Pseudoscleropodium purum (T.L. reports that Dicranum scoparium is also common here).

Scrub is extending at the western end of Dunstable Downs, but a worse example of scrub invasion of our downland is to be seen on Blow's Downs on the eastern side of the Dunstable Gap. Here the flora is again disappointing. From Blow's Downs the escarpment stretches to the outskirts of Luton but it, too, has suffered from scrub invasion, especially near to Luton where the hills have become a playground.

To the north of Luton are Warden and Galley Hills, preserved to some degree by the fact that they form part of a golf course. Galley Hill is the more interesting and here there has been a limited amount of grazing. On the top of the hill is a small pocket of Clay-with-Flints which gives rise to an unusual plant association. At the foot of both hills is some quantity of *Astragalus danicus*.

Habitat Study 66a   Galley Hill
Parish: Streatley   District: Lea   Grid Ref. 52/09202725
Alt. 500 ft.   Soil: Middle Chalk   pH 7.6
Surveyed: 1 July 1949 with W. M. Cornish; 14 April 1950; 25 Aug. 1950. This study was made about 200 yds. beyond the end of Dray's Ditches and about 30 yds. from the Icknield Way. Higher on the hill are considerable patches of *Bromus erectus*.

Abundant
Festuca ovina   Helictotrichon pratense

Frequent
Koeleria gracilis   Poterium sanguisorba   Pimpinella saxifraga

PLATE 15   RAILWAY CUTTING, AMPTHILL (Habitat Study 45b)

PLATE 16   SUTTON FEN (Habitat Study 49)

SANDY, TEMPSFORD AND OUSE VALLEY (Sandy is on the left of the picture)

Occasional

Asperula cynanchica	Festuca rubra	Leontodon hispidus
Carex flacca	Galium verum	Linum catharticum
Cirsium acaulon	Helianthemum nummu-	Plantago lanceolata
Crataegus monogyna	larium	Thymus pulegioides
8 ins.		

Rare

Achillea millefolium	Daucus carota	Plantago media
Anacamptis pyramidalis	Euphrasia nemorosa	Polygala oxyptera
Anthyllis vulneraria	Filipendula vulgaris	Ranunculus repens (1)
Astragalus danicus	Gentianella amarella (1)	Rosa canina 6 ins. (1)
Briza media	Gymnadenia conopsea	Scabiosa columbaria
Bromus erectus	Hieracium pilosella	Trifolium dubium (1)
Carlina vulgaris	Hippocrepis comosa	T. pratense
Centaurea nigra	Lotus corniculatus	Viburnum lantana 6 ins.
C. scabiosa	Medicago lupulina	(1)
Cerastium holosteoides		

Habitat Study 66b   Galley Hill
Parish: Streatley   District: Lea   Grid Ref. 52/092270
Alt. 600 ft.   Soil: Clay-with-Flints overlying Middle Chalk   pH 6.0
Surveyed: 1 July 1949 with W. M. Cornish; 14 April 1950; 25 Aug. 1950. This
study was made on the summit of the hill in the middle of the clay pocket.

Abundant

Helictotrichon pratense	Trifolium medium (L)

Frequent

Agrostis tenuis	Festuca ovina	Potentilla erecta
Calluna vulgaris (L)		

Occasional

Anthoxanthum odoratum	Filipendula vulgaris	Lotus corniculatus
(L)	Galium verum	Trisetum flavescens
Cirsium acaulon	Hypericum perforatum (L)	

Rare

Achillea millefolium	Crataegus monogyna 12 ins.	Luzula campestris
Anthyllis vulneraria	Cynosurus cristatus	Ononis repens (1)
Briza media	Dactylis glomerata	Plantago lanceolata
Bunium bulbocastanum (1)	Daucus carota	P. media
Campanula rotundifolia	Galium mollugo	Poterium sanguisorba
Carex flacca	Helianthemum nummu-	Sieglingia decumbens
Centaurea nigra	larium	Succisa pratensis
Cirsium arvense	Lathyrus pratensis (1)	Tragopogon minor
C. vulgare	Linaria vulgaris (L)	Trifolium pratense
Clinopodium vulgare	Lolium perenne	T. repens

To the north-east of Galley Hill is Telegraph Hill, which is at the
northern end of Lilley Hoo in Hertfordshire. Below it winds the
Icknield Way, by the side of which *Astragalus danicus* still persists.
Close by is Deacon Hill, which always repays study. *Anemone
pulsatilla* reappears here, and on disturbed ground *Iberis amara* is
almost certainly to be found. Among bushes at the foot of the hill
Lithospermum officinale, rare on the Bedfordshire chalk, is also to be
seen.

Our chalk scrub is also worthy of study. It is usually best developed where agriculture has been abandoned. The flora here is,
as a rule, different from that of the scrub which is developing so
rapidly on the hills. In both cases the scrub consists mainly of hawthorn and the various rose species, with rather less *Viburnum lantana*.

Habitat Study 67   Scrub, Galley Hill
Parish: Streatley   District: Lea   Grid Ref. 52/092274
Alt. 460 ft.   Soil: Middle Chalk   pH 7.6
Surveyed: 1 July 1949 with W. M. Cornish; 14 April 1950; 25 Aug. 1950. This
study was made about 50 yds. from the western side of Maulden Firs at about the
middle of the scrub. The site was ploughed up to or during the period of World
War I. It is now grazed occasionally by cattle.
Shrubs: Crataegus monogyna up to 12 ft. high and 10 ft. apart, Rosa canina (o).

Abundant

Festuca ovina	Helictotrichon pratense

Frequent

Anthyllis vulneraria	Koeleria gracilis	Poterium sanguisorba
Briza media	Pimpinella saxifraga	

Occasional

Asperula cynanchica	Festuca rubra	Linum catharticum
Centaurea nigra	Galium verum	Lotus corniculatus
Cirsium acaulon	Hieracium pilosella	Plantago lanceolata
Dactylis glomerata	Leontodon hispidus	Trisetum flavescens

Rare

Achillea millefolium	Euphrasia nemorosa	Prunella vulgaris
Anacamptis pyramidalis	Galium mollugo	Scabiosa columbaria
Bunium bulbocastanum	Gentianella amarella (1)	Thymus pulegioides
(1)	Medicago lupulina	Tragopogon minor
Carex flacca	Onobrychis viciifolia (1)	Trifolium dubium (1)
Centaurea scabiosa	Ononis repens (1)	T. pratense
Cerastium holosteoides	Plantago media	T. repens
Daucus carota (1)		

Bryophytes: Eurhynchium striatum, Pseudoscleropodium purum.

Scrub on downland has usually a greater proportion of chalk-hill
species. The study below gains additional interest from the inclusion
of *Blackstonia perfoliata*, which is common enough in scrub on the
Lower Chalk but is apparently absent on the Middle Chalk.

Habitat Study 68   Scrub, Barton
Parish: Barton   District: Ivel   Grid Ref. 52/097303
Alt. 425 ft.   Soil: Lower Chalk   pH 7.8
Surveyed: 2 July 1949 with H. B. Souster; 19 Aug. 1949; 7 April 1950. This study
was made at a point about 200 yds. from the lower end of the scrub and 30 yds.
from the path on its western side. The slope is slight, the patches between the
bushes well grazed by rabbits.
Shrubs: Crataegus monogyna up to 10 ft. high and 8 ft. apart, Rosa canina (o),
Viburnum lantana (o).

Abundant

Linum catharticum	Lotus corniculatus	Trisetum flavescens

Frequent

Poterium sanguisorba

Occasional

Blackstonia perfoliata	Hieracium pilosella	Polygala vulgaris
Carex flacca	Leontodon hispidus	Thymus pulegioides
Cirsium acaulon	L. taraxacoides	Viola hirta
Fragaria vesca	Origanum vulgare	

Rare

Agrimonia eupatoria	Centaurea nigra	Holcus lanatus
Agrostis stolonifera	C. scabiosa	Hypericum perforatum
Anacamptis pyramidalis	Clinopodium vulgare	Medicago lupulina
Bellis perennis (1)	Crepis capillaris	Pastinaca sativa (M)
Brachypodium sylvaticum	Dactylis glomerata	Plantago lanceolata
Bromus erectus	Euphrasia nemorosa	Potentilla reptans
Campanula glomerata	Festuca rubra	Prunella vulgaris
C. rotundifolia	Galium mollugo	Senecio erucifolius
Carlina vulgaris	Gentianella amarella	Trifolium repens

Beech is the common tree of the Chalk but beech woods are few. In some cases, as on the slopes of Totternhoe Knolls, they are obviously of recent origin. Normally, except in clearings, beech woods have a limited flora. *Cephalanthera damasonium* and *Monotropa hypopitys* each appear in a few places and *Ophrys insectifera* occurs on the margins of some woods. Without doubt the most interesting of the beech-wood species is *Epipactis phyllanthes*.

Bare banks under beech trees have their own moss flora, with *Barbula convoluta, B. recurvirostris, Brachythecium velutinum, Encalypta streptocarpa, Hypnum cupressiforme, Tortula subulata* and *Cirriphyllum crassinervium*.

Habitat Study 69   Leete Wood
Parish: Barton   District: Ivel   Grid Ref. 52/088294
Alt. 475 ft.   Soil: Middle Chalk   pH 7.8
Surveyed: 2 July 1949 with H. B. Souster; 7 April 1950; 29 Aug. 1950. This study was made on the brink of the hill at a point about 80 yds. from the south corner of the wood. The beech trees here are 60 ft. high, the only species evident in the shade being *Cephalanthera damasonium*. The species listed are from a clearing; the introduction of downland species in these conditions should be noted.
Shrubs, etc.: Bryonia dioica (r), Clematis vitalba (M), Crataegus monogyna 2 ft. (o), Fraxinus excelsior 1 ft. (r), Rosa canina (r), Rubus ? conjungens 2 ft. (r), Sambucus nigra 3 ft. (r), Tamus communis (r), Viburnum lantana 2 ft. (r).

Occasional

Festuca rubra	Mercurialis perennis (L)	Prunella vulgaris
Fragaria vesca (L)	Mycelis muralis	Sanicula europaea
Holcus lanatus	Poa pratensis	Viola hirta

Rare

Agrostis stolonifera	Clinopodium vulgare	Myosotis arvensis
Arctium minus (1)	Crepis capillaris	Plantago lanceolata
Arenaria serpyllifolia	Dactylis glomerata	Potentilla erecta
Brachypodium sylvaticum	Epilobium montanum	P. reptans
Campanula trachelium	× Festulolium loliaceum	Poterium sanguisorba
C. rotundifolia	Geum urbanum	Ranunculus acris
Carex flacca (M)	Hieracium pilosella	Reseda lutea (1)
Cephalanthera dama-	Hypericum hirsutum	Scleropoa rigida
sonium (M)	H. perforatum (M)	Silene vulgaris
Cerastium glomeratum	Leontodon autumnalis	Taraxacum sp.
C. holosteoides	L. taraxacoides	Trisetum flavescens
Chamaenerion angusti-	L. hispidus	Tussilago farfara (M)
folium	Linum catharticum	Verbascum nigrum (M)
Cirsium arvense	Medicago lupulina	Veronica chamaedrys

Bryophytes: Atrichum undulatum, Barbula convoluta, Bryum capillare, Camptothecium sericeum, Ceratodon purpureus, Encalypta streptocarpa, Neckera complanata.

A similar study was made at Maulden Firs which, notwithstanding its name and the fact that there are a few conifers on its edge, appears to be good native woodland. Maulden is eight miles away, which adds to the puzzle of its name.

Habitat Study 70   Maulden Firs
Parish: Streatley   District: Lea   Grid Ref. 52/093274
Alt. 500 ft.   Soil: Middle Chalk   pH 7.6
Surveyed: 1 July 1949 with W. M. Cornish; 14 April 1950; 25 Aug. 1950. This study was made half-way through the wood at 60 yds. from its western end. The beech trees are about 60 ft. high, up to 15 ft. apart, and have boles about 18 ins. in diameter. There were two ash trees (*Fraxinus excelsior*) and one hornbeam (*Carpinus betulus*) on the margin of the survey.

(a) normal woodland

Abundant
Hedera helix   Sanicula europaea

Frequent
Cephalanthera dama-   Viola reichenbachiana
   sonium

Rare
Achillea millefolium   Dactylis glomerata   P. pratensis
Brachypodium sylvati-   Festuca rubra   Sambucus nigra 12 ins.
   cum   Ligustrum vulgare 6 ins.   Tamus communis
Bromus erectus   Poa annua   Taraxacum officinale
B. sterilis (1)   P. nemoralis   Viburnum lantana 12 ins.

(b) clearing to east of above

Shrubs, etc.: Acer pseudoplatanus 1 ft. (1), Crataegus monogyna 6 ins. (r), Fraxinus excelsior 2 ft. (f), Ilex aquifolium 2 ft. (r), Ligustrum vulgare 18 ins. (f), Rosa canina (o), Rubus sp. (r).

Abundant
Hedera helix

Frequent
Tamus communis

Rare
Arrhenatherum elatius   Chamaenerion angusti-   Hypericum perforatum
Cephalanthera damaso-   folium
   nium   Heracleum sphondylium

Bryophytes: Brachythecium rutabulum, B. velutinum, Bryum capillare, Ceratodon purpureus, Hypnum cupressiforme.

A few caps of Boulder Clay occur on the Chalk and on them are sometimes small woods. In a wood near Streatley *Platanthera bifolia* has been found, and in hedgerows in the same neighbourhood is some quantity of *Ribes uva-crispa*.

The arable fields of the Chalk are of special botanical interest. *Sinapis alba* is the commonest weed but a number of the colonists of the cornfields are species normally rare in Britain. First place must be given to *Bunium bulbocastanum*, usually abundant in fields near Dunstable. *Galium tricorne*, by no means limited to the Chalk in Bedfordshire, is also common. Other weeds of the arable fields are *Valerianella rimosa*, *Ajuga chamaepitys*, *Torilis arvensis*, *T. nodosa*, *Iberis amara*, *Legousia hybrida* and various species of *Fumaria*. Mosses of such fields include *Barbula unguiculata*, *Phascum cuspidatum*, *Pottia truncata*, *P. intermedia* and *P. davalliana*.

Habitat Study 71   Arable field, Barton
Parish: Barton   District: Ivel   Grid Ref. 52/097296
Alt. 450 ft.   Soil: Glacial Gravel overlying Lower Chalk   pH 7.7
Surveyed: 2 July 1949 with H. B. Souster; 7 April 1950; 29 Aug. 1950. This study
was made in a field adjoining the large gravel patch, the subject of H.S. 61.

Acinos arvensis	Galium aparine	Ranunculus repens
Aethusa cynapium	Geranium dissectum	Reseda lutea
Alopecurus myosuroides	G. molle	Rumex crispus
Anagallis arvensis	G. pusillum	Senecio vulgaris
Aphanes sp.	Kickxia spuria	Sherardia arvensis
Arenaria serpyllifolia	Knautia arvensis	Silene vulgaris
Atriplex patula	Legousia hybrida	Sinapis alba
Carduus crispus	Linaria vulgaris	S. arvensis
Cirsium arvense	Melandrium album	Sonchus arvensis
Convolvulus arvensis	Myosotis arvensis	Stellaria media
Erodium cicutarium	Papaver rhoeas	Torilis nodosa
Euphorbia exigua	Plantago major	Tussilago farfara
Fumaria officinalis	Polygonum aviculare	Valerianella dentata
Galeopsis angustifolia	P. convolvulus	Veronica polita
G. tetrahit	Potentilla anserina	Viola arvensis

Habitat Study 72   Arable field, Totternhoe
Parish: Totternhoe   District: Ouzel   Grid Ref. 42/97902235
Alt. 400 ft.   Soil: Lower Chalk   pH 7.8
Surveyed: 21 June 1949; 27 Aug. 1949 with H. B. Souster; 19 April 1950. This
study was made in a field to the north of the castle mound.

Aethusa cynapium	Galeopsis angustifolia	Phleum nodosum
Alopecuroides myosu-	Galium tricorne	Poa compressa
roides	Geranium dissectum	P. trivialis
Anagallis arvensis	Knautia arvensis	Ranunculus arvensis
Arenaria serpyllifolia	Lapsana communis	R. repens
Bunium bulbocastanum	Legousia hybrida	Rhinanthus minor
Centaurea scabiosa	Linaria vulgaris	Scandix pecten-veneris
Cerastium holosteoides	Lithospermum arvense	Silene vulgaris
Chaenorhinum minus	Medicago lupulina	Sinapis alba
Cirsium arvense	Melilotus officinalis	S. arvensis
Convolvulus arvensis	Mentha arvensis	Tussilago farfara
Daucus carota	Myosotis arvensis	Valerianella dentata
Euphorbia exigua	Papaver argemone	Veronica persica
Fumaria officinalis	P. rhoeas	Viola arvensis

A similar study made in a field on the Middle Chalk revealed
fewer weeds, the crop at each visit being very clean.

Habitat Study 73   Arable field, Galley Hill
Parish: Streatley   District: Lea   Grid Ref. 52/093268
Alt. 575 ft.   Soil: Middle Chalk   pH 7.4
Surveyed: 1 July 1949 with W. M. Cornish; 14 April 1950. This study was made in
a field about 100 yds. to the east of H.S. 66b.

Anagallis arvensis	Knautia arvensis	P. reptans
Atriplex patula	Linaria vulgaris	Ranunculus repens
Bunium bulbocastanum	Lolium multiflorum	Reseda lutea
Centaurea scabiosa	Medicago lupulina	Rumex crispus
Cirsium acaulon	Melandrium album	Scandix pecten-veneris
Convolvulus arvensis	Poa compressa	Senecio vulgaris
Daucus carota	Polygonum aviculare	Silene vulgaris
Fumaria officinalis	Potentilla anserina	Sinapis arvensis

A number of places in this region support an adventive flora. By the railway, especially between Luton and Dunstable, *Linaria repens* is common with its hybrid with *L. vulgaris*. At Sewell also on the railway is *Minuartia tenuifolia* and *Lapsana intermedia*.

At Sundon is a large rubbish dump which receives refuse from London. Here may be found a strange assortment of plants ranging from garden throw-outs to such unusual adventives as *Bupleurum rotundifolium*, *Plantago ovata*, *Sisymbrium loeselii* and *Cannabis sativa*.

I.   The Clay-with-Flints and Lea Valley Region

This is a complex region in which for convenience I have had to include the upland clays overlying the Chalk, the river gravels and alluvium of the valleys, and some small exposures of chalk. Small as it is, it cannot be considered as a whole.

The Clay-with-Flints is in two detached areas to the east and west of the Lea Valley. In places the clay forms natural heathland, very similar to the Lower Greensand heaths, a small one lying in the part to the east of the valley at Chiltern Green. It is insignificant compared with the Hertfordshire commons in the same valley and a few miles distant, but has one small patch of *Viola canina* and a little *Calluna vulgaris*.

Habitat Study 74   Chiltern Green Common
Parish: Hyde   District: Lea   Grid Ref. 52/13371960
Alt. 575 ft.   Soil: Clay-with-Flints   pH 5.2
Surveyed: 25 June 1949; 30 July 1949; 17 April 1950; 29 Aug. 1950. This study was made on the western side of the dried-up pond at the north end of the common. Until it dried the pond had some quantity of *Peplis portula*.

(a) heath

Shrubs: Crataegus monogyna up to 20 ft. high spaced irregularly, Ilex aquifolium 15 ft. (1), Quercus robur 20 ft. (1), Rosa canina (r), Rubus sp. (r), Sambucus nigra 10 ft. (1).

Abundant
Agrostis tenuis   Galium saxatile

Frequent
Festuca rubra (L)   Holcus lanatus   Stellaria graminea
Hieracium pilosella

Occasional
Anthoxanthum odoratum   Centaurea nigra   Stachys officinalis
Campanula rotundifolia   Potentilla erecta   Rumex acetosella

Rare
Achillea millefolium   Cirsium acaulon   Koeleria gracilis
Agrostis stolonifera   Dactylis glomerata   Luzula campestris
Arrhenatherum elatius   Galium verum   Pimpinella saxifraga
Calluna vulgaris   Hypericum pulchrum   Viola canina

Bryophytes: Hylocomium splendens, Rhytidiadelphus squarrosus.

(b) dried-up pond

Abundant
Galium palustre (L)   Juncus effusus (L)

Frequent

Agrostis canina (L)
Glyceria fluitans (L)

Holcus lanatus (L)
Potentilla anserina (L)

Rumex acetosa (L)

Rare

Carex disticha
Centaurea nigra
Deschampsia cespitosa

Moehringia trinervia (1)
Plantago lanceolata
Poa pratensis

Ranunculus repens
Solanum dulcamara
Viola riviniana (M)

In the larger area to the west of the valley is more heathland. The best remaining portion is Whipsnade Heath, but parts of Studham Common, now mainly ploughed, are still interesting. Saunders recorded many species from Pepperstock, but there is some doubt as to whether his stations were not, in fact, in Hertfordshire.

Habitat Study 75   Whipsnade Heath
Parish: Whipsnade   District: Colne   Grid Ref. 52/01791784
Alt. 625 ft.   Soil: Clay-with-Flints   pH 7.4
Surveyed: 25 June 1949 with H. B. Souster; 12 Aug. 1949; 13 April 1950 with W. Durant. This study was made on the south-west side of the heath at a point five yards below the first pylon to the west of the main road. The heath here has a west aspect with a very slight slope and is well grazed by rabbits. The shrubs are clustered with open heath between them.
Shrubs: Corylus avellana 6 ft. (1), Crataegus monogyna 8 ft. (o), Prunus spinosa 5 ft. (1), Rosa canina (o).

Abundant

Festuca ovina

Linum catharticum

Lotus corniculatus

Frequent

Agrostis tenuis
Cynosurus cristatus

Helictotrichon pratense
Holcus lanatus

Koeleria gracilis (on ant
  hills)
Plantago lanceolata

Occasional

Cirsium acaulon
Festuca rubra
Galium verum

Lolium perenne
Luzula campestris
Medicago lupulina
  (native form)

Taraxacum sp.
Trifolium pratense
Trisetum flavescens

Rare

Achillea millefolium
Agrimonia eupatoria
Anthoxanthum odoratum
  (M)
Arenaria serpyllifolia
Arrhenatherum elatius
Bellis perennis
Briza media
Campanula rotundifolia
Carex flacca (M)

Cerastium holosteoides
Cirsium arvense (M)
C. vulgare (1)
Crepis capillaris
Dactylis glomerata
Erophila verna
Euphrasia nemorosa
Helianthemum nummu-
  larium
Hieracium pilosella

Lathyrus pratensis
Leontodon hispidus
Pimpinella saxifraga
Plantago media
Ranunculus repens
Rumex acetosa
Senecio erucifolius
Thymus pulegioides
Trifolium dubium
T. repens

Bryophyte: Pseudoscleropodium purum.

Apart from the heaths there are a few pastures on the Clay-with-Flints. The rough pasture surveyed below was possibly at one time arable land. It contains, though not within the area of the survey, *Gentianella germanica, Orchis morio* and *Ophioglossum vulgatum.*

Habitat Study 76   Pasture, Greencroft Barn
Parish: Studham   District: Colne   Grid Ref. 52/02311395
Alt. 580 ft.   Soil: Clay-with-Flints   pH 6.0
Surveyed: 25 June 1949 with H. B. Souster; 2 May 1950 with W. D. Coales and
H. B. Souster; 1 Sept. 1950. This study was made about 100 yards south-east of
the south-east corner of Greencroft Barn. The pasture has a northerly aspect, a
slight slope, and is grazed by cattle.

Abundant
Agrostis tenuis   Trifolium repens   Trisetum flavescens
  Frequent
Anthoxanthum odoratum   Cynosurus cristatus   Luzula campestris
Chrysanthemum leucan-   Lotus corniculatus   Trifolium pratense
  themum
  Occasional
Centaurea nigra   Plantago lanceolata   Taraxacum officinale
Hypochoeris radicata   Rumex acetosella
  Rare
Acer campestre 6 ins.   Festuca rubra   Prunella vulgaris (1)
Bellis perennis   Heracleum sphondylium   Ranunculus acris (1)
Cerastium holosteoides   Hieracium pilosella   Rosa canina 2 ft.
Crataegus monogyna 1 ft.   Holcus lanatus   Senecio jacobaea
Dactylis glomerata   Knautia arvensis   Tragopogon minor
Daucus carota   Linum catharticum

Bryophytes: Atrichum undulatum, Brachythecium rutabulum, Fissidens bryoides,
Mnium hornum, Physcomitrium pyriforme, Rhytidiadelphus squarrosus.

It is the woods of the Clay-with-Flints that command most
attention. In the portion to the east of the Lea Valley is an interesting belt of woodland. Spittlesea Wood must have been attractive
until it became incorporated first in Luton Sewage Farm and then
in the Vauxhall Works. On its eastern side Watson found a fine
bramble community. Further south are Dumb Hill Wood, George
Wood, Arden Dells Wood and Horsley's Wood. These make a fine
show in spring with wild cherry (*Prunus avium*), and produce also
Veronica montana and *Iris foetidissima*. Horsley's Wood is more heathy
than the rest and was made the subject of a study.

Habitat Study 77   Horsley's Wood
Parish: Hyde   District: Colne   Grid Ref. 52/13111910
Alt. 475 ft.   Soil: Clay-with-Flints   pH 5.2
Surveyed: 25 June 1949; 30 July 1949; 17 April 1950; 29 Aug. 1950. This study
was made at the first crossing of paths on the main ride of the wood on its western
side. The wood has suffered greatly from felling and subsequent planting.
Trees etc.: Betula pubescens 20 ft. (o), Castanea sativa 30 ft. (f), Corylus avellana
10 ft. (o), Rubus sp. (f).
                          (a) path, grazed by rabbits
  Abundant
Lysimachia nemorum (L)
  Frequent
Glechoma hederacea (L)   Potentilla anserina   Veronica officinalis
Juncus conglomeratus (L)   Prunella vulgaris
  Occasional
Agrostis tenuis   Rumex acetosella   Veronica serpyllifolia
Hypericum humifusum   Sagina procumbens   Viola canina (M)
Ranunculus repens   Stellaria graminea

Rare

Ajuga reptans	Fragaria vesca	P. pratensis
Aphanes microcarpa	Holcus lanatus	Potentilla reptans (1)
Centaurium minus	Juncus effusus (1)	P. sterilis (L)
Cerastium holosteoides	Leontodon hispidus	Rumex obtusifolius (1)
Cirsium palustre	Moehringia trinervia	Scrophularia nodosa (M)
Crepis capillaris	Myosotis discolor	
Deschampsia cespitosa	Poa annua	

Bryophytes: Atrichum undulatum, Ceratodon purpureus. Dicranella heteromalla, Eurhynchium praelongum.

(b) wood, adjoining above

Abundant

Chamaenerion angusti-   Pteridium aquilinum
folium

Frequent

Juncus conglomeratus   Viola reichenbachiana

Occasional

Lysimachia nemorum

Rare

Ajuga reptans	Juncus effusus (1)	Myosotis arvensis
Carex sylvatica	Luzula multiflora (1)	Oxalis acetosella
Fragaria vesca		

A short distance to the south of Horsley's Wood is the county boundary, but there are a few copses, one of which contains *Helleborus viridis*.

On the western side of the valley the woods are even more varied. On the higher ground in Luton Hoo Park there are woods with an abundance of *Symphytum tuberosum*, undoubtedly of garden origin. Kidney Wood, in the same park, has *Carex pallescens*, and in its woodland pools *Peplis portula*.

Towards and around Caddington are further woods, those nearer Luton having suffered a great deal from despoliation. They are usually coppiced with hazel, with which is often associated *Lathraea squammaria*. Folly Wood is probably the least spoiled.

Habitat Study 78   Folly Wood
Parish: Caddington   District: Colne   Grid Ref. 52/057201
Alt. 580 ft.   Soil: Clay-with-Flints   pH 6.7
Surveyed: 26 June 1949; 29 Aug. 1949; 28 April 1951; 16 Aug. 1952 with T.L.
This study was made 50 yards from the south-west corner of the wood at a point ten yards inside.
Trees, etc.: Acer campestre 10 ft. (r), Corylus avellana 10 ft. (o), Crataegus monogyna 6 ft. (r), Lonicera periclymenum (o), Prunus avium 60 ft. (f), Quercus robur 60 ft. (f), Rosa arvensis (o), Rubus sp. 4 ft. (f).

Abundant

Mercurialis perennis (L)

Frequent

Asperula odorata	Holcus lanatus	Viola riviniana
Glechoma hederacea	Milium effusum	

Occasional

Anemone nemorosa	Conopodium majus	Galeobdolon luteum
Brachypodium sylvati-	Deschampsia cespitosa	Stellaria holostea
cum	Endymion non-scriptus	Ranunculus ficaria
Bromus ramosus	Epilobium montanum	Torilis japonica
Carex sylvatica	Festuca gigantea	Vicia sepium

Rare

Arum maculatum	Geum urbanum	Potentilla sterilis
Cirsium palustre	Hypericum hirsutum	Scrophularia nodosa
Dactylis glomerata	Moehringia trinervia	Stachys sylvatica
Galium aparine	Poa pratensis	Veronica chamaedrys

Bryophytes: Cirriphyllum piliferum, Dicranella heteromalla, Eurhynchium praelongum, E. striatum, Fissidens bryoides, Mnium hornum, M. undulatum, Lopho-
colea heterophylla, Plagothecium denticulatum, Thamnium alopecurum.

More heathy conditions arise in the eastern part of Folly Wood
where *Pteridium aquilinum* is dominant in large patches and there is
some quantity of *Oxalis acetosella* and *Juncus conglomeratus*.

In the extreme south-west of the county is other varied woodland.
Two of the largest woods are the adjacent Deadmansey and Oldhill
Woods, the best area in the county for *Epipactis purpurata*. The road-
side between the woods is a good station for *Asperula odorata, Melica
uniflora* and *Hypericum dubium*. In Oldhill Wood is a large patch of
Helleborus viridis.

Habitat Study 79   Deadmansey Wood (1st Study)
Parish: Whipsnade   District: Colne   Grid Ref. 52/033167
Alt. 580 ft.   Soil: Clay-with-Flints   pH 7.6
Surveyed: 25 June 1949 with H. B. Souster; 17 Aug. 1949; 8 April 1950. This
study was made ten yards from the road along the second ride of the wood from
the Markyate-to-Whipsnade road.
Trees, etc.: Betula pendula 15 ft. (f), Carpinus betulus 14 ft. (1), Crataegus monogyna 6 ft. (o), Fagus sylvatica 20 ft. (1), Quercus robur 30 ft. (M), Rosa arvensis
5 ft. (o), Rosa canina 6 ft. (o), Rubus sp. 5 ft. (o), Salix caprea 12 ft. (r), Ulex
europaeus 3 ft. (r), Viburnum opulus 8 ft. (1).

(a) path

Abundant

Agrostis tenuis	Potentilla erecta

Frequent

Chamaenerion angusti-	Holcus lanatus	Pteridium aquilinum (M)
folium	Hypericum pulchrum (M)	
Galium saxatile		

Occasional

Ajuga reptans	Potentilla anserina	Veronica chamaedrys
Anthoxanthum odoratum	Prunella vulgaris	V. officinalis
Glechoma hederacea	Senecio erucifolius	Viola reichenbachiana
Juncus conglomeratus		

Rare

Centaurium minus	Hieracium pilosella (1)	Ranunculus repens
Cirsium arvense	Hypericum dubium (1)	Stellaria graminea
C. palustre	Hypochoeris radicata	Trifolium dubium
Clinopodium vulgare	Luzula multiflora	T. repens
Fragaria vesca	L. pilosa	Vicia angustifolia

Bryophytes: Mnium hornum, Plagiothecium denticulatum.

(b) wood, to west of above

Abundant

Pteridium aquilinum

Frequent

Lysimachia nemorum (L)

Occasional

Holcus lanatus   Lonicera periclymenum   Poa trivialis
Juncus conglomeratus

Rare

Carex sylvatica   Epilobium montanum   Potentilla anglica
Chamaenerion angusti-   Epipactis purpurata   Sanicula europaea
  folium   Lotus uliginosus   Scrophularia nodosa
Deschampsia cespitosa   Luzula multiflora   Veronica chamaedrys
Dryopteris filix-mas   L. pilosa   V. officinalis
Endymion non-scriptus   Mercurialis perennis   Vicia sepium

Habitat Study 80   Deadmansey Wood (2nd study)
Parish: Whipsnade   District: Colne   Grid Ref. 52/034167
Alt. 580 ft.   Soil: Clay-with-Flints   pH 5.4
Surveyed: 17 Aug. 1949; 8 April 1950; 7 July 1951. This study was made on a
continuation of the ride (H.S. 79) at a point where a deep dell (an earlier marling-
pit) is reached.
Trees, etc.: Betula pendula 30 ft. (r), Corylus avellana 10 ft. (o), Crataegus monogyna 2 ft. (r), Fagus sylvatica 60 ft. (1), Fraxinus excelsior 25 ft. (M), Lonicera
periclymenum 6 ft. (f), Quercus robur 60 ft. (r), Rosa arvensis 6 ft. (f), R. tomentosa 8 ft. (1), Rubus sp. (f), Thelycrania sanguinea 6 ft. (1), Viburnum opulus
8 ft. (1).

                                (a) path, to south of dell
Abundant

Holcus lanatus

Frequent

Agrostis tenuis   Plantago lanceolata   Potentilla anserina (L)
Centaurium minus   P. major (L)   P. sterilis
Euphrasia nemorosa   Poa annua   Veronica serpyllifolia
Odontites verna

Occasional

Agrimonia eupatoria   Poa trivialis   Sagina procumbens
Ajuga reptans   Prunella vulgaris   Veronica officinalis
Glechoma hederacea

Rare

Anthoxanthum odoratum   Dactylis glomerata   Malva moschata (M)
Bellis perennis   Hypericum perforatum   Potentilla erecta
Cerastium holosteoides   Juncus conglomeratus   Pteridium aquilinum (M)
Chamaenerion angusti-   J. effusus   Ranunculus repens
  folium   Linum catharticum   Senecio jacobaea
Cirsium arvense   Lolium perenne (1)   Trifolium dubium
C. palustre   Lotus uliginosus   Veronica chamaedrys

                                (b) south slope of dell
Abundant

Asperula odorata   Brachypodium sylvaticum

Frequent

Deschampsia cespitosa   Holcus lanatus   Viola reichenbachiana
Dryopteris filix-mas   Luzula pilosa

Occasional

Agrimonia eupatoria   Chamaenerion angusti-   Potentilla sterilis
Bromus ramosus   folium   Sanicula europaea

Rare

Arctium minus   Dactylis glomerata   Prunella vulgaris
Cirsium arvense   Hypericum pulchrum   Vicia sepium
Crepis capillaris   Primula vulgaris

Bryophytes: Brachythecium rutabulum, Hypnum cupressiforme, Mnium hornum.

Further west is yet another series of interesting woods. Long Wood, which has recently suffered by clearing, is the only station known to me for *Hordelymus europaeus*. It is mainly beech and probably lies on a chalk exposure. Ravensdell Wood and the small Elm Grove are, however, of the oak-hornbeam type so characteristic of the Clay-with-Flints. Each has some quantity of *Veronica montana*. Another contrast is to be seen at Greencroft Barn, where *Sorbus aria* appears to be native. It is a badly-kept wood on poor soil.

Habitat Study 81   Greencroft Barn
Parish: Studham   District: Colne   Grid Ref. 52/02251365
Alt. 580 ft.   Soil: Clay-with-Flints   pH 5.8
Surveyed: 25 June 1949 with H. B. Souster; 2 May 1950 with W. D. Coales and H. B. Souster; 1 Sept. 1950. This study was made on the northern fringe of the wood about 50 yards from its western end. The wood is badly overgrown and supports little ground vegetation.
Trees, etc.: Prunus avium 50 ft., abundant on fringe of wood, Corylus avellana 15 ft., abundant in interior, Crataegus monogyna 6 ft. (o), Fagus sylvatica 50 ft. (o), Fraxinus excelsior 40 ft. (o), Lonicera periclymenum 10 ft. (o), Rosa arvensis 4 ft. (r), Rubus sp. 2 ft. (r), Sambucus nigra 10 ft. (M).

Abundant

Mercurialis perennis (L)   Moehringia trinervia

Frequent

Ajuga reptans   Myosotis arvensis   Viola reichenbachiana
Endymion non-scriptus

Occasional

Primula vulgaris

Rare

Asperula odorata   Dryopteris filix-mas   Pteridium aquilinum
Anemone nemorosa   Fragaria vesca   Prunella vulgaris
Arum maculatum   Galeobdolon luteum   Ranunculus repens
Cerastium holosteoides Geum urbanum   Stellaria media
Chamaenerion angusti- Glechoma hederacea   Urtica dioica
   folium   Holcus lanatus   Veronica chamaedrys
Cirsium arvense   Potentilla sterilis   V. officinalis
Deschampsia cespitosa

Passing mention has been made of the trees of the Clay-with-Flints. Native here, but doubtfully so elsewhere in the county, are hornbeam (*Carpinus betulus*), white beam (*Sorbus aria*) and holly (*Ilex aquifolium*). Near Kensworth are some fine trees of *Ulmus plotii*.

To the roadside should be added *Crepis biennis*, abundant near Long Wood, and *Alchemilla vestita*, more common on the Clay-with-Flints than on the Lower Greensand. In Luton Hoo Park is a station for *Alchemilla xanthochlora*.

The Clay-with-Flints is well drained and, apart from a few ponds and fewer streams, there is no place to support a marsh or aquatic flora. Saunders recorded some aquatic species from Pepperstock but I have been unable to re-discover them.

The arable fields have weeds which merit some attention. *Scleranthus annuus* is frequent, and between Long Wood and Ravensdell Wood *Campanula rapunculoides* and *Geranium columbinum* grow in rough grassland and as weeds. At Luton Hoo the gardens produce unusual weeds in *Coronopus didymus* and *Rorippa sylvestris*.

Habitat Study 82   Arable field, Greencroft Barn
Parish: Studham   District: Colne   Grid Ref. 52/02201365
Alt. 580 ft.   Soil: Clay-with-Flints   pH 7.0
Surveyed: 25 June 1949 with H. B. Souster; 2 May 1950 with W. D. Coales and H. B. Souster; 1 Sept. 1950. This study was made in the field to the west of Greencroft Barn. The crop was reasonably clean.

Agrostemma githago	Melandrium album	Prunella vulgaris
Anagallis arvensis	Myosotis arvensis	Ranunculus repens
Aphanes arvensis	Odontites verna	Rumex acetosa
Cerastium glomeratum	Papaver rhoeas	R. crispus
C. holosteoides	Plantago lanceolata	Scleranthus annuus
Cirsium arvense	P. major	Senecio vulgaris
C. vulgare	Poa annua	Sinapis arvensis
Euphorbia exigua	P. pratensis	Trifolium repens
Galium aparine	Polygonum aviculare	Veronica arvensis
Geranium dissectum	P. persicaria	V. serpyllifolia
Matricaria maritima	Potentilla anserina	Vicia tetrasperma
Medicago lupulina	P. reptans	Viola arvensis

Any claim which might have been made to treat the Lea Valley as a separate region must be balanced against the growth of Luton which with its housing estates, factory sites and sewage works now occupies almost the whole valley. The river rises—it would be more true to say rose—near Leagrave, where there is a large area of river gravel. Within the valley are small caps of Boulder Clay that are often wooded. George Wood and Bramingham Wood are still sufficiently inaccessible or barred to the public to have some interest.

Habitat Study 83   George Wood
Parish: Streatley   District: Lea   Grid Ref. 52/07302715
Alt. 460 ft.   Soil: Boulder Clay overlying Middle Chalk   pH 6.2
Surveyed: 2 July 1949 with H. B. Souster: 14 April 1950; 22 Sept. 1950. This study was made at a point 100 yards along the main path from the north-east corner of the wood.
Trees, etc.: Acer campestre 12 ft. (r), Corylus avellana 10 ft. (o), Crataegus monogyna 8 ft. (o), Fraxinus excelsior 50 ft. (o), Lonicera periclymenum (o), Prunus avium 50 ft. (r), Rosa arvensis 6 ft. (1), R. canina 6 ft. (r), Rubus sp. 4 ft. (f), Sambucus nigra 8 ft. (1).

Abundant

Anemone nemorosa	Pteridium aquilinum (L)	Ranunculus ficaria
Mercurialis perennis (L)		

Frequent

Endymion non-scriptus	Glechoma hederacea	Veronica chamaedrys
Fragaria vesca	Holcus lanatus	Vicia sepium
Geranium robertianum		

Occasional

Ajuga reptans	Carex sylvatica	Filipendula ulmaria
Arum maculatum	Circaea lutetiana	Galium aparine (L)
Brachypodium sylvati-	Deschampsia cespitosa	Geum urbanum
cum	Epilobium montanum	Heracleum sphondylium
Bromus ramosus	Festuca gigantea	Poa trivialis

Rare

Adoxa moschatellina	Juncus effusus (1)	R. obtusifolius
Alopecurus pratensis	Moehringia trinervia	Scrophularia nodosa
Arctium minus	Myosotis arvensis	Solanum dulcamara
Cirsium palustre	Poa annua	Stachys sylvatica
Conopodium majus	Prunella vulgaris	Stellaria holostea
Dactylis glomerata	Ranunculus repens	Tamus communis
Galeopsis tetrahit	Rumex sanguineus	Viola reichenbachiana
Hypericum hirsutum		

Bryophytes: Brachythecium rutabulum, Eurhynchium praelongum, E. striatum.

Until comparatively recently there were many marshy places at Leagrave and Limbury, but these are rapidly disappearing. At Well Head, Leagrave, not the actual source of the Lea, is still a marshy depression.

Habitat Study 84   Well Head, Leagrave
Parish: Luton   District: Lea   Grid Ref. 52/06102475
Alt. 380 ft.   Soil: Alluvium overlying Middle Chalk   pH 6.3
Surveyed: 26 June 1949; 15 Aug. 1949; 8 May 1951.

(a) the southern end of the wettest part of the depression

Abundant

Agrostis stolonifera	Juncus inflexus	Poa pratensis

Frequent

Eleocharis palustris	Holcus lanatus	Potentilla anserina (L)
Galium palustre		

Occasional

Cirsium palustre	Equisetum palustre	R. repens
C. vulgare	Mentha aquatica	Rumex crispus
Epilobium parviflorum	Ranunculus acris	

Rare

Caltha palustris	Festuca pratensis	Ranunculus flammula
Cardamine pratensis	Filipendula ulmaria	Rumex conglomeratus
Cerastium holosteoides	Hypericum tetrapterum	R. obtusifolius
Cirsium arvense	Myosotis scorpioides	Vicia cracca
Epilobium hirsutum	Poa trivialis	

(b) stream bed to south of above—this has water only after periods of heavy rain

Abundant

Alopecurus geniculatus	Glyceria plicata (L)	Sparganium erectum (L)
Callitriche stagnalis	Lemna minor	

Frequent

Myosotis scorpioides	Ranunculus aquatilis
	subsp. pseudofluitans

Occasional

Epilobium hirsutum	Solanum dulcamara (M)

Rare

Apium nodiflorum	Mentha aquatica	Rumex conglomeratus
Arum maculatum (M)	Phalaris arundinacea	Urtica dioica

Leagrave Common was at one time one of the more interesting places in the county to the botanist. In the area of the study below Saunders found *Parnassia palustris* and I found *Sagina nodosa* there in 1918. As recently as 1940 there was some quantity of *Triglochin palustre* and until 1945 the leaves of *Menyanthes trifoliata*. A record was made of what is left.

Habitat Study 85   Leagrave Common
Parish: Luton   District: Lea   Grid Ref. 52/05972460
Alt. 380 ft.   Soil: Alluvium overlying Middle Chalk   pH 6.8
Surveyed: 26 June 1949; 15 Aug. 1949. This study was made in the extreme north-west corner of the marsh. Here it is low-lying and under water in wet seasons.

Abundant

Deschampsia cespitosa	Epilobium hirsutum (L)	

Frequent

Galium palustre	Holcus lanatus	Potentilla anserina

Occasional

Agrostis stolonifera	Festuca rubra	Juncus inflexus
Carex disticha	Galium uliginosum	Phleum pratense
Cirsium palustre		

Rare

Alnus glutinosa 6 ft. (1)	Fraxinus excelsior	Potentilla reptans (M)
Arrhenatherum elatius	6 ins. (1)	Ranunculus acris
Caltha palustris	Galium aparine	R. repens (M)
Carex hirta	Heracleum sphondylium	Rumex acetosa (M)
Cerastium holosteoides	Hypericum tetrapterum	R. crispus
Cirsium arvense	Mentha aquatica	Scrophularia aquatica
C. vulgare	Myosotis scorpioides	Solanum dulcamara
Dactylis glomerata	Poa pratensis	Urtica dioica (M)
Epilobium parviflorum	P. trivialis	Vicia cracca
Festuca pratensis		

At Limbury *Geranium pratense* and *Helleborus viridis* just survive but *Thalictrum flavum* has long been extinct. It is a much-fouled river which leaves Luton, but *Rumex maritimus* still grows by the lake in Luton Hoo Park as it does by the Luton Sewage Works at East Hyde with *Bidens cernua*. In a marshy meadow below the sewage works *Polygonum bistorta* holds its own. *Mimulus guttatus* was once abundant on this stretch of the river but disappeared a few years ago. The lake in Luton Hoo Park merits further study. Under trees by the lake-side *Iris foetidissima* is abundant.

Mycelis muralis makes a fine show on the wall of Luton Hoo Park, and until recently *Monotropa hypopitys* and *Cephalanthera damasonium* appeared in a small beech-wood on the road bordering the east side of the park. At Leagrave *Sambucus ebulus* is locally abundant by an old track.

There are many places in and near Luton which support an adventive flora but little worth special mention has been recorded. Between the Lea Valley and the Clay-with-Flints is a narrow outcrop of chalk where in suitable places *Petroselinum segetum* is especially plentiful.

Habitat Study 86   Arable field, Chiltern Green
Parish: Hyde   District: Lea   Grid Ref. 52/12781860
Alt. 480 ft.   Soil: Upper Chalk   pH 7.2
Surveyed: 25 June 1949; 17 April 1950; 29 Aug. 1950. This study was made in the
second field on the south side of the road from Chiltern Green Station.

Alopecurus myosuroides   Lithospermum arvense   Ranunculus repens
Anagallis arvensis   Matricaria maritima   Sherardia arvensis
Arenaria serpyllifolia   Medicago lupulina   Silene vulgaris
Atriplex patula   Myosotis arvensis   Sonchus arvensis
Centaurea scabiosa   Papaver rhoeas   S. asper
Convolvulus arvensis   Petroselinum segetum   Stellaria media
Geranium molle   Poa pratensis   Torilis arvensis
Knautia arvensis   Polygonum aviculare   Veronica persica

THE COMPOSITION OF THE FLORA

A native species may be defined as one which grows in a wild state,
usually with the same plant associates, provided that the place in
which it is growing is within the range of its continuous distribution
and that there is no evidence of its introduction from another place.
Too little is known of the past history of the plants of the county to
enable this definition to be applied too rigidly. Its wildest parts
have suffered some change at the hand of man and it is unlikely that
any species, except perhaps some micro-species of a genus such as
Rubus, had their origin in the county. Species accepted now without
hesitation as natives no doubt first entered the county under conditions which, if repeated now, would force one to consider them as
adventives; but to consider prehistoric plant-migrations in terms of
a present-day administrative county is academic.

With a few exceptions the native flora of the county is that of its
chalk hills, heaths, woods, marshes, rivers and riversides. Species
found normally elsewhere, even if they are native in other parts of
Britain, are doubtfully native in Bedfordshire. The places in which
native plants can survive in the county are becoming fewer,
though the county is in this respect better placed than most Midland
counties. The latest figures of land utilisation in the administrative
county are:

Woodland   .   .   .   .   12,180 acres (4%)
Arable land   176,853   ,,   (58.4%)
Permanent grass   .   .   .   63,726   ,,   (21%)
Rough grazing   3,111   ,,   (1%)
Urban and industrial sites   .   47,072   ,,   (15.6%)

Native species are still extending their range by natural means, for
it is unlikely that Abbot and other early botanists would not have
had their attention drawn to *Herminium monorchis*, *Aceras anthropophora*
and *Himantoglossum hircinum*, which have appeared within the last
fifty years. It should be noted, however, that the stations from which
each has been recorded were in every case disturbed chalk-
downland having a comparatively unstable flora.

Interesting also are those species which are native in other parts
of Britain but appear in Bedfordshire only in man-made habitats.

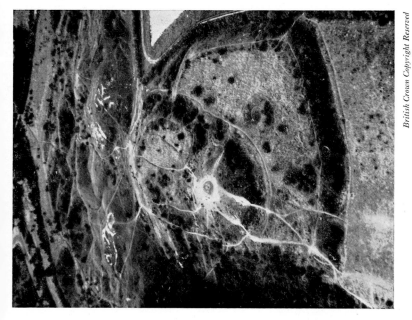

PLATE 19   TOTTERNHOE KNOLLS (Habitat Study 62)

PLATE 18   COW COMMON (Habitat Study 56)

These have a continuous distribution through native and man-made habitats until they reach us. They include most of the species limited to walls: *Ceterach officinarum, Asplenium adiantum-nigrum, A. ruta-muraria, Saxifraga tridactylites* and many of our mosses. They have a double status; but in Bedfordshire, where they would disappear if all the walls were pulled down, appear only as wall-denizens.

Similar problems arise with *Anthriscus sylvestris*, abundant on roadside verges in spring but also dominant in Stanbridge Wood, which is a rare case of an elm wood. Its status depends a great deal on that of the elm. *Anthyllis vulneraria*, without doubt native on the limestone in the west and north of Britain, appears in Bedfordshire only on regenerating chalk-downland or on disturbed soils on the Chalk and the Oolitic Limestone. Two problems arise here—how long is one to allow for a species to become native and what is established chalk-downland? Totternhoe Knolls is a case of downland still regenerating after probably 600 years. *Anthyllis vulneraria* is abundant there as it is in chalk quarries and railway banks of much lesser age. My own view, which is supported by J. F. Hope-Simpson, is that Knocking Hoe is virgin chalk downland and that the rest of the chalk hills were cleared by early man, reaching a comparatively stable vegetative condition many centuries ago. On this assumption one could consider as native those species which maintain themselves on Knocking Hoe, and hold in suspicion all the chalk-hill species which appear elsewhere.

This hypothesis would have been attractive thirty years ago when the hills were well grazed by sheep and a rabbit could scarcely keep alive, but the hills are now undergoing a great change. Hawthorn scrub is spreading, and *Bromus erectus*, which was previously restricted to hedge bottoms and cannot survive grazing, is now colonizing immense patches. It is kept down only in the vicinity of rabbit-warrens, and over-grazing by rabbits is allowing plant colonists to enter. A future botanist may very well wonder what was the native flora of the Chalk.

Like problems are presented by the flora of the greensand heaths, where *Calluna vulgaris* and some mosses and lichens could be considered the only native species. Areas not dominated by *Calluna* may be suspected of having been disturbed at some time. These areas contain some of the more interesting plant species of the county: *Teesdalia nudicaulis, Corydalis claviculata, Filago minima, Hypochoeris glabra*, etc.

Woodlands and marshes present fewer problems, but their flora can be changed immensely by the nature of the coppicing and by drainage.

Denizens proper may be defined as plant species which have every appearance of being native but about which there is sufficient evidence to suggest otherwise. A good example is *Acorus calamus*, which is plentiful with native species by the Ouse. Abbot, however, recorded it from one station only. Further evidence shows that it does not produce seed and is reproduced vegetatively. It is consequently

in a different category from *Herminium monorchis*. Similar to
Acorus is *Elodea canadensis*, the details of the introduction of which
are more accurately known. Some botanists consider that denizens
must be of long standing but I can see no difference in the status
of *Acorus*, a species of long standing in wild situations, and *Epilobium adenocaulon*, of very recent introduction. Shrubs, trees and
perennial herbs are in a different category from annuals. *Lycium
chinense* and *Symphoricarpos rivularis* have been considered to be
denizens but are more strictly garden escapes and in the same
category as a sunflower which escapes from a garden and lives only
one year. *Acaena anserinifolia* is established in a few places in the
county but only because it is a perennial. Few perennials introduced from outside become established in truly wild situations.

The bulk of the present Bedfordshire flora was probably colonist
in origin and a great deal of it is, strictly speaking, colonist still.
It is restricted to man-maintained habitats, and the species will con-
tinue only so long as man continues to use the land as he does now.
This colonist flora includes the weeds of arable land and most of
the species which grow by our roadsides. These habitats soon revert
to scrub if left untended, and the colonists quickly disappear[1].
Some native species appear also as colonists: *Aphanes microcarpa*
grows both in heathy pastures and in arable fields on light soils:
A. arvensis, on the other hand, appears only as a colonist. *Mentha
arvensis* appears in woodland rides and as a weed of arable land.
Is this a native species colonizing arable land or a colonist invading
a natural habitat? Probably the latter, as in woods it is usually
restricted to the ends of rides adjacent to arable fields.

Local colonists are few. Most of our colonists had their origin
elsewhere, some on the wind-swept raised beaches of the seashore,
others on East Anglian heaths; but many remain to be accounted
for. Colonists include some of the more interesting Bedfordshire
plant species. Notes on their status appear more appropriately in
the body of the Flora: see *Bunium bulbocastanum*, *Ajuga chamaepitys*,
Trifolium ochroleucon and *Ornithogalum pyrenaicum*.

Railroad colonists repay study, as their origin may be accounted
for, rarely the case with weeds of arable land. *Cerastium tetrandrum*
is abundant on ballast throughout the county and I have observed
it at every place where I have looked for it on the railway in other
counties. Similarly *Cochlearia danica* appears on railway ballast in
Bedfordshire and a number of inland counties. There can be no
doubt that both species have spread along the railway from the
seashore, where they are native. On the well-drained margins of
the tracks *Cerastium pumilum* is abundant, but less is known of its
distribution elsewhere. It may yet be linked with native stations.
Chaenorhinum minor and the allied *Linaria repens* are common as railway

---

[1] Peter Taylor has drawn my attention to the fact that this is invariably hawthorn scrub, the hawthorn spreading from the hedgerows where it was originally
planted. One wonders what was the nature of scrub before the Agrarian Revolu-
tion of the 18th century.

colonists, and in Bedfordshire they grow both on the railway and on rough chalky ground in places adjacent to it. *C. minor* is found on the railway throughout the county but *L. repens* is not found far from the Chalk.

A species which does not appear regularly in a given situation is known as a casual. Natives and colonists may appear as casuals also. Casuals appear usually in places with the least stable flora, such as railway sidings, gravel pits and rubbish dumps. When they appear on rubbish dumps for a number of years, as have *Conium maculatum, Descurainia sophia, Sisymbrium orientale* and *S. altissimum* on Sundon Rubbish Dump, they become colonists, for they are then playing an active part in transforming the dump into a more fertile site from which they themselves will ultimately disappear.

I have rejected the term alien, as its meaning is not clear. Many of the denizens, colonists and casuals are alien in the sense that they are not British natives, but it is surely beyond the scope of a local flora to enquire whether species are native in parts of Britain outside the area of its survey. For instance the wool adventives have all come from abroad yet a few are also natives of some part of Britain. Some have become established, but most are mere casuals. It is best to consider them to be adventives, a term that includes denizens, colonists and casuals introduced from abroad.

Wool manure, known more often as 'shoddy', is used extensively on light soils in the market-gardening areas of the county. It retains moisture and in decay adds to the organic content of the soil. It assures an early crop, a great advantage with this type of farming. Farmers use shoddy in varying quantities and rotations, a dressing of nine tons per acre every three years being not unusual. Shoddy is sold in two qualities: black shoddy, which is the waste products of the mills, contains no seeds, but grey shoddy, a residue of the preliminary scouring and combing processes, has a high seed content.

Wool is imported into Britain mainly from Australasia and South Africa, as the following table of average imports (1948–1951) in millions of tons shows:

Australia	364	} 77.3%
New Zealand	189	
South Africa	60	8.4%
India	14	} 3.8%
Pakistan	13	
Peru	4	
Chile	4	} 2.4%
Argentine	5	
Falkland Is.	4	
France	21	} 4%
Belgium	8	
Turkey	4	0.6%
Irish Republic	10	1.4%
Other countries	15	2.1%

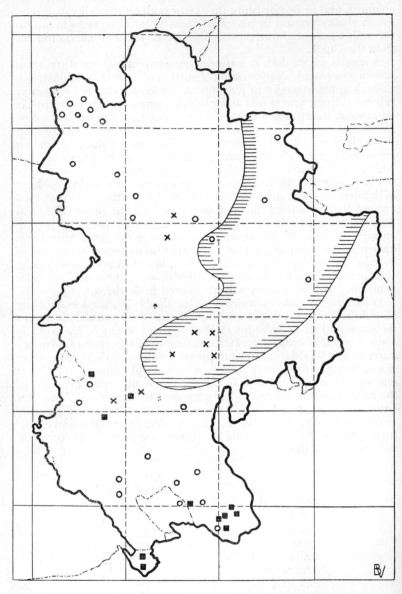

FIG. II

The hatched area is that in which wool manure is most used and the limit of
the distribution of *Medicago arabica* as a wool adventive. Its distribution as a
native species is shown with crosses. Circles show the distribution of *Geranium
pratense* and black squares that of *Veronica montana.*

The countries of origin of the wool adventives have no relationship to the sources of the wool. Most of them are cosmopolitan and had their origin in the Mediterranean area. This is the case with the various species of *Medicago* (usually the most evident in a crop of wool adventives) and *Trifolium* and with many of the grasses. Other cosmopolitan species include the South American *Xanthium spinosum* (the most well known of the wool adventives to farmers), *X. strumarium*, various species of *Bidens* and *Nicandra physalodes*.

From Australasia come *Acaena anserinifolia* and *Juncus pallidus*, both well established in a few places, various species of *Calotis* and *Chenopodium* and a number of grasses, and from South Africa come species of *Monsonia* and *Amaranthus*. One species comes from Tropical Africa.

The presence of the wool adventives raises many problems which it would be out of place to deal with here. It is sufficient to state that there is in Australasia and South Africa, from which most of the wool comes, already a large adventive flora, and any study of the British wool adventives must be made with this in mind. A number of the species do not flower in Britain, and there is no evidence yet of any becoming established weeds here as they apparently have in other countries.

The number of wool adventives so far recorded for the county is about 120, fewer than that listed by Hayward and Druce in *The Adventive Flora of Tweedside* (1919), but including a number not recorded by them. I hope elsewhere to deal more adequately with the wool-adventive flora when my studies of it in Bedfordshire and other parts of the country are more complete.

The present flora of the county consists of a number of native species which is diminishing slowly but being added to slightly as native species extend their range. To this, in the course of time, has been added a greater body of colonists and denizens that is neither increasing nor decreasing to any marked degree, and an increasing number of casuals, including the wool adventives, to which there would appear to be no limit.

Since botanists began to study the flora much has been lost. Below is a list of species which have not been recorded for more than 150 years—the date of the last certain record is given in brackets —most, it may be noted, species of heathy moorland:

Archidium alternifolium (1798)
Botrychium lunaria (1798)
Carex dioica (1798)
Caucalis latifolia (1798)
Centunculus minimus (1798)
Cicuta virosa (1798)
Colchicum autumnale (1798)
Dianthus deltoides (1690)
Dicranella rufescens (1801)
Drosera anglica (1801)
Equisetum hyemale (1798)
E. sylvaticum (1798)

Hammarbya paludosa (1798)
Hypericum elodes (1798)
Lycopodium clavatum (1798)
Narthecium ossifragum (1798)
Oenanthe silaifolia (1798)
Oxycoccus quadripetalus (1798)
Schoenus nigricans (1798)
Scirpus caespitosus (1798)
Scorpidium scorpioides (1798)
Splachnum ampullaceum (1798)
Thelypteris palustris (1798)
Tulipa sylvestris (1798)

A small list of species not recorded for more than 100 years reflects the small amount of work done on the botany of the county in the first half of the nineteenth century:

Caucalis daucoides (1841)   Epipactis palustris (1844)
Chrysosplenium alternifolium (1844)

A much greater number of species has not been recorded for more than fifty years. The list reflects the large amount of drainage carried out in the nineteenth century:

Anthoxanthum puelii (1890)   Philonotis fontana (1884)
Bryum inclinatum (1885)   Physcomitrella patens (1883)
B. murale (1885)   Orthotrichum lyellii (1882)
Campylium stellatum (1884)   O. tenellum (1892)
Carex diandra (1876)   O. pulchellum (1892)
C. pulicaris (1898)   Pyrola minor (1881)
C. × pseudo-axillaris (1876)   Ranunculus lingua (1881)
Crantoneuron commutatum (1882)   R. sardous (1889)
Drepanocladus sendtneri (1899)   Salvia pratensis (1890)
D. revolvens var. intermedius (1882)   Sedum dasyphyllum (1886)
Erica cinerea (c. 1880)   Seligeria calcarea (1903)
E. tetralix (c. 1880)   Sphagnum acutifolium (1882)
Fissidens viridulus (1878)   S. plumulosum (1885)
Fontinalis dolosa (1884)   Thesium humifusum (c. 1860)
Funaria meuhlenbergii (1885)   Tortula papillosa (1856)
Hydrocharis morsus-ranae (1886)   Urtica pilulifera (1884)
Myriophyllum alterniflorum (1889)

A few species, again mainly natives of marshy places, have not been recorded for more than 25 years:

Brachythecium populeum (1911)   Polytrichum urnigerum (1911)
Bryum pendulum (1906)   Potamogeton polygonifolius (1911)
Climaceum dendroides (1911)   Sagina nodosa (1918)
Oenanthe crocata (1911)   Salix pentandra (1911)
Parnassia palustris (1925)   S. repens (1911)
Pedicularis palustris (1926)   Sphagnum squarrosum (1911)
Pinguicula vulgaris (1921)

Further species have disappeared from their only known stations within the past 25 years:

Antennaria dioica (1936)   Geum rivale (1942)
Carex lepidocarpa (1953)   Drosera rotundifolia (1942)

It is too early to say that all these are now extinct; a number of species presumed so have been discovered while this Flora was being compiled.

PLANT DISTRIBUTION

A few species have been recorded for all neighbouring counties but not for Bedfordshire:

Cardamine amara, a riverside species and present in quantity in some neighbouring counties.

Sorbus torminalis, which may yet be found in our larger woods.

Cirsium dissectum, the most surprising absentee. Our marshes have been well-botanized and it is doubtful if it will be found.

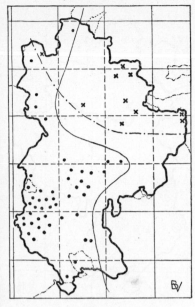

FIG. 12

Distribution of *Melandrium dioicum* (dots), a species frequent in the west of the county but absent in the east, and *Melampyrum cristatum* (crosses), an East Anglian species absent in the west of the county.

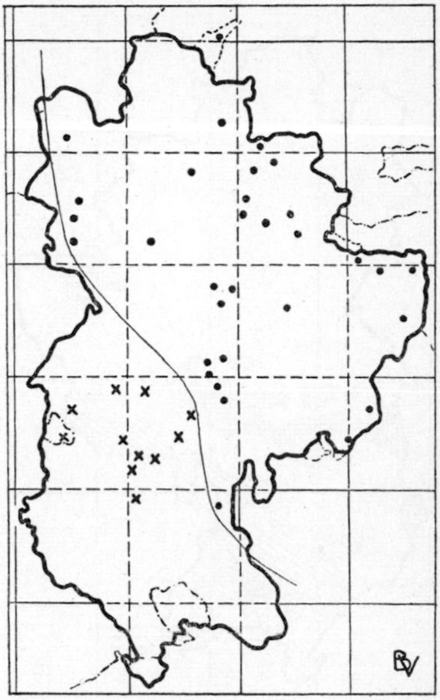

FIG. 13

Distribution of *Trifolium ochroleucon* (dots), another East Anglian species locally abundant in the east of the county, and *Chrysosplenium oppositifolium* (crosses), a species of marshy places associated with the Lower Greensand.

Chenopodium vulvaria, which will, no doubt, appear on one of our rubbish dumps.

Bromus interruptus, recorded for the county, with no details, by Druce.

Eleocharis multicaulis, *Scirpus pauciflorus* and *S. fluitans*, all very rare or extinct in neighbouring counties and which could have been overlooked by earlier Bedfordshire botanists.

A few well-established species are recorded for Bedfordshire but for no neighbouring counties: *Cerastium brachypetalum*, *Lathyrus tuberosus*, *Epilobium lanceolatum*, *Falcaria vulgaris*, *Lapsana intermedia* and *Lagarosiphon major*. All are of recent origin, except *Falcaria*, and no doubt await record in some neighbouring counties. To the list could be added many casuals and bryophytes[1] but their record in Bedfordshire reflects the greater attention paid to its botany and bears little relation to plant distribution.

Of the neighbouring counties Buckinghamshire has a flora most resembling that of Bedfordshire. It has been well studied, and the

[1] Bryophytes are for this reason excluded from this section as a whole.

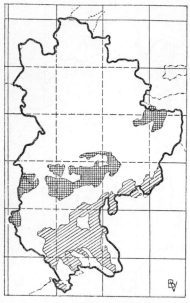

FIG. 14

Distribution of *Ornithopus perpusillus*
(cross hatching), a species of disturbed soils on the Lower Greensand (*Filago minima* has a similar
distribution), and *Asperula cynanchica* (oblique hatching), a common
species of chalk downland (*Hippocrepis comosa* and *Senecio campestris*
have a similar distribution).

FIG. 15

Distribution of *Calluna vulgaris*
(cross hatching), a species of heathy
places on the Lower Greensand
and Clay-with-Flints, and *Samolus
valerandi* (black), a species frequent
by the Ouse and appearing in a
few streams and marshy places
elsewhere.

Flora of Buckinghamshire, G. C. Druce (1926), gives a comparatively
recent account of it. Its weaknesses are that the reader is left to
guess whether certain species may be presumed to be extinct, that
there is too little respect shown by the author for the boundary of the
county, at least on the Bedfordshire and Hertfordshire side, and that
as much prominence is given to casuals and garden escapes as to
native species. It is nevertheless a most useful work. All the major
geological formations are common to both counties.

Cambridgeshire is likewise similar to Bedfordshire in having
some Lower Greensand and Chalk, but, unlike our county, it has
some fen country, and maritime species enter at Wisbech. Its flora
has been as well studied as that of any county, but the best account
is still Babington's *Flora of Cambridgeshire* (1860). Evans (1939) is
disappointing and incomplete, and Godwin (1938), while extremely
helpful, does not claim to account for the whole flora.

A good start was made with the flora of Hertfordshire by Webb
and Coleman, *Flora Hertfordiensis* (1849), which was followed by the

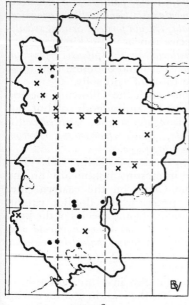

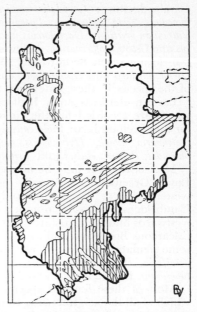

FIG. 16

Distribution of *Thalictrum flavum* (crosses), a species frequent by the Ouse and Ivel, and *Minuartia tenuifolia* (dots), a species of well-drained habitats on a variety of soils.

FIG. 17

Distribution of *Fissidens adianthoides* (vertical hatching), a species locally abundant on outcrops of the Chalk and Oolitic Limestone (*Camptothecium lutescens* and *Campylium chrysophyllum* have a similar distribution), and *Campylopus pyriformis* (oblique hatching), a species well distributed on the Lower Greensand.

ideal county flora in Pryor's *Flora of Hertfordshire* (1887). Since then no attempt has been made to collect together the mass of information made available by such workers as J. E. Little. The county has chalk in common with Bedfordshire, Buckinghamshire and Cambridgeshire, and many of the greensand species appear in Hertfordshire on the Clay-with-Flints.

Species of the Chalk common to the four counties but absent from the Northamptonshire limestones include *Fumaria parviflora, F. vaillantii, F. micrantha, Iberis amara, Bunium bulbocastanum, Himantoglossum hircinum* and *Herminium monorchis*. Calcifuge species common to Bedfordshire, Buckinghamshire and Hertfordshire but not recorded for other neighbouring counties include *Hypericum elodes, Vicia lathyroides* and *Carex curta*.

Western elements enter Bedfordshire with the following species recorded for both Buckinghamshire[1] and Bedfordshire but not for

[1] Many of these species appear also in Hertfordshire and Northamptonshire.

Cambridgeshire: *Corydalis claviculata*, *Viola palustris*, *Melandrium dioicum*, *Vicia sylvatica*, *Lathyrus montanus*, *Gentianella germanica*, *Vaccinium myrtillus*, *Dactylorchis incarnata*, *Luzula forsteri*, *Scirpus sylvaticus* and *Ceterach officinarum*. A few species appear in Buckinghamshire but do not reach Bedfordshire: *Scutellaria minor*, *Epipactis leptochila*, *Orchis militaris* and *Damasonium alisma*. It should be noted that most of the species in these categories are those of acid soils.

Eastern elements enter Bedfordshire with the following species, which are recorded for the county and Cambridgeshire[1] but not for Buckinghamshire: *Viola calcarea*, *Astragalus danicus*, *Seseli libanotis*, *Antennaria dioica*, *Hypochoeris maculata*, *Ajuga chamaepitys*, *Phleum phleoides*, *Primula elatior* and *Trifolium ochroleucon*.

Most of these, it may be noted, are species of alkaline soils. The Lower Greensand is generally more interesting botanically on the western side of the county and the Chalk on the eastern side. Cambridgeshire has a number of species which do not extend westwards into Bedfordshire. These include some species of the light calcareous Breckland soils, *e.g.*, *Silene otites* and *Veronica spicata* and some fenland species: *Viola stagnina*, *Lathyrus palustris*, *Selinum carvifolia*, *Peucedanum palustre*, *Senecio paludosus*, *S. palustris*, *Cirsium tuberosum*, *Lactuca saligna* and *Liparis loeselii*.

Most of these fenland species appear also in Huntingdonshire, which of all the neighbouring counties has the least in common with Bedfordshire. It has no Chalk and little Lower Greensand, and being small in area and population is probably the least botanized of all British counties. The only account of its flora is by Druce in *Victoria County History* (1926) and he would have been the first to admit that it was inadequate. Among the more interesting species which it has in common with Bedfordshire are *Primula elatior*, *Trifolium ochroleucon*, *Melampyrum cristatum* and *Ornithogalum pyrenaicum*. In addition to the fenland species listed above Huntingdonshire has *Myrica gale* and *Luzula pallescens*, which do not appear in Bedfordshire.

Northamptonshire has also little in common botanically with Bedfordshire. Some of the chalk-hill species appear also on the Northamptonshire limestones. I know no instance of a species recorded for Northamptonshire and not for Bedfordshire or a county adjoining it. There are a few species recorded for the Ouse in Northamptonshire, Bedfordshire, Huntingdonshire and in some cases Buckinghamshire, but not recorded for the Thames valley in Hertfordshire: *Ranunculus fluitans*, *Sium latifolium* and *Potamogeton praelongus*. There is a relatively recent account of the Northamptonshire flora by Druce, *Flora of Northamptonshire* (1930), but it leaves one with the impression that the author has gone back to the scene of his childhood in memory only. It is the work of a tired and aged man and lacks the inspiration of his earlier work.

From the above analysis of the distribution of some Bedfordshire plant species it will be seen that the more interesting problems of its

[1] Some of these species appear also in Hertfordshire.

botany rest in its borderline position between East Anglia and the
rest of England. Most of these problems have been already demon-
strated by Salisbury in The East Anglian Flora, *Trans. Norf. Norw.
Nat. Soc. 1932*, 191–263, and it is hoped that their further study in
various parts of this Flora may throw some additional light on
them.

One is often asked what are the most interesting plants entering
the county flora. My choice of ten is given with some hesitation,
with the number of vice-counties for which each species is recorded
added in brackets. Of greatest interest is *Bunium bulbocastanum* (4),
not only for its limited distribution but for its great abundance with
us. This must be followed by *Seseli libanotis* (4) and *Hypochoeris
maculata* (10). These are all species of the Chalk. *Carum carvi*, as a
rare native, must follow with *Onobrychis viciifolia*—vice-comital
distribution is in each case meaningless. The Boulder Clay adds
Primula elatior (8), *Ornithogalum pyrenaicum* (10), *Melampyrum cristatum*
(14) and *Trifolium ochroleucon* (16). Finally a place must be found
for one weed of arable land, *Melampyrum arvense*.

THE BOTANICAL DISTRICTS

These have many defects, for they vary in size and may give a
false impression of the actual distribution of species within the county.
The Chalk comes into four districts, with no relationship to the
distribution of chalk-hill species. But on the other hand the Lower
Greensand, which is in the Ouzel and Ivel districts, is divided
conveniently, and the large Ouse district reflects the uniformity of
the rather dull Ouse valley.

If a species is recorded for all the eight unequal and diverse
districts it can at least be assumed that it is well distributed in the
county. Indeed one of the more interesting facts which has emerged
from the use of the districts is that some species which one would
have assumed were well distributed in the county are not. The absence of some common species from certain districts is of greater
moment than the rarities they contain.

Botanical districts were first outlined for the county by Druce
in *Victoria County History* (1904) but his districts have had to be
emended:

1. Druce made an arbitrary division of the large Ouse District.
   I have preferred to use a watershed to introduce a new district, the Kym.
2. Druce combined the Colne and Lea catchment areas to make
   one district, the Lea. I have preferred to make them two to
   bring the county into line with Pryor's divisions for Hertfordshire.
3. Minor emendations had to be made in the boundaries of some
   districts, especially the Cam, which is larger than Druce considered it to be.

1. The Nene District

This very small district links with District VI (Nene B or Harper's Brook District) in Druce's *Flora of Northamptonshire*. It lies entirely within the Northern Uplands Region of this Flora.

It contains some profitable botanical country and five Habitat Studies were made in the district. Its best stations have proved to be Sharnbrook Summit (H.S. 1a, 1b) and Great Hayes Wood (H.S. 3a, 3b), but its arable fields are always worthy of study. The greatest rarity is *Cerastium brachypetalum* (known nowhere else in Britain), but otherwise there is no species limited to this district in the county.

Species recorded from all districts except this are:

Anacamptis pyramidalis	Galeobdolon luteum	Milium cffusum
Carex pallescens	Lotus uliginosus	Primula vulgaris
C. remota	Luzula pilosa	Stellaria holostea

Until recently it had been inadequately botanized. Druce recorded only one species, picked up from Newbould. Miss Kitchin gave me many records made in her childhood and was able to find most of the plants again. R. R. B. Orlebar, who knew the district well, was also helpful.

2. The Ouse District

This, the largest of the districts, links with the Ouse District in Druce's *Flora of Buckinghamshire* and *Flora of Northamptonshire*. It includes the whole of my Upper Ouse, Northern Oxford Clay and Southern Oxford Clay Regions, and parts of the Northern Uplands, Lower Greensand and Lower Ouse and Ivel Valley Regions. No less than 27 Habitat Studies were made within it.

Its most interesting features are the river and the meadows and marshes associated with it. There are, mainly by virtue of its size, a number of species limited to it:

Barbula cylindrica	Lilium martagon	Potamogeton praelongus
Cinclidotus fontinaloides	Nymphaea alba	Ranunculus fluitans
Eleocharis uniglumis	Orchis latifolia	Sium latifolium
Falcaria vulgaris	Orthotrichum cupulatum	Utricularia vulgaris

There is apparently no species recorded for all other districts still awaiting record here.

The district was well botanized in the past, for it includes the areas best known to Abbot, McLaren, Hillhouse and Hamson.

3. The Kym District

This small district links with District VI (Nene B or Harper's Brook District) in Druce's *Flora of Northamptonshire*. It lies entirely in the Northern Uplands Region of this Flora.

It is probably the least interesting district and only two Habitat Studies (2a, 2b) were made within it. There is no species recorded

here and for no other district, but the absences are considerable.
A closer search may reduce the list:

Carex caryophyllea	Raphanus raphanistrum	Veronica arvensis
Erigeron acris	Silene vulgaris	Valerianella dentata
Knautia arvensis	Torilis arvensis	V. locusta
Linaria vulgaris		

The large number of weeds of arable land in this list is surprising
in a district in which natural habitats are few.

It is the least botanized part of the county and for me proved the
most inaccessible.

4. The Ouzel District

This is, next to the Ivel District, the most interesting of all the
botanical districts. It links with the Ouzel District in Druce's
Flora of Buckinghamshire and included in it are parts of the Lower
Greensand, Gault and Chalk, Chalk Marl, and Chalk Regions of
this Flora. Twenty-four Habitat Studies were made within the
district.

It contains some of the best Lower Chalk at Totternhoe (H.S. 62),
marshes on the Chalk Marl (H.S. 56), Greensand heath, our best
wood—King's Wood, Heath and Reach (H.S. 30, 32), and New
Wavendon Heath (H.S. 35).

It has lacked a resident botanist but it is so easily accessible from
Luton that it has not suffered in this respect.

There are no species recorded for all other districts but awaiting
record here. A long list could be compiled of species limited to this
district.

5. The Ivel District

This is similar to the Ouzel District and more interesting only
because it is larger. It links with the Ivel District in Pryor's *Flora of
Hertfordshire*, and included in it are parts of the Lower Ouse and
Ivel Valleys, Lower Greensand, Gault and Chalk Marl, and Chalk
Regions of this Flora. No less than 53 Habitat Studies were made
within it. Many species are limited to this district and there is no
species recorded for all other districts and not here.

It is the best botanized district of the county, its central position
making it readily accessible. Both Crouchs lived in it and J. E.
Little's study of the district added much to the knowledge of the
botany of the county.

6. The Cam District

This very small district links with the Cam District in Pryor's
Flora of Hertfordshire and is included entirely in the Gault and Chalk
Marl Region of this Flora. It has limited possibilities botanically
and only one Habitat Study was made there.

Until recently it was inadequately botanized and Druce made no records from it. J. E. Little paid a few visits, and it was in this district that Laflin lived. It can now be said to be reasonably completely studied. It has some interesting species: *Melampyrum cristatum*, *Trifolium ochroleucon*, *Bupleurum rotundifolium* and *Orthodontium lineare*. *Primula elatior* is found here and in no other district.

There are many plant absences mainly because of its small size and the uniformity of its soils.

7. The Lea District

This small district links with the Lea District in Pryor's *Flora of Hertfordshire* and contains parts of the Chalk Region and Clay-with-Flints and Lea Valley Region of this Flora. It contains some interesting country, notwithstanding considerable urban development, and 14 Habitat Studies were made within it.

Two species are recorded here and for no other district: *Astragalus danicus* and *Alchemilla xanthochlora*. A few species recorded for all other districts are awaiting record here:

Rorippa microphylla	Kickxia elatine	Festuca arundinacea
Populus tremula	Anthemis cotula	

It is one of the best botanized districts of the county, as Saunders lived in it and it has been the centre of my work.

8. The Colne District

This is another small district and in many respects similar to the Lea District, which prompted Druce to combine them. It links with the Colne Districts in Druce's *Flora of Buckinghamshire* and Pryor's *Flora of Hertfordshire*, and includes parts of the Chalk Region and Clay-with-Flints and Lea Valley Region of this Flora. Nine Habitat Studies were made within it. It contains some interesting country but lacks water. Two species are recorded here and for no other district: *Sorbus aria*, as a native, and *Galium pumilum*. There are many plant absences, mainly aquatics.

Saunders knew the district well and it is one of the best botanized of the county.

THE PLAN OF THE FLORA

Sequence. This is that adopted in most modern floras. In the Liverworts (*Hepaticae*) it follows Wilson: *A Census Catalogue of the British Hepatics* (1930), and in the Mosses (*Musci*) Richards and Wallace: An Annotated List of British Mosses in *Trans. Brit. Bry. Soc.* 1950. For the higher plants it follows Clapham, Tutin and Warburg: *Flora of the British Isles* (1952). Minor deviations from the latter have been made within natural orders and genera, as my manuscript was prepared before the *Flora of the British Isles* was published, but cross-references to it appear throughout this Flora.

Type. Large type has been reserved for native species and for some denizens and colonists which are well established. Smaller type has been used for less well-established species, microspecies, subspecies, varieties, and for some plants of hybrid origin. It has also been used for charophytes, bryophytes and other sections of interest to the specialist rather than the general reader.

English names. As far as possible anglicized forms of Latin names have been avoided. Local names have been sought, the villages in which they are used being added in brackets. Names used by Batchelor (1808) are similarly indicated.

Synonyms. In addition to the Latin names used in the published works of Bedfordshire botanists, the names used in the following works are given:

A. Wilson, *A Census Catalogue of the British Hepatics*, Third Edition, 1930.

S. M. Macvicar, *The Student's Handbook of British Hepatics*, Second Edition, 1926.

P. W. Richards and E. C. Wallace, An Annotated List of British Mosses, *Trans. Brit. Bry. Soc.* 1950.

H. N. Dixon, *Student's Handbook of the British Mosses*, Third Edition, 1924.

V. F. Brotherus in Engler and Prantl, *Die Naturlichen Pflanzenfamilien*, Second Edition, 1924–5.

A. R. Clapham, T. G. Tutin and E. F. Warburg, *Flora of the British Isles*, 1952.

G. C. Druce, *British Plant List*, Second Edition, 1928.

G. C. Druce, *The Comital Flora of the British Isles*, 1932.

The London Catalogue of British Plants, Eleventh Edition, 1925.

A. R. Clapham, Check List of British Vascular Plants, *Journal of Ecology*, 1946.

Status. This is in accordance with the definitions given on pp. 128–134. When a species has a variable status it is shown in its highest category. A question mark preceding the status indicates that the author considers the species to be in this category but lacks definite evidence.

Distribution. The distribution given for neighbouring counties must be taken to refer to the species as a whole and not to subspecies or varieties, a note on which it may follow. The neighbouring counties must be interpreted in the broad sense; the author has not been able to check whether these records refer to the administrative county or the Watsonian vice-county.

Records. These are arranged in historical sequence in the botanical districts to which they refer. Punctuation is as follows: *colons* separate records from different sources; *semicolons* separate stations; *commas* separate additional information given for the station itself. Additional information following a station listed by another authority may be taken as added by the author. An exception is made to the above principles when the author has recorded a species from a common habitat (*e.g.*, arable fields) in a number of parishes; and in such cases the habitat and the various stations are separated by commas.

Citations. These are given in full only when they refer to a first record for the county. In other cases they are given in a form which may be checked fairly easily. Some records printed during the last years of the compilation of this Flora are not cited. When a date, but no citation, follows the record it indicates that the record is the first for the county, the date referring to its discovery.

First records. A printed record is given priority over records in manuscripts and herbaria, except when a long period of time has elapsed between the latter type of record and the printed record. I have accepted records made by Abbot in his annotated copy of *Flora Bedfordiensis* and in his correspondence with J. E. Smith, by W. Crouch and J. McLaren in their herbaria, and by Hamson in his manuscript flora, as each was 'recording' the species in what appeared to him the most permanent form. I differ from many botanists in that I consider a correctly-named, localized and dated herbarium specimen the most satisfactory of all records.

First evidence. This is reserved for cases where the first knowledge of the appearance of a species in the county is a localized but wrongly identified or unidentified herbarium specimen.

ABBREVIATIONS

* A specimen from this station in Herb. Luton Museum.

† A specimen from this station in Herb. Saunders.

! after a station. Indicates that the author has seen the species growing there. In the hepatics it refers to P. Taylor, in the mosses to T. Laflin.

PLATE 21   PEGSDON HILLS (with Pegsdon village in the foreground)

(Habitat Study 6)

! after the name of a recorder. Indicates that it is a joint record with the author.

v.c. 20 [Beds.], etc. A station in v.c. 20 Herts but now in the administrative county of Bedfordshire (see pp. 38–40).

B.E.C. Rep. Botanical Exchange Club Report.

Beds. Nat. Hist. Soc. Bedfordshire Natural History Society.

Bot. Guide. *Botanist's Guide*, D. Turner and L. W. Dillwyn, 1805.

Bot. Rec. Club Rep. Botanical Record Club Report.

B.P.L. Bedfordshire Plant List (*Trans. Beds. Nat. Hist. Soc.*, 1882).

Brit. Bry. Soc. British Bryological Society.

C.C. Charles Crouch.

C.R.O. List. List of plants made at Basmead by E. O. Squire (Beds. County Record Office).

C.T.W. Clapham, Tutin and Warburg, *Flora of the British Isles*, 1952.

E.B.B. (only in Rosa). E. B. Bishop.

E. Bot. *English Botany*, J. Sowerby and J. E. Smith, 1790–1814.

E. Bot. Supp. Supplement to *English Botany*, 1831–63.

E.M.-R. E. Milne-Redhead.

F.B. *Flora Bedfordiensis*, C. Abbot, 1798.

F.F.B. *Field Flowers of Bedfordshire*, J. Saunders, 1911.

Fl. Herts. *Flora of Hertfordshire*, A. R. Pryor, 1887.

Fl. Bucks. *Flora of Buckinghamshire*, G. C. Druce, 1926.

Herb. Herbarium.

Herb. J. S., etc. J. Saunders's Herbarium.

H.P. (mainly in *Rosa*). H. Phillips.

J. Bot. Journal of Botany.

J.E.L. J. E. Little.

J.E.L., B.E.C. The Ivel District of Bedfordshire, in *B.E.C. 1935 Rep.*, J. E. Little.

J.E.L., J. Bot. Notes on Bedfordshire Plants, in *J. Bot. 1919*, J. E. Little.

J.E.L., Diary. J. E. Little's field diaries (Hitchin Museum).

J.E.L., F.F.B. J. E. Little's annotated copy of *F.F.B.* (Hitchin Museum).

J.G.D. (only in *Bryophyta*). J. G. Dony.

J.H. *An Account of the Flora of Bedfordshire*, J. Hamson, 1906.

J.H. Notes. A manuscript Flora of Bedfordshire, J. Hamson.

J. McL. J. McLaren.

J.McL., B.M. (also J.McL., Herb. British Museum). A herbarium compiled by McLaren now in the British Museum (Natural History).

J.McL., L.M. (also J.McL., Herb. Luton Museum). A herbarium compiled by McLaren now in Luton Museum.

J.S. J. Saunders.

Marquand List. (See Bibliography—manuscripts.)

New Bot. Guide. *New Botanist's Guide*, W. C. Watson, 1835.

P.N.B. *Place-names of Bedfordshire and Huntingdonshire*, A. Mawer and F. Stenton, 1926.

Plantae Bedford. Plantae Bedfordiensis, C. Abbot, 1795.

P.T. P. Taylor.

Rep. Report.

R.M. (only in *Rosa*). R. Melville.

S.B.P.L. (also South Bedfordshire Plant List). On the Wild Flowers of South Bedfordshire, in *Trans. Beds. Nat. Hist. Soc.*, 1885, J. Saunders.

sp. An unidentified species of the genus indicated.

Subsp. Subspecies.

Top. Bot. *Topographical Botany*, H. C. Watson, 1873 4.

T.L. T. Laflin.

Trans. Transactions.

W.B.E.C. Rep. Watson Botanical Exchange Club Report.

W.C. Herb. W. Crouch's herbarium (cited in this form as it was previously so cited by Saunders).

W.F.B. *Wild Flowers of Bedfordshire*, J. Saunders, 1897–8.

W.F.S. Diary. Wild Flower Society Diary (see Bibliography—manuscripts).

W.H. 1875 (also Hillhouse List, 1875). A contribution towards a new Flora of Bedfordshire, *Trans. Beds. Nat. Hist. Soc.*, 1877, W. Hillhouse.

W.H. 1876 (also Hillhouse List, 1876). Bedfordshire Plant List for 1876, *Trans. Beds. Nat. Hist. Soc.*, 1878, W. Hillhouse.

W.I. Women's Institute.

W.W. (mainly in *Rubus*). W. C. R. Watson.

Var. Variety.

V.C.H. Botany, G. C. Druce, *Victoria County History* (Bedfordshire), 1904.

BIBLIOGRAPHY

PUBLISHED MATERIAL RELATING TO THE BEDFORDSHIRE FLORA

Abbot, C. *Flora Bedfordiensis*, 1798.

Adams, S. H. The microscopical structure of mosses (with list of Bedfordshire species); *Trans. Beds. Nat. Hist. Soc.*, 1882.

Batchelor, T. *A General View of Agriculture in the County of Bedford*, 1808.

Bishop, E. B. Notes on the Roses of Bedfordshire; *B.E.C. 1938 Rep.*, 1939.

Burtt, B. L. Lapsana intermedia in Britain; *Watsonia I*, 1950.

Brenan, J. P. M. and Chapple, J. F. G. The Australian Myriophyllum verrucosum Lindley in Britain; *Watsonia I*, 1949.

Corder, T. Claytonia perfoliata in Britain; *Phytologist IV*, 1852.

Crouch, C. Trifolium striatum L. in Beds.; *J. Bot. XXIX*, 1891.
Carex pulicaris on Chalk; *Ibid, XXXVI*, 1898.

Davis, F. See Paybody.

Dony, J. G. Additions and Emendations to Comital Flora records for v.c. 30 (Bedfordshire); *B.E.C. 1943–4 Rep.*, 1946.
What Bedfordshire Is; *Journ. Beds. Nat. Hist. Soc. I*, 1947.
Bedfordshire Naturalists: 1. William Crouch, *Ibid.*; 2. James Saunders, *Ibid., II*, 1948; 3. Charles Abbot, *Ibid., III*, 1949; 4. William Hillhouse, *Ibid., IV*, 1950.
Problems of distribution raised in the compilation of a county flora; in J. E. Lousley: *The study of the Distribution of British plants*, 1951.
Wool aliens in Bedfordshire; in J. E. Lousley: *The changing Flora of Britain*, 1953.

Druce, G. C. Plants of Bedfordshire; *J. Bot. XXV*, 1897.
Botany; *Victoria County History* (Bedfordshire), 1904.

Forbes, J. *Salictum Woburnense*, 1829.
Pinetum Woburnense, 1839.

Hamson, J. Ferns (Filices), Horsetails (Equisetaceae) and Fungi; in G. C. Druce: Botany, *Victoria County History* (Bedfordshire), 1904.
An Account of the Flora of Bedfordshire, 1906.

Hillhouse, W. On the surface geology and physical geography of Bedfordshire; *Trans. Beds. Nat. Hist. Soc., 1875–6*, 1877.
A contribution towards a new Flora of Bedfordshire [Hillhouse List, 1875]; *Ibid.*
Bedfordshire Plant List for 1876 [Hillhouse List, 1876]; *Ibid., 1876–7*, 1878.

Jackson, A. B. Bedfordshire plants; *J. Bot. LVIII*, 1920.

Linton, E. F. Rubi of Woburn Sands; *Ibid., XXXI*, 1893.
Bedfordshire Rubi; *Ibid.*

Little, J. E. The genus Populus in the Ivel basin; *Ibid.*, *LIV*, 1916.
Notes on Bedfordshire plants; *Ibid.*, *LVII*, 1919.
Euphorbia platyphyllos in Bedfordshire; *Ibid.*, *LXIX*, 1931.
The Ivel district of Bedfordshire; *B.E.C. 1935 Rep.*, 1936.

Mann, H. H. The weed herbage of a slightly acid soil; *Journal of Ecology*, *XXVII*, 1939.

Mann, H. H. and Barnes, T. W. The competition between barley and certain weeds under controlled conditions; *Annals of Applied Biology*, *XXXII*, 1945. Competition with Agrostis gigantea; *Ibid.*, *XXXVI*, 1949. Competition with Stellaria media; *Ibid.*, *XXXVII*, 1950. Competition with clover considered as a weed; *Ibid.*, *XXXIX*, 1952.

Milne-Redhead, E. Cerastium brachypetalum Pers. in Britain; *The Naturalist*, *822*, 1947.

Paybody, Miss. List of plants growing in the neighbourhood of Luton; in Davis, F.: *History of Luton*, 1855.

Pollard, J. Pinguicula vulgaris in Bedfordshire; *J. Bot. XIII*, 1875.

Pryor, A. R. Potamogeton praelongus in Bedfordshire; *Ibid.*, *XIII*, 1875.
On the occurrence of Medicago lappacea Lamk. in Bedfordshire . . . ; *Ibid.*, *XIV*, 1876.
Notes on the herbarium of Abbot; *Ibid.*, *XXIII*, 1881.

[Ransom, A.]. Bedfordshire Plant List for 1882; *Trans. Beds. Nat. Hist. Soc.*, 1882.

Riddelsdell, H. J. Rubus notes; *J. Bot. LXX*, 1932.

Saunders, J. The Mosses of South Bedfordshire; *Trans. Beds. Nat. Hist. Soc.*, 1882.
Bryological Note for South Beds.; *Midland Naturalist*, 1883.
Sphagnaces of the South Midlands; *Ibid.*
On the Flora of South Bedfordshire; *J. Bot. XXI*, 1883.
South Bedfordshire Mosses; *Ibid.*, *XXII*, 1884.
Bedfordshire Plants; *Ibid.*
Nitella mucronata in Beds.; *Midland Naturalist*, 1884.
On the Wild Flowers of South Bedfordshire; *Trans. Beds. Nat. Hist. Soc.*, 1885.
Additional Crytogamic notes on South Bedfordshire; *Ibid.*
Notes on Characeae gathered in Bedfordshire; *Ibid.*
Pinguicula vulgaris and Carum carvi in South Beds.; *J. Bot. XXIV*, 1886.
Notes on the flora of South Bedfordshire; *Ibid.*, *XXVII*, 1889.
Flora of the Ivel Valley, Bedfordshire; *Ibid.*
Bedfordshire and its Droseras; *Ibid.*, *XXVIII*, 1890.
Bedfordshire Plants; *Ibid.*, *XXXV*, 1897.
The Wild Flowers of Bedfordshire, a series of articles reprinted from *Bedfordshire Advertiser*, 1897–8.[1]

---

[1] Only two copies are known of this. Both are in Luton Museum and one is annotated by C. Crouch. There is a similar list of mosses of which there is only one copy and the origin of which is unknown.

On the shrinkage of the sources of the Upper Lea; *Trans. Herts. Nat. Hist. Soc.* 1903.

Mosses (Musci), scale mosses and liverworts (Hepaticae), and stoneworts (Characeae) in G. C. Druce: Botany, *Victoria County History* (Bedfordshire), 1904.

Field Flowers of Bedfordshire, 1911.

[Saunders, in addition to the above, wrote a number of papers on Mycetozoa, Witches' Broom, Geology, etc.]

Sinclair, G. *Hortus Gramineus Woburnensis*, 1816 (there were many subsequent editions containing additional material).

Smith, W. G. Bedfordshire plants; *J. Bot. XVI*, 1876, *XXIII*, 1885.

Verdcourt, B. The habitat of Cuscuta europaea L. in Britain; *Watsonia I*, 1950.

Watson, W. C. R. The brambles of Bedfordshire; *Journ. Beds. Nat. Hist. Soc. II*, 1948.

Wilmott, A. J. Orchis hircina in Bedfordshire; *J. Bot. LXX*, 1932.

Young, D. P. Studies in the British Epipactis; *Watsonia I*, 1949, *II*, 1952.

PRINCIPAL PUBLISHED WORKS DEALING WITH THE FLORA OF NEIGHBOURING COUNTIES

Webb, R. H. and Coleman, W. H. *Flora Hertfordiensis*, 1839.

Pryor, A. R. *A Flora of Hertfordshire* (edited by B. Daydon Jackson), 1887.

Hopkinson, J. Botany, *Victoria County History* (Hertfordshire), 1905.

The Natural History of the Hitchin Region (edited by R. L. Hine), 1934.

Transactions of Hertfordshire Natural History Society, 1879 continuing.

Druce, G. C. Botany, *Victoria County History* (Buckinghamshire), 1905.

Flora of Buckinghamshire, 1926.

Babington, C. C. *Flora of Cambridgeshire*, 1860.

Godwin, H. Botany, *Victoria County History* (Cambridgeshire), 1938.

Evans, A. H. *Flora of Cambridgeshire*, 1939.

Druce, G. C. Botany, *Victoria County History* (Huntingdonshire) 1926.

Dony, J. G. A contribution to the Flora of Huntingdonshire; *Watsonia I*, 1950.

Reports of Huntingdonshire Fauna and Flora Society, 1948 continuing.

Druce, G. C. Botany, *Victoria County History* (Northamptonshire) 1902.

Flora of Northamptonshire, 1930.

Journal of Northamptonshire Natural History Society, 1880 continuing.

MANUSCRIPTS RELATING TO THE FLORA OF
BEDFORDSHIRE

Abbot, C. Plantae Bedfordiensis, 1795. Linnean Society Library.
Hints to the Bedfordshire farming interest (prepared for Batchelor
T.: *A General View of Agriculture in the County of Bedford*). Bedford-
shire County Record Office, D.D.X. 135/4.
Correspondence with D. and S. Lysons. Brit. Mus. Add. MSS.
9804.
Annotated copy of *Flora Bedfordiensis*. Luton Museum.

Marsh, T. O. Collections for a biography of persons connected with
Bedfordshire. Brit. Mus. Add. MSS. 21. 067.
'Smith Correspondence'. This contains letters from C. Abbot,
J. Hemsted and T. O. Marsh. Linnean Society Library.

[Squire, E. O.]. List of indigenous plants found at Basmead
Manor (*c.* 1860). Bedfordshire County Record Office, D.D.S.Q.
185.

Crouch, C. Annotated copy of *Field Flowers of Bedfordshire*. Luton
Museum.
Annotated copy of *Wild Flowers of Bedfordshire*. Luton Museum.

Hamson, J. Flora of Bedfordshire, 5 volumes. Library of Bedford
Natural History and Archaeological Society.

'Marquand List'. A list of Bedfordshire bryophytes, dated 1911,
found in the papers of C. V. B. Marquand. I have checked the
writing with that of Saunders, Hamson, H. N. Dixon and
Marquand and find that it was not written by any of them.
Luton Museum.

Little, J. E. Field diaries. Hitchin Library.
Annotated copy of Field Flowers of Bedfordshire. Hitchin
Library.

Tremayne, L. J. Annotated copy of *London Catalogue*, Tenth Edition.
Luton Museum.

'Wild Flower Society Diary'. A list of Bedfordshire records made by
W.F.S. members. Mrs. R. H. Mortis (Hertford).

Laflin, T. Some observations on the mosses on a boulder-clay area
in East Bedfordshire and West Cambridgeshire, 1950. Copy in
Luton Museum.
Bedfordshire mosses, 1952. T. Laflin Library.

HERBARIA CONTAINING MAINLY BEDFORDSHIRE
MATERIAL

The herbaria which have been searched by the author for records
are shown *.

*Abbot, C. (*c.* 1794–1817), four volumes containing about 800
specimens. Arrangement: Linnean. Specimens not localized.
Placed on long loan at Luton Museum by Rev. H. H. Longuet
Higgins.

*Crouch, W. (1) (1842-5), four volumes containing about 600 specimens. Arrangement: alphabetical order of genera. Referred to in text as W.C. Herb. Luton Museum.

*Crouch, W. (2) (1842-5), one volume containing about 200 specimens. Arrangement: alphabetical order of genera. Referred to in text as Herb. Miss Crouch, as it is supposed to have been made by Crouch for his cousin. Luton Museum.

*McLaren, J. (1) (1864), four volumes containing about 700 specimens. Arrangement: Bentham. Stations given for rarer species. Luton Museum.

*McLaren, J. (2) (c. 1864-1888), details as above. British Museum (Natural History).

*Saunders, J. (1878-1926), fifteen volumes containing about 1,000 specimens. Arrangement: London Catalogue, Ninth Edition. Contains also specimens sent by J. McLaren and C. Crouch. Luton Museum.

*Bedfordshire Natural History Society (c. 1879-1888). Loose sheets, about 300 specimens, now packed in boxes and arranged by author to follow British Plant List. Contains specimens by J. Hamson, J. McLaren and J. Saunders. The original collection was large and must have been valuable, but was rendered almost worthless by insect infestation. Bedford Public Library.

*Cooper, E. (1856), a small collection of mosses including some collected by A. O. Black. Bedford Public Library.

*Day, G. H. (1901-1908), loose sheets, about 400 specimens. Arrangement: Hooker. Collection made mainly in the neighbourhood of Roxton. Library of Bedford Natural History and Archaeological Society.

*Luton Museum (1935 continuing). About 3,000 specimens of flowering and allied plants on loose sheets housed in cabinets and boxes. In course of arrangement in sequence of this Flora. Contains specimens by Mrs. Vipian, J. Saunders and C. Crouch in addition to those of author and his contemporaries. The bryophyte collection includes that of Saunders, with additions by P. Taylor. Luton Museum.

OTHER HERBARIA CONTAINING BEDFORDSHIRE MATERIAL

*J. Pollard: in course of arrangement to follow Flora of the British Isles. Hitchin Library.

Letchworth Museum (c. 1925 continuing): a collection of plants made within a twelve-mile radius of Letchworth.

*H. C. Watson: contains a number of specimens sent by Bedfordshire correspondents. Kew.

British Museum (Natural History): contains collections of H.Brown and D. M. Higgins and a number of specimens from miscellaneous sources.

Royal Botanic Gardens, Kew: contains bryophyte collection of
H. N. Dixon and specimens collected by E. Milne-Redhead and
A. B. Sampson as well as many specimens from miscellaneous
sources.

Cambridge University: contains J. E. Little's collection and T.
Laflin's Bedfordshire bryophyte collection.

Birmingham University: contains specimens collected by W.
Hillhouse.

J. E. Lousley (private herbarium): contains S. A. Chambers's
specimens.

P. Taylor (private): a good collection of ferns and fern allies.

THE FLORA OF BEDFORDSHIRE

CHAROPHYTA

Saunders was especially interested in this small group of plants and little work has been done on them recently. I wish to thank G. O. Allen who has named my gatherings.

CHARACEAE

Nitella Ag.

Nitella opaca Ag.
Rivers, ponds, etc.   Native
Rare: recorded for all neighbouring counties.
IVEL: Reed Pond, Sundon†; Pulloxhill; *F.F.B.*: Beckeringspark; *C.C.*: Ridgmont*; Maulden.
LEA: sources of the River Lea; *J.S. in Bot. Rec. Club 1881–2 Rep.*: Great Bramingham; *F.F.B.*
First record: Saunders, 1882.

N. mucronata Miq.
Rivers, ponds, etc.   Native
Rare: recorded for Cambs.
OUSE: near Bedford†*; *J.S. in Bot. Rec. Club 1881–2 Rep.*
IVEL: river, near Sandy; *F.F.B.*
First record: Saunders, 1882.

'I have found our Bedford river abounding in *Chara flexilis*, E. Bot. tab. 1070' wrote Abbot to J. E. Smith, 3 Sept. 1802. He may have intended *N. flexilis* Ag. which is recorded for Herts., Bucks. and Northants., but without further evidence the record cannot stand.

Tolypella Leonh.

Tolypella intricata Leonh.
Ponds   Native
Rare: recorded for all neighbouring counties.
OUSE: Lidlington; *C.C.*
OUZEL: The Litany; *F.F.B.*
IVEL: Sundon†; *F.F.B.*: Beckeringspark*; Pulloxhill*; *C.C.*
LEA: Little Bramingham; *J.S. in Bot. Rec. Club 1881–2 Rep.*
First record: Saunders, 1882.

T. glomerata Leonh.
Ponds   Native
Rare: recorded for Bucks., Hunts. and Northants.
OUSE: Bedford; *H. Davis in J. Bot. 1886*, 3.
OUZEL: near Woburn; *V.C.H.*
LEA: Icknield Hole, near Leagrave; *F.F.B.*
First record: H. Davis, 1886.

Chara L.

Chara vulgaris L.

Ponds, gravel pits, etc.   Native

Frequent: *E. Bot.* 336, original drawing is labelled 'July 8 1795, Revd. C. Abbot, Bedford'. Recorded for all neighbouring counties.

OUSE: Bedford; *J.S. in Bot. Rec. Club 1884–6 Rep.*: near Goldington†; *H. Davis in Trans. Beds. Nat. Hist. Soc. 1885*: Eaton Socon*; Sharnbrook; Eastcotts; Felmersham.

KYM: Melchbourne.

OUZEL: ponds near Watling Street†; *J.S. in Trans. Beds. Nat. Hist. Soc. 1885*: Totternhoe†; *F.F.B.*: Leighton Buzzard.

IVEL: Sundon; *F.F.B.*: College Farm, Pegsdon; *J.E.L., Diary*: Henlow*; Maulden*; Arlesey; Southill.

LEA: New Bedford Road; *J.S. in Bot. Rec. Club 1881–2 Rep.*: Bramingham*; Biscot; Limbury; *F.F.B.*

First record: common, Abbot, 1798; more certainly Saunders, 1882.

　　Var. **longibracteata** Kütz.

OUSE: Tempsford*; knot hole, Stewartby.

OUZEL: Leighton Buzzard.

LEA: river, near Biscot; *J.S. in Bot. Rec. Club 1881–2 Rep.*: New Bedford Road†; *Herb. J.S.*

　　Var. **papillata** Wallr.

NENE: Hinwick*; *the author in Watsonia I.(1950)* 262.

C. hispida L.

Ponds, etc.   Native

Rare: recorded for all neighbouring counties except Herts.

OUSE: Bedford; *Abbot's annotated F.B.*

KYM: Melchbourne*.

OUZEL: Totternhoe†*; *J.S. in Bot. Rec. Club 1881–2 Rep.*

IVEL: Maulden; *C.C.*

LEA: Limbury; *J.S. in Bot. Rec. Club 1881–2 Rep.*: Luton Hoo Lake; *F.F.B.*

First record: Abbot in letter to J. E. Smith, 2 Sept. 1802; more certainly Saunders, 1882.

C. globularis Thuill.

　　C. fragilis var. *hedwigii* (Bruz.) Kütz.

Ponds, etc.   Native

Not infrequent: recorded for Bucks. and Northants.

OUSE: Cardington Mill*; *E.M.-R.*!: Pavenham.

KYM: Pertenhall*.

LEA: Limbury†*; *J.S. in Trans. Beds. Nat. Hist. Soc. 1885*.

　　Var. **capillacea** (Thuill.) Zanev. (*C. fragilis* Desv.)

Many of the following records of *C. fragilis* may refer to *C. globularis* or *C. delicatula*. 'C. fragilis' is recorded for all neighbouring counties except Herts.

OUSE: near Bedford; *F.F.B.*

OUZEL: near Watling Street; Totternhoe†; *J.S. in Bot. Rec. Club 1881–2 Rep.*: near Battlesden; *F.F.B.*

IVEL: Flitwick†; *J.S. in Bot. Rec. Club 1881–2 Rep.*: Sharpenhoe; Pulloxhill; Parkfield Pond, Sundon; *F.F.B.*: Southill Park; *J.E.L., B.E.C.*

LEA: fountain, Luton Hoo gardens; *F.F.B.*

COLNE: Pepperstock*; *J.S.*

First record: for the aggregate, Saunders, 1882.

C. delicatula Ag.

Ponds   Native

Rare: recorded for Bucks. and Northants.

OUZEL: Rushmere*; *E.M.-R. and the author in B.E.C. 1943–4 Rep.* 814.

C. contraria Kütz. is recorded for v.c. 30 in *Comital Flora*, and in *V.C.H.* on the authority of Saunders from Totternhoe, but I cannot find the basis of the records. It is recorded for all neighbouring counties except Herts.

C. tomentosa L. was recorded in error from a spring near Stevington Church by Abbot in *Plantae Bedford.*, and from Clapham Springs in *Flora Bedfordiensis*.

BRYOPHYTA[1]

HEPATICAE

by P. Taylor

The Liverworts, for the most part, are plants of the more mountainous areas, and the number of species which could be expected to occur in Bedfordshire is small. The lists compiled by Abbot and Saunders were incomplete as is shown by recent work, done mainly in the south of the county. It is certain that still more species remain to be recorded and that the full distribution of those recorded is not yet known.

Some of the species recorded by Abbot but not found subsequently may have been listed by him in error. The same may apply to other species recorded for the county in the British Bryological Society's *Census Catalogue* but for which no authoritative record has yet been found. I wish to thank E. C. Wallace for lending me his annotated copy of the *Census Catalogue*, from which the distribution in neighbouring counties has been compiled.

RICCIACEAE

Riccia L.

Riccia glauca L.
Arable fields   Native
Local, but possibly overlooked. Recorded for Bucks. and Cambs.
COLNE: between Woodside and Caddington*; *P.T. in Trans. Brit. Bry. Soc.* 1949.

R. sorocarpa Bisch.
Arable fields, woodland rides, etc.   Native
More frequent than *R. glauca*, but also probably overlooked.
Recorded for Herts. and Bucks.
IVEL: Sutton; *T.L.*
OUZEL: King's Wood; field between Milton Bryan and Woburn.
LEA: Kidney Wood*.
COLNE: between Woodside and Caddington*; Deadmansey Wood.
First record: P.T. in *Trans. Brit. Bry. Soc.* 1949.

R. crystallina L.
Muddy edges of ponds, etc.   Native
Rare, but usually abundant when it occurs. Recorded for Herts., Cambs. and Northants.
OUZEL: Stockgrove; *R. D. Meikle*: Lower Drakelow Pond*.
First record: P.T. in *Trans. Brit. Bry. Soc.* 1949.

R. fluitans L.
Ricciella fluitans (L.) A. Braun
Floating on still water   Native
Rare or extinct. Recorded for Herts., Bucks. and Cambs.
LEA: Luton Hoo Park*; *J.S. in Victoria County History*, 1904.

[1] For the Bryophytes the exclamation mark refers to P. Taylor in the *Hepaticae* and to T. Laflin in the *Musci*. The source of Saunders's records are not given as they are supported by satisfactory herbarium specimens in every case.

MARCHANTIACEAE
Conocephalum Wigg.
Conocephalum conicum (L.) Dum.
Marchantia conica L.
Damp stonework and earth near water   Native
Not uncommon on weirs in the Ouse, rare elsewhere. Recorded for all neighbouring counties except Hunts.
OUSE: Clapham between Woods, *Abbot's annotated F.B.*: common on weirs in the river.
OUZEL: Sewell*; *Mrs. Twidell.*
First record: Abbot in letter to J. E. Smith, 3 Feb. 1799.

Lunularia Adans.
Lunularia cruciata (L.) Dum.
Damp stonework and earth   Native
Probably more common than the records suggest. Recorded for all neighbouring counties except Northants.
IVEL: Flitton; *E. C. Wallace in B.B.S. 1949 Rep.*: Steppingley*.
LEA: Wardown Park.
First record: E. C. Wallace, 1949.

Marchantia L.
Marchantia polymorpha L.
Damp soil, marshes, streamsides, etc.   Native
Probably not uncommon. Recorded for all neighbouring counties.
H.S. 40.
OUZEL: New Wavendon Heath, v.c. 24 [Beds.].
IVEL: Flitwick Moor!*; *J.S.*
LEA: Deodorizing Works, Luton*; *J.S.*
First record: Abbot, 1798.

JUNGERMANNIACEAE: ANACROGYNAE
Aneura Dum.
Aneura pinguis (L.) Dum.
Jungermannia pinguis L.
Marshes, ditches, etc.   Native
Rare, but considered common by Abbot. Recorded for Herts., Cambs. and Northants.
IVEL: Barton Springs*; *J.S.*
First record: Abbot, 1798.

A. multifida (L.) Dum.
Jungermannia multifida L.
Woods   Native
Considered common by Abbot but not recorded since. Recorded for Herts. and Bucks.
First record: Abbot, 1798.

Metzgeria Raddi
Metzgeria furcata (L.) Dum.
Jungermannia furcata L.
Trunks of trees   Native
Common: recorded for all neighbouring counties except Hunts.
OUZEL: Hockliffe*; *J.S.*
IVEL: Ampthill Park, where it was also found by Saunders*; *Plantae Bedford.*: Harlington*; *J.S.*: Fancott*; Barton*.
LEA: Limbury*; *J.S.*: foot of Warden Hill; *T.L.*
First record: Abbot, 1795.

Pellia Raddi
Pellia epiphylla (L.) Corda
Jungermannia epiphylla L.
Damp soil, ditches, etc.   Native

Local: recorded for all neighbouring counties.
OUZEL: King's Wood.
IVEL: Ampthill; Potton; *F.B.*: Flitwick Wood*; *J.S.*
First record: Abbot, 1798.

P. fabbroniana Raddi
 P. calycina Tayl.
Damp calcareous soil Native
Rare: recorded for all neighbouring counties except Hunts.
IVEL: Markham Hills*; *J.S.*: Barton.
First record: Saunders in *Victoria County History*, 1904.

Blasia pusilla L. is listed for the county in the *Marquand List* and in the *Census Catalogue of British Hepatics* probably in error for the following species.

Fossombronia Raddi
Fossombronia pusilla (L.) Dum.
 Jungermannia pusilla L.
Damp soil, woodland rides, etc. Native
Rare: recorded for Herts.
OUSE: Hazelwood Lane; *F.B.*
LEA: Kidney Wood*.
First record: Abbot, 1798.

F. wondraczeki (Corda) Dum.
Damp soil, fallow fields, woodland rides, etc. Native
Local: recorded for Bucks.
OUZEL: King's Wood*.
IVEL: between Woodside and Caddington*; Deadmansey Wood.
First record: P.T. in *Trans. Brit. Bry. Soc.* 1949.

JUNGERMANNIACEAE: ACROGYNAE
Alicularia Corda
Alicularia scalaris (Schrad.) Corda
Damp banks
Rare: recorded for Herts. and Bucks.
OUZEL: near Heath and Reach; *D. A. Reid*, 1952.
Lophozia Dum.
Lophozia turbinata (Raddi) Steph.
Damp exposed chalk Native
Common in suitable habitats. Recorded for Herts., Bucks. and Cambs.
OUZEL: Sewell*.
IVEL: Barton*; Pegsdon; Streatley.
First record: uncertain.

L. ventricosa (Dicks.) Dum.
Woods Native
Rare: recorded for Herts. and Bucks.
OUZEL: King's Wood; *E. C. Wallace*.

L. bicrenata (Schmid.) Dum.
Dry soil Native
Rare: recorded for Herts.
OUZEL: New Wavendon Heath, v.c. 24 [Beds.].
COLNE: Woodside*; *J.S.*
First record: Saunders in *Victoria County History*, 1904.
Sphenolobus (Lindb.) Steph.
Sphenolobus exsectiformis (Breidl.) Steph.
Damp woods Native
Rare: recorded for no neighbouring county.
OUZEL: King's Wood; *E. C. Wallace*.
Plagiochila Dum.
Plagiochila asplenioides (L.) Dum.
 Jungermannia asplenioides L.
Woods Native

Common: recorded for all neighbouring counties.
OUSE: Clapham Park; *Plantae Bedford*.
OUZEL: King's Wood.
IVEL: Streatley*; *J.S.*: Potton Wood; *T.L.*: Barton*.
CAM: Cockayne Hatley Wood; *T.L.*
COLNE: Deadmansey Wood.
First record: Abbot, 1795.

Lophocolea Dum.
Lophocolea bidentata (L.) Dum.
Jungermannia bidentata L.
Damp woods   Native
Common: recorded for all neighbouring counties except Hunts.
OUSE: Bromham; Oakley; Ravensden; *Marquand List*.
OUZEL: King's Wood.
IVEL: Barton*.
LEA: Marslets*; Luton Hoo*; Chiltern Green*; Daffodil Wood*; *J.S.*
COLNE: Woodside*; Pepperstock*; *J.S.*: Folly Wood; *T.L.*: Deadmansey Wood.
First record: Abbot, 1798.

L. heterophylla (Schrad.) Dum.
Damp woods   Native
Common: recorded for all neighbouring counties.
H.S. 78.
OUZEL: Woburn; King's Wood.
IVEL: Markham Hills*; *J.S.*: Barton; Fancott.
LEA: Luton Hoo*; Daffodil Wood; *J.S.*
COLNE: Deadmansey Wood; Folly Wood.
First record: Saunders in *Victoria County History*, 1904.

Chiloscyphus Corda
Chiloscyphus polyanthus (L.) Corda
Damp woods   Native
Rare: the record of *Jungermannia viticulosus* in *Flora Bedfordiensis* may refer to this species. Recorded for all neighbouring counties except Hunts.
OUZEL: King's Wood; *E. C. Wallace*.

Cephalozia Dum.
Cephalozia bicuspidata (L.) Dum.
Jungermannia bicuspidata L.
Damp woods   Native
Rare: but considered common by Abbot. Recorded for all neighbouring counties except Hunts.
OUZEL: King's Wood; *E. C. Wallace*.
First record: Abbot, 1798.

Cephaloziella starkii (Funck) Schiffn. is listed for v.c. 30 in *Census Catalogue of British Hepatics*, but it is not known on what authority. Recorded also for Herts. and Northants.

Calypogeia Raddi
Calypogeia fissa (L.) Raddi
Jungermannia fissa L.
Damp woods   Native
Rare: recorded for all neighbouring counties except Cambs.
OUZEL: King's Wood; *E. C. Wallace*.
IVEL: Ampthill Moor; *F.B.*
First record: Abbot, 1798.

Lepidozia Dum.
Lepidozia reptans (L.) Dum.
Damp woods on sandy soils   Native
Local: recorded for Herts., Bucks. and Northants.
OUZEL: Aspley Wood*; King's Wood*; *J.S.*
First record: Saunders in *Victoria County History*, 1904.

Diplophyllum Dum.
Diplophyllum albicans (L.) Dum.
Damp woods   Native
Rare: recorded for Herts., Bucks. and Northants.
OUZEL: King's Wood*; *J.S.*
First record: Saunders in *Victoria County History*, 1904.

Scapania Dum.
Scapania nemorosa (L.) Dum.
Damp woods   Native
Rare: recorded for Herts., Bucks. and Northants.
OUZEL: King's Wood; *N. Y. Sandwith.*
First record: N. Y. Sandwith in *Trans. Brit. Bry. Soc.* 1949.

Radula Dum.
Radula complanata (L.) Dum.
Jungermannia complanata L.
Damp tree-roots   Native
Local, considered common by Abbot. Recorded for all neighbouring counties.
OUSE: Bolnhurst; Ravensden; *Marquand List.*
OUZEL: near Hockliffe*; *J.S.*: King's Wood; *E. C. Wallace.*
LEA: Dallow Lane*; *J.S.*
First record: Abbot, 1798.

Madotheca Dum.
Madotheca platyphylla (L.) Dum.
Jungermannia platyphylla L., *Porella platyphylla* (L.) Lindb.
Damp walls, tree roots, etc.   Native
Local, considered common by Abbot. Recorded for all neighbouring counties.
OUSE: Biddenham; Oakley; *Marquand List*: Turvey; *E. C. Wallace.*
IVEL: Barton.
LEA: Dallow Lane*; Luton Hoo*; *J.S.*
COLNE: Pepperstock*; Chaul End*; Caddington; *J.S.*: Long Wood.
First record: Abbot, 1798.

Lejeunea Lib.
Lejeunea cavifolia (Ehrh.) Lindb.
Damp woods   Native
Rare: recorded for Bucks. and Cambs.
OUZEL: King's Wood; *E. C. Wallace.*

Frullania Raddi
Frullania dilatata (L.) Dum.
Jungermannia dilatata L.
Tree trunks   Native
Local: Abbot considered it to be common. Recorded for all neighbouring counties
except Northants.
OUSE: not uncommon around Bedford; Clapham; Bolnhurst; Ravensden;
Marquand List.
OUZEL: Hockliffe*; *J.S.*
IVEL: Toddington*; *J.S.*: Flitwick; *Marquand List.*
CAM: trunks of ash trees, Battle Brook; *T.L.*
LEA: Luton*; Limbury*; Biscot*; *J.S.*
First record: Abbot, 1798.

'*Jungermannia tamariscifolia*' was listed as common by Abbot. He probably intended
J. tamarisci L. (*Frullania tamarisci* (L.) Dum.), which has not been recorded subsequently, but is recorded for Herts. and Bucks.

ANTHOCEROTACEAE
Anthoceros L.
Anthoceros punctatus L.
Fallow fields   Native
Rare: recorded for Herts.
COLNE: Woodside*; *P.T. in Trans. Brit. Bry. Soc.* 1949.

MUSCI
by T. Laflin

The moss flora of Bedfordshire is not rich. The low altitude and lack of coastline precludes the occurrence of true mountain and maritime species. Bogs are few and limited in size. Although the range of species is limited there is much of interest, particularly in their distribution. Of the 210 species recorded for the county, 175 have been seen since 1947.

Saunders did valuable work in recording mosses, the more useful in that we have still his specimens. Many of his good stations have been destroyed or modified by time, but a large number of the mosses still grow where he found them.

Recent field work has been intensive. Alone, or in company with other botanists, I have visited many stations and have recorded in detail the species from 228 stations in varied habitats: 38 in deciduous woodland, 9 in coniferous woodland, 8 heaths, 16 scrub, 15 rough pastures, including chalk downs, 15 meadows, 19 arable fields, 16 hedgerows, 16 bogs and marshes, 23 rivers, streams and ponds, 25 quarries, pits and cuttings, and 28 miscellaneous habitats, including walls. J. G. Dony has collected material from 29 Habitat Studies not visited by me. J. P. M. Brenan, E. C. Wallace and P. Taylor, among others, have provided many valuable records.

SPHAGNACEAE
Sphagnum L. *Bog Mosses*

Sphagnum palustre L.
S. cymbifolium Ehrh.
Bogs   Native
Common in the bogs on the Lower Greensand. Recorded for all neighbouring counties except Cambs. and Hunts.
Ouzel: Mermaid's Pond!*; *J.S.*: New Wavendon Heath, v.c. 24 [Beds.]; *T.L.*
Ivel: Flitwick Moor!*; *J.S.*
H.S. 35, 36, 40.
First record: Abbot, 1798.

S. squarrosum Pers. ex Crome
Bogs   Native
Probably frequent in the 19th century but now extinct. Recorded for Herts.
Ouzel: Mermaid's Pond*; *J.S.*
Ivel: Flitwick Moor*; Westoning Moor*; *J.S.*
First record: Saunders in *Trans. Beds. Nat. Hist. Soc.* 1882.
　　Var. **laxum** Braithw.
　　S. teres Ångstr. var. *laxum* (Braithw.) Dix., *S. fimbriatum* Wils. var. *robustum* Braithw.
Ivel: Flitwick Moor*; *J.S. in Trans. Beds. Nat. Hist. Soc.* 1885.
　　Var. **imbricatum** Schimp.
Ivel: Flitwick Moor*; *J.S. in Trans. Beds. Nat. Hist. Soc.* 1885.

S. recurvum Beauv.
S. intermedium Hoffm.
Bogs   Native
Rare: the Warnstorfian segregate, **S. amblyphyllum** Russ., appears at New Wavendon Heath; *T.L.* Recorded for Bucks.
Ouzel: Mermaid's Pond*; *J.S.*: New Wavendon Heath, v.c. 24 [Beds.], see above.

PLATE 23   GALLEY HILL, STREATLEY

Maulden Firs (Habitat Study 70) and scrub (Habitat Study 67) are in the foreground and Galley Hill (Habitat Studies 66a and 66b) in the right-centre of the photograph

PLATE 24   GREAT EARTH-NUT (*Bunium bulbocastanum*) on the outskirts of Luton

PLATE 25   WELL HEAD, LEAGRAVE COMMON (Habitat Study 84)

IVEL: Flitwick Moor, where it is frequent!*; *J.S.*
First record: Saunders in *J. Bot. 1884.*

S. cuspidatum Ehrh. ex Hoffm. emend.
Bogs   Native
Frequent, usually in standing water, in bogs on the Lower Greensand. Recorded for Bucks.
H.S. 35.
OUZEL: Mermaid's Pond!*; *J.S.*: New Wavendon Heath, v.c. 24 [Beds.]; *T.L.*
IVEL: Flitwick Moor!*; *J.S.*
First record: Saunders in *Trans. Beds. Nat. Hist. Soc.* 1885.

S. subsecundum Nees   Native
I have found this locally abundant at New Wavendon Heath*. In addition to the species in its broad limits as conceived by Richards and Wallace there are many variants. I am unable to place these in the varieties adopted by Richards and Wallace as I have not had access to the papers on which their work is based. The Warnstorfian segregates I found at New Wavendon Heath were **S. obesum** (Wils.) Warnst., **S. subsecundum** Nees *sensu stricto*, **S. inundatum** Warnst. and **S. auriculatum** Schimp. The last named was by far the most common.

S. fimbriatum Wils.
Bogs   Native
Common in bogs on the Lower Greensand. Var. **validius** Card. is the common form, but vars. **intermedium** Russ. and **tenue** Grav. are also found at Flitwick. Not recorded for any neighbouring county.
H.S. 40.
OUZEL: Mermaid's Pond!*; *J.S.*: King's Wood*; *P.T.*
IVEL: Flitwick Moor!*; *J.S.*: Westoning Moor*; *J.S.*
First record: Saunders in *Trans. Beds. Nat. Hist. Soc.* 1882.

S. girgensohnii Russ.
Bogs   Native
Rare: recorded for Bucks.
OUZEL: Mermaid's Pond*; *J.S.*
IVEL: Flitwick Moor!*; *J.S.*
First evidence: Saunders in Herb. Luton Museum, 1882.

S. rubellum Wils. (*S. acutifolium* var. *rubellum* (Wils.) Russ.) was recorded by Saunders, with no details, in *V.C.H.* I have found no specimen to support the record. Recorded for Bucks.

S. nemoreum Scop.
  S. acutifolium Ehrh.
Bogs   Native
Probably extinct: recorded for Herts., Bucks. and Northants.
OUZEL: Aspley Heath Wood*; *J.S. in J. Bot. 1884.*

S. plumulosum Roell
  S. acutifolium var. *subnitens* (Russ. & Warnst.) Dix.
Bogs   Native
Probably extinct: recorded for Bucks.
IVEL: Flitwick Moor*; *J.S. in J. Bot. 1884.*

POLYTRICHACEAE
Atrichum Beauv.

Atrichum undulatum (Hedw.) Beauv.   *Undulate Hair Moss*
  Catharinea undulata (Hedw.) Web. & Mohr.
Woods, heaths and banks   Native
Common in woodland, frequent in scrub and on heaths, on acid or neutral soils. Occasional in woods elsewhere and in sandpits. Recorded for all districts and neighbouring counties.
H.S. 18, 19b, 23, 33–34b, 54, 69, 76, 77a.
First record: Abbot, 1798, as *Bryum undulatum.*

L

Polytrichum Hedw.   *Hair Mosses*

Polytrichum nanum Hedw.
  Pogonatum nanum (Hedw.) Beauv., *P. subrotundum* (Sm.) Lindb.
Heaths   Native
Occasional and now appearing also as a colonist. Recorded for all neighbouring
counties except Hunts.
OUZEL: Aspley*; *J.S.*
IVEL: Flitwick*; *J.S.*: Sandy Heath; Sutton Fen; *T.L.*
COLNE: Pepperstock*; Woodside*; Slip End*; Caddington*; *J.S.*
First record: Abbot, 1798, as *Polytrichum subrotundum*.

P. aloides Hedw.
  Pogonatum aloides (Hedw.) Beauv.
Heaths   ? Native
Rare on sandy heaths and in sandpits. It is most frequent on steep banks and
vertical faces and is possibly dependent on frequent erosion for survival. Re-
corded for all neighbouring counties except Hunts.
OUZEL: Aspley Heath*; King's Wood*; *J.S.*: New Wavendon Heath, v.c. 24
[Beds.]*; *P.T.*
IVEL: Flitwick Moor*; *J.S.*: Southill Park*; *J.McL.*: old sandpits, Clophill; *T.L.*
First record: Saunders in *J. Bot. 1884*.

P. urnigerum Hedw.
  Pogonatum urnigerum (Hedw.) Beauv.
Heaths and disused sandpits   Colonist
Rare and possibly extinct. Recorded for Northants.
IVEL: Southill*; *J.S.*: Shefford; *Marquand List*.
LEA: Luton Hoo*; *J.S.*
First record: Saunders in *J. Bot. 1884*.

P. piliferum Hedw.
Dry heaths and old sandpits   Native
Frequent on the Lower Greensand in the Ouse, Ouzel and Ivel districts. Recorded
for all neighbouring counties except Hunts.
H.S. 24, 47.
First record: Saunders in *Trans. Beds. Nat. Hist. Soc.* 1882.

P. juniperinum Hedw.
Heaths, woodland rides and old sandpits   Native
Common on acid soils and often locally dominant, especially on the Lower
Greensand. Recorded for all districts except Nene, Kym and Cam, and for all
neighbouring counties.
H.S. 22a, 22b, 27, 46, 47.
First record: Abbot, 1798.

P. gracile Sm.
Heaths, woods and old sandpits   Native
Occasional but widely distributed on the Lower Greensand, rare on the Clay-
with-Flints. Recorded for the Ouzel, Ivel and Colne districts and for Bucks.,
Cambs., Hunts. and Northants.
First evidence: Saunders in Herb. Luton Museum, 1887.

P. formosum Hedw.
  P. attenuatum Sm.
Heaths, woods and old sandpits   Native
Locally frequent on acid soils. Recorded for all districts except Nene, Kym and
Cam, and for all neighbouring counties except Hunts.
H.S. 19, 23, 30.
First record: Saunders in *Trans. Beds. Nat. Hist. Soc.* 1882.

P. commune Hedw.
Bogs   Native
Common in the bogs on the Lower Greensand. Recorded for all neighbouring
counties except Hunts.
H.S. 35, 40.

Ouzel: Mermaid's Pond!*; *J.S.*: New Wavendon Heath, v.c. 24 [Beds.]; *T.L.*
Ivel: Flitwick Moor!*; Sutton Fen; *J.S.*
First record: Abbot, 1798.

FISSIDENTACEAE
Fissidens Hedw.

Fissidens pusillus Wils. ex Milde
Stonework   Wall denizen
Rare and apparently limited to sandstone on locks and bridges of the Ivel. Recorded for Bucks. and Northants.
Ivel: Blunham Bridge (1950); lock, Ivel Navigation, Shefford; *T.L.*
    Var. **madidus** Spruce (? *F. minutulus* Sull. ex Braithw.)
Chalk blocks   Denizen
Occasional, but widely distributed. It grows usually in partial shade.
Ivel: Barton Cutting (1949)*; *P.T.*: Barton Hills; *T.L.*
Colne: Long Wood; Deadmansey Wood; *T.L.*

F. viridulus (Web. & Mohr) Wahlb. was recorded with no details by S. H. Adams in *Trans. Beds. Nat. Hist. Soc.*, 1882.

F. bryoides Hedw.
Woods, old scrub and banks   Native
Common throughout the county. It is also found occasionally in stubbles on calcareous soils, in sandpits, and on sandstone exposures on the Lower Greensand. Recorded for all districts and neighbouring counties.
H.S. 5, 76, 78.
First record: Abbot, 1798.

F. incurvus Starke ex Web. & Mohr
Margins of woods and banks   Native
Rare: recorded for Herts., Bucks., Cambs. and Northants.
Ivel: Barton Cutting; *J. P. M. Brenan.*
Cam: margin of Walk Wood, Cockayne Hatley; *T.L.*
First record: without locality, S. H. Adams in *Trans. Beds. Nat. Hist. Soc.* 1882.

F. crassipes Wils.
Stonework, wood, and tree roots in water   Wall denizen
Common on masonry, less so on logs, posts and tree roots, in and near water along the Ouse from Turvey to Great Barford, and the Ivel Navigation from Shefford to Stanford. Recorded for Herts., Bucks., Cambs. and Northants.
First record: Saunders in *Trans. Beds. Nat. Hist. Soc.* 1885.

F. exilis Hedw.
Wet rides in woods   Native
Rare: recorded for Herts., Cambs. and Northants.
Ouzel: King's Wood (1949)*; *P.T.*

F. taxifolius Hedw.
Woods, old scrub, banks, rough pasture and stubble   Native
Very common, especially on clay and chalk soils. It is rare along the flood valleys of the Ouse and Ivel and almost absent on the Lower Greensand. Recorded for all districts and neighbouring counties.
H.S. 1a–2b, 5, 34a, 34b, 52, 57, 62.
First record: Abbot, 1798, as *Hypnum taxifolium.*

F. cristatus Wils.
    F. decipiens De Not.
Chalk pastures   Native
Rare: recorded for Herts., Bucks., Cambs. and Northants.
Ivel: Pegsdon; *E. C. Wallace*: Barton Cutting; *J. P. M. Brenan*: Sharpenhoe*;
P.T.: Barton; *T.L.*
Lea: Warden Hills*; *J.S.*: Chiltern Green*; *J.S.*
First evidence: Saunders in Herb. Luton Museum, 1882.

F. adianthoides Hedw.

Pastures   Native

Common on the chalk downs and occasional in oolitic pastures. All the plants I have seen are of the small-celled 'chalk form' which bryologists now tend to place under *F. cristatus*. Recorded for all districts except Nene, Kym and Cam, and for all neighbouring counties except Hunts.

H.S. 59.

First record: Abbot, 1798, as *Hypnum adiantoides*.

ARCHIDIACEAE
Archidium Brid.
Archidium alternifolium (Hedw.) Schimp.

Bogs   Native

Probably extinct: recorded for Bucks. and Cambs.

OUSE: Stevington Bogs; *F.B.*, as *Phascum alternifolium*.

DICRANACEAE
Pleuridium Brid.
Pleuridium acuminatum Lindb.

  P. subulatum auct.

Woods; should be looked for on heaths   ? Native

Rare, in loamy woodland rides on acid soils. Recorded for all neighbouring counties.

OUZEL: King's Wood!*; *P.T.*

COLNE: Pepperstock*; *J.S.*

First record: common, Abbot, 1798, as *Phascum subulatum*.

P. subulatum (Hedw.) Lindb.

  P. alternifolium (Dicks.) Brid.

Sandpits   ? Colonist

Rare: recorded for Cambs. and Northants.

IVEL: old sandpit, Long Lane, Tingrith (1949)*; *J. P. M. Brenan and P.T.*

Ditrichum Hampe
Ditrichum flexicaule (Schleich.) Hampe

Chalk pastures   Native

Widespread but not common. Recorded for all neighbouring counties except Hunts.

OUZEL: Sewell!*; *P.T.*

IVEL: Barton!*; Markham Hills*; *J.S.*: Pegsdon; *T.L.*

LEA: Warden Hills*; Luton Hoo*; Luton Downs*; *J.S.*

First record: Saunders in *J. Bot. 1884*.

Ceratodon Brid.
Ceratodon purpureus (Hedw.) Brid.

In varied habitats   Native, but often a colonist

Common, being abundant on acid fallow land, common on acid gravels, ashes, fire sites and acid waste land. It is abundant on dry heaths, in acid open woodland, in old sandpits and on exposed sandstone. It appears occasionally on stumps, old thatch and wood, and on acid wall cappings. It is sometimes an epiphyte, especially on elder. Abundant on railway tracks, frequent at the base of beeches in beech hangers, rare on slate roofs. Recorded for all districts and neighbouring counties.

H.S. 1a, 1b, 11, 15, 22a, 22b, 24, 27, 36, 45a, 46, 47, 49, 69, 70, 77a.

First record: Abbot, 1798, as *Mnium purpureum*.

Seligeria Bruch, Schimp. & Guemb.
Seligeria paucifolia (Dicks.) Carruth.

Chalk exposures and blocks   Colonist

Frequent on loose chalk, especially on detached blocks in partial shade. Recorded for the Ouzel, Ivel, Lea and Colne districts and for Herts., Bucks. and Cambs.

First record: P. Taylor in *Trans. Brit. Bry. Soc.* 1949.

S. calcarea (Hedw.) Bruch, Schimp. & Guemb.
Vertical chalk faces   Colonist
Rare: recorded for Herts., Bucks. and Cambs.
IVEL: chalkpit, Barton Hill; *F.B.*
LEA: East Hyde (1903)*; *J.S.*
First record: Abbot, 1798, as *Bryum calcareum*.

Pseudephemerum (Lindb.) Hagen

Pseudephemerum nitidum (Hedw.) C. Jens.
Pleuridium axillare (Sm.) Lindb., *P. nitidum* (Hedw.) Rabenh., *Pseudephemerum axillare* (Sm.) Hagen.
Mud on pond margins and cart ruts   Colonist
Occasional on acid soils. Recorded for Herts., Bucks. and Northants.
OUZEL: Stockgrove; King's Wood; *T.L.*
IVEL: Flitwick*; Clophill*; Southill*; *J.S.*
LEA: Luton Hoo Park*; *J.S.*
COLNE: Deadmansey Wood; *T.L.*
First record: Saunders in *J. Bot. 1884.*

Dicranella Schimp.

Dicranella schreberiana (Hedw.) Dix.
D. schreberi (Hedw.) Schimp., *Anisothecium crispum* Lindb.
Bare wet earth by water   Colonist
Rare: recorded for Bucks. and Northants.
CAM: pond bank, Cockayne Hatley; *T.L.*
LEA: stream bank, Biscot*; *J.S.*
First record: Saunders in *J. Bot. 1884.*

D. varia (Hedw.) Schimp.
Anisothecium rubrum Lindb.
In varied habitats on bare earth   Colonist
Frequent on bare banks of calcareous clay near water and on bare compacted
chalk, rare in stubble on alkaline soils and on alkaline rock of coprolite quarries.
Recorded for all districts and neighbouring counties.
H.S. 1b.
First record: Abbot, 1798, as *Bryum simplex*.

D. rufescens (Sm.) Schimp.
Anisothecium rufescens (Sm.) Lindb.
Bare earth   Status not known
OUSE: Pickerings Closes; *Abbot in letter to J. E. Smith*: between Milton Ernest and
Radwell; *Abbot's annotated F.B.*, as *Bryum rufescens*.

D. cerviculata (Hedw.) Schimp.
Heathy places   Colonist
Rare, limited to the Lower Greensand. Recorded for Herts., Bucks. and Northants.
OUZEL: New Wavendon Heath, v.c. 24 [Beds.]; *T.L.*
IVEL: Ampthill; *F.B.*: Flitwick Moor*; *J.S.*
First record: Abbot, 1798, as *Bryum cerviculatum*.

D. heteromalla (Hedw.) Schimp.
Woods, heaths and similar places, and on rotten stumps   ? Native
Common on acid soils, also found frequently in old sandpits and on sandstone
exposures, around boles of beech trees and on rotten stumps. It also occurs rarely
as an epiphyte. Recorded for all districts except Kym, and for all neighbouring
counties.
H.S. 19, 23, 28, 49, 77a, 78.
First record: Abbot, 1798, as *Bryum heteromallum*.

Dichodontium Schimp.

Dichodontium pellucidum (Hedw.) Schimp.
Wet places   ? Colonist
Rare: not recorded for any neighbouring counties.

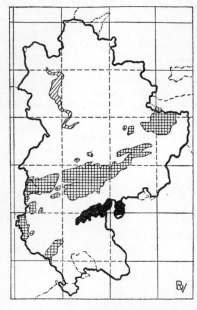

FIG. 18

Distribution of *Dicranum scoparium* (cross hatching) as a native species on greensand heaths and chalk downs (it occurs also as an epiphyte throughout the county), *Encalypta streptocarpa* (black), a species of beech woods on the Chalk, and *Encalypta vulgaris* (oblique hatching), a species of oolitic exposures and walls.

FIG. 19

Distribution of *Fissidens crassipes* (black), a species common by the Ouse and a short length of the Ivel, and *Dicranella heteromalla* (oblique hatching), a species of brown earths (*Ceratodon purpureus*, *Pohlia nutans*, *Polytrichum juniperinum*, *P. formosum*, *Mnium hornum* and *Bryum erythrocarpum* have a similar distribution, but in artificial habitats some have wider distributions).

NENE: step-like water channel, railway cutting, Wymington; *T.L.*
OUZEL: bog, Aspley Wood; *F.B.*
First record: Abbot, 1798, as *Bryum pellucidum.*
    Var. **flavescens** (Turn.) C. Jens. (*D. flavescens* (Turn.) Lindb.)
OUSE: Bedford; *Abbot in letter to J. E. Smith, 7 April 1801*, as *Bryum flavescens.*

Dicranoweissia Lindb.

Dicranoweissia cirrata (Hedw.) Lindb.
   Weissia cirrata Hedw.
Old wood, trees and masonry   Colonist
Epiphytic on trees, stumps, old wooden fences and bridges; occasional on walls, culverts and tombstones; rare on thatch. Well distributed and recorded for all districts, but most common on the Lower Greensand and the Chalk. Recorded for all neighbouring counties.
H.S. 27.
First record: Abbot, 1798, as *Bryum cirrhatum* (also '*Bryum dicksoni*' and '*Mnium cirrhatum*' in litt.)

Dicranum Hedw. *Fork Mosses*

Dicranum montanum Hedw.
Orthodicranum montanum (Hedw.) Loeske
Rotting stumps   Colonist
Locally abundant in a few woods on the Lower Greensand. Recorded for Herts., Bucks. and Northants.
OUZEL: Aspley Wood!*; *J.S.*: King's Wood, very common!*; *E. C. Wallace*.
IVEL: in old dense scrub, Barton Hills; *T.L.*
First record: Saunders in *J. Bot. 1884*.

D. flagellare Hedw.
Orthodicranum flagellare (Hedw.) Loeske
Rotting stumps   Colonist
Rare: not recorded for any neighbouring counties.
OUZEL: King's Wood, where it is locally common (1949)!*; *E. C. Wallace and P.T.*

D. majus Turn.
Woods   Native
Rare, limited to the Lower Greensand. Recorded for Herts., Bucks. and Northants.
OUZEL: King's Wood*; *E. C. Wallace and P.T.*
IVEL: Steppingley Firs*; *J.S.*: Shefford; *Marquand List.*
First record: Saunders in *J. Bot. 1884*.

D. bonjeani De Not.
D. palustre Bruch, Schimp. & Guemb.
Bogs on the Lower Greensand   Native
Rare: recorded for Herts., Bucks. and Northants.
OUZEL: Aspley Wood*; *J.S.*
IVEL: Clophill; Ampthill; *J.S.*: Flitwick Moor; *T.L.*
First record: Saunders in *J. Bot. 1884*.

D. scoparium Hedw.
Heaths and chalk downs   Native
Abundant on heaths on the Lower Greensand, locally common on chalk downs, occasional to frequent on rotten stumps, rare as an epiphyte. Recorded for all districts except Kym, and for all neighbouring counties.
H.S. 19, 22a, 23, 24, 27, 30, 47, 65.
First record: Abbot, 1798, as *Bryum scoparium*.

Campylopus Brid.

Campylopus fragilis (Turn.) Bruch, Schimp. & Guemb.
Greensand heaths   Native
Rare: not recorded for any neighbouring counties.
OUZEL: Aspley Wood; *Abbot's annotated F.B.*
IVEL: Shefford; *Marquand List*: Rowney Warren; Waterloo Wood, Sutton; Cox Hill, Sandy; *T.L.*
First record: Abbot, *c.* 1799, as *Bryum fragile*.

C. piriformis (K. F. Schultz) Brid.
Heaths, woods, and rotten stumps   Native
Frequent on dry heaths and in open woodland on the Lower Greensand in the Ouse, Ouzel and Ivel districts. Not previously recorded for the county. Recorded for all neighbouring counties.

C. flexuosus (Hedw.) Brid.
Tree stumps, heathy places and woods   Native
Locally abundant on the Lower Greensand from Shefford westwards. Saunders recorded it without a station from the 'chalk scarp' but not found there since. Recorded for Herts., Bucks., Cambs. and Northants.
OUZEL: Aspley!*; *J.S.*: King's Wood!*; New Wavendon Heath, v.c. 24 [Beds.]!*; *P.T.*: dry heath, Heath and Reach; *T.L.*
IVEL: Rowney Warren; *T.L.*
First record: Saunders in *J. Bot. 1884*.

Leucobryum Hampe
Leucobryum glaucum (Hedw.) Schimp.
Woods   Native
Limited to the western part of the Lower Greensand. Recorded for Herts., Bucks.,
Cambs. and Northants.
OUZEL: Aspley Wood, where it was also found by Saunders*; *F.B.*: King's
Wood!*; *P.T.*: Stockgrove; *T.L.*
IVEL: Ampthill Heath; *F.B.*
First record: Abbot, 1798, as *Bryum glaucum.*

ENCALYPTACEAE

Encalypta Hedw.
Encalypta vulgaris Hedw.   *Extinguisher Moss*
Oolitic walls and exposures   Denizen
Widespread in the north of the county. Recorded for all neighbouring counties
except Hunts.
OUSE: Oakley Church wall, where it has since been found by P.T.*; *H. Davis*:
walls, Biddenham; *Marquand List*: railway cutting, Milton Ernest*; *P.T.*
First record: Abbot, 1798, as *Bryum extinctorium.*

E. streptocarpa Hedw.
   E. contorta Hoppe
Chalky banks under beeches   Native
Locally frequent: recorded for Bucks. and Northants.
H.S. 69.
IVEL: Barton Leete (1948)!*; *J.G.D.*: Pegsdon Common; *T.L.*

POTTIACEAE

Tortula Hedw.   *Screw Mosses*
Tortula ruralis (Hedw.) Crome
   Barbula ruralis Hedw.
Walls, embankments and disused chalkpits   Colonist
Frequent: recorded for all districts and neighbouring counties.
H.S. 1a, 1b, 45a.
First record: Abbot, 1798, as *Bryum rurale.*

T. intermedia (Brid.) Berk.
   Barbula intermedia (Brid.) Milde, *T. montana* (Nees) Lindb.
Walls and rock exposures   Colonist
Frequent, especially on limestone masonry and oolitic exposures. Recorded for all
districts and neighbouring counties.
First record: Saunders in *Trans. Beds. Nat. Hist. Soc.* 1882.

T. laevipila (Brid.) Schwaegr.
   Barbula laevipila (Brid.) Bruch, Schimp. & Guemb.
Tree trunks and stumps   Denizen
A widespread epiphyte, particularly on isolated trees or stumps and trees in
hedgerows. Recorded for all districts and neighbouring counties.
First record: Saunders in *Trans. Beds. Nat. Hist. Soc.* 1882.
      Var. **laevipiliformis** (De Not.) Limpr.
In similar places to the type, but rarer.
OUSE: Roxton; *H. N. Dixon in Brit. Bryol. Soc. 1931 Rep.*
LEA: Limbury*; *J.S.*

T. papillosa Wils. ex Spruce
Tree trunks   Denizen
Possibly extinct: recorded for Herts., Bucks., Cambs. and Northants.
OUSE: Sharnbrook; ? *E. Cooper in Herb. Bedford Public Library.*
KYM: Knotting (1856); *A. O. Black in Herb. Bedford Public Library.*

T. latifolia (Bruch) Hartm.
  T. mutica (K. F. Schultz) Lindb.
Walls and trees by water   Denizen
Very common near water on the Ouse above Bedford, rare on the Ivel and Flit.
Recorded for all neighbouring counties.
First evidence: J. Hamson in Herb. Luton Museum, 1882.

T. subulata Hedw.
  Barbula subulata (Hedw.) Beauv.
Calcareous banks and walls   ? Native
Frequent on bare chalk banks beneath beeches, rare on alkaline banks and on
walls. Not known on sandy banks (a common habitat in many areas) since
Saunders's time. Recorded for all neighbouring counties except Hunts.
OUSE: Oakley Church wall; *C. C. Foss.*
IVEL: Steppingley Firs*; Ampthill*; *J.S.*: common in suitable places on the
  Chalk.
CAM: Wrestlingworth; *T.L.*
First record: Abbot, 1798, as *Bryum subulatum.*

T. muralis Hedw.
  Barbula muralis (Hedw.) Web. & Mohr
Masonry and loose bricks   Wall denizen
Common on masonry, loose stones and bricks, and also found occasionally on
exposed oolitic and coprolitic rock. Recorded for all districts and neighbouring
counties.
First record: Abbot, 1798, as *Bryum murale.*
  Var. **aestiva** (Beauv.) Brid.
IVEL: Everton Church (1892); *H. N. Dixon* in Herb. Kew.

Aloina Kindb.

Aloina ambigua (Bruch, Schimp. & Guemb.) Limpr.
  Barbula ambigua Bruch, Schimp. & Guemb., *Tortula ambigua* (Bruch, Schimp. &
Guemb.) Ångstr., *A. ericaefolia* Kindb.
Limestone walls and calcareous exposures   Colonist
Common in the north of the county, locally frequent on compacted chalk in the
south. It grows best in dry, sunny situations. Recorded for all districts except
Kym and Cam and for Bucks., Cambs. and Northants.
First record: Abbot, 1798, as *Bryum rigidum.*

A. aloides (Schultz) Kindb.
  Tortula aloides (Schultz) De Not.
Calcareous exposures   Colonist
Rare: recorded for all neighbouring counties except Hunts.
OUSE: on coprolitic rock in old quarry, Houghton Park (1951); *T.L.*

Pterygoneurum Jur.

Pterygoneurum ovatum (Hedw.) Dix.
  Pottia cavifolia Bruch, Schimp. & Guemb., *Tortula pusilla* (Lindb.) Mitt., *Ptery-
goneurum cavifolium* (Bruch, Schimp. & Guemb.) Jur.
Rubbly limestone walls   Wall denizen
Widespread, but rare, in the Nene and Ouse districts. Recorded for all neigh-
bouring counties.
First record: Abbot, 1798, as *Bryum ovatum.*

Pottia Fuernr.

Pottia lanceolata (Hedw.) C. Muell.
Exposed rocks   Native
Common on oolitic exposures in the Nene and Ouse districts, rare on the Chalk in
the Ivel district. Recorded for all neighbouring counties except Hunts.
First record: Abbot, 1798, as *Bryum lanceolatum.*

P. intermedia (Turn.) Fuernr.
Fallow ground   Short-term colonist
Common: recorded for all districts and neighbouring counties.
First record: Saunders in *J. Bot. 1884.*

P. truncata (Hedw.) Fuernr.
  P. truncatula (Sm.) Lindb.
Fallow ground   Short-term colonist
Common: recorded for all districts and neighbouring counties.
First record: Abbot, 1798, as *Bryum truncatulum.*

P. davalliana (Sm.) C. Jens.
  P. rufescens (Schultz) Warnst., *P. minutula* (Schleich.) Fuernr.
Fallow ground   Short-term colonist
Frequent on alkaline soils in all districts. Recorded for all neighbouring counties
except Hunts.
First record: Saunders in *J. Bot. 1884.*

P. bryoides (Dicks.) Mitt. was recorded from Clapham Springs by Abbot as
Phascum bryoides in his annotated *Flora Bedfordiensis.* It is also listed for v.c. 30 in
Census Catalogue (1926), but the authority is not known. Recorded for Herts.,
Cambs. and Northants.

P. recta (Sm.) Mitt.
  Phascum rectum Sm.
Calcareous banks   Colonist
Rare: recorded for Herts., Bucks., Cambs. and Northants.
CAM: Battle Brook, Wrestlingworth; *T.L.*
LEA: Warden Hills*; Luton Downs*; *J.S.*
First record: Saunders in *J. Bot. 1884.*

Phascum Hedw.

Phascum curvicollum Hedw.
Calcareous soils   Colonist
Rare: recorded for Herts., Bucks., Cambs. and Northants.
OUSE: Clapham; *F.B.*
IVEL: Barton Hills*; *P.T.*
First record: Abbot, 1798.

P. cuspidatum Hedw.
Fallow fields   Short-term colonist
Common on alkaline soils in all districts. Recorded for all neighbouring counties.
First record: Abbot, 1798, as *Phascum acaulon.*
      Var. **piliferum** (Hedw.) Hook. & Tayl. (*P. piliferum* Hedw.)
OUSE: Clapham Hill; *F.B.*
      Var. **curvisetum** (Sm.) Nees, Hornsch. & Sturm (*P. curvisetum* Sm.)
This is credited to Abbot by Braithwaite in *Brit. Moss Flora* 1884, 190, but no
mention of it can be found in Abbot's writings.

Acaulon C. Muell.

Acaulon muticum (Brid.) C. Muell.
Fallow fields   Short-term colonist
Rare in fields on acid soils on the Lower Greensand, but Abbot considered it
'common on garden walks'. Recorded for Herts., Bucks. and Northants.
IVEL: near Sutton Cross-roads; *T.L.*
First record: Abbot, 1798, as *Phascum muticum.*

Cinclidotus Beauv.

Cinclidotus fontinaloides (Hedw.) Beauv.
Walls and bases of trees in water   Denizen
Common in or near water along the Ouse from Turvey to Great Barford. Re-
corded for Bucks. and Northants.
First record: Abbot, 1798, as *Fontinalis minor.*

A record of *C. riparius* (Host) Arn. (*C. nigricans* (Brid.) Dix.) from Barford Bridge
in Herb. Bedford Public Library is not supported by a specimen and must be
regarded as suspect.

Barbula Hedw.
Barbula convoluta Hedw.

In varied habitats   Colonist
Common on compacted gravel, especially in disused gravel pits, on compacted
waste ground, ash paths, railway tracks and rubble of walls: frequent on calcareous rock exposures and floors of disused sandpits. Widespread on chalk downs
and bare banks in beech woods. Recorded for all districts and neighbouring
counties.
H.S. 1a, 1b, 11, 27, 69.
First record: Abbot, 1798, as *Bryum convolutum*.

B. unguiculata Hedw.

In varied habitats   Colonist
Common on alkaline soils, rubbly walls, calcareous rock exposures, compacted
gravels, ash paths and waste ground. It appears also in some acid fallows. Re-
corded for all districts and neighbouring counties.
H.S. 5, 11, 27.
First record: Abbot, 1798, as *Bryum unguiculatum*.

B. revoluta Brid.

Calcareous walls and rocks   Colonist
Rare: recorded for Bucks., Hunts. and Northants.
OUSE: coprolitic rock, Houghton Park; *T.L.*
IVEL: walls, Silsoe (1948); *E. C. Wallace*.

B. fallax Hedw.

Calcareous walls, rock and earth   Colonist and denizen
Common on limestone walls, oolitic exposures and compacted chalk: rare on
compacted alkaline clays. Recorded for all districts and neighbouring counties.
First record: Abbot's annotated *Flora Bedfordiensis*, c. 1799, as '*Bryum imberbe*'.

B. reflexa (Brid.) Brid.
  B. recurvifolia (Wils.) Schimp.

Chalk pits   Colonist
Rare: recorded for Northants.
IVEL: old chalk pit, Barton Hills*; *J.S.*: Barton Cutting; *J. P. M. Brenan*.
First evidence: Saunders in Herb. Luton Museum, 1885.

B. rigidula (Hedw.) Mitt.

Walls   Wall denizen
Rare: recorded for all neighbouring counties except Herts.
OUSE: Turvey (1951); *E. C. Wallace*.
KYM: Knotting*; *J.G.D.*
COLNE: Stockwood Park wall; *T.L.*

B. trifaria (Hedw.) Mitt.
  Didymodon luridus Hornsch., *B. lurida* (Hornsch.) Lindb.

Stonework   Denizen
Rare by water along the Upper Ouse and the Ivel. Recorded for Bucks., Cambs.
and Northants.
OUSE: Bromham Bridge; *T.L.*: Harrold Bridge*; *J.G.D.*
IVEL: lock, Ivel Navigation, Shefford (1951); *T.L.*

B. tophacea (Brid.) Mitt.
  Trichostomum tophaceum Brid., *Didymodon tophaceus* (Brid.) Jur.

Rubbly walls   Colonist
Occasional throughout the county. Recorded for all districts except Kym and
Colne, and for Bucks., Cambs. and Northants.
First record: Abbot, 1798, as '*Bryum trifarium*' (Wilson in *Bryologia Britannica* (1854)
114, confirms that a specimen of Abbot's '*B. trifarium*' from Clapham Springs is
Barbula tophacea).

B. cylindrica (Tayl.) Schimp.

Muddy places   Colonist
Rare: recorded for all neighbouring counties.
OUSE: silt on base of willow by the river, Stevington (1952); *T.L.*

B. vinealis Brid.
Sandstone walls   Denizen
Rare: recorded for all neighbouring counties except Cambs.
OUSE: Oakley Bridge; *T.L.*
IVEL: Wrest Park! (1948); *E. C. Wallace*: Pegsdon; *T.L.*

B. recurvirostris (Hedw.) Dix.
    B. rubella (Hueben.) Lindb., *Didymodon rubellus* (Hueben.) Bruch, Schimp. &
Guemb.
Bare chalk banks and rubbly walls   Colonist
Locally frequent in beech woods on the Chalk: occasional on walls. Recorded for
all districts except Kym and Cam, and for all neighbouring counties except Hunts.
First record: Saunders in *Trans. Beds. Nat. Hist. Soc.* 1882.

Eucladium Bruch, Schimp. & Guemb.

Eucladium verticillatum (With.) Bruch, Schimp. & Guemb.
    Weissia verticillata With.
Stonework by water   ? Denizen
Rare: recorded for Cambs. and Northants.
IVEL: detritus on stonework of lock, Ivel Navigation, Shefford (1951); *T.L.*

Tortella (C. Muell.) Limpr.

Tortella tortuosa (Hedw.) Limpr.
    Barbula tortuosa (Hedw.) Web. & Mohr, *Trichostomum tortuosum* (Hedw.) Dix.
Chalk downs   Native
Rare. Recorded for Bucks. and Cambs.
OUZEL: Dunstable Downs*; *R. Rogers.*
IVEL: Barton Hills*; *J.S.*: north side of Deacon Hill; *W. M. Cornish.*
First record: Saunders in *Victoria County History*, 1904.

Pleurochaete squarrosa (Brid.) Lindb. was recorded in the *Marquand List* from
Bolnhurst. Without further evidence it must be suspect.

Trichostomum Bruch

Trichostomum sinuosum (Wils.) Lindb.
    Barbula sinuosa (Wils.) Braithw.
Walls and stumps   Native
Rare, and usually found near water. Recorded for all neighbouring counties
except Hunts.
OUSE: walls, Turvey, where it has also been found by E. C. Wallace; *H. N. Dixon.*
IVEL: lock, Ivel Navigation, Shefford; *T.L.*
CAM: tree stumps and roots, Battle Brook, Wrestlingworth; *T.L.*
First evidence: H. N. Dixon, 1906, in Herb. Kew.

Weissia Hedw.

Weissia controversa Hedw.
    W. viridula Brid.
Sandy banks, old sand and gravel pits and broken earth   Colonist
Widely distributed but not common. Recorded for all districts and for all neigh-
bouring counties except Hunts.
First record: Abbot, 1798, as *Bryum virens* (also '*B. viridulum*' in litt.).

W. microstoma (Hedw.) C. Muell.
    Hymenostomum microstomum (Hedw.) R.Br.
Chalk downs   Native
Frequent from Pegsdon to Houghton Regis but not appearing further west. Re-
corded for the Ouzel and Ivel districts and for Herts., Bucks., Cambs. and
Northants.
H.S. 59.
First record: P.T. in *Trans. Brit. Bry. Soc.* 1949.
First evidence: Saunders in Herb. Luton Museum, 1882.

W. multicapsularis (Sm.) Mitt.
Fallow ground   Short-term colonist
Rare or extinct. Abbot in his annotated *F.B.* says 'common on molehills and arable land'. Not recorded for any neighbouring counties.
OUSE: rare, cart ruts, Clapham Park Wood; *F.B.*
IVEL: Ampthill, from which it is credited to Abbot by Braithwaite, *British Moss Flora* (1885) 232, but I can find no mention of it in Abbot's writings.
First record: Abbot, 1798, as *Phascum sphaerocarpon* and *P. multicapsulare*, which he evidently regarded as two different species, though they are the same.

W. crispa (Hedw.) Mitt.
Phascum crispum Hedw., *Astomum crispum* (Hedw.) Hampe
Chalk downs and calcareous banks   ? Native
Occasional on the chalk downs, appearing also rarely on calcareous clay banks. Recorded for all neighbouring counties.
OUSE: Clapham Pastures; *Abbot's annotated F.B.*
OUZEL: Totternhoe Knolls*; *J.G.D.*: Dunstable Downs; *T.L.*
IVEL: Knocking Hoe; *T.L.*
CAM: Cockayne Hatley; *T.L.*
H.S. 62.
First record: Abbot, *c.* 1799.

Leptodontium Hampe

Leptodontium flexifolium (Sm.) Hampe
Heaths   Native
Rare: recorded for Bucks. and Northants.
IVEL: Rowney Warren (1949); *T.L.*

GRIMMIACEAE
Grimmia Hedw.

Grimmia apocarpa Hedw.
Walls   Wall denizen
Rare: recorded for all neighbouring counties.
OUSE: Clapham; St. Mary's Church, Bedford; *F.B.*
IVEL: Ampthill; *F.B.*: Northill; *Marquand List.*
CAM: concrete block, Cockayne Hatley; *T.L.*
LEA: bridge, Luton Hoo*; *J.S.*
First record: Abbot, 1798, as *Bryum apocarpum.*

G. pulvinata (Hedw.) Sm.
Walls   Wall denizen
Common on masonry. Recorded for all districts and neighbouring counties.
First record: Abbot, 1798, as *Bryum pulvinatum.*

Rhacomitrium Brid.

Rhacomitrium heterostichum (Hedw.) Brid.
Stonework   Casual
Rare: recorded for Northants.
OUZEL: imported rocks on embankment, road cutting, Hockliffe (1952); *P.T. and T.L.*

R. lanuginosum (Hedw.) Brid.
R. hypnoides Lindb.
Stonework   ? Native in Abbot's time
Rare: not recorded for any neighbouring county.
H.S. 1b.
NENE: rocks on railway embankment, Sharnbrook Summit (wrongly recorded as *R. canescens* in *Trans. Beds. Nat. Hist. Soc.* 1948)!*; *J.G.D.*
OUZEL: rocks on embankment, road cutting, Hockliffe; *P.T. and T.L.*
IVEL: fir grove, Warden; *F.B.*
First record: Abbot, 1798, as *Bryum hypnoides.*

FUNARIACEAE

Funaria Hedw.   *Cord Mosses*
Funaria hygrometrica Hedw.
Variable habitats on earth   Short-term colonist
Common on fire sites, ashes and compacted waste ground, especially railway
sidings, frequent on soil in flower pots, occasional on walls and on clay banks,
rare in fallow fields. Recorded for all districts and neighbouring counties.
H.S. 1a, 24, 40, 49, 64.
First record: Abbot, 1798, as *Mnium hygrometricum.*

F. muehlenbergii Turn.
  F. dentata Crome, *F. calcarea* Wahlb.
Chalk downs   ? Native
Rare: not recorded for any neighbouring county.
LEA: on chalk, near Luton, 1885, a scrap in his herbarium but not detected by
Saunders*; *J.S.*

F. fascicularis (Hedw.) Schimp. is of doubtful record by Saunders in *Trans. Beds.
Nat. Hist. Soc.* 1885. His herbarium contains two specimens so named—one is a
Hertfordshire gathering, the other, from Flitwick, is immature, *cf. Pottia intermedia.*

An alien species, with excurrent nerve, but sterile, found by *T.L.* on railway sidings
at Tempsford, 1950, growing with wool adventives, *cf.* the Australian *F. cuspidatum*
Hook. f. & Wils.

Physcomitrium Brid.
Physcomitrium pyriforme (Hedw.) Brid.
Muddy places   Short-term colonist
Frequent on bare, peaty mud on the Lower Greensand; occasional by the sides of
rivers and ditches, especially on mud after flooding or thrown up after cleaning out
watercourses. Recorded for all districts and neighbouring counties.
H.S. 15, 37, 53b, 76.
First record: Abbot, 1795, as *Bryum pyriforme.*

Physcomitrella Bruch, Schimp. & Guemb.
Physcomitrella patens (Hedw.) Bruch, Schimp. & Guemb.
Muddy places   Short-term colonist
Rare: recorded for all neighbouring counties except Hunts.
OUSE: around Bedford; *Abbot* (see below).
IVEL: Southill Park*; *J.S.* in *J. Bot. 1884.*
First record: W. Wilson in *Bry. Brit.* (1854) 38, wrote that some of the specimens
of *Phascum multicapsulare* in Herb. Dawson Turner sent by Abbot from around
Bedford are this.

EPHEMERACEAE

Ephemerum Hampe
Ephemerum serratum (Hedw.) Hampe
  Phascum serratum Hedw., *P. stoloniferum* Sm.
Fallow land   Short-term colonist
Rare: recorded for Herts., Bucks., Cambs. and Northants.
OUSE: Caldwell Enclosures; *J. Hemsted in Abbot's annotated F.B.*
IVEL: Long Lane, Tingrith*; *J. P. M. Brenan and P.T.*
First record: J. Hemsted, *c.* 1799.

SPLACHNACEAE

Splachnum Hedw.
Splachnum ampullaceum Hedw.   *Bottle Moss*
Bogs   ? Native
Extinct: recorded for Cambs. and Northants.
IVEL: Ampthill; *F.B.*

TETRAPHIDACEAE
Tetraphis Hedw.
Tetraphis pellucida Hedw.
  Georgia pellucida (Hedw.) Rabenh.
Rotting stumps, stools and sandy banks   Colonist
Occasional in western part of the county. Recorded for all neighbouring counties.
OUZEL: frequent on the Lower Greensand.
IVEL: occasional on the Lower Greensand west of Clophill.
CAM: Potton Wood; *T.L.*
First record: Abbot in letter to J. E. Smith, 5 May 1798, as *Mnium pellucidum.*

BRYACEAE
Orthodontium Schwaegr.
Orthodontium lineare Schwaegr.
  O. gracile var. *heterocarpum* Wats.
Tree stumps   Colonist
Locally frequent: recorded for Herts., Bucks. and Cambs.
OUZEL: King's Wood*; *R. D. Meikle.*
IVEL: Potton Wood, common; Sandy Heath; Rowney Warren; *T.L.*
CAM: Walk Wood, Cockayne Hatley, common (1947); *T.L.*
COLNE: Folly Wood*; *E. C. Wallace.*

Leptobryum Wils.
Leptobryum pyriforme (Hedw.) Wils.
Bare peat and ashes   Short-term colonist
Abundant on bare peat in the Flit Valley and at Sutton Fen, where the peat is
drying out after inundation, but variable in frequency from season to season.
Occasional on peaty mud elsewhere on the Lower Greensand; locally frequent on
ashes and on soil in flower pots. Recorded for all districts except Nene, Kym and
Colne, and for all neighbouring counties.
H.S. 37.
First record: Saunders in *J. Bot.* 1884.

Pohlia Hedw.
Pohlia nutans (Hedw.) Lindb.
  Webera nutans Hedw.
Sandy places, peat and rotten stumps   Native
Common on the ground and on rotten stumps on the Lower Greensand, rare
elsewhere. Occasionally amongst sphagna in bogs. Recorded for all districts
except Kym, and for all neighbouring counties.
H.S. 1b, 22b–24, 36, 40, 42, 47.
First record: Abbot's annotated *Flora Bedfordiensis, c.* 1799, as *Bryum sericeum.*

P. cruda Hedw. (*Webera cruda* (Hedw.) Bruch) was recorded by Saunders in *V.C.H.*
from Hawnes but his specimen is *P. nutans.*

P. annotina (Hedw.) Loeske
  Webera annotina (Hedw.) Bruch
Woods, along rides   ? Native
Rare in loamy rides in woods on acid soils. Recorded for Northants.
OUSE: St. Cuthbert's; road to Milton Ernest; *Abbot's annotated F.B.*
OUZEL: rides, King's Wood; *T.L.*
COLNE: rides, Deadmansey Wood; *T.L.*
First record: Abbot, 1798, as '*Mnium scandens*'.

P. proligera Lindb.
  Webera proligera (Lindb.) Kindb.
Sandpits   Colonist
Rare: recorded for Northants.
IVEL: old sandpit, Silsoe (1949)*; *J. P. M. Brenan and P.T.*

P. delicatula (Hedw.) Grout
Webera carnea (Brid.) Schimp., *Mniobryum carneum* (Brid.) Limpr.
Wet clay-banks   Colonist
Common on bare, wet clay or chalk banks of rivers, streams and ditches;
occasional on bare soil of woodland rides, embankments, cuttings, banks and old
gravel pits.
Recorded for all districts and neighbouring counties.
H.S. 6, 57.
First record: Saunders in *J. Bot. 1884*.

Bryum Hedw.

Bryum pendulum (Hornsch.) Schimp.
Very rare: recorded for Herts.
OUSE: Turvey (1906); *H. N. Dixon in Herb. Kew*.
First record: without locality, *Victoria County History*, 1904.

B. inclinatum (Brid.) Bland.
Rare: recorded for Bucks. and Northants.
IVEL: Flitwick*; *J.S. in V.C.H.*

B. pseudotriquetrum (Hedw.) Schwaegr.
  B. ventricosum Hook. & Tayl.
Marshy places   Native
Considered common by Abbot; it is rare or extinct now. Recorded for Herts.,
Cambs. and Northants.
OUSE: Cox's Pits*; *J. Hamson*: Great Barford Bridge; *E. Cooper*.
OUZEL: bog, Heath and Reach*; *J.S.*
IVEL: Flitwick Moor*; Westoning Moor*; *J.S.*
First record: Abbot, 1798, as *Mnium triquetrum*.
        Var. **bimum** (Brid.) Rich. & Wall. (*B. bimum* (Brid.) Turn.)
Bogs   Native
Occasional on peat in wet places and common on peat in fern and orchid houses.
Recorded for Cambs., Hunts. and Northants.
OUSE: by Great Barford Bridge*; *J.G.D.*
IVEL: Flitwick Moor!*; *J.S.*
First record: Saunders in *V.C.H.*

B. affine (Bruch) F. W. Schultz
Rare: recorded for Northants.
OUZEL: cement works, Houghton Regis (H.S. 64)*; *J.G.D.*

B. caespiticium Hedw.
In varied habitats   Colonist
Frequent on rubbly walls and mortar, on rubble, chalk waste, waste ground, ashes,
gravel paths, floors of disused gravel pits and calcareous rock exposures; occasional
as a stump epiphyte and on chalk banks, especially in beech woods. Recorded for
all districts and neighbouring counties.
H.S. 1a, 1b, 63b.
First record: Abbot, 1798, as *Mnium caespiticium*.

B. argenteum Hedw.   *Silver Moss*
Walls and waste places   ? Colonist
Common on compacted waste ground, paths, railway tracks, walls, masonry and
in the angles of pavements; occasional on stumps, rotting woodwork, fire sites and
slate roofs; rare on natural rock. Recorded for all districts and neighbouring
counties.
First record: Abbot, 1798.
        Var. **lanatum** (Beauv.) Bruch, Schimp. & Guemb.
Walls   Wall denizen
Rare: recorded for Hunts. and Northants.
NENE: Podington*; *P.T.*
OUSE: Great Barford Bridge*; *J.G.D.*

B. bicolor Dicks.
B. atropurpureum Web. & Mohr
Bare earth   Colonist
Not generally common, but widely distributed and locally abundant on all soils,
usually associated with compacted ground. It appears occasionally on fire sites,
old wood and exposed sandstone rock. Recorded for all districts and neighbouring
counties except Hunts.
First record: Saunders in *J. Bot. 1884.*

B. murale Wils.
Walls   Wall denizen
Rare or extinct: recorded for Bucks. and Northants.
Lea: Luton Hoo*; *J.S.*
Colne: Stockwood Park wall*; *J.S.*
First record: Saunders in *Victoria County History*, 1904.

B. erythrocarpum Schwaegr.
Sandy soils   Colonist
Widely distributed on the Lower Greensand and Clay-with-Flints. Recorded for
the Ouzel, Ivel, Lea and Colne districts and for Bucks., Cambs. and Northants.
First evidence: Saunders, 1883, in Herb. Luton Museum.

B. capillare Hedw.
In varied habitats   Colonist
Common on rubbly walls and mortar; a frequent epiphyte on stumps, ash stools
and on elders; frequent on chalk banks in beech woods; occasional on chalk
downs, natural calcareous rock, thatch and waste ground. Recorded for all
districts and neighbouring counties.
H.S. 1a, 1b, 5, 27, 45a, 47, 69, 70.
First record: Abbot, 1795, as *Mnium capillare.*

Rhodobryum (Schimp.) Limpr.
Rhodobryum roseum (Hedw.) Limpr.
Bryum roseum (Hedw.) Sm.
Sandy banks   Native
Considered common by Abbot, it is certainly rare now. Recorded for all neigh-
bouring counties except Hunts.
ouzel: Aspley Heath*; *C. F. Boultbee.*
Ivel: Lidlington*; *C.C.*: between Eversholt and Steppingley*; *P.T.*
First record: Abbot, 1798, as *Mnium proliferum.*

MNIACEAE
Mnium Hedw.
Mnium hornum Hedw.
In varied habitats   ? Native
Common on the ground and on rotten stumps in woodland on acid soils; occasional
on banks in beech woods, on stumps and at the base of trees in alkaline soil areas;
a rare epiphyte. Recorded for all districts and neighbouring counties.
H.S. 31, 33, 42, 76, 78, 79a, 80b.
First record: Abbot, 1798.

M. cuspidatum Hedw.
Woods   ? Native
Rare: Abbot recorded it as common, no doubt in error for *M. longirostrum* which he
did not record. Recorded for Herts., Bucks., Cambs. and Northants.
Ouzel: Aspley Wood (1948)*; *P.T.*
Ivel: Potton Wood; *T.L.*

M. longirostrum Brid.
M. rostratum (Sm.) Roehl.
In varied terrestrial habitats   Native
Common on the Lower Greensand in scrub, woodland rides, old pits and banks;
frequent in shade on the Chalk; occasional elsewhere in woodland and old pasture.

M

Rare in the north of the county. Recorded for all districts except Kym, and for all neighbouring counties.
H.S. 1a, 1b, 33, 43.
First record: Abbot, 1798, as *M. cuspidatum*.

M. affine Bland.
Woodland rides   Native
Uncommon but widespread in the county. Recorded for all districts and for all neighbouring counties except Hunts.
H.S. 34b.
First record: Saunders in *J. Bot. 1884*.

M. seligeri (Jur. ex Lindb.) Limpr.
Marshy places   Native
Rare: the distribution in neighbouring counties is not known. It is regarded as synonymous with *M. affine* var. *elatum* Bruch, Schimp. & Guemb., but the two have been much confused in the past.
OUSE: water meadow, Eaton Socon; *T.L.*
IVEL: Flitwick Moor (1888); *J.S. in Herb. Braithwaite.*

M. undulatum Hedw.
Woods and shady places   Native
Common in woodland and appearing also in scrub. It is rare in old pasture on calcareous clays, the Chalk and Clay-with-Flints, and generally on the Lower Greensand. Recorded for all districts and neighbouring counties.
H.S. 2b, 34b, 43, 78.
First record: Abbot, 1798.

M. punctatum Hedw.
Wet places   Native
Rare on the Lower Greensand and Clay-with-Flints. Recorded for all neighbouring counties.
OUZEL: New Wavendon Heath, v.c. 24 [Beds.]*; *P.T.*: King's Wood; *T.L.*
IVEL: Ampthill; Southill; *F.B.*: Clophill*; *J.S.*: Stratford; *T.L.*
LEA: Luton Hoo*; *J.S.*
First record: Abbot, 1798.

AULACOMNIACEAE
Aulacomnium Schwaegr.

Aulacomnium palustre (Hedw.) Schwaegr.
Bogs   Native
Locally frequent in bogs on the Lower Greensand. Recorded for all neighbouring counties except Hunts. and Cambs.
H.S. 40.
OUZEL: Aspley; *F.B.*: Mermaid's Pond; New Wavendon Heath, v.c. 24 [Beds.]; *T.L.*
IVEL: Ampthill Bogs; Potton Bogs; *F.B.*: Flitwick Moor!*; *J.S.*
LEA: Limbury Marsh*; *J.S.*
First record: Abbot, 1798, as *Mnium palustre.*

A. androgynum (Hedw.) Schwaegr.
In varied habitats   Colonist
Common on the ground, on stumps, on exposed sandstone and occasionally as an epiphyte on the Lower Greensand. An occasional epiphyte on elders on the Chalk and rare on stumps elsewhere. Recorded for all districts except Nene and Kym and for all neighbouring counties.
H.S. 15, 23, 27, 47, 49.
First record: Abbot, 1798, as *Bryum androgynum* (also as *Mnium androgynum* in litt.).

BARTRAMIACEAE
Bartramia Hedw.

Bartramia pomiformis Hedw.   *Apple Moss*
Sandy banks   Native
Rare and limited to the Lower Greensand and Clay-with-Flints. Recorded for all neighbouring counties except Hunts.

OUZEL: Aspley*; *J.S.*: Woburn*; *P.T.*
IVEL: Clophill; Potton; Southill; *F.B.*: Flitwick*; Steppingley Firs*; *J.S.*: Oak
Spinney, Sutton; *T.L.*
COLNE: Pepperstock*; *J.S.*: Whipsnade*; *P.T.*
First record: Abbot, 1795, as *Bryum pomiforme.*

Philonotis Brid.

Philonotis fontana (Hedw.) Brid.
Bogs   Native
Probably extinct: recorded for all neighbouring counties except Hunts.
OUZEL: Heath and Reach*; Totternhoe Mead*; *J.S.*
IVEL: Ampthill; *F.B.*: Flitwick Moor*; *J.S.*
First record: Abbot, 1798, as *Mnium fontanum.*

ORTHOTRICHACEAE

Zygodon Hook. & Tayl.

Zygodon viridissimus (Dicks.) R.Br.
Epiphyte and on walls   Colonist
An occasional to frequent epiphyte, especially in hedgerows and elder scrub. It
occurs also on sandstone walls. Recorded for all districts and neighbouring
counties.
First record: without locality, Saunders in *Victoria County History*, 1904.

Orthotrichum Hedw.

Orthotrichum anomalum Hedw. var. **saxatile** (Wood) Milde
Walls   Wall denizen
Widespread and sometimes frequent on limestone walls. Recorded for all districts
and neighbouring counties.
First record: Saunders in *Victoria County History*, 1904, as *O. saxatile* Wood.

O. cupulatum Brid.
Limestone masonry near water   Colonist
Common in its known stations. Recorded for Northants.
OUSE: Oakley Bridge (1952); *T.L. and J.G.D.*: Bromham Bridge (1952); *T.L.*

O. affine Brid.
Epiphyte   Colonist
An occasional to frequent epiphyte, especially on elder; rare on stonework.
Recorded for all districts and neighbouring counties.
H.S. 27.
First record: Abbot, 1798, as *Bryum striatum.*

O. striatum Hedw. (*O. leiocarpum* Bruch, Schimp. & Guemb.) was recorded by
Abbot as *Bryum striatum*, but his description appears to fit *O. affine.* A specimen
by Saunders in Herb. Luton Museum labelled '*O. leiocarpum*, Pepperstock, 1883'
is *O. lyellii.*

O. lyellii Hook. & Tayl.
Epiphyte in woods   ? Native
Once frequent in woods on the Clay-with-Flints, but now possibly extinct.
Recorded for all neighbouring counties.
OUSE: Ravensden; *Marquand List.*
LEA: Luton Hoo*; *J.S.*
COLNE: Pepperstock*; Caddington*; *J.S.*
First record: Saunders in *Trans. Beds. Nat. Hist. Soc.* 1882.

O. tenellum Bruch
Epiphyte on elders   ? Native
Rare: recorded for Bucks. and Northants.
LEA: Luton Hoo (1892)*; *J.S. in Herb. Luton Museum.*

O. winteri Schimp. (*O. pulchellum* var. *winteri* (Schimp.) Braith.) was recorded by Saunders in *J. Bot. 1897* in error. I have compared Saunders's specimen with Schimper's type at Kew and find it to be *O. pulchellum*. This is confirmed by Dixon in *Student's Handbook of British Mosses* (1924) 281, although the specimen (*O. pulchellum*) in his herbarium is named *O. winteri*.

O. pulchellum Brunton
Epiphyte on elders   ? Native
Rare or extinct: recorded for Bucks.
LEA: Luton Hoo (1892)*; *J.S. in Herb. Luton Museum* (as *O. pulchellum* var. *winteri*).

O. diaphanum Brid.
Epiphyte   Colonist
A widespread but not common epiphyte, usually in hedgerows, on isolated trees, or in elder scrub. It occurs occasionally on stumps and walls. Recorded for all districts and all neighbouring counties.
First record: Saunders in *Trans. Beds. Nat. Hist. Soc.* 1882.

Ulota Brid.

Ulota crispa (Hedw.) Brid.
   U. ulophylla Broth.
Epiphyte on elders
Rare and limited to the Chalk. Recorded for Bucks., Cambs. and Northants.
IVEL: Barton Hills (1949)*; *P.T.*
COLNE: Long Wood*; *P.T.*

FONTINALACEAE

Fontinalis Hedw.

Fontinalis antipyretica Hedw.
Rivers and ponds   Native
Common, attached to wood, masonry or stream-bed stones. Recorded for all districts and neighbouring counties.
H.S. 6.
First record: Abbot, 1798.

F. dolosa Card. ex Dix.
Ponds   ? Native
Probably extinct: not recorded for any neighbouring county.
OUSE: Bedford, a very doubtful record; *Marquand List.*
LEA: Limbury (1882)*; *J.S. in J. Bot. 1897.*

CLIMACIACEAE

Climacium Web. & Mohr

Climacium dendroides (Hedw.) Web. & Mohr
Marshy places and pastures   Native
Probably extinct: recorded for Herts., Cambs. and Northants.
IVEL: Ampthill; Southill; Clophill; *F.B.*: Flitwick Moor*; *J.S.*
LEA: Limbury*; Luton*; *J.S.*
First record: Abbot, 1795, as *Hypnum dendroides.*

CRYPHAEACEAE

Cryphaea Mohr

Cryphaea heteromalla (Hedw.) Mohr
   C. arborea (Beauv.) Lindb.
Epiphyte   Colonist
Possibly extinct; but should be looked for on elders on the Chalk. Recorded for all neighbouring counties except Hunts.
OUSE: Clapham Park Wood; *F.B.*: Turvey; *C. L. Higgins in J. Bot. 1884.*
First record: Abbot, 1798, as *Sphagnum arboreum.*

LEUCODONTACEAE
Leucodon Schwaegr.
Leucodon sciuroides (Hedw.) Schwaegr.
Epiphyte
Considered common by Abbot, but rare now. Recorded for all neighbouring counties.
IVEL: Shefford; *Marquand List*: Knocking Hoe; *E. C. Wallace*.
LEA: Chalgrave*; Limbury*; Luton Hoo*; *J.S.*
First record: Abbot, 1795, as *Hypnum sciuroides*.

Antitrichia Brid.
Antitrichia curtipendula (Hedw.) Brid.
Of doubtful record. Saunders' record in *J. Bot. 1884* is in error as his specimens are *Cirriphyllum crassinervium*. Recorded for Herts. and Northants.
OUZEL: Woburn, suspect without further evidence; *Abbot's annotated F.B.*, as *Hypnum curtipendulum*.

NECKERACEAE
Neckera Hedw.
Neckera crispa Hedw.
Chalk downs   Native
Locally common; recorded for Herts. and Bucks.
IVEL: Markham Hills*; *J.S.*: Barton Hills, also found here by P.T.*; *H. N. Dixon*: Sharpenhoe Clappers*; Pegsdon!*; *P.T.*
LEA: Warden Hill; *T.L.*
First record: Saunders in *Trans. Beds. Nat. Hist. Soc.* 1882.

N. complanata (Hedw.) Hueben.
Tree trunks and chalk downs   Native
An occasional to frequent epiphyte, especially on ash on calcareous soils. It occurs sometimes on the ground in beech woods, on chalk downs and in overgrown chalk pits. Recorded for all districts and neighbouring counties.
H.S. 69.
First record: Abbot, 1798, as *Hypnum complanatum*.

Homalia (Brid.) Bruch, Schimp. & Guemb.
Homalia trichomanoides (Hedw.) Bruch, Schimp. & Guemb.
Epiphyte
Occasional to frequent on calcareous soils especially on ash stools. Recorded for all districts and neighbouring counties.
First record: Saunders in *Trans. Beds. Nat. Hist. Soc.* 1882.

Thamnium Bruch, Schimp. & Guemb.
Thamnium alopecurum (Hedw.) Bruch, Schimp. & Guemb.
Porotrichum alopecurum (Hedw.) Mitt.
Woods, masonry and stones   Native
Common in woods on calcareous soils; occasional on brickwork, limestone blocks, chalk blocks and flints in woodland or overgrown chalk pits: rare at the base of dense hedges on calcareous clays. Recorded for all districts and neighbouring counties.
H.S. 18, 20, 52, 78.
First record: Abbot, 1795, as *Hypnum alopecurum*.

LESKEACEAE
Leskea Hedw.
Leskea polycarpa Hedw.
Near water   ? Native
Common by the Ouse above Bedford, rare below Bedford and by the Ivel. Found commonly on tree roots, bases of trees and wooden piles, and occasionally on masonry and river banks. Recorded for the Ouse and Ivel districts and for all neighbouring counties.
First record: Abbot in undated letter to J. E. Smith, as *Hypnum polycarpum*.

THUIDIACEAE
Anomodon Hook. & Tayl.
Anomodon viticulosus (Hedw.) Hook. & Tayl.
In varied habitats   Native
Occasional on chalk downs and chalk banks in beech woods. It appears also as
an epiphyte occasionally on the Chalk and rarely on calcareous clay. It is also
rare on brickwork near water. Recorded for all districts except Nene and Kym,
and for all neighbouring counties except Hunts.
First record: Abbot, 1798, as *Hypnum viticulosum*.

Thuidium Bruch, Schimp. & Guemb.
Thuidium abietinum (Hedw.) Bruch, Schimp. & Guemb.
  Hypnum abietinum Hedw., *Abietinella abietina* (Hedw.) C. Muell.
Chalk downs   Native
Probably extinct: recorded for Herts., Cambs. and Northants.
IVEL: Barton Hills; *F.B.*

T. hystricosum Mitt.
  Abietinella hystricosa (Mitt.) Broth.
Chalk downs   Native
Rare: recorded for Bucks.
IVEL: Deacon Hill (1948)!*; *P.T.*

T. tamariscinum (Hedw.) Bruch, Schimp. & Guemb.
Woods   Native
A frequent ground moss in woodland, but confined on the Lower Greensand to the
more loamy woods in the west of the county. It is apparently absent on the Chalk,
although it is found on the Clay-with-Flints. Recorded for all districts and
neighbouring counties.
H.S. 2b, 3b, 19, 20, 30, 34a, 34b, 51a.
First record: Saunders in *Trans. Beds. Nat. Hist. Soc.* 1882.

T. philiberti Limpr.
Chalk banks   Native
Rare: recorded for all neighbouring counties except Hunts.
OUZEL: Sewell (1949)*; *P.T.*

HYPNACEAE
Cratoneuron (Sull.) Roth
Cratoneuron filicinum (Hedw.) Roth
  Hypnum filicinum Hedw., *Amblystegium filicinum* (Hedw.) De Not.
Marshy places   Native
Common in calcareous marshes and springs, frequent by rivers and streams,
occasional by ponds in disused gravel pits. It is apparently absent from the
Lower Greensand. Saunders found it in the Lea and Colne districts, but it is
absent there now. Recorded for all districts and neighbouring counties.
H.S. 10a–10c.
First record: Abbot, 1798.
      Var. **fallax** (Brid.) Moenkm. (var. *vallisclausae* (Brid.) Wils.)
IVEL: Harlington Ponds*; *J.S. in Herb. Luton Museum*, 1883.

C. commutatum (Hedw.) Roth
  Hypnum commutatum Hedw., *C. glaucum* (Lam.) C. Jens.
Marshy places   Native
Rare: recorded for Northants.
OUZEL: Hockliffe*; *J.S.*
IVEL: Flitwick Moor*; Markham Hills*; *J.S.*
First record: Saunders in *Trans. Beds. Nat. Hist. Soc.* 1882.

Campylium (Sull.) Mitt.
Campylium stellatum (Hedw.) Lange & C. Jens.
  Hypnum stellatum Hedw.
Marshes   Native
Rare or extinct: recorded for all neighbouring counties except Hunts.

OUZEL: Heath and Reach*; Totternhoe Mead*; *J.S.*
IVEL: Ampthill; *F.B.*: Flitwick Moor*; *J.S.*
LEA: Limbury Marsh*; *J.S.*
First record: Abbot, 1798.

C. chrysophyllum (Brid.) Bryhn
Hypnum chrysophyllum Brid.
Pastures   Native
Frequent in pastures and on grassy banks on the Oolite and Chalk. Recorded for all districts except Kym and Cam, and for Bucks., Cambs. and Northants.
H.S. 63b.
First record: Saunders in *Trans. Beds. Nat. Hist. Soc.* 1882.

C. elodes (Spruce) Broth.
Hypnum elodes Spruce
Marshy pastures   Native
Rare or extinct: recorded for Cambs.
OUSE: Ravensden; *Marquand List.*
OUZEL: Totternhoe*; *J.S.*
First evidence: Saunders in Herb. Luton Museum, 1883.

Leptodictyum (Schimp.) Warnst.
Leptodictyum riparium (Hedw.) Warnst.
Hypnum riparium Hedw., *Amblystegium riparium* (Hedw.) De Not.
In or near water   Native
Common on masonry, wood and tree roots, at about water level, by rivers, streams and ponds. It is occasionally submerged. Recorded for all districts and neighbouring counties.
H.S. 43.
First record: Abbot, 1798.

Amblystegium Bruch, Schimp. & Guemb.
Amblystegium serpens (Hedw.) Bruch, Schimp. & Guemb.
Hypnum serpens Hedw.
In varied habitats   Native
Common in woods, especially on fallen wood, except on the Lower Greensand; frequent on hedgerow and woodland stumps, on brickwork near water and at shady bases of walls. It is an epiphyte on elders, especially on the Chalk and Lower Greensand, and on willows by the larger rivers. It occurs also on compacted gravel in old gravel pits, on chalk banks in beech woods and on stones on embankments and cuttings. Recorded for all districts and neighbouring counties.
H.S. 1a, 1b, 5, 63b.
First record: Abbot, 1798.

A. juratzkanum Schimp.
In varied habitats in moist places   ? Native
Widespread, but not common, on masonry and wood near water, at bases of shaded walls and on tree stumps. Recorded for Bucks., Cambs., Hunts. and Northants.
OUSE: marsh, Stevington; by river, Eaton Socon; *T.L.*
KYM: wall, Knotting Church*; *J.G.D.*
IVEL: Barton Cutting; *J. P. M. Brenan*: Blunham Bridge; base of elder, Barton Hills; *T.L.*
LEA: Luton Hoo*; *J.S.*
First evidence: Saunders in Herb. Luton Museum, 1883.

A. kochii Bruch, Schimp. & Guemb.
Leptodictyum trichopodium var. *kochii* (Bruch, Schimp. & Guemb.) Broth.
Rare: recorded for Bucks. and Northants.
CAM: pondside, Cockayne Hatley (1949); *T.L.*

A. varium (Hedw.) Lindb.
Rare: recorded for Bucks., Cambs. and Northants.
OUSE: pondside, Wyboston (1950); *T.L.*

Drepanocladus (C. Muell.) Roth

Drepanocladus aduncus (Hedw.) Warnst.
Hypnum aduncum Hedw., *H. kneiffii* Schimp.
Near water   Native
Locally common in ponds and by ditches on calcareous clays and marls, occa-
sionally by the Ouse. Recorded for all districts and for all neighbouring counties.
H.S. 14b.
First record: Abbot, 1798.

D. sendtneri (Schimp.) Warnst.
Hypnum sendtneri Schimp.
Probably extinct: not recorded for any neighbouring county.
OUZEL: Totternhoe Mead*; *J.S. in J. Bot. 1884*.

D. fluitans (Hedw.) Warnst.
Hypnum fluitans Hedw.
Ditches   Native
Locally frequent: recorded for all neighbouring counties except Hunts.
OUSE: ponds, Goldington Road, Bedford; *F.B.*
OUZEL: Totternhoe*; *J.S.*
IVEL: Flitwick Moor!*; *J.S.*
LEA: Luton*; Limbury*; *J.S.*
First record: Abbot, 1798.

D. exannulatus (Bruch, Schimp. & Guemb.) Warnst.
Hypnum exannulatum Bruch, Schimp. & Guemb.
Bogs   Native
Rare: not recorded for any neighbouring county.
OUZEL: Mermaid's Pond; *T.L.*
IVEL: Flitwick Moor*; *J.S.*
First record: Saunders in *J. Bot.* 1884.

D. revolvens (Sm.) Warnst. var. **intermedius** (Lindb.) Rich. & Wall.
Hypnum intermedium Lindb., *D. intermedius* (Lindb.) Warnst.
Rare or extinct: recorded for Northants.
OUZEL: Totternhoe (1882); *J.S. in Herb. H. N. Dixon.*

Hygrohypnum Lindb.

Hygrohypnum luridum (Hedw.) Jennings (*Hypnum palustre* Brid., *Hygrohypnum
palustre* (Brid.) Loeske) is of doubtful record for the county. The record of Saunders
from *V.C.H.* is in error; his specimen is *Eurhynchium murale*. A record from Flitwick
in the *Marquand List* is unlikely and unconfirmed.

Acrocladium Mitt.

Acrocladium stramineum (Brid.) Rich. & Wall.
Hypnum stramineum Brid., *Calliergon stramineum* (Brid.) Kindb.
Bogs   Native
Rare, growing usually in association with *Sphagnum*.
Recorded for Bucks.
OUZEL: Mermaid's Pond; *T.L.*
IVEL: Flitwick Moor!*; *J.S.*
First record: Saunders in *J. Bot.* 1884.

A. cordifolium (Hedw.) Rich. & Wall.
Hypnum cordifolium Hedw., *Calliergon cordifolium* (Hedw.) Kindb.
Marshy places   Native
Limited to the Lower Greensand: recorded for Herts., Hunts. and Northants.
OUZEL: Mermaid's Pond; *J.S.*: King's Wood!*; *P.T.*
IVEL: Flitwick Moor!*; Westoning Moor*; *J.S.*
First record: Saunders in *Trans. Beds. Nat. Hist. Soc.* 1882.

A. cuspidatum (Hedw.) Lindb.
  Hypnum cuspidatum Hedw., *Calliergonella cuspidata* (Hedw.) Loeske
Pastures and wet places   Native
Common in calcareous marshes, wet pastures, moist banks, moist rides in woods,
chalk pastures and downs, by ponds, rivers, streams and ditches. Less common on
the Lower Greensand than elsewhere. Recorded for all districts and neighbouring
counties.
H.S. 2b, 10a–10c, 34a, 34b, 37, 51a, 53b, 57, 63b.
First record: Abbot, 1798.

Isothecium Brid.

Isothecium myurum (Brid.) Brid.
  Eurhynchium myurum (Brid.) Dix., *I. viviparum* Lindb.
Epiphyte   ? Native
An occasional to frequent epiphyte and found also on stumps in woodland. It is
occasional on flints and rare on the ground in woods. Recorded for all districts
and neighbouring counties.
First record: Saunders in *Trans. Beds. Nat. Hist. Soc.* 1885.

I. myosuroides (Brid.) Brid.
  Hypnum myosuroides Brid., *Eurhynchium myosuroides* (Brid.) Schimp.
Epiphyte   Native
An occasional to common epiphyte and found also on stumps in woodland. It is
also occasional on oolitic blocks and flints but is rare on the ground in woods.
Recorded for all districts and neighbouring counties.
H.S. 19, 20, 33, 51a, 52.
First record: Abbot, 1795.

Camptothecium Bruch, Schimp. & Guemb.

Camptothecium sericeum (Hedw.) Kindb.
  Hypnum sericeum Hedw., *Homalothecium sericeum* (Hedw.) Bruch, Schimp. &
Guemb.
In varied habitats   ? Native
Common on masonry and on trees and stumps, especially on isolated and hedgerow
trees, but also common in woodland. It is also epiphytic on elders and is occa-
sional on bare chalk banks in beech woods. Recorded for all districts and neigh-
bouring counties.
H.S. 69.
First record: Abbot, 1798.

C. lutescens (Hedw.) Bruch, Schimp. & Guemb.
  Hypnum lutescens Hedw.
Pastures, chalk downs, grassy banks and scrub   Native
Common in likely habitats on the Oolite and Chalk. Recorded for all districts
except Kym and Cam, and for all neighbouring counties.
H.S. 61, 63b.
First record: Abbot, 1798.

Brachythecium Bruch, Schimp. & Guemb.

Brachythecium albicans (Hedw.) Bruch, Schimp. & Guemb.
In varied habitats   ? Native
Frequent in compacted sand of tracks and pits and in short turf on the Lower
Greensand and glacial gravels. Saunders found it also on thatch. Recorded for
the Ouse, Ouzel and Ivel districts and for all neighbouring counties.
H.S. 27.
First record: Saunders in *J. Bot.* 1884.

B. glareosum (Bruch) Bruch, Schimp. & Guemb.
Grassy banks   Native
Rare, but widespread, on calcareous soils. Recorded for Bucks., Cambs. and
Northants.

OUSE: Houghton Conquest; *T.L.*
OUZEL: Puddle Hill*; *J.S.*
IVEL: near Clophill*; *J.S.*: Barton; *T.L.*
LEA: Stopsley*; *J.S.*
First record: Saunders in *J. Bot.* 1884.

B. salebrosum (Web. & Mohr) Bruch, Schimp. & Guemb.
Probably extinct: recorded for Cambs. and Northants.
OUSE: Sheerhatch Wood; *F.B.*, as *Hypnum plumosum.*

B. rutabulum (Hedw.) Bruch, Schimp. & Guemb.
Hypnum rutabulum Hedw.
In varied habitats   Native
Common on waste ground, in pastures and bogs, on banks and roadside verges,
in woodland and scrub, on stumps, stools and fallen wood, and on walls and other
masonry. A frequent epiphyte. Recorded for all districts and neighbouring counties.
H.S. 1a–2b, 10a–11, 15, 27, 34a, 37, 38, 42, 43, 52, 53b, 57, 61, 63b, 64, 70, 76, 80b, 83.
First record: Abbot, 1798.

B. velutinum (Hedw.) Bruch, Schimp. & Guemb.   *Velvet Feather Moss*
Hypnum velutinum Hedw.
In varied habitats   Native
A frequent stump-moss, particularly in hedgerows. It appears also at the base of
hawthorn in scrub, on banks under hedges on the Lower Greensand and on chalk
banks in beech woods. It is occasional on the ground in woodland and on stones
on embankments and cuttings, rare on rotting wood, walls and as an epiphyte on
elders. Recorded for all districts and neighbouring counties.
H.S. 1a, 1b, 5, 45a, 70.
First record: Abbot, 1798.

B. populeum (Hedw.) Bruch, Schimp. & Guemb. is of doubtful record for the county.
Dixon's specimen from Woburn Sands in Herb. Kew refers to Bucks. Records
from Bolnhurst and Cardington in the *Marquand List* are unlikely and unconfirmed.

Cirriphyllum Grout

Cirriphyllum piliferum (Hedw.) Grout
Eurhynchium piliferum (Hedw.) Bruch, Schimp. & Guemb.
Woods, scrub and banks   Native
Frequent on the ground in woods and scrub, and on banks and on the sides of
ditches. It is rare on the Lower Greensand. Recorded for all districts and neigh-
bouring counties.
H.S. 2b, 20, 34a, 42, 55a, 78.
First record: Saunders in *Trans. Beds. Nat. Hist. Soc.* 1882.

C. crassinervium (Tayl.) Loeske & Fleisch.
Eurhynchium crassinervium (Tayl.) Bruch, Schimp. & Guemb.
Beech woods   Native
Locally common in a few stations on the Chalk. Recorded for all neighbouring
counties except Cambs. and Hunts.
IVEL: Barton Leete!*; *J.S.*
LEA: Hoo Wood*; *J.S.*
COLNE: Long Wood; *T.L.*
First record: Saunders in *Victoria County History*, 1904.

Eurhynchium Bruch, Schimp. & Guemb.

Eurhynchium striatum (Hedw.) Schimp.
In varied habitats   Native
Common on the ground in woods on calcareous soils, less common in woods on
acid soils, occasional in old scrub and rare on shaded slopes of chalk downs. It is
less common on the Lower Greensand than elsewhere. Recorded for all districts
and neighbouring counties.
H.S. 3b, 20, 33–34b, 67, 78, 83.
First record: Saunders in *Trans. Beds. Nat. Hist. Soc.* 1882.

E. praelongum (Hedw.) Hobk.
   Hypnum praelongum Hedw., *Oxyrrhynchium praelongum* (Hedw.) Warnst.
In varied habitats   Native
Very common on the ground in woodland and scrub and extending to the bases
of trees and stumps; common also at the bases of hedges, on banks and the sides
of ditches, in rough pasture and on shady waste ground. It is frequent in marshes
and occasional at the bases of walls and in stubble. Recorded for all districts and
neighbouring counties.
H.S. 2a, 2b, 19, 20, 23, 30, 33–34b, 40, 51a, 52, 53b, 54, 55a, 77a, 78, 83.
First record: Abbot, 1798.

E. swartzii (Turn.) Curn.
   Oxyrrhynchium swartzii (Turn.) Warnst.
Pastures, marshes, chalk banks and scrub   Native
Frequent on banks amongst grass and occasional in open rides and clearings in
woods on calcareous clays; common in rough pastures and scrub, on banks on the
Chalk and in calcareous marshes and springs. It is rare at the bases of walls.
Recorded for all districts and neighbouring counties.
H.S. 1a, 1b, 34a, 34b, 52, 55a.
First record: Saunders in *J. Bot.* 1884.

E. riparioides (Hedw.) Jennings
   Hypnum rusciforme Brid., *Platyhypnidium rusciforme* (Brid.) Fleisch., *E. rusciforme*
(Brid.) Milde, *Rhynchostegium ruscifolium* (Sm.) Bruch, Schimp. & Guemb.
Aquatic   ? Native
Common on stones, masonry, tree roots and wood in running water. Recorded for
all districts and neighbouring counties.
First record: Abbot, 1798.

E. murale (Hedw.) Milde
   Rhynchostegium murale (Hedw.) Bruch, Schimp. & Guemb.
Masonry   Wall denizen
Widespread, but not common. It grows usually in moist and shaded places.
Recorded for all districts except Kym, and for Herts., Bucks., Cambs. and
Northants.
First record: Saunders in *J. Bot.* 1884.

E. confertum (Dicks.) Milde
   Rhynchostegium confertum (Dicks.) Bruch, Schimp. & Guemb.
In varied habitats   Native
Widespread and frequent on stumps, a frequent epiphyte on *Salix fragilis* by rivers,
and elders on the Chalk. It is occasional on shady walls and rare on the ground
at hedge bases. Recorded for all districts and neighbouring counties.
H.S. 1b, 27.
First record: Saunders in *Trans. Beds. Nat. Hist. Soc.* 1882.

E. megapolitanum (Bland.) Milde
   Rhynchostegium megapolitanum (Bland.) Bruch, Schimp. & Guemb.
Rocks and banks   Native
Rare: recorded for Herts., Bucks. and Cambs.
OUSE: oolitic block, Woodcraft; *T.L.*
IVEL: chalk bank, Barton Leete (1949)*; *P.T. and T.L.*

Rhynchostegiella Limpr.

Rhynchostegiella pallidirostra (A. Braun) Loeske
   Hypnum pumilum Wils., *Oxyrrhynchium pumilum* (Wils.) Broth., *Eurhynchium pumilum*
(Wils.) Schimp.
Near water   Native
Rare. A record by Saunders from Barton in Herb. Luton Museum, det. A.
Boswell, is in error. Recorded for Bucks., Cambs. and Northants.
OUSE: tree root near water, Woodcraft (1952); *T.L.*

R. curviseta (Brid.) Limpr.
  Eurhynchium curvisetum (Brid.) Husn.
On detritus on masonry by water   ? Native
Rare: recorded for Northants.
OUSE: Tempsford Mill (1950); *T.L.*

R. tenella (Dicks.) Limpr.
  Eurhynchium tenellum (Dicks.) Milde, *Rhynchostegium tenellum* (Dicks.) Bruch,
Schimp. & Guemb., *Rhynchostegiella algiriana* (Brid.) Broth.
In varied habitats   ? Native
Widespread but rare on masonry, at bases of trees, on elders, on flints and chalk
blocks. Recorded for Bucks., Cambs. and Northants.
OUSE: near Bedford; *S. H. Adams*.
IVEL: Sandy (1878)*; *S. H. Adams*: Blunham Bridge*; *T.L.*
CAM: elm trees, Cockayne Hatley; *T.L.*
COLNE: Long Wood; Folly Wood; *T.L.*
First record: S. H. Adams in *Trans. Beds. Nat. Hist. Soc.* 1882.

Entodon C. Muell.

Entodon orthocarpus (La Pyl.) Lindb.
  Cylindrothecium concinnum (De Not.) Schimp.
Calcareous banks and chalk downs   Native
Rare and limited to the Chalk and Oolite. Recorded for Bucks., Cambs. and
Northants.
OUSE: Colworth; *E. Cooper*.
IVEL: Markham Hills*; *J.S.*: old chalk pit, Barton; *T.L.*
First evidence: E. Cooper in Herb. Bedford Public Library, 1856.

Pseudoscleropodium (Limpr.) Fleisch.

Pseudoscleropodium purum (Hedw.) Fleisch.
  Hypnum purum Hedw., *Brachythecium purum* (Hedw.) Dix.
Pastures, banks and scrub   Native
Common on the ground in open parts of woods, in scrub, on heaths, chalk downs,
banks, and at the sides of ditches. Found also occasionally in old pastures. Re-
corded for all districts and neighbouring counties.
H.S. 2a, 2b, 22b, 23, 27, 28, 45a, 49, 59, 61, 62, 65, 67, 75.
First record: Abbot, 1798.

Pleurozium Mitt.

Pleurozium schreberi (Brid.) Mitt.
  Hypnum schreberi Brid.
Heaths and acid woods   Native
Frequent on heaths, occasional in old sandpits and woodland rides on the Lower
Greensand. Also found occasionally on acid gravels. Recorded for the Ouse,
Ouzel, Ivel, Lea and Colne districts and for all neighbouring counties except
Hunts.
H.S. 23, 27, 47, 49.
First record: Abbot, 1798, as *Hypnum parietinum*.

Isopterygium Mitt.

Isopterygium elegans (Hook.) Lindb.
  Plagiothecium elegans (Hook.) Sull., *P. borrerianum* Spruce
Sandy banks   ? Native
Occasional and usually growing in light shade. It is apparently limited to the
western part of the Lower Greensand and the Clay-with-Flints. Recorded for
Herts., Bucks. and Northants.
OUZEL: Woburn*; *A. B. Jackson*: Aspley Wood*; *R. D. Meikle and P.T.*: King's
Wood*; *E. C. Wallace, R. D. Meikle and P.T.*: Bakers Wood; *T.L.*
IVEL: Clophill*; *P.T.*
COLNE: Whipsnade Common; *P.T. and T.L.*
First record: A. B. Jackson in Herb. Luton Museum, 1909.

Plagiothecium Bruch, Schimp. & Guemb.
Plagiothecium denticulatum (Hedw.) Bruch, Schimp. & Guemb.
Hypnum denticulatum Hedw.
In varied habitats   Native
Common on the ground on sandy banks, at bases of trees and on stumps on the
Lower Greensand, Clay-with-Flints and glacial gravels. It appears occasionally
in marshes on the Lower Greensand and on stumps and bases of trees elsewhere.
Recorded for all districts except Kym, and for all neighbouring counties.
H.S. 78, 79.
First record: Abbot, 1798.
    Var. **aptychum** (Spruce) Dix.
IVEL: Flitwick Moor (H.S. 40); Cox Hill, Sandy; *T.L.*
COLNE: Pepperstock*; *J.S. in J. Bot. 1884.*

P. silvaticum (Brid.) Bruch, Schimp. & Guemb.
In varied habitats   Native
Growing usually with the previous species, but much less common. A record of
Saunders of var. *succulentum* (Wils.) Husn. from Aspley in Herb. Luton Museum,
1883, is in error; his specimen is a form of *P. denticulatum.* Recorded for all districts
except Nene and Kym, and for all neighbouring counties.
First evidence: Saunders in Herb. Luton Museum, 1882.

P. undulatum (Hedw.) Bruch, Schimp. & Guemb.
    Hypnum undulatum Hedw.
Acid banks   Native
Rare: recorded for Herts., Bucks. and Northants.
OUZEL: Aspley*; *S. H. Adams*: King's Wood*; *E. C. Wallace, R. D. Meikle and
P.T.*
IVEL: Steppingley Firs*; Barton Hills*; *Mrs. Twidell*: sandpit, Silsoe*; *P.T.*
First record: Mrs. Twidell in *Beds. Nat. Hist. Soc.* 1885.

Hypnum Hedw.
Hypnum cupressiforme Hedw.
In varied habitats   Native
Very common on trees, stumps, fallen wood and on the ground in woodland, on
heaths, in marshes, on chalk downs and banks, and on walls and roofs. It is a
very variable species: varieties recognized by modern taxonomists and found in
the county are listed below. Recorded for all districts and neighbouring counties.
H.S. 3b, 5, 19, 22a, 22b, 27, 31, 40, 47, 51a, 52, 70, 80b.
First record: Abbot, 1798.
    Var. **resupinatum** (Wils.) Schimp. (*H. resupinatum* Wils.)
Common on trees and stumps in all districts.
    Var. **filiforme** Brid.
LEA: Luton Hoo*; *J.S.*
    Var. **ericetorum** Bruch, Schimp. & Guemb.
Common at the base of *Calluna* and on banks on heaths on the Lower Greensand
and the Clay-with-Flints. Recorded for all districts except Nene, Kym and Cam.
H.S. 24.
    Var. **tectorum** Brid.
Once frequent on old thatch and occasional on walls, but not found recently.
Recorded for the Ouzel, Ivel, Lea and Colne districts.
    Var. **lacunosum** Brid. (var. *elatum* Bruch, Schimp. & Guemb.)
Occasional on chalk downs. Recorded for the Ouzel, Ivel and Lea districts.

H. patientiae Lindb.
    Breidleria arcuata (Lindb.) Loeske
Woods   Native
Occasional in rides in loamy woods on the Clay-with-Flints and western part of the
Lower Greensand. Recorded for Herts. and Bucks.
OUZEL: King's Wood!*; *E. C. Wallace, R. D. Meikle and P.T.*
LEA: Luton Hoo*; *J.S.*: Kidney Wood!; *E. C. Wallace.*
COLNE: Slip End*; Pepperstock*; *J.S.*: Deadmansey Wood; *T.L.*
First record: Saunders in *J. Bot.* 1884.

H. pratense Koch (*Breidleria pratensis* (Koch) Loeske) was recorded by Saunders from Slip End in Herb. Luton Museum, 1885, in error for *H. patientiae*. *H. pratense* is not usually regarded as a British species.

Ptilium crista-castrensis (Hedw.) De Not. (*Hypnum crista-castrensis* Hedw.) was recorded by Abbot in error for *Ctenidium molluscum*. It was corrected by him in his annotated copy of *Flora Bedfordiensis*.

Ctenidium (Schimp.) Mitt.

Ctenidium molluscum (Hedw.) Mitt.
   Hypnum molluscum Hedw.
Pastures, chalk downs and woods   Native
Common on downs, banks and rough pastures on the Chalk; frequent in chalk scrub and woodland, and in rough pastures and banks on the Oolite. It is also occasionally found in woodland on calcareous clays. Recorded for all districts and neighbouring counties.
H.S. 34b, 59, 63a, 63b.
First record: Abbot's annotated *Flora Bedfordiensis, c.* 1799.

Rhytidiadelphus (Lindb.) Warnst.

Rhytidiadelphus triquetrus (Hedw.) Warnst.
   Hypnum triquetrum Hedw., *Hylocomium triquetrum* (Hedw.) Bruch, Schimp. & Guemb.
Woods, scrub and chalk downs.   Native
Frequent in scrub and woodland on calcareous soils and on shaded slopes of chalk downs. Recorded for all districts and neighbouring counties.
H.S. 23, 51a.
First record: Abbot, 1798.

R. squarrosus (Hedw.) Warnst.
   Hypnum squarrosum Hedw., *Hylocomium squarrosum* (Hedw.) Bruch, Schimp. & Guemb.
Pastures, banks, woods and marshes   Native
Frequent on banks, in rough pastures, woodland, scrub and bogs; particularly frequent in scrub on calcareous clays and chalk and in short rabbit-grazed turf on acid heaths. Recorded for all districts and neighbouring counties.
H.S. 49, 74a, 76.
First record: Abbot, 1798.

Hylocomium Bruch, Schimp. & Guemb.

Hylocomium splendens (Hedw.) Bruch, Schimp. & Guemb.
   Hypnum splendens Hedw., *H. proliferum* Brid., *Hylocomium proliferum* (Brid.) Lindb.
Pastures and heaths   Native
Occasional on acid heaths and chalk downs, particularly on acid cappings. Recorded only from the Lower Greensand, Chalk and Clay-with-Flints in the Ouse, Ouzel, Ivel, Lea and Colne districts. Recorded for all neighbouring counties except Hunts.
H.S. 74a.
First record: Abbot, 1798.

PTERIDOPHYTA

LYCOPODIACEAE

LYCOPODIUM L.

Lycopodium clavatum L. C.T.W. 1.1.4.   *Stag's-horn Moss*
Heathy places   Native
Probably extinct: recorded for all neighbouring counties except
Hunts.
IVEL: Potton Heath; *F.B.*
COLNE: Birchin Grove, Pepperstock, but the station was probably
a few yards within v.c. 20, Herts*; *J.S. in J. Bot. 1907*, 451.
First record: Abbot, 1798.

L. inundatum L. C.T.W. 1.1.2.   *Marsh Club-moss*
Heathy bogs   Native
Rare: recorded for all neighbouring counties except Hunts.
OUSE: Ampthill; *F.B.*
OUZEL: New Wavendon Heath, a single plant, v.c. 24 [Beds.]!*; *P.T.*
First record: Abbot, 1798.

EQUISETACEAE

EQUISETUM L.

Equisetum arvense L. C.T.W. 4.1.10.   *Field* or *Common Horsetail*
Arable land, waste places, etc.   Colonist
Common: recorded for all neighbouring counties.
H.S. 1b, 11, 13, 17, 44, 64, [58b].
First record: Abbot, 1798.

E. telmateja Ehrh. C.T.W. 4.1.11.   *Great Horsetail*
  E. maximum Lam.
Marshes, damp banks, etc.   Native
Local: this is probably the *E. fluviatile* of Abbot, which was con-
sidered common by him. Recorded for all neighbouring counties.
H.S. 10c, 43.
OUSE: Stevington, plentiful below the church!; *J.McL.*, *L.M.*:
Bedford; Pavenham!; *J.H.*: Cranfield; King's Wood.
KYM: roadside north of Riseley; *E.M.-R.*!: spinney between
Keysoe and Little Staughton.
OUZEL: Milton Bryan!; *W.H. 1875*: Woburn; *V.C.H.*: Battlesden
Park.
IVEL: Southill; *J.McL.*, *B.M.*: Barton, near the springs!†*; *S.B.P.L.*:
Washer's Wood, both in wood and marsh adjoining it; marsh at foot
of Cooper's Hill*; Fancott; plentiful in marsh, Long Lane, Tingrith.
First record: Abbot, 1798.

E. sylvaticum L. C.T.W. 4.1.8.   *Wood Horsetail*
Woods, etc.   Native
Apparently extinct, but it may reappear as it did on a railway bank
at Hitchin in 1932. Recorded for all neighbouring counties except
Hunts.
IVEL: Ampthill pleasure gardens; *Plantae Bedford.*: Ampthill; *F.B.*
First record: Abbot, 1795.

E. fluviatile L. C.T.W. 4.1.6.   *Smooth Horsetail*
    E. limosum L.
Wet places   Native
Frequent, but considered common by both Abbot and Saunders.
Recorded for all neighbouring counties.
H.S. 37, 39a, 40.
NENE: Hinwick Lodge.
OUSE: Kempston; Husborne Crawley; Oakley; Radwell.
KYM: Melchbourne Park.
OUZEL: Horsemoor Farm.
IVEL: near Toddington; *E. Forster in Bot. Guide*: Chalton†; *S.B.P.L.*:
R. Ivel, Sandy; *F.F.B.*: Flitwick, plentifully on the Moor!*; *J.H.*:
Gravenhurst; Flitton Moor.
LEA: Limbury†; *Herb. J.S.*
First record: Abbot, 1798.

E. palustre L. C.T.W. 4.1.7.   *Marsh Horsetail*
Wet places   Native
Common, but considered rare by Abbot. Recorded for all districts
except Colne, and for all neighbouring counties.
H.S. 10a, 10c, 12, 14b, 15, 38, 39a, 41, 56a, 84a.
First record: Abbot, 1798.

E. hyemale L. C.T.W. 4.1.1.   *Rough Horsetail*
Bogs   Native
Extinct: recorded for Cambs.
IVEL: Ampthill Moor; *Plantae Bedford.*: Potton Marshes; *F.B.*
First record: Abbot, 1795.

OSMUNDACEAE

OSMUNDA L.

Osmunda regalis L. C.T.W. 5.1.1.   *Royal Fern*
Damp woods on acid soils   Native
Rare: recorded for Bucks. where it may still survive, extinct in
Cambs. and Northants.
OUSE: Aspley Wood; *F.B.*: New Wavendon Heath, v.c. 24 [Beds.], a
few small plants, probably relics of a native colony; Stockgrove,
probably planted*.
IVEL: Tingrith Park, almost certainly planted; *J.S., J. Bot. 1889.*
First record: Abbot, 1798.

POLYPODIACEAE

PTERIDIUM Scop.

Pteridium aquilinum (L.) Kuhn C.T.W. 7.1.1.   *Bracken*
Pteris aquilina L., *Eupteris aquilina* (L.) Newman
Heaths, woods, walls, etc.   Native
Common in woods on heaths on light soils. When it appears on
walls it presents a very different appearance, but can usually be
recognised by the complete absence of scales on the stalk. Recorded
for all neighbouring counties.
H.S. 22a–24, 27, 31c, 36, 43, 47, 49, 53b, 54, 77b, 79a–80a, 81, 83.
NENE: railway wall, Wymington.
OUSE: Harrold; Tempsford*; *G. H. Day*: Odell Great Wood;
R. R. B. Orlebar: Sharnbrook; Pavenham.
KYM: Swineshead Church; Kimbolton Station, v.c. 30 [Hunts.].
OUZEL and IVEL: common on the Lower Greensand; rare on the
Chalk by Barton Cutting; *P.T.*
LEA: common near Chiltern Green; frequent on walls in Luton.
COLNE: common on the Clay-with-Flints.
First record: Abbot, 1798.

BLECHNUM L.

Blechnum spicant (L.) Roth C.T.W. 7.5.1.   *Hard Fern*
Osmunda spicant L.
Marshes, damp woods, heaths, etc.   Native
Rare and limited to acid soils: recorded for all neighbouring
counties.
H.S. 35.
OUZEL: Aspley†; *S.B.P.L.*: New Wavendon Heath, v.c. 24 [Beds.]*.
IVEL: Potton Marshes, where it was still found by McLaren; *F.B.*:
Flitwick Moor, where it grew on the sides of deep ditches, but not
appearing recently!†; *S.B.P.L.*
LEA: Luton Hoo†; *S.B.P.L.*
First record: Abbot, 1798.

PHYLLITIS Hill

Phyllitis scolopendrium (L.) Newman C.T.W. 7.6.1.
                                              Hart's-tongue Fern
Asplenium scolopendrium L., *Scolopendrium vulgare* Sm.
Old walls, hedgebanks, etc.   Native
Frequent on walls, where it can often be traced to cultivated
plants near by; rare elsewhere. Recorded for all neighbouring
counties.
NENE: Wymington; *P.T.*!: Hinwick.
OUSE: Bromham Library steps; *Plantae Bedford.*: Cardington;
J.McL., B.M.: Elstow; *J.McL., L.M.*: Newnham; *W. Davis in J.H.*:
Elstow; *J.H.*: Lidlington; *C.C.*: Hawnes; *C. C. Foss.*

N

Kym: Pertenhall Church; Kimbolton Station, v.c. 30 [Hunts.].
Ouzel: Houghton Hall; *E.M.-R.*: sides of lakes, Woburn Park.
Ivel: Eversholt; *F.B.*: Ampthill; *W.H. 1875*: Toddington; *S.B.P.L.*:
Potton Station; *P.T.*!: Harlington; Barton Church; *P.T.*: Ickwell
Bury; *C. C. Foss*!
Lea: Luton Hoo!; *S.B.P.L.*: East Hyde*; Leagrave.
Colne: hedgebank, Whipsnade, the only station known not on
brickwork!; *P.T.*: Caddington Church.
First record: Abbot, 1795.

ASPLENIUM L.

Asplenium trichomanes L. C.T.W. 7.7.4.   *Maidenhair Spleenwort*
Old walls   Wall denizen
Rare: the status of this and other ferns now found growing on old
walls is difficult to decide. They can be traced westwards through
equally artificial habitats to natural habitats in the West country,
but in many cases with us they have been introduced from gardens.
Recorded for all neighbouring counties.
Ouse: Stafford Bridge; *Plantae Bedford.*: Pavenham Church;
Abbot's annotated F.B.: Cardington; *J.McL.*, *L.M.*: Turvey;
G. H. Day.
Ouzel: Heath and Reach†; *P.T.*: sides of lakes, Woburn Park*.
Ivel: Potton Station; *P.T.*!: Ampthill Station.
Lea: Luton Hoo†*; *H. Catt in S.B.P.L.*: Leagrave Station.
First record: Abbot, 1795.

A. adiantum-nigrum L. C.T.W. 7.7.1.   *Black Spleenwort*
Old walls   Wall denizen
Rare: recorded for all neighbouring counties.
Nene: Wymington; *P.T.*!
Ouse: Stafford Bridge; *Plantae Bedford.*: Willington; *P.T.*!
Souldrop*.
Kym: Swineshead Church.
Ouzel: Leighton Buzzard*; *B. Verdcourt.*
Lea: East Hyde; *P.T.*: Leagrave Station; *S. P. Rowlands*!
First record: Abbot, 1795.

A. ruta-muraria L. C.T.W. 7.7.6.   *Wall Rue*
Old walls   Wall denizen
Frequent: recorded for all neighbouring counties.
Nene: Wymington; *P.T.*!
Ouse: Bromham Library steps; *Plantae Bedford.*: Bedford; *J.H.*:
Chellington; Turvey!; Harrold; *G. H. Day*: Stevington; *P.T.*:
Bromham Bridge.
Ivel: Hawnes Park; *C. C. Foss*: Ampthill Station.
Lea: Luton Hoo†; *H. Catt in S.B.P.L.*: East Hyde; Putteridge
Park Wall; Leagrave Station.
Colne: wall, Stockwood Park*.
First record: Abbot, 1795.

CETERACH Garsault

Ceterach officinarum DC. C.T.W. 7.8.1.   *Rusty-back Fern*
Old walls   Wall denizen
Rare: recorded for Herts., Bucks. and Northants.
Ouse: Stevington!*; *P.T.*
Ivel: Ampthill Station*.
Lea: Chiltern Green Station!; *P.T.*: Leagrave Station*.
First record: P.T. in *B.E.C. 1946–7 Rep.* 322.

ATHYRIUM Roth

Athyrium filix-femina (L.) Roth C.T.W. 7.9.1.   *Lady Fern*
  Polypodium filix-femina L.
Damp woods, rarely on walls, etc.   Native
Frequent: recorded for all neighbouring counties.
H.S. 53b.
Nene: Great Hayes Wood; *P.T.*!
Ouse: Ampthill Park; *H. B. Souster*!
Ouzel: Aspley Wood†*; *J.McL.*, *L.M.*: King's Wood; *P.T.*!: New
Wavendon Heath, v.c. 24 [Beds.]; *P.T.*
Ivel: Potton Marshes; *F.B.*: Flitwick Moor*; *J.S.*: Cooper's Hill*;
P.T.!: Daintry Wood; *P.T.*: Moors Plantation; abundant, Becker-
ingspark Moor; Washer's Wood.
Lea: Horsley's Wood!*; Luton Hoo!; Leagrave Station; Whitehill
Wood; *P.T.*
Colne: Elm Grove Wood; Deadmansey Wood; Ravensdell Wood;
P.T.
First record: Abbot, 1798.

CYSTOPTERIS Bernh.

Cystopteris fragilis (L.) Bernh. C.T.W. 7.10.1.   *Brittle Bladder Fern*
Walls   Wall denizen
Rare: recorded for Herts. and Northants. This, with the other ferns recorded
from Leagrave Station, presents interesting problems of status. The wall is kept
perpetually damp with the exhaust from locomotives, which solves the problem
of their survival, but whether they have been introduced by means of the railroad,
or whether this station can be regarded as a link in discontinuous distribution
westwards, is a problem which merits some study.
Lea: Leagrave Station*; *P.T. in Watsonia II (1951)* 61.

DRYOPTERIS Adans.

Dryopteris filix-mas (L.) Schott C.T.W. 7.12.1.   *Male Fern*
  Polypodium filix-mas L., *Lastrea filix-mas* (L.) C. Presl
Woods, rarely on old walls, etc.   Native
Most frequent in woods on light soils but occasionally found in
woods on the Boulder Clay and Chalk. Recorded for all districts
and neighbouring counties.
H.S. 18, 30a, 55b, 79b, 80b, 81.
First record: Abbot, 1798.

D. borreri Newman C.T.W. 7.12.2.   *Golden-scaled Male Fern*
   D. filix-mas var. *paleacea* Druce p.p.
Woods   Native
Rare and found only where there is some quantity of *D. filix-mas*.
Recorded for all neighbouring counties except Cambs.
NENE: Great Hayes Wood!; *P.T.*
OUZEL: King's Wood!*; *P.T.*
LEA: Luton Hoo*; *P.T.*!: Horsley's Wood; *P.T.*
COLNE: Ravensdell Wood; *P.T.*
First record: P.T. in *B.E.C. 1946–7 Rep.* 322.

D. spinulosa (O. F. Muell.) Watt C.T.W. 7.12.6.
   Lastrea spinulosa (O. F. Muell.) C. Presl   *Narrow Buckler Fern*
Damp woods, marshes, etc.   Native
Rare: recorded for all neighbouring counties.
H.S. 40, 42, 49, 53b.
OUZEL: King's Wood; *P.T.*
IVEL: Flitwick, in association with and with intermediates with
D. dilatata!†*; Sutton Fen!; *W.F.B.*: Westoning Moor; *C.C.*: Moors
Plantation; Washer's Wood; Beckeringspark Moor.
LEA: Luton Hoo; *P.T.*!: Horsley's Wood; *P.T.*
COLNE: Kidney Wood; *P.T.*
First record: *Newbould MS.*, 1874.

D. × *uliginosa* (Newman) Kuntze ex Druce (*Lastrea uliginosa*
Newman) was recorded by J. McLaren (*J.McL., B.M.*) from Flitwick in error—see author's note in *B.E.C. 1943–4 Rep.* 813.

D. dilatata (Hoffm.) A. Gray   *Broad Buckler Fern*
   Lastrea dilatata (Hoffm.) C. Presl, *D. austriaca* auct. C.T.W. 7.12.7.
Damp woods   Native
Frequent. This was not recorded by Abbot but the plate he cites in
his record of *Polypodium cristatum* L. (*Dryopteris cristata* (L.) A. Gray)
from Aspley Wood and Potton Marshes represents *D. dilatata*. *D.
cristata* is recorded for Hunts. and *D. dilatata* for all neighbouring
counties.
H.S. 33, 35, 42, 49.
NENE: Great Hayes Wood; *P.T.*: Hinwick.
OUSE: Odell Dungee*; *G. H. Day*: Sheerhatch Wood; *P.T.*!: King's
Wood.
OUZEL: Aspley Wood, as *D. spinulosa*; *J.McL., L.M.*: Woburn; *V.C.H.*:
King's Wood!; New Wavendon Heath, v.c. 24 [Beds.]!; *P.T.*
IVEL: Potton Marshes; *J.McL., L.M.*: Flitwick Wood; *J.S., J. Bot.
1889*: Southill; *J.E.L., J. Bot.*: Flitwick Moor!*; *J.E.L., Diary*: Clop-
hill; Silsoe; *P.T.*: Moors Plantation; Sutton Fen; Washer's Wood.
LEA: Horsley's Wood; Luton Hoo; Whitehill Wood; *P.T.*
COLNE: Deadmansey Wood!; Elm Grove!; *P.T.*: Long Wood;
S. P. Rowlands!
First certain record: J. McLaren, 1864.

POLYSTICHUM Roth

Polystichum aculeatum (L.) Roth   *Prickly Shield Fern*
   Polypodium aculeatum L., *Polystichum lobatum* (Huds.) C. Presl
C.T.W. 7.13.2.
Damp shady banks   Native
Rare: recorded for all neighbouring counties.
IVEL: Eversholt, where it was still found by McLaren; *Plantae Bedford.*: Flitwick; *S.B.P.L.*: Chicksands; *C.C. in J. Bot. 1889*: Steppingley!; *P.T.*
LEA: Bramingham; *J.S., J. Bot. 1883*: Chiltern Green, where it has not been seen since 1948!; *P.T.*
COLNE: Whipsnade!*; *P.T.*: Caddington*; *L. J. Margetts.*
First record: Abbot, 1795.

THELYPTERIS Schmidel

Thelypteris oreopteris (Ehrh.) Slosson C.T.W. 7.14.1.
                                                       Mountain Fern
   Lastrea montana Moore, *Dryopteris oreopteris* (Ehrh.) Maxon
Heaths   Native
Rare: Listed for v.c. 30 in *Comital Flora*, but I know of no basis for its inclusion. Recorded for all neighbouring counties except Hunts.
OUZEL: New Wavendon Heath, v.c. 24 [Beds.]!*; *P.T. in B.E.C. 1946-7 Rep.* 322.

T. palustris Schott C.T.W. 7.14.2.   *Marsh Fern*
   Polypodium thelypteris (L.) Weis, *Lastrea thelypteris* (L.) Bory, *Dryopteris thelypteris* (L.) A. Gray
Bogs   Native
Extinct: recorded for all neighbouring counties except Herts.
IVEL: Potton Marshes; *F.B.*

GYMNOCARPIUM Newman

Gymnocarpium dryopteris (L.) Newman   *Oak Fern*
   Phegopteris dryopteris (L.) Fée, *Dryopteris dryopteris* (L.) Christ, *Thelypteris dryopteris* (L.) Slosson C.T.W. 7.14.4.
Walls   Wall denizen
Rare: recorded for Herts. and Bucks.
LEA: Leagrave Station*; *P.T.*!

G. obtusifolium (Schrank) Schwarz   *Limestone Polypody*
   G. robertianum (Hoffm.) Newman, *Phegopteris robertiana* (Hoffm.) A. Braun, *Dryopteris robertiana* (Hoffm.) C. Chr., *Thelypteris robertiana* (Hoffm.) Slosson C.T.W. 7.14.5.
Walls   Wall denizen
Rare: recorded for Bucks. and Northants.
LEA: Leagrave Station*; *P.T.*! *in Watsonia I (1951)* 61.

POLYPODIUM L.

Polypodium vulgare L. C.T.W. 7.15.1.   *Polypody Fern*
Old walls, banks, etc.   Native
Considered common by Abbot, it is now rare. Recorded for all neighbouring counties.

NENE: Hinwick; *P.T.*!: Farndish.
OUSE: old wall, Turvey.
KYM: Pertenhall Church*.
OUZEL: Aspley Heath in woodland, v.c. 24 [Beds.]; *P.T.*
IVEL: Sandy; *J.McL.*, *B.M.*: Silsoe, on bank!†*; *Herb. J.S.*:
brickwork, Ickwell Bury; *C. C. Foss*!: old walls, Westoning, Flitwick*,
Potton.
COLNE: near Zouche's Farm, v.c. 20 [Beds.]; *Fl. Herts.*: bank,
Whipsnade, now gone!; *P.T.*
First record: Abbot, 1798.

OPHIOGLOSSACEAE

BOTRYCHIUM Sw.

Botrychium lunaria (L.) Sw. C.T.W. 10.1.1.   *Moonwort*
   Osmunda lunaria L.
Dry pastures   Native
Probably extinct: recorded for all neighbouring counties except
Hunts.
OUSE: Oakley West Field, under the furze; *Plantae Bedford.*

OPHIOGLOSSUM L.

Ophioglossum vulgatum L. C.T.W. 10.2.1.   *Adder's Tongue*
Pastures, wet pastures, woodland drives, etc.   Native
Frequent: an interesting species with a wide range of habitats from
waterlogged meadows to almost pure chalk. It is much eaten by
cattle: recorded for all neighbouring counties.
NENE: Hinwick; *R. R. B. Orlebar*: Great Hayes Wood*.
OUSE: Fenlake; *Abbot's annotated F.B.*: near Sheerhatch Wood;
J.McL., *L.M.*: Wilstead; Hillfoot Farm; *J.McL.*, *B.M.*: Bushmead;
D. Martyn in J.H. Notes: Ravensden; *A. Ransome in J.H. Notes*:
Turvey; *Mr. Fernaby in J.H. Notes*: Bolnhurst; *I. Evans in J.H. Notes*:
Harrold*; *G. H. Day*: Brown's Wood; *Mrs. Soper*: Souldrop;
West Wood; *P.T.*!
KYM: Dean; *M. Dalton.*
OUZEL: Aspley; *C.C.*: Thorn; *S.B.P.L.*: Houghton Regis; *E.M.-R.*:
Totternhoe Knolls!; *P.T.*: Salford Wood; *J. Burrell*: Cow Common.
IVEL: Cainhoe; *Herb. Miss Crouch*: Sundon†; *S.B.P.L.*: Toddington;
Mrs. Horley in J.H. Notes: Flitwick; *M. L. Berrill in J.H. Notes*:
Ridgmont; Clophill; *C.C.*: Barton Leete; Sharpenhoe; *P.T.*
CAM: Potton Wood; *B. Verdcourt*: Cockayne Hatley Wood.
LEA: Biscot†; *S.B.P.L.*
COLNE: Caddington; *S.B.P.L.*: Stockwood; *G. K. Long*: Studham,
v.c. 20 [Beds.]*.
First record: Abbot, 1798.

SPERMATOPHYTA
GYMNOSPERMAE

The pinetum at Woburn Park is one of the finest and oldest in the country and there is a smaller but equally old one at Flitwick Manor. Most parks also have a number of planted coniferous trees. A full list of the species still growing in the county, with references to Forbes's *Pinetum Woburnensis*, would be beyond the scope of this work.

PINACEAE

PINUS L.

Pinus sylvestris L. C.T.W. 11.6.1. *Scots Pine*
Woods, plantations, roadsides, etc. Planted
Planted extensively on the Lower Greensand and less so on the Chalk. It is just possible that it may be native on the Lower Greensand, where seedlings are frequently seen. Recorded for all districts and neighbouring counties.
H.S. 27, 45b, 47.
First record: Herb. McLaren, Luton Museum, 1864.

LARIX Mill.

Larix decidua Mill. C.T.W. 11.5.1. *Larch*
  L. larix (L.) Karst.
Hedgerows, plantations, etc. Planted
Frequently planted and often self-sown. Recorded for all districts and neighbouring counties.
H.S. 27.

CUPRESSACEAE

JUNIPERUS L.

Juniperus communis L. C.T.W. 12.1.1. *Juniper*
Downland Native
Probably extinct: recorded for Herts., Bucks. and Northants.
IVEL: Warden; *F.B.*: between Warden and Southill, probably planted†; *J.McL. in Herb. J.S.*: one bush on Barton Hills, where I have failed to find it; *J.S. in J. Bot. 1889.*
First record: Abbot, 1798.

TAXACEAE

TAXUS L.

Taxus baccata L. C.T.W. 13.1.1. *Yew*
Churchyards, old gardens, etc. Planted
Frequent, but in no place appearing to be wild. Recorded for all districts and with varying degrees of status for all neighbouring counties.
First record: Abbot in letter to J. E. Smith, 3 Sept. 1802.

ANGIOSPERMAE : DICOTYLEDONES

RANUNCULACEAE

CLEMATIS L.

Clematis vitalba L. C.T.W. 14.10.1.   *Traveller's Joy,*
                                                      Old Man's Beard
Hedgerows, wood borders, chalk hills, etc.   Native
Frequent as a climbing plant on calcareous soils but absent from the
Lower Greensand. It is most common on the Chalk, where a pros-
trate form occurs on rough ground. Recorded for all districts and
neighbouring counties.
H.S. 63b, 69.
First record: Abbot, 1798.

THALICTRUM L.

Thalictrum flavum L. C.T.W. 14.15.1.   *Meadow Rue*
Riversides, water meadows, etc.   Native
A feature of some of the Ouse water meadows, rare elsewhere.
Recorded for all neighbouring counties.
OUSE: Cox's Pits; *J.H. Notes*; Fenlake; *B.P.L.*: Cardington Mill;
J.McL., B.M.: Harrold; *G. H. Day*: Oakley; *A. W. Guppy*: Bromham;
Stevington; Sharnbrook!*; *B. B. West*: Castle Mills; *H. B. Souster*!:
meadows, Felmersham, Biddenham; frequent in meadows between
Bromham and Kempston.
OUZEL: meadows, Leighton Buzzard; *Mr. Piffard in S.B.P.L.*
IVEL: Gravenhurst; *W.C. Herb.*: Biggleswade; *Mrs. Cotton in J.H.
Notes*: meadow near Campton Mill.
LEA: Limbury, as var. *riparium* Jord.—the site is now built over, I
saw it last in 1931!*; *F.F.B.*
First record: Abbot, 1798.

A specimen in Herb. R. H. Webb collected at Flitwick in 1841 was named by
Pryor *T. minus* ? var. *saxatile*, see *J. Bot.* 1876. Little in *J.E.L., F.F.B.* refers this to
T. calcareum Jord., but there is no evidence that he had seen the specimen. It is
unlikely that any of the forms of *T. minus* L. would appear at Flitwick except as
garden escapes.

ANEMONE L.

Anemone pulsatilla L. C.T.W. 14.9.4.   *Pasque Flower*
Chalk downs   Native
Rare: one of our more attractive plant species. Recorded for all
neighbouring counties.
H.S. 59.
IVEL: Barton Hills, where it occurs sparingly!†; *F.B.*: Pegsdon Hills,
? Deacon Hill, where it occurs in small quantity!; *I. Brown in Herb.*
Watson: Sundon Hills; *J.McL., B.M.*: Streatley, plentiful on
Smithcombe Hill!; *B.P.L.*: Table Hill; *P.T.*: Sharpenhoe Clappers;
B. Garratt: Knocking Hoe, plentifully*.
LEA: Luton Downs, probably Warden Hill; *F.B.*
First record: Abbot, 1798.

A. nemorosa L. C.T.W. 14.9.1.   *Wood Anemone*
Woods   Native
Common, especially in woods on the Boulder Clay.   J. E. Little
recorded var. *purpurea* DC. from Cockayne Hatley Wood in his
diary. Recorded for all districts and neighbouring counties.
H.S. 2b, 18, 19b, 20b, 32, 33, 34a, 34b, 51a, 52, 54, 55a, 78, 81, 83.
First record: Abbot, 1798.

ADONIS L.

Adonis annua L. C.T.W. 14.12.1.   *Pheasant's Eye*
   A. autumnalis L.
Cornfields, rough pastures, etc.   Colonist
Rare: but apparently once more common. Abbot wrote in his
annotated copy of *Flora Bedfordiensis* 'Mrs. Abbot has often brought
this elegant plant in from the cornfields and it would have been
inserted in the body of the work, but for the general suspicion of
its being not indigenous'. Recorded for all neighbouring counties
except Hunts.
OUSE: Pickerings; St. Leonards; *Abbot's annotated F.B.*
IVEL: field, border of Silsoe Park; *Herb. Miss Crouch*: Pegsdon!*;
Ruth Bott in J.E.L., B.E.C.
LEA: waste ground, Old Bedford Road, Luton, garden escape*;
K. Goodwin.
First record: Abbot in letter to J. E. Smith, 3 Sept. 1802.

MYOSURUS L.

Myosurus minimus L. C.T.W. 14.13.1.   *Mouse-tail*
Arable land   Colonist
Rare, but possibly overlooked. It is apparently established in the
same field for a number of years. Recorded for all neighbouring
counties.
OUSE: Ford End; Bedford field; *Plantae Bedford.*: Biddenham; Fen-
lake; *F.B.*: Chapel End; *J.McL., L.M.*: Cardington; *W.H. 1875*:
Chawston; Roxton; *G. H. Day*: Putnoe; *J.H.*: Kempston; *B. B. West.*
OUZEL: Salford; *V.C.H.*
IVEL: Ridgmont†; *W.C. Herb.*: Silsoe; *C.C. in F.F.B.*: Steppingley;
C.C.: Flitwick; *S.B.P.L.*: Wilbury, where it was found for a number
of years; *R. Morse.*
COLNE: Blow's Downs, v.c. 20 [Beds.]; *H. B. Sargent.*
First record: Abbot, 1795.

RANUNCULUS L.

Ranunculus acris L. C.T.W. 14.11.1.   *Upright Crowfoot*
Marshy places   Native
Frequent. It is very variable, but no serious work has been done on
its forms in the county: 'var. *steveni* (Andrz.)' was recorded from Flit-
wick in *V.C.H.* Recorded for all districts and neighbouring counties.
H.S. 5, 8–10c, 14b, 21, 23, 38, 45b, 56b, 57b, 69, 76, 84a, 85.
First record: Abbot, 1798.

R. repens L. C.T.W. 14.11.2.   *Creeping Crowfoot*
Arable land, woodland rides, ditches, etc.   Native
A very common but variable species growing as a prostrate form on
dry arable land but as a tall sub-erect form in ditches and woods.
Recorded for all districts and neighbouring counties.
H.S. 3a, 4, 8–10a, 10c, 12, 14b, 15, 19a, 20a, 34a, 34b, 37, 38, 41, 43,
52–53b, 55b, 56a, 57b, 58b, 62, 64, 66a, 71–73, 74b, 75, 77a, 79a,
80a, 81–84a, 85, 86.
First record: Abbot, 1798.

R. bulbosus L. C.T.W. 14.11.3.   *Buttercup, Bulbous Crowfoot*
Meadows   Native
Common in meadows but rarely seen elsewhere. A depauperate and
pale-flowered form is found on chalk downland. Recorded for all
districts and neighbouring counties.
H.S. 1a, 16, 63a.
First record: Abbot, 1798.

R. arvensis L. C.T.W. 14.11.5.   *Corn Crowfoot, Starveacre,*
                                           Scratch-bur (Batchelor)
Cornfields   Colonist
Frequent and seen at its best on calcareous soils, but not limited to
them. Recorded for all districts and neighbouring counties.
H.S. 72.
First record: Abbot, 1798.

R. sardous Crantz C.T.W. 14.11.6.   *Pale Hairy-crowfoot*
  R. philonotis Ehrh.
Cornfields, waste places, etc.   Colonist
Rare or extinct: both Abbot and McLaren considered it common. Recorded for
Herts., Cambs. and Hunts. and doubtfully for Northants. I saw it in Hunts. as
recently as 1948.
OUSE: clover field near Cardington†; *J.McL.*
IVEL: Cainhoe; *W.C. Herb.*: Ridgmont; *C.C. in J. Bot. 1889.*
LEA: moist meadows, a doubtful record; *Davis.*
First record: Abbot, 1798.

R. parviflorus L. C.T.W. 14.11.7.   *Small-flowered Crowfoot*
Dry banks, arable fields, etc.   Native
Rare and possibly decreasing. Recorded for all neighbouring
counties except Hunts.
OUSE: Elstow; Ford End; Kempston; *Plantae Bedford.*: Bromham;
Biddenham; *Abbot's annotated F.B.*: Cardington†; Goldington;
J.McL. in J. Bot. 1884: Colmworth, extending over three fields on
glacial gravel and growing as a luxuriant plant on banks and in a
stunted form on the edge of arable fields*.
OUZEL: Eggington; *Mrs. Tindall in W.F.B.*
IVEL: Tingrith†*; *C.C. in W.F.B.*
First record: Abbot, 1795.

R. auricomus L. C.T.W. 114.11.8.   *Goldilocks*
Woods, shady places, etc.   Native
Frequent and well distributed. It is very variable in its leaf forms
and apetalous forms are common. Recorded for all districts except
Cam, and for all neighbouring counties.
H.S. 52.
First record: Abbot, 1798.

R. lingua L. C.T.W. 14.11.9.   *Greater Spearwort*
Ponds, riversides, etc.   Native
Probably extinct: recorded for all neighbouring counties.
OUSE: Oakley West Field; Goldington; *Plantae Bedford.*: near
Sharnbrook; *J.McL., L.M.*
LEA: Reed Pond, Sundon, where I have looked for it in vain†; *S.B.P.L.*
First record: Abbot, 1795.

R. flammula L. C.T.W. 14.11.10.   *Lesser Spearwort*
Marshes   Native
Frequent and well distributed. A somewhat variable species:
var. *serratus* DC., with large flowers and serrated leaves, is found
on Cow Common, var. *ovatus* Pers., with rounded leaves, is
recorded from Flitwick, see *B.E.C. 1905 Rep.* 798, and var. *radicans*
Noltke, with no details, in *W.H. 1876.* Recorded for all districts
except Cam, and for all neighbouring counties.
H.S. 15, 35, 37, 56a, 84a,
NENE: Wymington Scrub.
OUSE: Oakley; Stevington; *F.B.*: Bromham; *J.H.*
OUZEL: Aspley; *Abbot's annotated F.B.*: Woburn; *W.H. 1875*: pond,
King's Wood; Horsemoor Farm; Woburn; Cow Common; Stock-
grove; New Wavendon Heath, v.c. 24 [Beds.].
IVEL: Ampthill; *F.B.*: Maulden Bogs; *W.C. Herb.*: Potton Marshes;
J.McL., L.M.: Biggleswade; *W.H. 1875*: Flitwick!†*; *W.F.B.*:
Warren Farm, Sandy; *J.E.L., F.F.B.*: Flitton Moor!; *V. H.Chambers*:
Westoning Moor; Biggleswade Common.
LEA: Limbury; Marslets; *W.F.B.*: Leagrave Marsh!†; *Herb. J.S.*
COLNE: Deadmansey Wood!; *V. H. Chambers*: pond, Pepperstock.
First record: Abbot, 1798.

R. sceleratus L. C.T.W. 14.11.13.   *Celery-leaved Crowfoot*
Ponds, streamsides, etc.   Native
Common: recorded for all districts and neighbouring counties.
H.S. 13, 14a.
First record: Abbot, 1798.

R. hederaceus L. C.T.W. 14.11.14.   *Ivy-leaved Crowfoot*
Muddy places   Native
Rare and found most often on the Lower Greensand. Procumbent
forms on the mud are most usual, but floating forms are seen occa-
sionally. Recorded for all neighbouring counties.
OUSE: Oakley; *Plantae Bedford.*: Cardington; *J.McL., B.M.*

OUZEL: Aspley; *Plantae Bedford.*: Husborne Crawley; *C.C.*: Horse-moor Farm*; Totternhoe Mead.
IVEL: Ampthill; *Plantae Bedford.*: Maulden; *W.C. Herb.*: Potton; *J.McL.*, *L.M.*: Flitwick!†; Chorton (? Chalton)†*; *S.B.P.L.*: Clophill†; *Herb. J.S.*: Steppingley; Ridgmont; *C.C.*: Biggleswade Common; Warren Farm, Sandy; *J.E.L.*, *J. Bot.*: Ampthill Park; *A. E. and E. M. Ellis*: foot of Cooper's Hill!*; *W. D. Coales*.
First record: Abbot, 1795.

R. fluitans Lam. C.T.W. 14.11.18. *Water Crowfoot*
Rivers *Native*
Limited to the Ouse and its main tributaries, where it is frequent. Recorded for all neighbouring counties except Herts.
OUSE: Cardington; *J.McL.*, *B.M.*: Bedford†; *S.B.P.L.*: Tempsford*; Great Barford; Eaton Socon.
OUZEL: Leighton Buzzard†; *Herb. J.S.*
IVEL: near Sandy; *W.F.B.*: Shefford†; *Herb. J.S.*: Biggleswade; *J.E.L.*, *Diary*.
First record: *Hillhouse List, 1876*.

R. circinatus Sibth. C.T.W. 14.11.19.
Rivers, ponds, etc. *Native*
Frequent and occurring usually in stagnant water. Recorded for all neighbouring counties.
H.S. 7.
OUSE: above Bedford†; *R. A. Pryor in J. Bot. 1875*, 212: Cardington; *J.McL.*, *B.M.*: gravel pit, Cople*; Oakley Bottom.
OUZEL: near Aspley†; *Mrs. Twidell in S.B.P.L.*: Battlesden Lake!†; *W.F.B.*: Lower Drakelow Pond*.
IVEL: near Shefford!†; *W.F.B.*: Cainhoe; *C.C.*: Southill Lake; *J.E.L. in W.B.E.C. 1913 Rep.*
First record: R. A. Pryor, 1875.

R. aquatilis L. C.T.W. 14.11.21. *Water Crowfoot*
The treatment by British botanists of this and the two preceding species has varied greatly. The treatment followed here is that of the *Flora of the British Isles*. I wish to thank R. W. Butcher for naming my material.

Subsp. **aquatilis**
R. heterophyllus Weber
Ponds *Native*
Frequent except in the south of the county. Recorded for all districts except Lea and Colne and for all neighbouring counties.

Var. **submersus** (Bab.) Clapham
OUZEL: Totternhoe*.
IVEL: Maulden; Westoning.
CAM: Cockayne Hatley.

Subsp. **radians** (Rével) Clapham
Ponds *Native*
Rare: recorded for all neighbouring counties except Northants.
OUSE: near Wootton church.
IVEL: gravel pit, Long Lane, Tingrith!*; *V. H. Chambers*: pond, Sundon.
First record: the author in *B.E.C. 1943–4 Rep.* 806.

Subsp. **peltatus** (Schrank) Syme
Ponds   Native
Frequent, but variable: recorded for all districts except Nene and Cam, and for all neighbouring counties.
OUSE: Putnoe; *J.McL., B.M.*: gravel pit, Cople; *E.M.-R.*: near Wootton Church*.
KYM: Riseley*; *C. C. Foss.*
OUZEL: pond near King's Wood!*; *B.P.L.*
IVEL: Sundon; *B.P.L.*: Clophill*; near Westoning Church*.
LEA: Chiltern Green*; Limbury*; *J.S.*
COLNE: Slip End†; *S.B.P.L.*: Slip End; *J.S. in J. Bot. 1889*: Woodside!*; *W.F.B.*: near Caddington Church; near Greencroft Barn, v.c. 20 [Beds.]*.
First record: *Hillhouse List, 1876.*

Subsp. **pseudofluitans** (Syme) Clapham
Streams   Native
Apparently limited to chalky streams in the south of the county. Saunders recorded subsp. *pseudofluitans* from Heath and Reach in *Bot. Rec. Club 1883 Rep.*, and a form linking it with *R. fluitans* from Leighton Buzzard in *W.F.B.* Both records are to be doubted, as his herbarium contains only *R. fluitans* from Leighton Buzzard. Recorded for all neighbouring counties.
IVEL: Barton Springs†; *J.S. in J. Bot. 1889*: Cadwell Farm; *J.E.L. in W.B.E.C. 1922–3 Rep.*: R. Hiz, Arlesey.
LEA: Limbury†; *S.B.P.L.*: Biscot; New Mill End!*; *W.F.B.*: Leagrave* (H.S. 84).
COLNE: R. Gade, Studham*.
First record: Saunders, 1885.

R. trichophyllus Chaix C.T.W. 14.11.20.   *Water Crowfoot*
Subsp. **trichophyllus**
Ponds   Native
Common: recorded for all districts and neighbouring counties. This would be considered by some botanists as no more deserving of specific rank than the subspecies of *R. aquatilis*.
H.S. 9, 14b, 21.
First record: W. W. Newbould in *Topographical Botany*, 1873.

Subsp. **drouetii** (F. W. Schultz ex Godr.) Clapham
Ponds   Native
Rare: recorded for all neighbouring counties except Hunts.
OUZEL: Aspley†; *S.B.P.L.*: pond, Milton Bryan*.
LEA: Limbury†; Biscot; *W.F.B.*
First record: W. W. Newbould in *Topographical Botany*, 1873.

R. baudotii Godr. was recorded by Saunders from Sundon in *S.B.P.L.* and, probably on his authority, in *Top. Bot. Ed. II* (*J. Bot. 1905*) and in *Comital Flora.* He did not include it in later lists and there is no specimen to support the record. It is unlikely that it would appear in the county.

R. ficaria L. C.T.W. 14.11.23.   *Lesser Celandine*
Damp and shady places   Native
Common, but able to withstand only limited competition and rarely found in turf. The common form in the county has no axillary bulbils, *cf.* var. *fertilis* of the *Flora of the British Isles*; but forms with axillary bulbils have been observed at Great Hayes Wood; Bidwell; Blackgrove Wood and Wardown. Recorded for all districts and neighbouring counties.
H.S. 2b, 6, 8, 9–10c, 13, 18, 19b, 20b, 32, 33, 51a, 52, 53a, 54, 55a, 57b, 78, 83.
First record: Abbot, 1798.

CALTHA L.

Caltha palustris L. C.T.W. 14.1.1.   *Kingcup, Marsh Marigold*
Marshy places   Native
Common in marshes and waterlogged meadows throughout the
county: var. *guerangerii* (Bor.) Lamotte was recorded by Saunders
from Limbury in *W.F.B.* and by J. E. Little from Cadwell, v.c. 20
[Beds.], in *J.E.L., F.F.B.* Recorded for all districts and neighbouring
counties.
H.S. 8, 14b, 38, 39a, 41, 84a, 85.
First record: Abbot, 1798.

HELLEBORUS L.

Helleborus viridis L. C.T.W. 14.3.2.   *Bear's-foot, Green Hellebore*
Woods, damp meadows, etc.   ? Native
Rare, but evenly distributed in the county. Recorded for all neigh-
bouring counties.
OUSE: Goldington; *Plantae Bedford.*: pastures near King's Wood;
J.McL., L.M.: Stagsden, presumably Astey Wood, where it still
survives!; *J.H.*: Turvey; *J.H. Notes*: Felmersham; *G. H. Day.*
KYM: Keysoe; Little Staughton!; *L. W. Wilson.*
OUZEL: spinney near Bidwell, plentiful!; *W.F.B.*
IVEL: Cadwell, v.c. 30 [Herts.]; *I. Brown in Herb. Watson.*
LEA: meadows, Limbury, once plentiful now just surviving!†*;
B.P.L.: wood, East Hyde!*; *W.F.B.*
COLNE: Whipsnade Wood; *W. G. Smith in J. Bot. 1876*: Oldhill
Wood!; *C. Dunham.*
First record: Abbot, 1795.

H. foetidus L. C.T.W. 14.3.1.   *Setterwort, Stinking Hellebore*
Woods, shady places, etc.   Denizen
Rare, but very well established in a number of places, usually near to houses.
E. Bot. tab. 613, Feb. 1799, was drawn from a specimen sent by Abbot. Recorded
for all neighbouring counties.
NENE: hedgerow, Podington!; *R. R. B. Orlebar.*
OUSE: Stevington; *Plantae Bedford.*: Bromham; Stagsden, where it still grows in
and near White's Wood!*; *F.B.*: Basmead; *C.R.O. List*: Hill Plantation; *J.McL.,
L.M.*: Eaton Socon, Harrold, *G. H. Day*: Salem Thrift; *R. Townsend*: Goldington;
L. W. Wilson.
KYM: Dean; *G. H. Day.*
OUZEL: rectory garden, Stanbridge; *H. B. Souster.*
IVEL: Higham Gobion, in spinney near Church!*; *W.C. Herb.*: spinney near
Vicarage, Clophill!; *R. L. R. Bearman*: spinney, Ickwell Bury; *C. C. Foss!*
LEA: garden escape, Luton*; near Limbury†; *S.B.P.L.*: garden escape, Dunstable
Cemetery; *W. D. Coales.*
First record: Abbot, 1795.

AQUILEGIA L.

Aquilegia vulgaris L. C.T.W. 14.14.1.   *Wild Columbine*
Woods, cleared woodland, etc.   ? Native
Rare, but often well established. Recorded for all neighbouring counties.
OUSE: Elstow; Stevington; *Abbot's annotated F.B.*: Basmead; *C.R.O. List.*
OUZEL: Aspley Wood, with every appearance of being a garden escape!; *V. H.
Chambers.*

IVEL: Barton Leete, apparently well established, as it was found here by
Saunders†*; *F.B.*: near Pegsdon (? Beds.); *H. Brown*: Potton Wood; *T.L.*: beech
woods on Streatley Hills.
LEA: Dallow Farm; *Davis*: garden escape, New Bedford Road, Luton; *G. H.
Latchmore in W.F.B.*
First record: Abbot, 1798.

DELPHINIUM L.

Delphinium gayanum Wilmott   *Larkspur*
    D. ajacis auct. C.T.W. 14.7.1.
Cornfields, waste places, etc.   Garden escape
Rare: it was probably a frequent cornfield weed in earlier times but it is now a
plant of waste places. The taxonomy of larkspurs has been much confused and I
have taken Abbot's *D. consolida* to refer to this species. Recorded for all neigh-
bouring counties.
OUSE: Bedford; *F.B.*: Pavenham; *J.H.*: Biddenham; *A. W. Guppy*: gravel pit,
Eaton Socon.
IVEL: field on border of Silsoe Park; *Herb. Miss Crouch*: clover field, Potton;
J.McL., L.M.: Harlington†; *Mr. Anstee in J.S., J. Bot. 1889*: roadside between
Barton and Gravenhurst; *H. B. Souster*: rubbish dump, Sundon*.
LEA: cornfield near Lilley (? Beds.); *J.S. in Herb. Beds. Nat. Hist. Soc.*
First record: Abbot, 1798.

BERBERIDACEAE

BERBERIS L.

Berberis vulgaris L. C.T.W. 16.1.1.   *Barberry*
Hedgerows   Native
A not infrequent shrub in the north of the county. It was at one
time probably common but was destroyed in great quantity in the
early years of the nineteenth century in the belief that it caused the
spreading of wheat rust. Contemporary botanists scorned this
belief, but it has since proved to be well founded. Thomas Batchelor
in his *General View of Agriculture in the county of Bedford* (1808) gives
interesting evidence of its distribution and wholesale destruction in
his day. Abbot, it is surprising to note, considered it rare. Recorded
for all neighbouring counties.
NENE: Podington; *A. W. Guppy*: Wymington.
OUSE: Clapham Lane; Milton Ernys; *F.B.*: Pavenham; Felmer-
sham; Sharnbrook; Bletsoe!; Oakley; *Batchelor*: Manor Lane,
Cardington; *J.McL., L.M.*: Fenlake; *J.McL., B.M.*: Carlton!;
B.P.L.: Honey Hills; *J.H.*: Biddenham; Clapham; *J.H. Notes*:
Roxton; Chellington; Odell; *G. H. Day*: Stagsden; *E. Proctor*.
KYM: Knotting; Melchbourne; *Batchelor*.
OUZEL: Woburn; *Batchelor*: Battlesden; *H. E. Pickering*.
IVEL: Ridgmont; *Batchelor and W.C. Herb.*: Steppingley; *S.B.P.L.*
CAM: Wrestlingworth.
LEA: Luton; *Batchelor*.
First record: Abbot, 1798.

NYMPHAEACEAE

NUPHAR Sm.

Nuphar lutea (L.) Sm. C.T.W. 17.2.1.   *Yellow Water-lily*
Nymphaea lutea L.
Rivers, ponds, etc.   Native
Common and native in the Ouse and its main tributaries; frequent
in lakes and large ponds, but less certainly native here. Recorded
for all neighbouring counties.
H.S. 14a.
Ouse: common in the river.
Ouzel: Leighton Buzzard!; *J.H. Notes*: river, Heath and Reach.
Ivel: common in the Ivel and Flitt.
Lea: pond, Luton Park; *Davis*: East Hyde, no doubt native here!;
W.F.B.
First record: Abbot, 1798.

NYMPHAEA L.

Nymphaea alba L. C.T.W. 17.1.1.   *White Water-lily*
Castalia alba (L.) Wood
Deep slow-flowing rivers, lakes, etc.   Native
Common in the Ouse and frequent in the Ouzel, but apparently
absent in the Ivel. It is occasionally planted in ornamental lakes.
Recorded for all neighbouring counties.
H.S. 6, 7, 13.
Ouse: common in the river.
Ouzel: Woburn, I have seen it only in the lakes in Woburn
Park!; *S.B.P.L.*: river, Heath and Reach; *H. B. Souster*: lake,
Stockgrove.
Cam: lake, Cockayne Hatley.
First record: Abbot, 1798.

CERATOPHYLLACEAE

CERATOPHYLLUM L.

Ceratophyllum demersum L. C.T.W. 18.1.1.   *Hornwort*
Rivers, lakes, etc.   Native
Common in the Ouse, frequent elsewhere. Recorded for all neigh-
bouring counties.
H.S. 7, 14a.
Ouse: common in the river.
Ouzel: Battlesden Lake*; *J. E. Dandy*!
Ivel: Pulloxhill†; *C.C.*: Southill Lake!; *J.E.L.*, *J. Bot.*: pond,
Fancott; lakes, Flitwick Park, Wrest Park, Ickwell Bury.
Lea: The Hyde; *R. A. Pryor in J. Bot. 1876*: Luton Hoo Lake!*;
S.B.P.L.
First record: Abbot in letter to J. E. Smith, 3 Sept. 1802.

C. submersum L. C.T.W. 18.1.2.
Ponds   Native
Rare, but possibly overlooked. Recorded for Bucks., Hunts. and
Northants. and doubtfully for Cambs.
IVEL: shallow pool near Flitwick, 1887*; *J.S. in Herb. H. N. Dixon*
(I am indebted to N. D. Simpson for drawing my attention to this
and sending me a duplicate specimen.)

PAPAVERACEAE

PAPAVER L.

Papaver rhoeas L. C.T.W. 19.1.1.   *Common Poppy*
Arable fields, waste places, etc.   Colonist
A common but very variable species. Forms with reddish hairs on
the stems, *cf.* var. *pryorii* Druce, have been observed at Sundon*,
Rowney Warren* and Studham, and var. *chelidonioides* Kuntze
was recorded from Clifton; H. Phillips in *B.E.C. 1934 Rep.* Recorded
for all districts and neighbouring counties.
H.S. 12, 16, 17, 50, 64, 71, 72, 82, 86.
First record: Abbot, 1798.

P. dubium L. C.T.W. 19.1.2.   *Smooth long-headed Poppy*
Cornfields, waste places, etc.   Colonist
Common: recorded for all districts and neighbouring counties.
H.S. 16, 44, 50.
First record: Abbot, 1798 (but see note to the following species).

P. lecoqii Lamotte C.T.W. 19.1.3.   *Babington's Poppy*
Waste places, railway banks, etc.   Colonist
This species needs more study. Babington (*Fl. of Cambs.* 1860) considered it to be
the common species in Cambridgeshire and *P. dubium* to be rare. The species as
understood in the nineteenth century was long-fruited with dark yellow latex and
narrow acuminate segments to the leaves. This fits the specimen of '*P. dubium*' in
Abbot's herbarium which Pryor, Newbould and Hillhouse passed as *P. lecoqii*,
and was no doubt the concept of the species as recorded for the county, from
between Luton and Hexton, by T. B. Blow in *Bot. Rec. Club 1876 Rep.*, by Pryor,
from near Dunstable, in *J. Bot. 1876*, and by Hillhouse, from Arlesey and Sandy
(W.H. 1875). Further records were made by Saunders, Druce and Little. More
recent opinion has been that *P. dubium* may have yellow latex and acuminate
segments to the leaves, and that *P. lecoqii* differs from it in the shape of the capsules
and is a very rare species. I cannot follow the various authorities in distinguishing
the species by the relative lengths of the stigma rays, as I have found plants showing
the same differences in one capsule. The only instance I know in the county of
P. lecoqii with all the essential characters is on a railway bank at Flitwick, where it
was also recorded and collected by Saunders!†*. The plants here are always
galled, with some capsules becoming spherical in shape. A specimen in Herb.
McLaren, British Museum, labelled 'railway bank', is also correctly identified as
this. Long-fruited poppies with dark yellow latex are more common than those
with white or colourless latex, and have been recorded for all districts.

P. argemone L. C.T.W. 19.1.5.   *Rough long-headed Poppy*
Rough ground, waste places, etc.   Colonist
Frequent, especially as a railway weed, where it grows on ballast

o

and between the metals. Recorded for all districts except Nene, Kym and Cam, and for all neighbouring counties.
H.S. 44, 72.
First record: Abbot, 1798.

P. hybridum L. C.T.W. 19.1.4.   *Rough round-headed Poppy*
Cornfields, waste places, etc.   Colonist
Rare: appearing at its best as a cornfield weed and, usually only in a depauperate state, in waste places. Unlike *P. argemone* it is rarely established. Abbot considered it to be common. Recorded for all neighbouring counties.
NENE: railway bank, Wymington.
OUSE: Elstow; *Abbot's annotated F.B.*: Cox's Pits; *J.McL., B.M.*: Biddenham; *J.McL. in Herb. J.S.*: cornfield, Roxton*; *V. H. Chambers*!: rubbish dump, Clapham; *E.M.-R.*!: roadside, Wyboston; *H. B. Souster*: roadside, Willington.
OUZEL: Dunstable Downs†; *W. G. Smith in J. Bot. 1885*: arable field, Houghton Regis.
IVEL: Pegsdon; *C.C.*: cornfield, Arlesey*; *E.M.-R.*
LEA: Luton, 1935; *W.F.S. Diary.*
First record: Abbot, 1798.

CHELIDONIUM L.

Chelidonium majus L. C.T.W. 19.5.1.   *Greater Celandine*
Roadsides, railway banks, etc.   Colonist
Common, but rarely seen far from houses. It may have been grown originally in gardens for medicinal purposes. Recorded for all districts and neighbouring counties.
First record: Abbot, 1798.

FUMARIACEAE
CORYDALIS Medic.

Corydalis claviculata (L.) DC. C.T.W. 20.1.2. *Climbing Fumitory*
    Capnoides claviculata (L.) Kuntze
Heaths   Native
Frequent over a large part of the Lower Greensand, and growing most plentifully in cleared woodland or in places where the soil has been disturbed. Recorded also for Bucks.
OUZEL: Aspley Wood; *W.H. 1875*: Woburn!; *G. C. Druce in B.E.C. 1914 Rep.*: Heath and Reach!; *J.S. in Bot. Rec. Club 1880 Rep.*: New Wavendon Heath, v.c. 24 [Beds.]!; *V. H. Chambers.*
IVEL: heath, Clophill!†*; *J.McL., L.M.*: Ampthill!; *S.B.P.L.*: Millbrook; *W.F.B.*: Maulden! *C.C.*
First record: J. McLaren, 1864.

C. lutea (L.) DC. C.T.W. 20.1.3.   *Yellow Fumitory*
    Capnoides lutea (L.) Gaertn.
Old walls   Wall colonist
Rare: recorded for all neighbouring counties except Cambs.
OUSE: Cotton End; *B.P.L.*: Cardington; *J.McL.*: Elstow†; *J.McL., B.M.*: Odell; *B. B. West*: Harrold; Turvey; Tempsford; *G. H. Day*: Millbrook; *L. J. Tremayne.*
IVEL: Clophill!; *B. Verdcourt*: Ampthill*.
LEA: rubbish dump, Dunstable; East Hyde!; *H. B. Souster.*
First record: Abbot in letter to J. E. Smith, 3 Sept. 1802.

FUMARIA L.

Fumaria officinalis L. C.T.W. 20.3.8.   *Fumitory*
Arable fields, waste places, etc.   Colonist
Common: recorded for all districts and neighbouring counties.
H.S. 64, 71, 72, 73.
First record: Abbot, 1798.
 Var. **wirtgenii** Hausskn.
OUZEL: Totternhoe.
IVEL: Barton Hills!*; *J.E.L., J. Bot.*: Pegsdon!*; *J.E.L., F.F.B.*
LEA: Luton*; Leagrave.
 Var. **elegans** Pugsl.
IVEL: Shefford*.

F. micrantha Lag. C.T.W. 20.2.7.
Arable fields   Colonist
Rare, and mainly limited to the Chalk. Recorded for Herts.,
Bucks. and Cambs.
IVEL: near High Down; *R. A. Pryor in J. Bot. 1876*, 22: Barton†;
S.B.P.L.: Mead Hook; *C.C.*: Clophill*.
LEA: Luton; Dunstable; *R. A. Pryor in J. Bot. 1876*: Limbury; *J.S.,
J. Bot. 1889*: Leagrave; *W.F.B.*
COLNE: between Kensworth and Half Moon, etc. v.c. 20 [Beds.];
Fl. Herts.
First record: R. A. Pryor, 1876.

F. vaillantii Lois. C.T.W. 20.2.9.
Arable fields   Colonist
Rare, and apparently limited to the Chalk. Recorded for Herts.,
Bucks. and Cambs.
OUZEL: Totternhoe*.
IVEL: near High Down; *R. A. Pryor in J. Bot. 1876*, 22: Barton
Hills!*; *C.C. in J. Bot. 1889.*
LEA: about Dunstable; *R. A. Pryor in J. Bot. 1876*: Luton: *D. M.
Higgins in W.B.E.C. 1900–1 Rep.*: Galley Hill; *J.E.L., F.F.B.*
COLNE: Isle of Wight Farm; hills, south-west of Dunstable, v.c. 20
[Beds.]; *Fl. Herts.*
First record: R. A. Pryor, 1876.

F. parviflora Lam. C.T.W. 20.2.10.
Arable fields   Colonist
Rare: like the preceding apparently limited to the Chalk and
usually found in association with it. Recorded for Herts., Bucks. and
Cambs.
IVEL: near High Down; *T. B. Blow in Bot. Loc. Rec. Club. 1876 Rep.*:
Barton Hills†; *S.B.P.L.*
LEA: Leagrave; *F.F.B.*
First record: T. B. Blow, 1876.
 Var. **acuminata** Clav.
IVEL: Barton Hills; *J.E.L., J. Bot.*: Noon Hill!*; *V. H. Chambers.*
LEA: Galley Hill; *J.E.L., B.E.C.*

CRUCIFERAE

RORIPPA Scop.

Rorippa nasturtium-aquaticum (L.) Hayek   *Watercress*
   Sisymbrium nasturtium-aquaticum L., *Nasturtium officinale* R.Br.,
C.T.W. 21.39.1, *Radicula nasturtium-aquaticum* (L.) Britten & Rendle
Streams, ponds, ditches, etc.   Native
Common, and appearing most frequently in calcareous water.
Recorded for all districts except Colne and for all neighbouring
counties.
First record: for the aggregate, Abbot, 1798.

R. microphylla (Boenn.) Hyland.
   Nasturtium microphyllum (Boenn.) Reichb., C.T.W. 21.39.2.,
N. uniseriatum Howard & Manton
Streams, ponds, ditches, etc.   Native
More common and widely distributed than the preceding, with
which it often grows. Recorded for all districts except Lea, and for
all neighbouring counties.
H.S. 6, 8, 13, 14b, 56a.
First record: the author in *B.E.C. 1946–7 Rep.* 283.
First evidence: Cainhoe, W. Crouch Herb., 1844.

R. × sterilis Airy Shaw (R. microphylla × nasturtium-aquaticum)
OUSE: pond, Souldrop, with both parents!*; *E.M.-R.*
IVEL: Barton; *D. M. Higgins,* 1899, in Herb. Brit. Mus. (see *Watsonia I (1950)* 232).

R. sylvestris (L.) Bess. C.T.W. 21.40.1.   *Creeping Yellow-cress*
   Sisymbrium sylvestre L., *Nasturtium sylvestre* (L.) R.Br., *Radicula
sylvestris* (L.) Druce
Damp meadows, etc.   Native
Frequent by the Ouse, rare elsewhere. Records by Saunders from
Luton in *J. Bot. 1883*, and New Mill End in *F.F.B.*, are to be
doubted, as his herbarium specimens are represented by *R. islandica*.
It is a species which can survive in dry situations, when it assumes a
straggling form which it is sometimes difficult to recognize. Abbot
considered it to be common. Recorded for all neighbouring counties.
OUSE: Pavenham; *J.H. in Herb. Beds. Nat. Hist. Soc.*: Newnham;
J.H.: Bedford*; *C.C.*: Bromham; Stevington*; Sharnbrook; Great
Barford; rubbish dump, Cople.
COLNE: plentiful as weed in garden, Luton Hoo*.
First record: Abbot, 1798.

R. islandica (Oeder) Borbás C.T.W. 21.40.2.   *Marsh Watercress*
   Sisymbrium terrestre Curt., *Nasturtium palustre* (L.) DC., *Radicula
islandica* (Oeder) Druce
Damp meadows, waste places, etc.   Native
Considered rare by Abbot and other early botanists, it has probably
increased. It is now found frequently on rubbish dumps and railway
sidings. Recorded for all districts except Nene and Cam. I have

seen it only at Kimbolton Station, v.c. 30 [Hunts.], in the Kym district. Recorded for all neighbouring counties.
H.S. 7.
First record: Abbot, 1798.

R. amphibia (L.) Bess. C.T.W. 21.40.3. *Great Yellow-cress*
   Sisymbrium amphibium L., *Nasturtium amphibium* (L.) R.Br., *Radicula amphibia* (L.) Druce
Riversides   Native
Frequent on the banks of the larger rivers. Recorded for all neighbouring counties.
H.S. 6, 7, 8.
OUSE: Elstow; Goldington†; *F.B.*: Newnham!; Clapham; Honey Hills; Cox's Pits; *J.H. Notes*: Biddenham!; *C.C.*: Oakley!; *A. W. Guppy*; Tempsford; *V. H. Chambers*!: Bromham; Stevington; Pavenham; Felmersham; Great Barford; Eaton Socon; Radwell.
OUZEL: Leighton Buzzard; Billington; *V. H. Chambers*: Heath and Reach.
IVEL: Arlesey; Clifton; Shefford!; *J.E.L., J. Bot.*
LEA: New Mill End!; *J.S., J. Bot. 1883*: Luton Hoo; *E.M.-R.*!
First record: Abbot, 1798.

BARBAREA R. Br.

Barbarea vulgaris R.Br. C.T.W. 21.35.1.   *Winter Cress, Yellow Rocket*
   Erysimum barbarea L., *B. barbarea* (L.) MacMill.
Streamsides, ditches, roadsides, etc.   Native
Common and well distributed in the county, but more plentiful in wet places. It is a variable species, and J. E. Little and Druce recorded var. *silvestris* Fr., var. *campestris* Fr. and var. *transiens* Druce. Recorded for all districts and neighbouring counties.
First record: Abbot, 1798.
      Var. **arcuata** (Opiz) Fr.
OUZEL: Salford; *G. C. Druce.*
IVEL: Ivel Navigation, Shefford.

ARABIS L.

Arabis hirsuta (L.) Scop. C.T.W. 21.37.4.   *Hairy Rock-cress*
Dry banks   Native
Rare: limited to previously disturbed soils on the Chalk and Oolitic Limestone. Recorded for all neighbouring counties except Herts.
NENE: railway cutting, Wymington*.
OUSE: railway cutting, Souldrop*; *P.T.*!
OUZEL: limited on top of Dunstable Downs!*; *W. D. Coales.*
LEA: Lynches*; *H. Salmon and D. B. Slope in B.E.C. 1945 Rep.* 51.
First record: the first printed record is Salmon and Slope's, but Verdcourt (*Watsonia I (1950)* 245) claims that he found it at Dunstable Downs in 1943; Coales, however, knew it there still earlier.

CARDAMINE L.

Cardamine pratensis L. C.T.W. 21.33.1. *Cuckoo-flower, Lady's Smock*
Marshes, meadows, damp woods, etc.   Native
Common in all likely habitats. Recorded for all districts and neigh-
bouring counties.
H.S. 8, 10a, 14b, 34a, 34b, 37–39a, 84a.
First record: Abbot, 1798.

C. hirsuta L. 21.33.5.   *Hairy Bitter-cress*
Gardens, woodland rides, old walls, etc.   Native
Frequent and widely distributed in the county but awaiting record
in the Cam district. Recorded for all neighbouring counties.
H.S. 45a, 45b.
First record: Abbot, 1798.

C. flexuosa With.   *Wood Bitter-cress*
Shady places, etc.   Native
Rare, and found growing in similar situations to *C. hirsuta*, but
generally in greater shade. Recorded for all neighbouring counties.
OUZEL: Wavendon Heath, v.c. 24 [Beds.]; *P.T.*
IVEL: Shefford; *J.E.L., J. Bot.*: abundant on mud cleared from
lakes, Silsoe Park, 1950*.
LEA: Luton; *J.E.L., F.F.B.*: Wardown Park*.
First record: J. E. Little, 1919.

ALYSSUM L.

Alyssum alyssoides (L.) L. C.T.W. 21.27.1.   *Small Alison*
  A. calycinum L.
Dry banks, arable fields, etc.   Colonist
Rare: recorded for all neighbouring counties.
OUSE: Biddenham, 1864; *J.McL. in J. Bot. 1884*: Oakley; *F. W. Crick in Herb. Beds.
Nat. Hist. Soc.*
IVEL: railway embankment, Flitwick, where it is well established!*; *G. D. Nicholls.*
First record: clover fields; Herb. McLaren, Luton Museum, 1864.

DRABA L.

Draba muralis L. C.T.W. 21.30.4.   *Wall Whitlow Grass*
Waste ground, gardens, etc.   Colonist
Rare, but probably increasing. Most recent records are for plants introduced from
gardens, although the species itself is not a garden plant. Recorded for Herts.
OUSE: Hunter's Nursery; *A. B. Sampson.*
KYM: Kimbolton†; *J. Hamson in J. Bot. 1889.*
OUZEL: well established in gravel pit, Leighton Buzzard*.
LEA: 'gathered by the Revd. Mr. Abbot in flower on April 14 last on the Wardon
Hills near Barton in the Clay'; *E. Bot. tab. 912, Aug. 1801*: Vicarage Garden,
Luton; *M. Brown in W.F.S. Diary*: public gardens, Dunstable!*; *E.M.-R.*
First record: Abbot, 1801.

EROPHILA DC.

Erophila verna (L.) Chevall. C.T.W. 21.31.1.   *Whitlow Grass*
  Draba verna L.
Dry places, old walls, etc.   Native
Common. This is a very variable species of which I have made many
gatherings, but hesitate to name. Various names have been given

to Bedfordshire specimens in the past, see J. E. Little in *W.B.E.C.
1914 Rep.*, G. C. Druce in *B.E.C. 1929 Rep.*, etc. Many of these
names may be proved to be synonymous, and in any case are not
upheld by present-day taxonomists. The most modern treatment of
the genus is by Ö. Winge: Taxonomic and Evolutionary Studies in
Erophila, *Comptes Rendus des travaux du Laboratoire Carlberg* (1940).
Following this all my gatherings, except one, fall into his *E. duplex.*
The remaining gathering is from a mixed population to which V. H.
Chambers drew my attention at Flitwick Moor in 1941. Much of
this fits in every detail Winge's *E. quadruplex*, but needs to be studied
cytologically before a definite determination can be made. The
specimens were examined by A. J. Wilmott who could find no
material with which to compare them nor a name from the then
accepted treatment to apply to them. *Erophila verna* has been
recorded for all districts and neighbouring counties.
H.S. 1a, 44, 45a, 47, 56b, 61, 75.
First record: Abbot, 1798.

COCHLEARIA L.

Cochlearia danica L. C.T.W. 21.23.5.   *Danish Scurvy-grass*
A railway weed   Colonist
This increased on the railway in the war years, when labour was not available to
clear the track. It has persisted for a number of years and will probably survive on
the ballast outside the metals. Recorded for Herts. and Northants.
OUSE: Willington*.
IVEL: Flitwick; *J. Ounsted.*
LEA: East Hyde*; in some quantity on about half-a-mile of the track, two miles
south of Luton L.M.R. station.
First record: the author in *Watsonia I (1949)* 38.

HESPERIS L.

Hesperis matronalis L. C.T.W. 21.45.1.   *Dame's Violet*
Waste places   Garden escape
Frequent and often established. Recorded for Herts., Bucks. and Northants.
OUSE: Goldington; Millbrook; *J.H. Notes*: Sharnbrook; *H. B. Souster.*
OUZEL: Woburn Sands; *V.C.H.*
IVEL: Ridgmont*; *C.C.*: Flitwick*; old chalk workings, Barton.
LEA: Luton Hoo†; *S.B.P.L.*: Leagrave; *W.F.S. Diary*: rubbish dump, Luton.
COLNE: Woodside.
First record: Saunders, 1885.

ARABIDOPSIS (DC.) Heynh.

Arabidopsis thaliana (L.) Heynh. C.T.W. 21.50.1.   *Thale Cress*
  Arabis thaliana L., *Sisymbrium thalianum* (L.) Gay
Dry banks, arable fields, etc.   Native
Common, especially on the Lower Greensand, but rare or absent on
heavy clay soils. It was considered rare by Abbot, and the original
drawing for *E. Bot.* tab. 901, July 1801, is marked with his name.
Recorded for all neighbouring counties, but awaiting record for
districts Kym and Cam.
H.S. 22b, 25.
First record: Abbot, 1798.

SISYMBRIUM L.

Sisymbrium officinale (L.) Scop. C.T.W. 21.49.1. *Hedge Mustard*
  Eysimum officinale L.
Waste places, roadsides, etc.   Colonist
Common: recorded for all districts and neighbouring counties.
H.S. 16, 48, 50.
First record: Abbot, 1798.
  Var. **leiocarpum** DC.
A well-marked form, with glabrous fruits, and perhaps deserving specific rank.
OUSE: railway siding, Bedford.
IVEL: rubbish dump, Sundon*; *R. C. L. Burges*: railway siding, Ampthill; roadside,
Clophill.
LEA: Wardown Park.

S. altissimum L. C.T.W. 21.49.4.   *Tall Rocket*
  S. pannonicum Jacq.
Waste places   Colonist
A recent introduction which has become established in a number of places.
Recorded for all neighbouring counties.
OUSE: Milton Ernest Mill; *E. M. Langley in J.H. Notes*: Thurleigh; *J.H. Notes*:
Bedford, plentiful in and near the town!; *L. W. Wilson*: Tempsford*; *B. Reynolds*.
OUZEL: Leighton Buzzard.
IVEL: between Henlow and Shefford; *J.E.L., B.E.C.*: Ampthill, as '*S. sophia*'*;
C.C.: rubbish dump, Sundon; old gravel pit, Meppershall.
LEA: frequent around Luton*.
First record: E. M. Langley, 1907.

S. orientale L. C.T.W. 21.49.3.   *Eastern Rocket*
  S. columnae Jacq.
Waste places, arable fields, etc.   Colonist
A newcomer which has increased rapidly. It occurs frequently with wool adven-
tives, and there is no doubt that wool manure is one of the agencies of its intro-
duction. Recorded for all districts except Cam and Colne, and for all neighbouring
counties.
First record: J. E. Little in *B.E.C. 1936 Rep.*

DESCURAINIA Webb & Berth.

Descurainia sophia (L.) Webb ex Prantl C.T.W. 21.52.1.   *Flixweed*
  Sisymbrium sophia L.
Waste places   Colonist
Considered to be common by Abbot, but it is rare now. Recorded for all neigh-
bouring counties.
OUSE: Cardington Mill*; *J.McL. in Bot. Rec. Club 1884–6 Rep.*: gravel pit, Eaton
Socon; *J. Russell*.
IVEL: rubbish dump, Sundon, where it appears to be well established*.
First record: Abbot, 1798.

ALLIARIA Scop.

Alliaria petiolata (Bieb.) Cavara & Grande C.T.W. 21.48.1.
  Garlic Mustard, Jack-by-the-Hedge, Satan's Flowers (Westoning)
  Erysimum alliaria L., *Sisymbrium alliaria* (L.) Scop.
Hedgerows, wood borders, etc.   Native
Common: recorded for all districts and neighbouring counties.
H.S. 6, 7, 13, 32.
First record: Abbot, 1798.

ERYSIMUM L.

Erysimum cheiranthoides L. C.T.W. 21.46.1.   *Treacle Mustard*
Arable fields, waste places, etc.   Colonist
Common on the Lower Greensand, but usually only a temporary
weed elsewhere. It appears to thrive best on soils previously re-
claimed from marsh or bog. Recorded for all neighbouring
counties.
H.S. 16, 44.
OUSE: Willington*; *L. W. Wilson*: Blunham; *G. H. Day*: Houghton
Conquest!; *C. C. Foss.*
OUZEL and IVEL: common on the richer soils.
CAM: Wrestlingworth.
LEA: south of Luton!†; *S.B.P.L.*: railway sidings, Luton.
COLNE: Whipsnade.
First record: Clophill, Herb. W. Crouch, 1843.

CAMELINA Crantz

Camelina sativa (L.) Crantz C.T.W. 21.51.1.   *Gold of Pleasure*
Waste places   Colonist or Casual
Rare: recorded for all neighbouring counties.
OUSE: Cardington Mill; *J.McL.*, *B.M.*
OUZEL: near Salford; *G. C. Druce in B.E.C. 1902 Rep.*
IVEL: Ridgmont*; *J. F. Crouch*: Flitton Moor; *C.C.*: rubbish dump, Sundon,
where it appears regularly*; *E.M.-R.*!
LEA: south of Luton†; *S.B.P.L.*: Dunstable; *S. A. Chambers in F.F.B.*
First record: J. F. Crouch, 1864, spec. in Herb. Luton Museum.

SINAPIS L.

Sinapis arvensis L. C.T.W. 21.4.1.   *Charlock*
  Brassica arvensis (L.) Rabenh. non L.
Arable fields, waste places, etc.   Colonist
Common, but usually replaced on the Chalk by *S. alba*. It is some-
what variable and there are many records for var. **orientalis** (L.)
Koch & Ziz. Recorded for all districts and neighbouring counties.
H.S. 7, 12, 58b, 71–73, 82.
First record: Abbot, 1798.

S. alba L. C.T.W. 21.4.2.   *White Mustard*
  Brassica alba (L.) Rabenh.
Arable fields, waste places, etc.   Colonist
Common on the Chalk and on heavy clay soils, but rare on sandy
and gravelly soils. Recorded for all districts and neighbouring
counties.
H.S. 12, 63b, 64, 71, 72.
First record: Abbot, 1798.

DIPLOTAXIS DC.

Diplotaxis tenuifolia (L.) DC. C.T.W. 21.6.2.   *Perennial Wall-rocket*
Waste places   Colonist
Rare, but usually well established. Recorded for all neighbouring counties.

Ivel: rubbish dump, Sundon*; railway siding, Shefford.
Ouzel: railway siding, Totternhoe.
Lea: waste ground, Dallow Road!*; *H. B. Souster*: railway siding, Dunstable
North Station.
First record: the author in *B.E.C. 1943–4 Rep.* 806.

D. muralis (L.) DC. C.T.W. 21.6.1.   *Wall-rocket, Stinkweed*
Railway tracks, etc.   Colonist
Common on ballast on the railway, but seen also on clinkers around cement works,
brick factories, etc., and at the base of walls: var. *babingtonii* Syme was recorded
from Cardington station by E. M. Langley in *J.H. Notes* and from the railway at
Woburn Sands in *V.C.H.* Recorded for all districts except Kym and Cam, and
for all neighbouring counties.
First record: T. Corder in *Beds. Times*, 2 July 1861.

CORONOPUS Boehm.

Coronopus squamatus (Forsk.) Aschers. C.T.W. 21.15.1.
                                                          Swine-cress
Cochlearia coronopus L., *Coronopus procumbens* Gilib., *C. coronopus* (L.)
Karst.
Cultivated fields, waste places, etc.   Colonist
Common, and thriving best on well-manured ground, such as the
bottom of dung-heaps. Recorded for all districts and neighbouring
counties.
First record: Abbot, 1798.

C. didymus (L.) Sm. C.T.W. 21.15.2.   *Lesser Swine-cress*
Senebiera didyma (L.) Pers.
Cultivated fields, waste places, etc.   Colonist
Rare: this appears as a weed of arable land, and is sometimes well established as
such. It is introduced also with us as a wool adventive as it was on Tweedside.
Recorded for Herts., Bucks. and Hunts.
Ouse: Cardington*; *J.McL., in J. Bot. 1884.*
Ivel: arable field, Flitwick; *M. Oldfield*: railway siding, Flitwick*.
Lea: waste ground, Biscot; *H. B. Souster*: waste ground, Luton Hoo; *E.M.-R.*
Colne: plentifully in Luton Hoo gardens*.
First record: J. McLaren, 1884.

CAPSELLA Medic.

Capsella bursa-pastoris (L.) Medic. C.T.W. 21.21.1.
Thlaspi bursa-pastoris L., *Bursa pastoris* Weber   *Shepherd's Purse*
Waste places, arable land, etc.   Native
Very common. In the early thirties there were attempts to divide
this into a number of species and, from gatherings made in the
county by H. Phillips and Druce, Almquist recognised a number of
species, including one new to science. For accounts of these the
reader is referred to *B.E.C. 1930 Rep.* 238, 334, and *1932 Rep.* 90.
Recorded for all districts and neighbouring counties.
H.S. 9, 12, 16, 17, 22a, 25, 34a, 44, 48, 50.
First record: Podington, O. St. John Cooper in *Collections towards the
history and antiquities of Bedfordshire*, 1783.

LEPIDIUM L.

L. campestre (L.) R. Br. C.T.W. 21.14.2.   *Pepperwort Field-cress*
Cultivated fields, waste places, etc.   Colonist
Not infrequent: recorded for all neighbouring counties.

NENE: railway cutting, Wymington.

OUSE: Bedford; *Abbot in letter to J. E. Smith, 2 Nov. 1804*: Cardington Mill; *J. McL. in Herb. Beds. Nat. Hist. Soc.*: Lidlington; *W.F.B.*: cornfield, Marston Thrift*; *V. H. Chambers*: Chellington; *G. H. Day*: riverside, Felmersham.

KYM: Newton Gorse; *C. C. Foss.*

OUZEL: railway bank, Aspley Guise.

IVEL: Pulloxhill*; *C.C.*: cornfield near Ampthill Station.

LEA: Sundon; *S.B.P.L.*: near Luton; *D. M. Higgins*: waste ground, Limbury.

COLNE: arable field, Caddington; *H. B. Souster.*

First record: Abbot, 1804.

L. smithii Hook. (*L. heterophyllum* auct.), C.T.W. 21.14.3, is of doubtful occurrence in the county, as it is often and easily confused with forms of the preceding species. A specimen collected by C. Crouch at Ridgmont* is *Cardaria draba*. It would be useful to know of satisfactory specimens to support the records of Hamson from Goldington, Castle Mills by E. M. Langley, Box End by R. Townsend, and Harlington by S. A. Chambers in *F.F.B.* It is recorded for all neighbouring counties.

CARDARIA Desv.

Cardaria draba (L.) Desv. C.T.W. 21.16.1.   *Hoary Cress*
   Lepidium draba L.
Waste places, arable fields, etc.   Colonist
A comparatively recent introduction, which is now one of the worst weeds on calcareous soils, especially in the east of the county. Recorded for all districts and neighbouring counties.
First record: Kempston Mill, Thomas Corder in *Beds. Times*, 3 July 1861.

THLASPI L.

Thlaspi arvense L. C.T.W. 21.19.1.   *Field Penny-cress*
Arable land, waste places, etc.   Colonist
Common and probably increasing. It varies considerably in size according to its habitat. Recorded for all districts and neighbouring counties.
First record: Abbot, 1798.

T. perfoliatum L., C.T.W. 21.19.2, was recorded in the name of C. Crouch from Kitchen End in *W.F.B.*, but his herbarium specimen is *T. arvense*. It is listed for v.c. 30 in *Comital Flora*, but I know of no other evidence of its appearance. Recorded for Bucks.

IBERIS L.

Iberis amara L. C.T.W. 21.18.1.   *Wild Candytuft*
Chalk pits, arable land, rough pastures, etc.   ? Native
Essentially a species of disturbed soils, and limited to a small area of the Chalk. A record from the railway at Turvey in *Hillhouse List, 1875*, no doubt refers to a garden variety or *I. umbellata*, as does a record by G. A. Battcock from Cardington, seen and named *I. amara* var. *ruficaulis* Lejeune by Hamson, see *J.H. Notes*. Recorded for Herts., Bucks. and Cambs., and doubtfully for Northants.

OUZEL: *V.C.H.* with no details.

IVEL: Deacon Hill!; *I. Brown in Herb. Watson*: near High Down; *R. A. Pryor in J. Bot. 1876*: Barton Hills!†; *S.B.P.L.*: plentiful in old

chalk pit, Knocking Hoe*, and in arable land and on disturbed downland nearby.
First record: Isaac Brown, 1839.

TEESDALIA R. Br.

Teesdalia nudicaulis (L.) R. Br. C.T.W. 21.20.1. *Shepherd's Cress*
Iberis nudicaulis L.
Borders of heaths, dry banks, etc.   Native
Limited to the Lower Greensand, where it is locally abundant. It cannot withstand much competition and is consequently seen at its best on greensand exposures not yet colonized with bracken or heather. Recorded for all neighbouring counties, but doubtfully for Northants.
H.S. 22b, 47.
OUZEL: Aspley; *F.B.*: plentiful on open greensand around Heath and Reach.
IVEL: Ampthill!; Clophill!; *F.B.*: Steppingley Fox Covert; *W.C. Herb.*: Ridgmont*; *J. F. Crouch 1864*: Southill Park; *J.McL., B.M.*: Millbrook†; *J.McL. in Herb. J.S.*: sandstone rock, Maulden Green End!; *M. L. Berrill in J.H. Notes*: Flitwick, plentiful by roadside!; *W.F.B.*: Rowney Warren!; Maulden Wood!; between Sandy and Potton!; *J.E.L., J. Bot.*: between Shefford and Southill; near Galley Hill, Sutton!; Portobello Farm, Sutton!; Sandy Heath!; *J.E.L. Diary*: Bunker's Hill.
First record: Abbot, 1798.

RAPHANUS L.

Raphanus raphanistrum L. C.T.W. 21.8.1.   *Wild Radish*
Arable fields, waste places, etc.   Colonist
Common on calcareous soils but less so on sandy and gravelly soils. Yellow forms are not infrequent, and are to be seen consistently at Biddenham, Oakley, etc. Recorded for all districts and neighbouring counties.
First record: Abbot, 1798.

RESEDACEAE
RESEDA L.

Reseda lutea L. C.T.W. 22.1.2.   *Wild Mignonette*
Chalk downs, rough ground, etc.   Native
Frequent in disturbed ground on calcareous soils. It is more common on the Chalk than elsewhere and is most conspicuous there in the vicinity of rabbit warrens, and in disused chalk pits. Elsewhere it appears in brickpits and on railway sidings. It is very plentiful on waste ground in, and around, Bedford. Recorded for all neighbouring counties and all districts except Cam, but rare in the Kym district.
H.S. 63b, 69, 71, 73.
First record: Abbot, 1795.

R. luteola L. C.T.W. 22.1.1.   *Dyer's Rocket, Dyer's Weed, Weld*
Rough pastures, old walls, etc.   Native
In similar habitats to the preceding but more widely distributed.
It also appears, but sparingly, on the Lower Greensand and other
acid soils. Recorded for all districts and neighbouring counties.
H.S. 16, 25, 60, 61, 64.
First record: Abbot, 1798.

VIOLACEAE

VIOLA L.

Viola odorata L. C.T.W. 23.1.1.   *Sweet Violet*
Shady places, banks, etc.   Native
Common and well distributed. A very variable species with a
number of named varieties, the distribution of which in the county
has still to be studied. S. M. Walters, in *B.E.C. 1943–4 Rep.*, wrote
that he had observed var. **dumetorum** (Jord.) Rouy in a number
of places in the county, and named a gathering of mine from
between Ampthill and Ridgmont as this. Recorded for all districts
and neighbouring counties.
First record: Abbot, 1798.

V. hirta L. C.T.W. 23.1.2.   *Hairy Violet*
Pastures   Native
Common in pastures on calcareous soils and especially so on the
Chalk. A white-flowered form appears on the tunnel baulk at
Wymington. Considered rare by Abbot. Recorded for all districts
and neighbouring counties.
H.S. 1a–2a, 3a, 56b, 60, 68, 69.
First record: Abbot, 1798.

V. × **permixta** Jord. (*V. hirta* × *odorata*)
A very variable hybrid and probably overlooked.
OUSE: rectory garden, Houghton Conquest*.
IVEL: Flitwick Manor*.
CAM: Eyeworth*.
LEA: New Mill End Road; *S.B.P.L.*

V. calcarea (Bab.) Gregory
    V. hirta subsp. *calcarea* (Bab.) E. F. Warb. C.T.W. 23.1.2.
Chalk hills   Native
Rare; but there seems little to distinguish this from dwarfed late-
flowering specimens of *V. hirta*.
IVEL: Barton Hills, 1886†; *J.S. see E. S. Gregory in J. Bot. 1904*, 186.

V. palustris L. C.T.W. 23.1.9.   *Marsh Violet*
Marshes   Native
Rare, and limited to marshes in the middle of the county: *E. Bot.*
tab. 444, Jan. 1798 was used by Abbot in *Flora Bedfordiensis*.
Recorded for Bucks.

H.S. 40.
OUZEL: *V.C.H.*, with no details; but it is interesting to note that
Druce gives Heath as a station in *F. of Bucks*.
IVEL: 'in a standing pit a fourth part of a mile from Biggleswade in
the way to St. Neots'; *Ray's Synopsis*: Ampthill, where it still grows
sparingly in a marsh at the foot of Cooper's Hill!†*; *Plantae Bedford.*:
Flitton Marshes; *J.McL.*, *L.M.*: Westoning, where it grows in
limited quantity on Westoning Moor!*; Flitwick, where it is likewise limited, on the edge of a *sphagnum* bog in Folly Wood!; *C.C.*
First record: *Viola palustris stellata*, in *Indiculus Plantarum Dubiarum*,
an appendix to Ray's *Synopsis Ed. III*, 1724.

V. reichenbachiana Jord. ex Bor. C.T.W. 23.1.5. *Pale Wood Violet*
    V. sylvestris Lam. p.p.
Woods, shady places, etc.   Native
Common, and well distributed, but more frequent in woods on clay
soils. Recorded for all districts and neighbouring counties.
H.S. 2b, 5, 18, 19b, 33–34b, 52, 70a, 77b, 79a, 80b, 81, 83 [20b,
51a, 53b].
First record: *Hillhouse List, 1875*.

V. riviniana Reichb. C.T.W. 23.1.4.   *Common* or *Wood Violet*
Woods   Native
This is the large-flowered woodland violet, commonly known as the
dog violet. It is more common and less restricted in its habitats
than the preceding species. Recorded for all districts and neigh-
bouring counties.
H.S. 3b, 18, 30b, 32, 34b, 52, 54, 55a, 74b, 78.
First record: *Hillhouse List, 1876*.

V. canina L. C.T.W. 23.1.6.   *Heath Violet*
Heaths, dry open spaces, etc.   Native
Frequent in suitable places on the Lower Greensand and Clay-with-
Flints. All our Bedfordshire material is referable to subspecies
canina of the *Flora of the British Isles*. This is itself variable, but the
Bedfordshire forms appear to be only habitat ones. Early records of
V. canina refer to *V. riviniana*. Recorded for all neighbouring
counties.
H.S. 28, 74a, 77a.
OUZEL: Heath and Reach!†*; *S.B.P.L.*: near Brickhill; *G. C. Druce
in B.E.C. 1914 Rep.*: Horsemoor Farm*; Stockgrove.
IVEL: Maulden; *J.McL.*, *L.M.*: plentiful, Rowney Warren.
LEA: Chiltern Green, locally plentiful on the Common!†*; *S.B.P.L.*:
sparingly, Galley Hill.
COLNE: Studham Common!; *W.F.B.*: Whipsnade Common!*;
V. H. Chambers.
First record: J. McLaren, 1864.

V. rupestris Schmidt, C.T.W. 23.1.3, is recorded in *Comital Flora* for Herts., Beds.,
Bucks. and Northants.; but I cannot trace the source of the Beds. record.

POLYGALA 223

V. arvensis Murr. C.T.W. 23.1.12.   *Heartsease, Wild Pansy*
Cultivated fields, waste places, etc.   Colonist
Common: this, with *V. tricolor* L., was considered to be critical by
earlier botanists. Both have been recorded for the county, but I
have not seen any plants in Bedfordshire to compare with *V. tricolor*
as it appears on the East Anglian heaths. R. D. Meikle has referred
a specimen collected by me at Houghton Regis to *V. arvensis* ×
tricolor, and it is possible that this is to be recorded elsewhere in
the county. Recorded for all districts and neighbouring counties.
H.S. 13, 22b, 44, 71, 72, 82.
First record: Abbot, 1798, as *V. tricolor*.

V. lutea Huds., C.T.W. 23.1.10, was recorded, almost certainly in error, in *Hillhouse List, 1875*, on the authority of A. Poulton.

POLYGALACEAE

POLYGALA L.

Polygala vulgaris L. C.T.W. 24.1.1.   *Milkwort*
Pastures   Native
Abundant on the Chalk, and appearing with less certainty on the
Oolite and Boulder Clay. Recorded for all neighbouring counties.
H.S. 1a, 59, 60, 68.
NENE: tunnel baulk, Wymington*.
OUSE: Clapham; *F.B.*: Cardington; *Newbould MS.*: Colworth;
W.H. 1875: Bedford, Oakley; *J.H. Notes*: Twin Woods, Cleat Hill;
Colesdon, plentifully!; Salem Thrift; *A. W. Guppy*: Lady Wood,
Harrold; Dungee Corner; Pavenham!; *G. H. Day*: tunnel baulk,
Souldrop.
KYM: Dean; *M. Dalton*.
OUZEL: Aspley, possibly *P. serpyllifolia*; *F.B.*: Totternhoe Knolls!;
V. H. Chambers: Dunstable Downs.
IVEL: plentiful on the Chalk, records on other formations include:—
Ampthill; Sandy; both possibly *P. serpyllifolia*; *F.B.*: Clophill Heath,
mixed gathering; *W.C. Herb.*: Warden; *Newbould MS.*: pasture near
Washer's Wood*; railway bank, south of Shefford.
LEA: Limbury Marsh†; *J.S.*: Galley Hill!; *J.E.L., Diary*: Warden
Hill.
COLNE: Caddington; *R. A. Pryor*: Pepperstock; *D. M. Higgins in
W.B.E.C. 1899–1900 Rep.*: Whipsnade Zoo.
First record: Abbot, 1798.

P. oxyptera Reichb. C.T.W. 24.1.2.
  P. dubia Bellynck
Chalk hills   Native
Frequent on the Chalk, but easily confused with *P. vulgaris*.
Saunders's records of *P. serpyllacea* are to be referred to this species.
Recorded for all neighbouring counties except Hunts., but doubt-
fully for Northants.
H.S. 56b, 62, 66a.

OUZEL: Totternhoe, frequent on the Knolls!†*; *W.F.B.*: Cow Common.
IVEL: Wayting Hill; *J.E.L. in B.E.C. 1930 Rep.*: Sundon Hills.
LEA: Warden Hill!†; *W.F.B.*: plentiful on Galley Hill.
COLNE: Whipsnade Common!*; *V. H. Chambers.*
First record: G. C. Druce in *Bot. Rec. Club 1878 Rep.*

P. serpyllifolia Hose C.T.W. 24.1.3.   *Heath Milkwort*
  P. serpyllacea Weihe
Heathy places   Native
Rare and limited to the Lower Greensand and Clay-with-Flints.
Recorded for all neighbouring counties.
H.S. 23, 38.
OUZEL: Heath and Reach; Woburn, *V.C.H.*: sandy rides, King's Wood*; Horsemoor Farm, Woburn; New Wavendon Heath, v.c. 24 [Beds.].
IVEL: Clophill Heath, a mixed sheet as *P. officinalis*; *W.C. Herb.*: frequent on drier parts of Flitwick Moor.
COLNE: drive, Deadmansey Wood!*; *V. H. Chambers*: Studham Common.
First record: G. C. Druce, 1904.
First evidence: W. Crouch, 1842.

HYPERICACEAE

HYPERICUM L.

Hypericum hirsutum L. C.T.W. 25.1.12.   *Hairy St. John's Wort*
Woods and shady places   Native
Probably the most common of our St. John's Worts. It thrives best in boulder clay woods, but is by no means limited to them. Recorded for all districts and neighbouring counties.
H.S. 2a, 3a, 19a, 20a, 34a, 34b, 51a, 51b, 69, 78, 83.
First record: Abbot, 1798.

H. pulchrum L. C.T.W. 25.1.11.   *Heath St. John's Wort*
Heathy woods, commons, etc.   Native
Frequent on the Lower Greensand and Clay-with-Flints. Recorded for all neighbouring counties except Hunts.
H.S. 74a, 79a, 80b.
OUZEL: Aspley Wood!; *Plantae Bedford.*: Woburn; *W.H. 1875*: Heath and Reach; *W.F.B.*: frequent in the rides in King's Wood.
IVEL: Ampthill; Potton; *F.B.*: Rowney Warren; *J.McL., L.M.*: Clophill; *J.McL., B.M.*: Green End, Maulden; *M. L. Berrill in J.H. Notes.*
LEA: Chiltern Green, frequent on the Common!; *W.F.B.*
COLNE: Pepperstock!†*; *Davis*: Deadmansey Wood!; *V. H. Chambers*: Studham Common; Folly Wood, Caddington.
First record: Abbot, 1795.

H. tetrapterum Fr. C.T.W. 25.1.8   *Marsh St. John's Wort*
 H. quadrangulum L., *H. acutum* Moench
Marshes, wet meadows and riversides   Native
Common in suitable habitats throughout the county, but awaiting
record from the Colne district. Recorded for all neighbouring
counties.
H.S. 3a, 10c, 19a, 20a, 34a, 34b, 41, 53a, 57a, 84a, 85.
First record: Abbot, 1798.

H. dubium Leers C.T.W. 25.1.6.   *Imperforate St. John's Wort*
 H. quadrangulum auct.
Heathy woods   Native
Rare, but possibly often overlooked owing to its superficial resem-
blance to *H. perforatum*. Recorded for all neighbouring counties
except Hunts.
H.S. 79a.
OUZEL: woods near Woburn; *W.H. 1875*: Aspley Wood; *C.C.*: near
Little Brickhill; *A. B. Jackson*: King's Wood*; *E.M.-R.*!; roadside
near Potsgrove; *A. J. Wilmott*: Stockgrove.
IVEL: Silsoe; *W.F.S. Diary*.
LEA: Luton; *T. Vaux in F.B.*
COLNE: Oldhill Wood!; *W. D. Coales*: Caddington; *L. J. Margetts*:
Byslip Wood*.
First record: Abbot, 1798.

H. perforatum L. C.T.W. 25.1.5.   *Common St. John's Wort*
Hedgebanks, roadsides, etc.   Native
Common and well distributed throughout the county. Recorded
for all districts and neighbouring counties.
H.S. 2a, 25, 26, 53b, 58b, 63a, 63b, 66b, 68, 69, 70b, 80a.
First record: Abbot, 1798.

H. humifusum L. C.T.W. 25.1.9.   *Trailing St. John's Wort*
Heaths, woods on heathy soils, etc.   Native
Frequent and well distributed in the county. Recorded for all
neighbouring counties.
H.S. 25, 77a.
NENE: Wymington Scrubs; *V. H. Chambers*!
OUSE: Stagsden; *R. Townsend*: Marston Thrift!; *V. H. Chambers*.
OUZEL: Aspley, frequent in Aspley Wood!; *F.B.*: Woburn, in many
places!; *W.H. 1875*: King's Wood; Rushmere.
IVEL: Ampthill; *Plantae Bedford.*: Potton; *F.B.*: Clophill Warren;
W.C. Herb.: Keeper's Warren; *J.McL.*, *B.M.*: Maulden Wood!;
M. L. Berrill in J.H. Notes: Rowney Warren*; Simpson's Hill
Plantation.
LEA: Chiltern Green, plentiful in Horsley's Wood!†; *S.B.P.L.*
COLNE: near Pepperstock (? Beds.); *H. Brown*: Deadmansey
Wood!*; *V. H. Chambers*: Kidney Wood.
First record: Abbot, 1795.

P

H. elodes L. C.T.W. 25.1.14. *Bog St. John's Wort*
Marshes Native
Extinct: recorded for Bucks. and Cambs.
IVEL: Potton Marshes; *F.B.*

CISTACEAE

HELIANTHEMUM Mill.

Helianthemum chamaecistus Mill. C.T.W. 26.1.2.
Cistus helianthemum L., *H. nummularium* auct. *Common Rockrose*
Dry calcareous pastures Native
Common on the Chalk, rare on oolitic exposures and the Boulder
Clay. Recorded for all neighbouring counties.
H.S. 59–61, 65–66b, 75.
OUSE: Stevington; *Abbot's annotated F.B.*: Salem Thrift; *A. W. Guppy*.
OUZEL: common on the Chalk.
IVEL: common on the Chalk: Shefford Hill; *J. McL.*, *B.M.*: Rowney
Warren; *J.E.L.*, *F.F.B.*
LEA and COLNE: common on the Chalk.
First record: Abbot, 1798.

CARYOPHYLLACEAE

DIANTHUS L.

Dianthus armeria L. C.T.W. 30.7.1. *Deptford Pink*
Railway banks, etc. Colonist
Rare: recorded for Herts., Bucks. and Hunts.
LEA: Luton†; Leagrave!†*, where it is well established; *D. M. Higgins in W.B.E.C.*
1891–2 Rep.
First record: Abbot in letter to J. E. Smith, 2 Nov. 1802.

D. deltoides L. C.T.W. 30.7.2. *Maiden Pink*
Dry pastures Native
Extinct, but it is a species which could be overlooked and suitable
habitats should be searched for it. Recorded for, but extinct in,
Herts., Bucks. and Cambs.
IVEL: 'on Sandy hills, not far from an ancient Roman camp';
Ray's Synopsis Ed. II, 198: between Everton and Gamlingay; *Plantae
Bedford*.
First record: J. Ray, 1690.

SAPONARIA L.

Saponaria officinalis L. C.T.W. 30.9.1. *Soapwort*
Roadsides, waste places, etc. Colonist
Frequent, and usually well established. Both single and double-flowered forms
have been observed. Recorded for all districts except Kym, and for all neighbouring counties.
First record: Abbot, 1798.

SILENE L.

Silene vulgaris (Moench) Garcke.
Bladder Campion, White Bee Hen, Pudding Bag (Willington)
Cucubalus behen L., *S. cucubalus* Wibel C.T.W. 30.1.1, *S. angustifolia*
(Mill.) Schinz & Thell. non Poir.
Arable land, waste places, etc. Colonist
Common in most of the county, but rare in the north-east: 'var.
puberula Jord.' is recorded from Barton in *V.C.H.* Recorded for all
districts and neighbouring counties.
H.S. 1b, 61, 69, 71–73, 86.
First record: Abbot, 1798.

MELANDRIUM Röhl.

Melandrium noctiflorum (L.) Fr. C.T.W. 30.3.1.
Silene noctiflora L. *Night-flowering Campion*
Arable fields, waste places, etc. Colonist
Frequent as a weed of arable land on light clay soils and as an
adventive in waste places. Recorded for all neighbouring counties.
E. Bot. tab. 291, Dec. 1795, was drawn from a specimen sent by
Abbot.
OUSE: Oakley West Field; *F.B.*: Basmead; *C.R.O. List*: Biddenham;
J.McL., *B.M.*: arable fields, Milton Ernest; near Hanger Wood;
Old Race Course, Bedford; *C. C. Foss*: arable field, Souldrop*;
E.M.-R.
KYM: arable field, Knotting*; *C. C. Foss*!
IVEL: Barton Hills; *S.B.P.L.*: arable field, Stotfold; *H. B. Souster*!:
arable field, Gravenhurst*; *E.M.-R.*!: rubbish dump, Sundon*;
railway sidings, Biggleswade, Arlesey.
LEA: Leagrave; Luton; *J.S.*, *J. Bot. 1889*.
COLNE: rough ground, Farley Estate; *H. B. Souster*.
First record: Abbot, 1795.

M. album (Mill.) Garcke C.T.W. 30.3.3. *White Campion*
Lychnis alba Mill.
Waste places, roadsides, etc. Colonist
Common: recorded for all districts and neighbouring counties.
H.S. 16, 25, 29, 44, 47, 48, 54, 71, 73, 82.
First record: Red and White Lychnis, rare, Abbot, 1798.

M. dioicum (L.) Coss. & Germ. *Red Campion*
Lychnis dioica L., *M. rubrum* (Weigel) Garcke C.T.W. 30.3.2.
Roadsides, wood borders, etc. Native
Common on the western part of the Lower Greensand, and frequent
in other parts of the county except on the eastern and southern
borders. It is rare on the Chalk. Hybrids with *M. album* are not
infrequent, especially in the neighbourhood of Woburn, where
M. dioicum is more common than elsewhere in the county. Recorded
for all neighbouring counties, except Cambs., but it is rare in Hunts.
H.S. 31b.

NENE: Podington; *W. Kitchin.*
OUSE: Millbrook; *W.H. 1875.*
KYM: Dean!; *M. Dalton.*
OUZEL: common on the Lower Greensand.
IVEL: common on the Lower Greensand west of the Ivel.
LEA: Bramingham; *W.F.B.*: Limbury!†; *Herb. J.S.*
COLNE: Whipsnade; *W. D. Coales*: near Zouche's Farm, v.c. 20
[Beds.]; *H. B. Souster.*
First record: Abbot, 1798.

LYCHNIS L.

Lychnis flos-cuculi L. C.T.W. 30.4.1.   *Ragged Robin*
Marshes, damp woods, etc.   Native
Common in all likely habitats in the county. Recorded for all
districts and neighbouring counties.
H.S. 7, 10c, 14b, 19a, 34a, 37, 39a, 41, 43, 53a, 53b.
First record: Abbot, 1798.

AGROSTEMMA L.

Agrostemma githago L. C.T.W. 30.5.1.   *Corn Cockle*
  Lychnis githago (L.) Scop.
Cornfields   Colonist
Frequent, but uncertain in its appearance and possibly diminishing.
A white-flowered form has been observed at Studham. Recorded
for all districts except Cam, and for all neighbouring counties.
H.S. 82.
First record: Abbot, 1798.

CERASTIUM L.

Cerastium tetrandrum Curt. C.T.W. 30.12.9. *Dark-green Mouse-ear Chickweed*
Railway tracks   Colonist
Plentiful on the railway, and growing mainly between the metals. It no doubt
spread in the war years when labour on the railway was scarce and tracks were less
frequently relaid, and it is already beginning to diminish. It is, because of the
absence of the railway, not recorded for the Kym, Cam and Colne districts. I have
found it in all neighbouring counties.
First record: E. Milne-Redhead in *B.E.C. 1943–4 Rep.* 806.

C. pumilum Curt. C.T.W. 30.12.10.   *Curtis's Mouse-ear Chickweed*
Railway tracks and banks   Colonist
Plentiful along considerable stretches of the railway, and usually growing at the
side of the cinder tracks outside the permanent way and on banks and embankments with a ballast foundation. It is more likely to be a permanent member of
the flora than is *C. tetrandrum.* Abbot recorded it from Aspley and Ampthill,
unlikely stations in his day, and his herbarium specimen is *C. semidecandrum* which
still grows at both places. A sketch of '*C. pumilum*', drawn from material sent by
Abbot but not used for *English Botany*, appears to represent a form of *C. glomeratum*
(see *J. Bot. 1903*). I have found *C. pumilum* in some quantity on the railway in
Herts. and Northants.
OUSE: common on the railway between Oakley and Sharnbrook; *E.M.-R.!*:
common on the railway south of Bedford.
IVEL: common on the railway between Sundon and Harlington* and to the south
of Henlow.
LEA: plentifully north of Leagrave.
First certain record: E. Milne-Redhead and the author in *Watsonia I (1950)* 246.

C. brachypetalum Pers. C.T.W. 30.12.8.
Dry railway banks   Colonist
Plentiful for about a mile in the two large railway cuttings at Wymington*, and
not yet recorded elsewhere in Britain. Its status is uncertain.
First record: E. Milne-Redhead in *The Naturalist, 1947*, 95–6.

C. semidecandrum L. C.T.W. 30.12.11.
Little Mouse-ear Chickweed
Heaths, railway banks, etc.   Native
Frequent on the Lower Greensand, but to be found also on glacial
gravels and well-drained railway banks and sidings. It shows con-
siderable variability in size and form. Recorded for all neighbouring
counties.
H.S. 27, 48.
NENE: railway cutting, Wymington!*; *E.M.-R.*
OUSE: Cardington†; *J.McL. in Herb. J.S.*: waste ground, Cople;
E.M.-R.
OUZEL: Aspley, as *C. pumilum*!; *F.B.*: Heath and Reach; *V.C.H.*:
IVEL: Ampthill!†*, as *C. pumilum*, it is common on Cooper's Hill;
F.B.: Maulden!; *W.C. Herb.*: between Clophill and Ampthill; *T. B.
Blow in Herb. Watson*: near Shefford; *J.McL., L.M.*: Southill;
J.McL., B.M.: Flitwick*; *W.F.B.*: Wilbury Hill; *J.E.L. in W.B.E.C.
1912–3 Rep.*: Rowney Warren!; Sandy!; *J.E.L., J. Bot.*: Westoning*.
First record: Abbot, 1798, but his herbarium specimen is repre-
sented by *C. holosteoides* and his specimen of *C. pumilum* is represented
by this species; more certainly, W. Crouch, 1845.

C. glomeratum Thuill. C.T.W. 30.12.7.
C. viscosum auct.   *Sticky Mouse-ear Chickweed*
Arable fields, waste places, railway banks, etc.   Colonist
Common and growing at its best in well-drained habitats. Recorded
for all districts and neighbouring counties.
H.S. 23, 24, 27, 44, 53b, 54, 69, 82.
First record: Abbot, 1798.

C. holosteoides Fr.   *Mouse-ear Chickweed*
C. vulgatum auct. C.T.W. 30.12.6.
Meadows, rough pastures, waste places, etc.   Native
Common, and growing in more varied habitats than *C. glomeratum*.
It is probably native in meadows, where it is a regular constituent
of the flora, but elsewhere it is a colonist. A variable species, but
it is doubtful if any of its forms deserve varietal rank. Recorded for
all districts and neighbouring counties.
H.S. 2a, 4, 9, 10a, 11, 14b, 16, 22b, 23, 25–29, 31a, 37, 41, 47, 53a,
55b, 56b, 61, 63a, 66a, 67, 69, 72, 75–77a, 80a, 81, 82, 84a, 85.
First record: Abbot, 1798.

C. arvense L. C.T.W. 30.12.2.   *Field Mouse-ear Chickweed*
Edges of arable fields, disturbed soils, etc.   Native
A species of unusual distribution in the county. It occurs frequently
on disturbed soils on the Chalk and Clay-with-Flints, and occa-

sionally also on rough ground on the Lower Greensand and on gravels. Recorded for all neighbouring counties.
Ouse: Ford End!; Kempston; *F.B.*: Cox's Pits!; *J.McL.*, *L.M.*: Biddenham; *J.McL.*, *B.M.*: Cardington; *J.McL. in J.H. Notes*: Honey Hill Fields.
Ouzel: Totternhoe!*; *W.F.B.*: Dunstable!; *F.F.B.*: it is frequent here on the Chalk.
Ivel: Barton Hill!; *F.B.*: Pegsdon!; Cadwell; *J.E.L.*, *F.F.B.*: roadside near Sandy.
Lea: Warden Hills!*; *S.B.P.L.*: frequent on gravelly soils at Leagrave and Limbury.
Colne: Blow's Downs!; *Fl. Herts.*: chalk quarry, Mount Pleasant; edge of arable field, Studham.
First record: Abbot, 1798.

MOENCHIA Ehrh.

Moenchia erecta (L.) Gaertn., Mey. & Scherb. C.T.W. 30.16.1.
Upright Chickweed
Sagina erecta L., *Cerastium erectum* (L.) Coss. & Germ.
Sandy heaths   Native
Very rare, but possibly overlooked. Recorded for all neighbouring counties.
Ivel: Clophill; Ampthill; *Plantae Bedford.*: Maulden Moor; *W.C. Herb.*: Deepdale; *J.McL.*, *L.M.*: Flitwick†*; *J.McL.*, *J. Bot. 1884*: Rowney Warren; *J.E.L.*, *Diary*.
First record: Abbot, 1795.

MYOSOTON Moench

Myosoton aquaticum (L.) Moench C.T.W. 30.13.1.
Water Chickweed
Cerastium aquaticum L., *Stellaria aquatica* (L.) Scop.
Riversides, ditches, etc.   Native
Common by all the larger rivers of the county. Recorded for all districts except Nene, Cam and Colne, and for all neighbouring counties.
H.S. 6, 13.
First record: Abbot, 1798.

STELLARIA L.

Stellaria media (L.) Vill. C.T.W. 30.14.2.
Alsine media L.   *Common Chickweed, Chickenweed*
Arable land, waste places, etc.   Colonist
Common, but very variable. Recorded for all districts and neighbouring counties.
H.S. 1a, 6, 12, 16, 17, 22a, 22b, 24–27, 29, 34a, 44, 48–50, 54, 71, 81, 86.
First record: Abbot, 1798.

S. apetala Ucria C.T.W. 30.14.3.   *Lesser Chickweed*
  S. boraeana Jord.
Sandy pastures   Native
Probably not infrequent on the Lower Greensand.
IVEL: Ampthill!*; Flitwick; *V.C.H.*: open sandy pasture, Maulden;
E.M.-R.
First record: G. C. Druce, 1904.

S. neglecta Weihe C.T.W. 30.14.4.   *Greater Chickweed*
Shady pastures   Native
Probably more common than the records indicate, as it is super-
ficially like robust forms of *S. media*. Recorded for all neighbouring
counties.
OUSE: Kempston, spec. in Herb. Birmingham Univ.; *W.H. 1875*.
OUZEL: Aspley Woods; *W.F.B.*
IVEL: pasture on verge of Flitwick Moor*; *E.M.-R.*
First record: W. Hillhouse, 1875.

S. holostea L. C.T.W. 30.14.5.   *Greater Stitchwort*
Waysides, edges of woods, etc.
Common over the greater part of the county, but awaiting record
in the Nene district. Recorded for all neighbouring counties.
H.S. 31a, 32, 42, 78, 83.
First record: Abbot, 1798.

S. palustris Retz. C.T.W. 30.14.6.   *Marsh Stitchwort*
  S. dilleniana Moench non Leers, *S. glauca* With.
Marshes, wet meadows, etc.   Native
Rare: recorded for all neighbouring counties.
H.S. 38, 39a.
OUSE: Ford End; *F.B.*: Fenlake; *J.McL.*, *L.M.*: Cardington;
J.McL., *B.M.*: Bromham Park; *V. H. Chambers*.
IVEL: Ampthill Bogs; *F.B.*: Sandy; *W.H. 1875*: Flitton Moor;
W.C. Herb.: Flitwick Moor!†*; *J.H. in W.F.B.*: Shefford; *J.E.L.*,
J. Bot.
First record: Abbot, 1798.

S. graminea L. C.T.W. 30.14.7.   *Lesser Stitchwort*
Heathy pastures, etc.
Common on the lighter soils throughout the county. Recorded for
all districts except Cam, and for all neighbouring counties.
H.S. 22a, 23, 28, 37, 38, 39b, 41, 49, 74a, 77a, 79a.
First record: Abbot, 1798.

S. alsine Grimm C.T.W. 30.14.8.   *Bog Stitchwort*
  S. uliginosa Murr.
Marshy places, etc.   Native
Frequent: recorded for all neighbouring counties.
H.S. 14a, 37, 41.

OUSE: Kempston; *F.B.*: Marston; *Ada Stimson in J.H. Notes*: damp meadows, Radwell; *N.Y. Sandwith*: meadow, Lidlington; marsh, Ampthill Park; meadow, Eaton Socon.
OUZEL: Heath and Reach; *S.B.P.L.*: meadows, Tilsworth; Horsemoor Farm; meadow, Husborne Crawley.
IVEL: Ampthill Bogs, it still grows at the foot of Cooper's Hill!*; *F.B.*: Clophill; *W.C. Herb.*: Potton Marshes; *J.McL., L.M.*: Flitwick, frequent on the Moor!; *S.B.P.L.*: Flitton!; Maulden; *J.H.*: Westoning!; *F.F.B.*: Sandy; Biggleswade; *J.E.L., F.F.B.*: marsh near Cainhoe Castle; *H. B. Souster*!: marsh Tingrith; Horsemoor Farm.
LEA: Marslets†; *S.B.P.L.*: meadow, East Hyde.
COLNE: grassy rides, Oldhill Wood*.
First record: Abbot, 1798.

MOEHRINGIA L.

Moehringia trinervia (L.) Clairv. C.T.W. 30.21.1.
Arenaria trinervia L.   *Three-nerved Sandwort*
Woods, shady places, etc.   Native
Well distributed and common throughout the county. Recorded for all districts and neighbouring counties.
H.S. 27, 49, 54, 74b, 77a, 78, 81, 83.
First record: Abbot, 1798.

ARENARIA L.

Arenaria serpyllifolia L. C.T.W. 30.22.1. *Thyme-leaved Sandwort*
Arable fields, waste places, etc.   Colonist
Common and usually a feature of unstable habitats. Recorded for all districts and neighbouring counties.
H.S. 22b, 28, 44, 45a, 56b, 61, 69, 71, 72, 75, 86.
First record: Abbot, 1798.

A. leptoclados (Reichb.) Guss. C.T.W. 30.22.2.
                                           Lesser Thyme-leaved Sandwort
Dry fields, old walls, etc.   Colonist
Less common and not easily distinguishable from the preceding, with which it frequently grows. Awaiting record for the Cam district; but recorded for all neighbouring counties.
H.S. 61.
First record: R. A. Pryor in *J. Bot. 1876*.

MINUARTIA L.

Minuartia tenuifolia (L.) Hiern C.T.W. 30.18.4.
Arenaria tenuifolia L.   *Fine-leaved Sandwort*
Old walls, railway banks, railway sidings, etc.   Colonist
A species which is apparently increasing with the increase of suitable well-drained habitats. Recorded for all neighbouring counties except Hunts.

Ouse: Newnham Wall; *F.B.*: railway embankment, Souldrop*;
P.T.!: railway wall, Moor End, Felmersham; *H. B. Souster*!:
railway siding, Bedford.
Ouzel: Totternhoe; railway bank, Sewell!*; *E.M.-R.*
Ivel: Hawnes; *Abbot's annotated F.B.*: near Pegsdon Road (? Beds.);
H. Brown: Barton Hills, it grows sparingly near Ravensburgh Castle
on glacial gravel!; *J.E.L.*, *F.F.B.*: railway bank, Harlington; *P.T.*!:
railway sidings, Ampthill, Harlington*; Sundon Rubbish Dump.
Lea: railway siding, Chaul End*; *H. B. Souster*!
First record: Abbot, 1798.

SAGINA L.

Sagina nodosa (L.) Fenzl C.T.W. 30.17.9.   *Knotted Pearlwort*
   Spergula nodosa L.

Marshes   Native
Possibly extinct. Recorded for all neighbouring counties.
Ouse: Stevington; *Plantae Bedford*.
Ivel: Ampthill; *F.B.*: Gravenhurst Moor; *W.C. Herb.*: Flitwick,
where it was also seen by Saunders*; *J.McL.*, *B.M.*
Lea: Leagrave Marsh, where it was plentiful with *Triglochin palustre*
and *Menyanthes trifoliata*, but I have not seen it here since 1926!†*;
F.F.B.
First record: Abbot, 1795.

S. ciliata Fr. C.T.W. 30.17.2.   *Ciliate Pearlwort*
Dry places, old walls, etc.   Native
Frequent, but easily overlooked. Recorded for all neighbouring
counties.
H.S. 22b.
Nene: old wall, Hinwick.
Ouse: old wall, Stevington*; *E.M.-R.*
Ouzel: on open greensand, Heath and Reach*.
Ivel: Cainhoe, as *S. apetala*; *W.C. Herb.*: Ampthill; *V.C.H.*: Shefford;
E.M.-R.: old wall, Tingrith*; station yard, Sandy*; Rowney
Warren.
Cam: old wall, Eyeworth*.
Lea: Luton†; *S.B.P.L.*
Colne: chalk quarry, Mount Pleasant.
First record: Saunders, 1885.
First evidence, W. Crouch, 1845.

S. filicaulis Jord. C.T.W. p. 313.
Dry places, old walls, etc.   Wall denizen
Rare, unless overlooked; recorded for Herts., Bucks. and Hunts.
Ouse: old wall, Eaton Socon, 1936; *A.J. Wilmott in B.E.C. 1945 Rep.*: station yard,
Bedford*; *D. P. Young*.
Kym: old wall, Melchbourne*; railway siding, Kimbolton, v.c. 30 [Hunts.].
Lea: railway siding, Luton*.
First record: A. J. Wilmott, 1945.

S. apetala Ard. C.T.W. 30.17.1.   *Common Pearlwort*
Dry places, old walls, etc.   Native
More common than the two preceding species, with which it often
grows. Recorded for all neighbouring counties.
H.S. 27.
NENE: old wall, Hinwick.
OUSE: Bedford; *C. C. Foss*: railway wall, Souldrop*; Felmersham
Bridge*.
KYM: Knotting Church*; Dean Church*; railway sidings, Kim-
bolton, v.c. 30 [Hunts.].
OUZEL: Heath and Reach; *N. Y. Sandwith*: Leighton Buzzard
Church.
IVEL: Ampthill; *Plantae Bedford.*: Rowney Warren!; *J.McL.*,
B.M.: Shillington; *J. Pollard*: Pulloxhill; *C.C.*: Flitwick; old wall,
Tingrith*.
LEA: East Hyde; *R. A. Pryor in J. Bot. 1876*: Luton; *W.F.B.*
COLNE: Market Street, v.c. 30 [Herts.]; *Plantae Bedford.*: dry bank,
Deadmansey Wood*.
First record: Abbot, 1795.

S. procumbens L. C.T.W. 30.17.4.   *Procumbent Pearlwort*
Heaths, woodland rides, lawns, etc.   Native
Common and well distributed in the county, but more common on
sandy soils than elsewhere. Recorded for all districts and neigh-
bouring counties.
H.S. 3a, 19a, 20a, 22b, 23, 26, 27, 34a–35, 53b, 54, 77a, 80a.
First record: Abbot, 1798.

SPERGULA L.

Spergula arvensis L. C.T.W. 30.23.1.   *Corn Spurrey, Beggar Weed*
                                                             (Batchelor)
Arable fields, gravel pits, etc.   Colonist
Common in arable fields on the lighter soils. It is variable and
was at one time considered as comprising two species, but these
are, no doubt quite rightly, reduced to varietal rank in the *Flora
of the British Isles*. These, var. **arvensis** (var. *vulgaris* (Boenn.)
Mert. & Koch) and var. **sativa** (Boenn.) Mert. & Koch, are
apparently equally common with us. Recorded for all neighbouring
counties.
H.S. 44.
OUSE: gravel pit, Eaton Socon; *E.M.-R.*!: arable field, Oakley;
H. B. Souster!
OUZEL and IVEL: common on the Lower Greensand.
COLNE: Pepperstock†; *Herb. J.S.*: arable fields, Studham, Kens-
worth, v.c. 20 [Beds.].
First record: Abbot, 1798.

SPERGULARIA (Pers.) J. & C. Presl

Spergularia rubra (L.) J. & C. Presl C.T.W. 30.24.1.
Arenaria rubra L.   *Sand Spurrey*
Heaths, etc.   Native
Locally abundant in heathy places, especially on previously dis-
turbed soils on the Lower Greensand. Rare on the Clay-with-
Flints. Recorded for all neighbouring counties.
H.S. 22a, 24, 27.
OUZEL: Aspley; *J.McL.*, *L.M.*: common around Heath and Reach*.
IVEL: Potton!; Sandy!; Ampthill!; *F.B.*: Cainhoe; *W.C. Herb.*:
Clophill; *J.McL.*, *L.M.*: Flitwick!*; *S.B.P.L.*: Maulden!; *M. L.
Berrill in J.H. Notes*: Westoning!; *W.F.B.*: Southill; Tingrith;
Sutton!; *J.E.L.*, *F.F.B.*: Biggleswade; *J.E.L.*, *Diary*: Rowney
Warren.
COLNE: Pepperstock; *S.B.P.L.*: heathy ground, Whipsnade; *W. D.
Coales*: on clinkers, Luton Hoo Gardens.
First record: Abbot, 1798.

ILLECEBRACEAE
SCLERANTHUS L.

Scleranthus annuus L. C.T.W. 30.31.1.   *Annual Knawel*
Heathy pastures, arable fields, etc.   Colonist
Frequent on the Lower Greensand and Clay-with-Flints. Recorded
for all neighbouring counties.
H.S. 82.
OUSE: Millbrook; *W.H. 1875*: Bromham; *A. W. Guppy.*
OUZEL: Woburn; *W.H. 1875*: Heath; *V.C.H.*: Aspley; New Waven-
don Heath, v.c. 24 [Beds.].
IVEL: Steppingley; *W.C. Herb.*: Sandy; *W.H. 1875*: Flitwick!*;
Ampthill!; *J.H. Notes*: Rowney Warren!; Southill; Maulden!;
J.E.L., *F.F.B.*: Appley End.
LEA: Streatley; Chiltern Green; Stopsley*.
COLNE: Whipsnade; near Ravensdell Wood, v.c. 20 [Beds.].
First record: Abbot, 1798.

PORTULACACEAE
MONTIA L.

Montia fontana L. C.T.W. 31.1.   *Blinks*
Marshy places, damp pastures, etc.   Native
Rare. I wish to thank S. M. Walters who has named all my
material. It is difficult to know into which subspecies to place
Abbot's records from Ampthill Moor in *Plantae Bedfordiensis* and
Potton Marshes in *Flora Bedfordiensis* and a record from Flitwick in
V.C.H. Montia is recorded for all neighbouring counties except
Hunts., but the distribution of the subspecies in them is not known.
First record: Abbot, 1795.

Subsp. **chrondrosperma** (Fenzl) S. M. Walters.
M. verna auct. C.T.W. 31.1.1.
IVEL: Maulden Bogs; *Herb. Miss Crouch*: disused sandpit, Sutton!*; *T.L.*: turf, Westoning Moor!*; *W. D. Coales and H. B. Souster.*

Subsp. **intermedia** (Beeby) S. M. Walters
M. lusitanica Sampaio C.T.W. 31.1.3.
Appearing usually in wetter places than the preceding.
H.S. 37.
OUZEL: Heath and Reach†*; *Herb. J.S.*, 1893: plentiful in marsh and stream, Horsemoor Farm*.
IVEL: Flitton*; *C.C.*

Subsp. **variabilis** S. M. Walters
M. rivularis auct.
OUSE: marsh, Ampthill Park*; *H. B. Souster!*

CLAYTONIA L.

Claytonia perfoliata Donn ex Willd. C.T.W. 31.2.1.
Shady places, banks, etc.   Denizen
Frequent on the Lower Greensand, a casual elsewhere. Recorded for all neighbouring counties.
H.S. 44, 47.
OUZEL: Leighton Buzzard, frequent in a number of places near the town!; *R. C. Chambers in W.F.B.*: Heath and Reach; Aspley Guise; Woburn.
IVEL: Ampthill, where it is still a feature of sandy banks and shady places!†; *T. Corder in The Phytologist, 1852*, 485: between Ampthill and Maulden!; *J.McL., L.M.*: Maulden!; *S.B.P.L.*: Pulloxhill; *J. Pollard*: Flitwick, especially common near the church!; *C.C.*: Old Warden; *J.E.L., J. Bot.*: Moors Plantation!; *H. B. Souster*: Sandy; Westoning*; Potton; Bunker's Hill; rubbish dump, Sundon.
LEA: Limbury; *F.F.B.*: railway siding, Luton.
First record: T. Corder, 1852, the first for Britain; but there is an incorrectly-named specimen in Herb. Watson collected at Gorleston, Suffolk, in 1837.

CHENOPODIACEAE

CHENOPODIUM L.

I have had considerable help from J. P. M. Brenan who has not only named my material, but made a number of visits to the county.

Chenopodium bonus-henricus L. C.T.W. 34.1.1.
Good King Henry
Roadsides, waste places, etc.   Colonist
Frequent, especially near towns and villages, where it is well established. Recorded for all neighbouring counties.
NENE: Wymington*.
OUSE: Elstow; Wilstead; St. Cuthberts; Goldington; *F.B.*: Cardington; *J.McL., B.M.*: Bedford; *J.H. Notes*: Houghton Conquest; *C.C.*: Harrold; *G. H. Day*: Sharnbrook; Felmersham, in the neighbourhood of which it is common.

OUZEL: Totternhoe!*; *W.F.B.*
IVEL: Hawnes; *F.B.*: Cainhoe; *W.C. Herb.*: Steppingley; *C.C.*:
Shefford; *J.E.L.*, *F.F.B.*: Sundon; *H. B. Souster*!
LEA: R. Lea, south of Luton; *S.B.P.L.*: Someries Castle†; *W.F.B.*:
Dallow Lane!; *V. H. Chambers.*
First record: Abbot, 1798.

C. polyspermum L. C.T.W. 34.1.2.   *Many-seeded Goosefoot, All-seed*
Arable fields, waste places, etc.   Colonist
Frequent, especially on well-manured soils. Recorded for all
neighbouring counties.
H.S. 34a.
NENE: Podington.
OUSE: Cotton End; Cardington; *J.McL., in J. Bot. 1884*: Souldrop*;
Marston Mortaine; Wilstead Wood.
KYM: Pertenhall; Swineshead, near Worley's Wood.
OUZEL: Woburn Park; Tilsworth.
IVEL: Cainhoe, as *C. murale*; *W.C. Herb.*: Silsoe; *W.F.B.*: Pulloxhill!;
C.C.: Clophill; Streatley; Flitwick*; Westoning*.
CAM: Cockayne Hatley; *T.L.*: Thistly-grounds Farm.
LEA: Luton Hoo!†*; *J. Catt. in Bot. Rec. Club 1884–6 Rep.*: railway,
Luton!; *H. B. Souster*: East Hyde.
COLNE: gardens, Luton Hoo.
First record: J. McLaren, 1884.
First evidence: W. Crouch, 1843.

C. album L. C.T.W. 34.1.4.   *Fat Hen, White Goosefoot,*
                                             Wild Spinach (Batchelor)
Arable fields, waste places, etc.   Colonist
Common, especially on lighter soils, but very variable. J. P. M.
Brenan has referred various gatherings of mine to f. *cymigerum*
(Koch) Schinz & Thell. Recorded for all districts and neighbouring
counties.
H.S. 12, 16, 17, 44, 50.
First record: Abbot, 1798.
    C. album × **reticulatum**
IVEL: Toddington*; Westoning*.
LEA: Markyate, v.c. 20 [Beds.]*.

C. reticulatum Aellen C.T.W. 34.1.5.
Arable fields   Colonist or Casual
Probably more common than the records indicate. Recorded for Northants.
NENE: Hinwick.
OUSE: Eaton Socon*.

C. suecicum J. Murr C.T.W. 34.1.6.
    C. viride auct.
Arable fields   Colonists or Casual
Rare: there are many records by Saunders and J. E. Little of *C. viride*, a name
usually applied by earlier botanists to *C. album* f. *cymigerum.*
OUZEL: Tilsworth*.

C. ficifolium Sm. C.T.W. 34.1.10.   *Fig-leaved Goosefoot*
Arable fields, waste places, etc.   Colonist
Rare, but considered common by Abbot. Recorded for all neighbouring counties.
OUZEL: Woburn; *H. K. Airy Shaw.*
IVEL: Arlesey; *J.E.L., B.E.C.*: Shefford*.
First record: J. E. Little, 1936.

C. murale L. C.T.W. 34.1.12.   *Nettle-leaved Goosefoot*
Arable fields, waste places, etc.   Colonist
Rare, and probably introduced recently as a wool adventive. Abbot, no doubt
mistakenly, considered it common. There are no specimens to support records in
J.H. Notes by E. M. Langley from Willington and Ivor Evans from Newnham.
Recorded for Herts., Bucks. and Cambs.
OUSE: railway yard, Bedford.*
IVEL: old garden, Shefford*; *J. E. Lousley*!: Ampthill, both on railway sidings and
in arable fields*; arable fields, Maulden; railway siding, Flitwick.
First record: uncertain.

C. urbicum L. C.T.W. 34.1.13.   *Upright Goosefoot*
Waste places   Casual
Rare: Abbot recorded it from St. Cuthbert's and has a correctly-named specimen;
but sent *C. rubrum* in mistake for it to Sowerby. McLaren's specimens are correctly
named; but there are no specimens to support records of C. Crouch, Ivor Evans,
Hamson and L. W. Wilson. Recorded for all neighbouring counties except Hunts.
OUSE: on a dung-hill, Cardington, specimens also in Herb. B.M.; *J.McL., B.M.*
First certain record: Herb. J. McLaren, Luton Museum, 1864.

C. hybridum L. C.T.W. 34.1.14.   *Maple-leaved Goosefoot, Sowbane*
Arable fields, waste places, etc.   Colonist
Rare, and probably introduced recently as a wool adventive. Recorded for all
neighbouring counties.
OUSE: Bedford, where it appears in some quantity on the railway sidings!*;
F.B.: Willington; *E. M. Langley in J.H. Notes.*
OUZEL: arable field, Woburn*.
IVEL: Sandy; *J. E. Little in W.B.E.C. 1912-3 Rep.*: Ampthill; *A. E. Ellis in B.E.C.
1939-40 Rep.*: arable field, Maulden; *H. B. Souster*!: arable field, Hollington*.
First record: Abbot, 1798.

C. rubrum L. C.T.W. 35.1.15.   *Red Goosefoot*
Arable fields, waste places, etc.   Colonist
Common, but very variable in appearance. It is most plentiful on well-manured
soils, and is especially abundant on rubbish dumps. Abbot considered it to be
rare, but it is doubtful if he knew the *Chenopodia* well. Recorded for all districts and
neighbouring counties.
First record: Abbot, 1798.

C. glaucum L. was recorded by Abbot in a letter to J. E. Smith, 3 Sept. 1802, and
from Bedford in *J.H. Notes.* It needs confirmation. Recorded for Herts. and Bucks.

ATRIPLEX L.

The varieties of the species of this genus were named for me by A. J.
Wilmott.

Atriplex patula L. C.T.W. 34.3.2.   *Common Orache*
Arable fields, waste places, etc.   Colonist
Common but very variable. Recorded for all districts and neigh-
bouring counties.
H.S. 6, 7, 17, 71, 73, 86.
First record: Abbot, 1798.

Var. **erecta** (Huds.) Lange
Ouzel: Woburn Sands; *R. A. Pryor.*
Ivel: Arlesey; *J.E.L., F.F.B.*
Lea: Luton; *R. A. Pryor*: Great Bramingham; *S.B.P.L.*: Luton*.
    Var. **angustissima** Gren. & Godr.
Ouse: Cardington, as *A. deltoidea*; *J.McL., B.M.*
Ouzel: Woburn Sands; *W. Moyle-Rogers.*
Lea: Dallow Road*.
    Var. **linearis** Moss & Wilmott
Nene: Wymington*.
Kym: Pertenhall*.
Ouzel: Leighton Buzzard*.
Colne: Slip End*.

A. hastata L. C.T.W. 34.3.3.   *Hastate Orache*
    A. deltoidea Bab.
Arable fields, waste places, etc.   Colonist
Less common than *A. patula*, but appearing plentifully on dung-heaps, etc. Recorded for all districts except Nene, Ouzel, and Cam and for all neighbouring counties.
H.S. 7, 13.
First record: Abbot, 1798.

TILIACEAE

TILIA L.

Tilia platyphyllos Scop. C.T.W. 35.1.1.   *Large-leaved Lime*
Roadsides   Planted
This occurs only as a planted tree in the county, where it may be more common than the records indicate. Recorded similarly for Herts., Bucks. and Northants.
Ouse: Eaton Socon.
Ouzel: Eggington!*; *E.M.-R.*: Woburn*.
Ivel: Barton Springs*.
Colne: near Greencroft Barn, v.c. 20 [Beds.]; *E.M.-R.*
First record: the author in *B.E.C. 1943–4 Rep.* 807.

T. × vulgaris Hayne (*T. cordata* × *platyphyllos*) C.T.W. 35.1.3.   *Common Lime*
    T. europaea L. nom. ambig.
Roadsides, woods, etc.   Planted
This is the commonly planted lime of the county. It was considered rare by Abbot, but, while his herbarium specimen is correctly named, the stations he gives are more likely to be those of *T. cordata*. Recorded for all districts and neighbouring counties.
First record: Abbot, 1798; but more certainly, Aspley Wood; J. McLaren, Herb. Luton Museum, 1864.

T. cordata Mill. C.T.W. 35.1.2.   *Small-leaved Lime*
Woods   Native
This is our only native lime, and occurs as such in a few woods on the Lower Greensand. Elsewhere it is found, especially near houses and in parks, as a planted tree. It is probably native also in Bucks., and is recorded from Herts., Hunts. and Northants., where its status should be studied.

Ouse: Clapham Park, planted.
Ouzel: Aspley!; *V.C.H.*: King's Wood, where it is plentiful*;
A. J. Wilmott!
Ivel: Chicksands Wood, where E. Milne-Redhead and the author
found it still growing plentifully!*; *J. Aubrey*.
Lea: Luton, planted; Leagrave, planted.
Colne: near Greencroft Barn, v.c. 20 [Beds.], probably planted!;
E.M.-R.: Luton Hoo, planted.
First record: 'there are woods, where are thousands *e.g.* at Chick-
sands (Sir Osborne's) and in other woods thereabouts', J. Aubrey
in letter to John Ray, 1691 (*Ray's Correspondence*, 1848, 237).

MALVACEAE
MALVA L.

Malva moschata L. C.T.W. 36.1.1.   *Musk Mallow*
Edges of woods, roadsides, etc.   Native
Frequent, but considered rare by Abbot. It is one of the more
attractive of our wild flowers, and white-flowered forms which are
seen occasionally, *e.g.* at Barton Springs, are no less attractive than
the normal pink forms. Recorded for all districts and neighbouring
counties.
H.S. 80a.
First record: Abbot, 1798.

M. sylvestris L. C.T.W. 36.1.2.   *Common Mallow*
Waste places, arable land, etc.   Colonist
The commonest of our mallows. It is a frequent weed of arable land
on the Lower Greensand. Recorded for all districts and neighbour-
ing counties.
H.S. 7.
First record: Abbot, 1798.

M. neglecta Wallr. C.T.W. 36.1.3.   *Dwarf Mallow*
  M. rotundifolia auct.
Arable land, roadsides, waste places, etc.   Colonist
Frequent, especially near farmhouses; and possibly introduced as a
wool adventive and in other ways. Recorded for all districts and
neighbouring counties.
First record: Abbot, 1798.

LINACEAE
LINUM L.

Linum catharticum L. C.T.W. 37.1.3.   *Purging Flax*
Pastures   Native
Common on pastures on calcareous soils throughout the county.
A double-flowered form has been seen at Eaton Socon. Recorded
for all districts and neighbouring counties.
H.S. 1a, 21, 45b, 56b, 59–66a, 67–69, 75, 76, 80a.
First record: Abbot, 1798.

Radiola linoides Roth, C.T.W. 37.2.1, *R. radiola* (L.) Karst. *All-seed*, is recorded, with no details, in *Hillhouse List, 1876*, and with a question-mark in *Comital Flora*. Bedfordshire is well within the range of the native species, and it would be interesting to have a more definite record. It is recorded for Herts. and Bucks., and doubtfully for Cambs.

GERANIACEAE

GERANIUM L.

Geranium pratense L. C.T.W. 38.1.1.   *Meadow Cranesbill*
Meadows, roadside verges, etc.   Native
Frequent in the west of the county, but less so in the east. It is one of the more attractive species of the flora and often makes a fine show in old meadow land. Recorded for all neighbouring counties.
NENE: a marked feature of this small district.
OUSE: Oakley!; Cardington Mill; *Abbot's annotated F.B.*: Fenlake; Pavenham; *W.H. 1875*: Harrold!; *J.H. Notes*: Tempsford!; Eaton Socon!; *L. W. Wilson*: Bromham; *A. W. Guppy*; Souldrop.
OUZEL: Woburn!; *C.C.*: Totternhoe!*; *W.F.B.*: Bidwell; *E.M.-R.*: Stanbridge; Sewell; Battlesden; Sheep Lane.
IVEL: Sharpenhoe; *W.F.B.*: Biggleswade; *C. C. Foss*: between Stotfold and Astwick; *H. B. Souster*.
LEA: Limbury!†*; Nether Crawley; *S.B.P.L.*: near Luton; *H. Brown*: Farley Road; *Davis*: Leagrave!; *H. B. Souster*.
COLNE: 'by the Market Street Road, Luton, 1841'; *J. Foster in Herb. Watson*: Caddington; *L. J. Margetts*.
First record: Abbot, 1798.

G. pyrenaicum Burm. f. C.T.W. 38.1.9.   *Mountain Cranesbill*
Roadsides, waste ground, etc.   Colonist
Common near villages and towns, where it is usually well established. It has probably increased, as it was considered rare by Abbot and to be on the increase by Hillhouse. Recorded for all districts except Nene, Kym and Cam, and for all neighbouring counties.
First record: Abbot, 1795.

G. columbinum L. C.T.W. 38.1.10.   *Long-stalked Cranesbill*
Open pastures, dry banks, etc.   Native
Rare, and apparently limited to well-drained soils. Recorded for all neighbouring counties.
OUSE: Cox's Pits; Biddenham; *F.B.*: Kempston Park; *W.C. Herb.*
IVEL: Gravenhurst; *W.C. Herb.*: Pegsdon; *J. Pollard in W.F.B.*: Barton Hills, where it is frequent on glacial gravel near Ravensburgh Castle!*; *J.E.L., F.F.B.*
COLNE: near Long Wood, v.c. 20 [Beds.], in rough arable land; *E.M.-R.*!; Deadmansey Wood; *P.T.*
First record: Abbot, 1798.

G. dissectum L. C.T.W. 38.1.11.   *Cut-leaved Cranesbill, Crowsfoot*
Arable fields, waste places, etc.   Colonist
Common: recorded for all districts and neighbouring counties.
H.S. 4, 7, 9, 55b, 71, 72, 82.
First record: Abbot, 1798.

G. molle L. C.T.W. 38.1.13.   *Dove's-foot Cranesbill, Plum Pudding*
(Westoning)
Roadsides, arable fields, waste ground, etc.   ? Native
Common and well distributed. A variable species, both in the size
of the plants and size and colour of the flowers: forms with smooth
carpels, *cf.* var. *aequale* Bab., are not infrequent. Recorded for all
districts and neighbouring counties.
H.S. 22b, 25, 27, 44, 48, 61, 71, 86.
First record: Abbot, 1798.

G. rotundifolium L. C.T.W. 38.1.12.   *Round-leaved Cranesbill*
Railway banks, arable fields, waste places, etc.   Colonist
Rare, but usually well established. Recorded for all neighbouring
counties except Hunts.
NENE: railway bank, Wymington.
OUSE: Bedford; *Abbot's annotated F.B.*: Cox's Pits!†; *J.McL., L.M.*:
Cardington†; *J.McL., in Herb. J.S.*: roadside, Fenlake; *H. B.
Souster*: railway banks, Souldrop, Sharnbrook; gravel pit, Eaton
Socon.
OUZEL: allotments, Heath and Reach*; *S. P. Rowlands and W. D.
Coales*!
LEA: railway bank, Leagrave!†*; *W.F.B.*: Old Bedford Road,
Luton; *P.T.*
COLNE: near Caddington; *Fl. Herts.*
First record: Abbot, in letter to J. E. Smith, 3 Sept. 1807.

G. lucidum L. C.T.W. 38.1.15.   *Shining Cranesbill*
Old walls, etc.   Wall denizen
Rare: Saunders found this growing in a swampy place among trees at Marslets,
where it was apparently well established. Recorded for all neighbouring counties
except Hunts.
OUSE: old walls, Elstow; *Plantae Bedford.*: Cauldwell; *F.B.*: walls, Bedford; *J.McL.,
L.M.*: old walls, St. Leonards; *J.McL., B.M.*: Cardington; *J.McL. in J.H. Notes*:
Bedford Cemetery*; *A. W. Guppy.*
IVEL: old wall, Ickwell Bury!*; *C. C. Foss.*
LEA: Luton Park; *Davis*: Marslets, see note above†; *S.B.P.L.*: pavements at a
nursery, Stopsley; *G. K. Long.*
First record: Abbot, 1795.

G. pusillum L. C.T.W. 38.1.14.   *Small-flowered Cranesbill*
  G. parviflorum Curt.
Dry pastures, arable fields, etc.   Native
Not as common as *G. molle*, but well distributed. Recorded for all
neighbouring counties.
H.S. 22b, 25, 27, 44, 71.
NENE: plentiful on railway banks.
OUSE: Elstow; *C. C. Foss*: Milton Ernest; Ford End; common on
railway banks.
OUZEL and IVEL: plentiful on sandy soils.
LEA: Dallow Lane, Luton†; *Herb. J.S.*: old sewage works, Luton.
First record: Abbot, 1798.

G. robertianum L. C.T.W. 38.1.16. *Herb Robert*
Roadsides, waste places, old walls, etc. Colonist
Common: recorded for all districts and neighbouring counties.
H.S. 1a, 1b, 2b, 3a, 5, 25, 32, 42, 43, 52, 55b, 83.
First record: Abbot, 1798.

ERODIUM L'Hérit.

Erodium moschatum (L.) L'Hérit. C.T.W. 38.2.2. *Musk Storksbill*
Sandy fields, railway sidings, etc. Wool adventive
Rare until recently, when it has appeared as one of the more frequent of our wool
adventives. *E. Bot.* tab. 902, July 1801, was drawn from a specimen sent by
Abbot. Recorded for all neighbouring counties except Bucks.
Ouse: frequent as a wool adventive around Eaton Socon and north of Sandy.
Ivel: Ampthill; *C.A. in E. Bot. 1801*: Eversholt†; *J.McL., B.M.*: too many stations
as a wool adventive to merit listing.
First record: Abbot, 1801.

E. cicutarium (L.) L'Hérit. C.T.W. 38.2.3. *Hemlock Storksbill,*
Wild Musk (Batchelor)
Rough sandy and gravelly pastures, arable fields, etc. Native
One of the more difficult species in the Bedfordshire flora. As a
native species it occurs well distributed on the Greensand, in gravelly
pastures and arable fields on the river gravels of the Ouse around
Bedford and of the Lea around Limbury, and on glacial gravels,
especially those overlying the Chalk. It is variable, and J. E. Little
gave many stations for the var. *pimpinellifolium* (With.) Sm. which
I have not studied. More recently it has become one of our most
common wool adventives and as such shows even greater variability.
Recorded for the Ouse, Ouzel, Ivel and Lea districts and for all
neighbouring counties.
H.S. 22b, 25, 27, 44, 48, 50, 61, 71.
First record: Abbot, 1798.

OXALIDACEAE

OXALIS L.

Oxalis acetosella L. C.T.W. 39.1.1. *Wood-sorrel*
Woods Native
In woods throughout the county, but more frequent in those on the
Lower Greensand and Clay-with-Flints. Recorded for all neigh-
bouring counties.
H.S. 77b.
Ouse: Ravensden; *F.B.*: Thurleigh; *Abbot's annotated F.B.*: Lidling-
on; *W.H. 1875*: Cardington; *J.McL. in J.H. Notes*: Harrold;
G. H. Day: Odell; B. B. West:
Ouzel: common in most of the woods on the Lower Greensand.
Ivel: Eversholt; *F.B.*: Rowney Warren; *J.E.L., F.F.B.*: foot of
Water Gutter Hill, Streatley, on chalk; *V. H. Chambers*: Maulden
Wood*; Washer's Wood; Daintry Wood.
Lea and Colne: common in most woods on the Clay-with-Flints.
First record: Abbot, 1798.

BALSAMINACEAE

IMPATIENS L.

Impatiens capensis Meerb. C.T.W. 40.1.2.   *Orange Balsam*
  I. biflora Walt.
Riversides   Denizen
A recent introduction, which has increased rapidly along our larger rivers. Recorded for all neighbouring counties, where it has likewise increased, except Northants.
H.S. 6.
OUSE: common above Bedford.
OUZEL: Leighton Buzzard!*; *V. H. Chambers*: Billington.
First record: V. H. Chambers in *B.E.C. 1943–4 Rep.* 807.

I. parviflora DC. C.T.W. 40.1.3.   *Small Balsam*
Old yards, shady places, etc.   Colonist
Rare, but usually well established. Recorded for all neighbouring counties.
OUSE: Bromham Hall, 1917, apparently still there; *J.H. Notes*: old wood yard, Tempsford!*; *A. E. Baxter*: gravel pit, Eaton Socon.
OUZEL: old garden, Leighton Buzzard.
IVEL: well established, Deadman's Wood, Maulden; Long Lane, Toddington; *H. B. Souster*: Flitwick Moor, where it was established from 1917 to 1923*.
First record: the author in *B.E.C. 1943–4 Rep.* 807.

I. glandulifera Royle C.T.W. 40.1.4.   *Indian Balsam, Policeman's Helmet*
Riversides   Denizen
A recent introduction, which is likely to increase. Recorded for all neighbouring counties.
IVEL: near Tempsford*; *E.M.-R.!*: Blunham Bridge, abundant!; *I. J. Allison*.
LEA: Marslets, not established; *C. Swain*.
First record: E. Milne-Redhead and the author in *B.E.C. 1945 Rep.* 54.

ACERACEAE

ACER L.

Acer pseudoplatanus L. C.T.W. 41.1.1.   *Sycamore*
Woods, roadsides, etc.   Denizen
A common tree of comparatively recent origin in the county. It was considered rare by Abbot, but is now well distributed in the county and reproducing itself, but nowhere appearing to be native. Recorded for all districts and neighbouring counties.
H.S. 3b, 5, 64, 70b.
First record: Abbot, 1798.

A. campestre L. C.T.W. 41.1.3.   *Maple*
Woods, hedgerows, etc.   Native
A common shrub of hedgerows and well-coppiced woods, but rarely reaching the stature of a tree. Druce recorded vars. *hebecarpum* DC. and *leiocarpon* Wallr. in *V.C.H.* Recorded for all districts and neighbouring counties.
H.S. 1b, 2b, 5, 18, 20b, 32–34a, 51–53a, 76, 78, 83.
First record: Abbot, 1798.

AQUIFOLIACEAE

ILEX L.

Ilex aquifolium L. C.T.W. 44.1.1. *Holly*
Woods, hedgerows, etc. Native
Holly occurs throughout the county and is recorded for all districts, except Nene and Cam. It is native and plentiful on the Clay-with-Flints. Here it is usually a bush, but often develops into a tree. It is also possibly native on the Lower Greensand in the Ouzel district, and it is interesting to note that Abbot gave Aspley as his only station. It would be difficult to single out native stations in this part of Bedfordshire, as holly is more extensively planted here than elsewhere in the county. Recorded as a native in Herts., Bucks. and Cambs., and as a planted tree in Hunts. and Northants.
H.S. 70b, 74a.
First record: Abbot, 1798.

CELASTRACEAE

EUONYMUS L.

Euonymus europaeus L. C.T.W. 45.1.1. *Spindle-tree*
Woods, hedgerows, etc. Native
Well distributed in the county, and growing usually on the Chalk and Boulder Clay. Abbot and Hillhouse recorded it from Aspley and I have seen it on the Lower Greensand, near Pinfold Farm, at Woburn. Recorded for all districts and neighbouring counties.
H.S. 2b.
First record: Abbot, 1795.

BUXACEAE

BUXUS L.

Buxus sempervirens L. C.T.W. 46.1.1. *Box*
Supposed at one time to be native on the chalk hills. It is now frequently planted in parks and hedges. Recorded, with some doubt, as native for Herts. and Bucks. OUZEL: 'in plenty on the chalk hills near Dunstable', on the authority of Woodward; *Stoke's edition Withering's Natural Arrangement, 1787.*

RHAMNACEAE

RHAMNUS L.

Rhamnus catharticus L. C.T.W. 47.1.1. *Buckthorn*
Hedgerows, etc. Native
Frequent, and with a similar distribution to *Euonymus europaeus*, but I know no station on the Lower Greensand. Abbot considered it to be rare. Recorded for all districts and neighbouring counties.
H.S. 52.
First record: 'in the hedges between Dunstable and St. Albans, Mr. Newton'; *Ray's Synopsis Ed. III*, 1724, 465.

FRANGULA Mill.

Frangula alnus Mill.   C.T.W. 47.2.1.   *Alder-buckthorn*
Rhamnus frangula L.
Woods   Native
Very rare, and limited to the Lower Greensand. Recorded for all
neighbouring counties.
OUZEL: Aspley Wood, where it was frequent near Mermaid's
Pond until the wood was cleared in 1940!†*; *Plantae Bedford.*:
Woburn, ? same station; *W.H. 1875*: in limited quantity, Stock-
grove*.
IVEL: Everton, ? error; *Plantae Bedford.*: Eversholt; *F.B.*
First record: Abbot, 1795.

LEGUMINOSAE

GENISTA L.

Genista anglica L.   C.T.W. 49.3.2.   *Petty Whin*
Heaths   Native
Extinct. Records by McLaren (*J. Bot. 1884*) from Clapham Wood
and Davis from Sundon were probably in error. Recorded for
Bucks., Cambs. and Northants.
OUZEL: Heath and Reach; *W.H. 1875*.
IVEL: Ampthill Heath; *Plantae Bedford.*: Everton Heath; *J.McL.*,
L.M.: Maulden, Heath and Moor; *W.C. Herb.*
First record: Abbot, 1795.

G. tinctoria L.   C.T.W. 49.3.1.   *Dyer's Greenweed*
Rough pastures   Native
Rare, but well established in a few places on calcareous soils in the
north of the county. Recorded for all neighbouring counties.
H.S. 1a.
NENE: railway cutting, Wymington, where it makes a fine show; site
of disused airfield, Hinwick.
OUSE: Stevington; Clapham; *F.B.*: Exeter Wood†*; *J.McL.*, *B.M.*:
Cotton End; *J.McL. in J.H. Notes*: Wilden; *E. Laxton in J.H. Notes*:
Twin Wood; *A. B. Sampson*: Chawston; *J.H. Notes*: Colesdon
Grange!*; *A. G. Nash*: Roxton; *G. H. Day*.
IVEL: Southill; *J.McL. in Herb. Beds. Nat. Hist. Soc.*: Tingrith;
H. E. Pickering.
First record: Abbot, 1795.

ULEX L.

Ulex europaeus L.   C.T.W. 49.4.1.   *Gorse, Furze*
Heaths, rough pastures, etc.   Native
Common on the Lower Greensand and Clay-with-Flints, but absent
over large parts of the county. The appearance of gorse in place
names, *e.g.* Newton Gorse, in a part of the county where the plant
does not grow, may be deceptive (see note on *Ononis spinosa*).
Recorded for all neighbouring counties.

H.S. 22a, 24, 27–29, 45a–46, 79.
NENE: side of Great Hayes Wood; *G. H. Day*.
OUSE: Clapham; *Abbot's annotated F.B.*: Harrold, one bush; *G. H.
Day*: Bromham; *A. W. Guppy*: Pavenham.
OUZEL and IVEL: common on the Lower Greensand.
LEA and COLNE: common on the Clay-with-Flints.
First record: Abbot, 1798.

SAROTHAMNUS Wimm.

Sarothamnus scoparius (L.) Wimm. ex Koch C.T.W. 49.5.1.

Cytisus scoparius (L.) Link   *Broom*
Sandy places   Native
Frequent in open places on the Lower Greensand, but not as
common as the preceding. Broom, spelt Brume in *Domesday Book*,
probably owes its name to an abundance of the species. Recorded
for all neighbouring counties.
H.S. 22a, 47, 48.
OUSE: railway sidings, Bedford, well established.
OUZEL and IVEL: frequent in all suitable places.
LEA: Stockwood Lawn; *Davis*: Chiltern Green, probably planted
here!; *S.B.P.L.*
First record: Abbot, 1798.

ONONIS L.

Ononis repens L. C.T.W. 49.6.1. *Restharrow, Cammock* (Batchelor)
Rough pastures, roadsides, etc.   Native
Frequent on calcareous soils. A feature of old chalk quarries and
roadside verges on the Chalk but rarely seen on well-established
downland. Recorded for all districts and neighbouring counties.
H.S. 66b, 67.
First record: Abbot, 1798.

O. spinosa L. C.T.W. 49.6.2.   *Prickly Restharrow, Cranfield Clover*
Rough pastures   Native
More common than the preceding and especially abundant on
basic clays, from which *O. repens* is usually absent. Batchelor said
that it was usually known as gorse in the county. White-flowered
forms are not infrequent. Recorded for all districts and neighbouring counties.
H.S. 1b, 56b.
First record: Abbot, 1798.

MEDICAGO L.

Medicago falcata L. C.T.W. 49.8.1.   *Yellow* or *Sickle Medick*
Waste places   Colonist
Rare, but usually well established. Recorded for all neighbouring counties except
Hunts.
OUSE: Cox's Pits*; *L. W. Wilson*: gravel pit, Eaton Socon*; Castle Mills; *H. B.
Souster*!
IVEL: near Flitwick Mill†; *R. C. Chambers in W.F.B.*: waste ground, Henlow Mill;
A. K. Smith: near Holme Mills*.
First record: Abbot, *Hints to the Bedfordshire Farming Interest*, 1808.

M. sativa L. C.T.W. 49.8.2.   *Lucerne, Alfalfa*
Roadsides, waste places, etc.   Escape from cultivation
Frequent and often well established in rough pastures on the Chalk. In common
with many fodder crops it is very variable. Recorded for all districts and neigh-
bouring counties.
H.S. 58b.
First record: Abbot, 1798.

M. lupulina L. C.T.W. 49.8.3.   *Black Medick, Nonsuch*
Pastures, roadsides, waste places, etc.   ? Native
Common. This occurs in a number of forms, the most interesting
being a dwarfed procumbent one as a constituent of closely-grazed
turf on light soils, *e.g.* the Lower Greensand and Clay-with-Flints.
Elsewhere the normal form is sub-erect and is found frequently in
rough pastures, chalk quarries, etc. Another form is a relic of
cultivation, *cf.* var. *willdenowiana* Koch, and is frequently seen on
roadsides and field borders. Recorded for all districts and neighbouring counties.
H.S. 1a–2a, 3a, 4, 13, 21, 22b, 25, 45b, 56b, 58b, 61, 63a, 64, 66a,
67–69, 72, 73, 75, 82, 86.
First record: Abbot, 1798.

M. arabica (L.) Huds. C.T.W. 49.8.6.   *Spotted Medick,*
                                                              Calvary Clover
Pastures, arable fields, railway sidings, etc.   Native
This has every appearance of being native in pastures on light soils,
but appears also as one of our more common wool adventives.
In the latter category a robust form, obviously used as a fodder crop,
is frequent. *E. Bot.* tab. 1616, July 1806, was drawn from material
sent by Hemsted. Recorded for all neighbouring counties.
OUSE: Bedford; *J. Hemsted*: Cardington*; *J.McL. in Bot. Rec.
Club 1884–6 Rep.*: Elstow; *J.McL., B.M.*: common as a wool
adventive.
OUZEL: pasture, Eggington; *E.M.-R.*!: Milton Bryan*; *H. E.
Pickering*.
IVEL: Clophill!†; *J. H. Crouch in S.B.P.L.*: Silsoe, frequent in
pastures in Wrest Park!; Pulloxhill; *C.C.*: Sandy!; *J.E.L., F.F.B.*:
Tingrith; *J.E.L., Diary*: pastures, Maulden; common as a wool
adventive.
CAM: arable field, Wrestlingworth.
First record: J. Hemsted, 1806.

MELILOTUS Mill.

Melilotus altissima Thuill. C.T.W. 49.9.1.   *Tall Melilot*
Hedgerows, roadsides, etc.   Native
Frequent, but much less common than *M. officinalis*. Recorded for
all districts and neighbouring counties.
First record: uncertain.
First evidence: Herb. J. McLaren, Luton Museum, 1864.

M. officinalis (L.) Pall. C.T.W. 49.9.2. *Common Melilot*
Trifolium officinale L., *M. arvensis* Wallr.
Waste places, rough pastures, etc. Colonist
The common melilot of the county. It thrives best on calcareous
soils and makes a fine show on railway banks and spoil banks in
chalk and clay quarries. It is difficult to know whether early records
refer to this or the preceding species. Recorded for all districts and
neighbouring counties.
H.S. 1a, 1b, 45a, 58b, 64, 72.
First record: Podington; O. St. John Cooper in *Collections towards the
history and antiquities of Bedfordshire*, 1783; more certainly, Herb.
Miss Crouch, 1844.

M. alba Medic. C.T.W. 49.9.3. *White Melilot*
Waste places, rough pastures, etc. Colonist
A species which has increased. It also thrives best on calcareous soils and is often
well established on clay soils. Recorded for all neighbouring counties.
OUSE: Ravensden; *B. B. West*: Bedford!; *A. W. Guppy*: Harrold; Chimney
Corner, well established!; *C. C. Foss*: gravel pits, Eaton Socon, Cople.
KYM: Keysoe; *B. B. West*.
OUZEL: Leighton Buzzard; *V.C.H.*: near A.C. Sphinx Works, Dunstable; *W. D.
Coales*.
IVEL: between Henlow Station and Shefford!; *J.E.L., Diary*: Chalton; *W. D.
Coales*: Toddington; *H. B. Souster*: railway siding, Ampthill.
LEA: Leagrave; *C. C. Foss*: Limbury*; arable field, Warden Hills; waste ground,
Luton; sewage works, East Hyde.
COLNE: Blow's Downs; *L. J. Margetts*.
First record: G. C. Druce, 1904.

TRIFOLIUM L.
Trifolium medium L. C.T.W. 49.10.3. *Zigzag Clover*
Pastures, woodland rides, etc. Native
Frequent: this species sometimes dominates considerable patches of
rough pasture, especially in the north of the county. Abbot con-
sidered it rare; Saunders did not know it in the county, but it is
possible that he may have overlooked it. Recorded for all neigh-
bouring counties.
H.S. 1a–2a, 66b.
NENE: tunnel baulk and cutting, Wymington!*, where it is especially
common; *R. R. B. Orlebar*.
OUSE: Bromham; Milton Hill; *F.B.*: Twin Woods; *J.McL. in Herb.
Beds. Nat. Hist. Soc.*: Harrold*; *G. H. Day*.
KYM: West Wood, common on main drive.
OUZEL: Milton Bryan*; *H. E. Pickering*: Leighton Buzzard!; *W. D.
Coales*: roadside, Aspley Wood; *H. B. Souster*: roadsides, Heath and
Reach, Potsgrove.
IVEL: woods, Eversholt; *J.McL., L.M.*: drives, Maulden Wood;
H. B. Souster: between Millbrook and Ridgmont; *L. J. Tremayne*:
rides, Potton Wood.
CAM: rough pasture, Cockayne Hatley.
LEA: Galley Hill, in clay pocket with *Calluna*; *J. F. Hope-Simpson*!
COLNE: ride, Deadmansey Wood!; *W. D. Coales*.
First record: Abbot, 1798.

T. pratense L. C.T.W. 49.10.1. *Common* or *Meadow Clover*
Pastures, roadside verges, etc. Native
Common, both as a native species and as a relic of cultivation.
Many forms are used as fodder plants. An attractive white-
flowered form has been observed at Stevington. Recorded for all
districts and neighbouring counties.
H.S. 1a, 2a, 4, 8–10a, 14b, 21, 37, 53a, 56b, 58b, 63a–67, 75, 76.
First record: Abbot, 1798.

T. ochroleucon Huds. C.T.W. 49.10.2. *Sulphur Clover*
Pastures, roadside verges, etc. Native
One of the botanical features of the county. It is abundant on road-
side verges, usually with a Boulder Clay foundation, in the north and
east of the county. Podington would appear to be its western limit
in Britain. *E. Bot.* tab. 1224, Oct. 1803, was drawn from a specimen
sent by Abbot. Recorded for all neighbouring counties except Bucks.
NENE: Podington!; *R. R. B. Orlebar.*
OUSE: Everton; Clapham; *F.B.*: Park Lane (? Cardington†);
J.McL., L.M.: Bedford; *J.McL. in Bot. Rec. Club Rep. 1884–6*:
Turvey; *R. Townsend*: Roxton!; Harrold; *G. H. Day*: Colesdon*;
L. W. Wilson: Thurleigh; *A. W. Guppy*: Colmworth; Wilstead;
Wyboston; Eaton Socon.
KYM: Keysoe!; *C. C. Foss*: Bolnhurst; *H. B. Souster*: near Kimbolton
station, v.c. 30 [Hunts.].
IVEL: Potton; *Plantae Bedford.*: Clophill!†*; *C.C.*: Stotfold; *J.H.
Notes*: Wilbury Hill!; *J.E.L., J. Bot.*: Higham Gobion; Silsoe; it
makes an especially fine show on the roadside between Clophill and
Wilstead.
CAM: Dunton; Cockayne Hatley.
LEA: Barton Hill Farm Road; *P.T.*
First record: 'in chalky meadows and pastures'; Hudson, *Flora
Anglica, Ed. II*, 1778, 326.

T. arvense L. C.T.W. 49.10.7. *Hare's-foot Clover*
Rough pastures, arable fields, etc. Native
Common on the Lower Greensand, frequent in gravel pits, but rarely
found in well-established pasture. It appears also on railway sidings
and in arable fields with wool aliens and no doubt is also introduced
with wool manure. Var. *strictius* Mert. & Koch was recorded from
Maulden by J. E. Little and M. Brown in *B.E.C. 1931 Rep.* and var.
longisetum (Boiss. & Bal.) Boiss. from Eaton Socon gravel pits by J. E.
Lousley in *Watsonia I (1949)* 41. Recorded for all neighbouring
counties.
H.S. 23, 48.
OUSE: Biddenham; *Abbot's annotated F.B.*: Oakley; *J.H. in Herb.
Beds. Nat. Hist. Soc.*: railway siding, Bedford; *C. C. Foss*: gravel pits,
Eaton Socon, Cople; brick pit, Brogborough.
OUZEL and IVEL: common in all likely stations.
First record: Abbot, 1798.

T. scabrum L. was recorded by Abbot in *Plantae Bedford.* from between Potton and Gamlingay, and in *Flora Bedfordiensis* from Sandy and Ampthill; but his herbarium specimen is too immature to be determined. Specimens named *T. scabrum* from Flitwick in *J.McL., L.M.* and from Biddenham in *J.McL., B.M.* and in *Herb. J.S.* are represented by *T. striatum.* To this species must also be referred C. Crouch's record of *T. scabrum* in *W.F.B.* from Limbury. *T. scabrum* is recorded for Cambs. and Northants.

T. subterraneum L. C.T.W. 49.10.11.   *Subterranean Clover*
Sandy pastures, arable fields, etc.   Native
As a native species this is very rare and is limited to a few heathy pastures on the Lower Greensand. It is also one of the more common wool adventives and as such is very variable, the most marked form being var. *oxaloides* (Bunge) Rouy, a common fodder crop in Australia (see J. E. Lousley, *Watsonia I (1949)* 118). The records given below are for the native species. Recorded for all neighbouring counties except Northants.
IVEL: Clophill Warren, where it was also found by C. Crouch†; *Plantae Bedford.*: Ampthill; *F.B.*: Everton Heath; *J.McL., L.M.*: Rowney Warren!*; *J.McL., B.M.*: Maulden Green End; *M. L. Berrill in J.H. Notes*: Chicksands; *H. and D. Meyer.*
First record: Abbot, 1795.

T. striatum L. C.T.W. 49.10.8.   *Soft* or *Knotted Clover*
Rough pastures   Native
In similar situations to and often growing in association with *T. arvense.* Recorded for all neighbouring counties except Hunts. H.S. 25.
OUSE: Ford End; *Plantae Bedford*: Biddenham, where it was also found by McLaren†; *F.B.*: gravel pits, Eaton Socon, Willington.
OUZEL and IVEL: common in all likely stations: Wilbury Hill; *H. B. Souster.*
LEA: Limbury, as *T. scabrum*†; *C.C. in Herb. J.S.*
First record: Abbot, 1795.

T. fragiferum L. C.T.W. 49.10.17.   *Strawberry Clover*
Pastures   Native
Not uncommon on clay soils, but rare or absent on the Chalk, Lower Greensand and gravels. Recorded for all neighbouring counties.
H.S. 14b.
NENE: Podington; *R. R. B. Orlebar*: Wymington.
OUSE: Clapham; *W.H. 1875*: Park Lane; *J.McL., B.M.*: Hammer Hill Range; *M. L. Berrill in J.H. Notes*: Roxton; Harrold; *G. H. Day*: Bromham; *B. B. West*: Elstow; *V. H. Chambers*: Lidlington; near Marston Thrift; Souldrop; Cranfield; Wootton; Stevington; Eaton Socon.
KYM: Dean; *B. B. West*: Knotting; Swineshead.
OUZEL: Sheep Lane; *H. E. Pickering*: Milton Bryan; Little Billington!; *V. H. Chambers*: near Blackgrove Wood; *H. B. Souster!*
IVEL: Clophill; Barton; *C.C. in J. Bot. 1889*: Kitchen End; *C.C.*: Stondon; *R. Long in J.E.L., J. Bot.*: Arlesey; Henlow; *J.E.L.*,

J. Bot.: Meppershall; *J.E.L.*, *F.F.B.*: Upper Gravenhurst; *V. H. Chambers*: Harlington; Tingrith; Long Lane, Toddington.
CAM: Cockayne Hatley; *J.E.L.*, *Diary*: Dunton!; *H. and D. Meyer*: Wrestlingworth.
LEA: Streatley; Leagrave Marsh; *V. H. Chambers*.
First record: Abbot, 1798.

T. campestre Schreb. C.T.W. 49.10.18.   *Hop-trefoil*
    T. procumbens auct.
Dry pastures, roadside verges, railway banks, etc.   Native
A common and attractive species that often dominates considerable patches in suitable habitats. Recorded for all districts and neighbouring counties.
H.S. 1a, 1b, 25, 44, 45b.
First record: Abbot, 1798.

T. dubium Sibth. C.T.W. 49.10.20.   *Lesser Yellow-trefoil*
    T. minus Relhan
Heaths, dry pastures, roadside verges, etc.   Native
Common, especially on heaths on the Lower Greensand and Clay-with-Flints, but growing wherever there is no competition with taller plants. Stunted specimens are often mistaken for *T. micranthum*.
Recorded for all districts and neighbouring counties.
H.S. 9, 11, 16, 22b, 23, 25–28, 44, 45b, 48, 61, 66a, 67, 75, 79a, 80a.
First record: Abbot, 1798.

T. micranthum Viv. C.T.W. 49.10.21.   *Slender Trefoil*
    T. filiforme L. nom. ambig.
Closely grazed pastures   Native
Not infrequent on heathy and gravelly soils. Recorded for all neighbouring counties.
H.S. 23, 26, 27.
OUSE: Biddenham; *F.B.*: Cardington†; *J.McL. in J. Bot. 1884*: Millbrook*; *C.C.*: open pasture, Tempsford*.
OUZEL: Aspley; Leighton Buzzard; *V.C.H.*: grassy ride, King's Wood.
IVEL: Cainhoe; *W.C. Herb.*: Silsoe*; Ampthill; *C.C.*: open pasture, Westoning*; *E.M.-R.*: drive, Simpson's Hill Plantation; open pasture, Tingrith; Rowney Warren.
LEA: Highfield, Luton; *M. Brown in W.F.S. Diary*.
COLNE: edge of Studham Common*; *E.M.-R.*!
First record: Abbot, 1798.

T. hybridum L. C.T.W. 49.10.15.   *Alsike Clover*
Roadside verges, waste places, etc.   Colonist
Common, but usually an escape from cultivation. Proliferating forms are frequently seen. A robust form, *cf. T. elegans* Savi, was recorded from Aspley in *Hillhouse List, 1875*, and was abundant for a time in the gravel pits at Eaton Socon* and Langford. *T. hybridum*, which was considered rare by Abbot, is recorded for all districts and neighbouring counties.
H.S. 13, 21, 64.
First record: Abbot, 1798.

T. repens L. C.T.W. 49.10.16. *White* or *Dutch Clover*
Pastures, roadside verges, etc. Native
Very common as a native species, and frequently sown in leys.
Recorded for all districts and neighbouring counties.
H.S. 2a, 3a, 4, 7, 8, 10a, 10b, 11, 14b, 16, 23, 25, 26, 28, 29, 44, 53a,
56a, 61, 63a, 64, 66b, 67, 68, 75, 76, 79a, 82.
First record: Abbot, 1798.

ANTHYLLIS L.

Anthyllis vulneraria L. C.T.W. 49.11.1. *Kidney-vetch,*
Ladies' Fingers
Calcareous soils Native
Common on the Chalk where the soil has at some time been dis-
turbed, and frequent on the Oolite in similar circumstances. It is
absent on the chalk downs where the turf is well established and is
to be seen at its best on road and rail cuttings. On the site of
Totternhoe Castle it is abundant. Abbot, it is interesting to note,
considered it to be rare. Recorded for all neighbouring counties.
H.S. 62–67.
NENE: railway cutting, Wymington.
OUSE: Oakley; Cox's Pits; *F.B.*: Colesdon; *A. W. Guppy*: railway
cutting, Souldrop.
OUZEL, IVEL, LEA and COLNE: common on the Chalk where the
soil has been disturbed.
First record: Abbot, 1798.

LOTUS L.

Lotus uliginosus Schkuhr C.T.W. 49.12.3. *Large Birdsfoot-trefoil*
L. major auct.
Marshes, damp woods, meadows, etc. Native
Frequent in most of the likely places in the county. The common
form is pubescent, but glabrous forms, *cf.* var. *glaber* Bréb., are not
uncommon in marshy places. Recorded for all districts except
Nene, and for all neighbouring counties.
H.S. 2a, 10a, 14b, 15, 19a, 23, 35, 37–39a, 40–43, 79b, 80a.
First record: *Hillhouse List*, 1875.

L. corniculatus L. C.T.W. 49.12.1. *Birdsfoot-trefoil*
Pastures, roadsides, etc. Native
Very common on all soils: recorded for all neighbouring counties.
H.S. 1a, 2a, 11, 21, 25, 28, 29, 45b, 51b, 56b, 59–63b, 65–68, 75, 76.
First record: Abbot, 1798.

L. tenuis Waldst. & Kit. ex Willd. C.T.W. 49.12.2. *Slender Birdsfoot-trefoil*
Pastures Colonist
Rare and growing on a great variety of soils. In all places where I have seen it the
soil appears to have been disturbed at some time. The colony at Wymington
shows considerable variability. Recorded for all neighbouring counties.
NENE: Podington; *Newbould MS.*: railway cutting, Wymington*.

Ouse: Harrowden Hill†; *J.McL., B.M.*: Cardington; *J.McL. in Bot. Rec. Club 1884–6 Rep.*: Twin Wood; *G. C. Druce in B.E.C. 1920 Rep.*: Turvey; *W. W. Mason in B.E.C. 1928 Rep.*: gravel pits, Eaton Socon*; Cople.
Ouzel: Aspley; *J.McL., L.M.*: Leighton Buzzard, where it grows on the edge of a sandpit!*; *A. Templeman in B.E.C. 1920 Rep.*
Ivel: Haynes; *Edward Forster in Herb. B.M.*: Ridgmont*; *C.C.*: Pegsdon; *J.E.L., B.E.C.*: roadside, Fancott*.
Cam: Cockayne Hatley; *J.E.L., F.F.B.*
First record: Edward Forster, 1849.

ASTRAGALUS L.

Astragalus danicus Retz. C.T.W. 49.16.1.   *Purple Milk-vetch*
Chalk hills   Native
Rare: one of the more interesting Bedfordshire plant species. There is a similarity in its two stations in the county, as both are on the Middle Chalk and on the verges of old roads across the hills. It is probable that Gerard's early record refers to Warden Hills, as may a record by Woodward from hills near Dunstable in *Bot. Guide*, 1805. Recorded for Herts., Cambs. and Northants.
H.S. 66a.
Lea: near Pegsdon, possibly the north end of Lilley Hoo!, where it appears in a limited area; *H. Brown*: Warden Hills!†*, where it extends along the foot of the hills to Galley Hill; *J.S. in J. Bot. 1883*.
First record: 'upon Barton hills fower miles from Lewton', *Gerard's Herbal*, 1596, 1062.

A. glycyphyllos L. C.T.W. 49.16.3.   *Wild Licorice, Milk-vetch*
Railway banks, hedgebanks, etc.   Colonist
Rare, and usually a species of disturbed ground on a great variety of soils. As with so many species now a feature of railway banks, Abbot considered it rare. Recorded for all neighbouring counties.
Nene: railway cutting, Wymington*.
Ouse: Bromham; Oakley; *F.B.*: Stevington; *Abbot's annotated F.B.*: Clapham; *A. B. Sampson*: Mowsbury Hill; Twin Wood; *J.H. Notes*: Turvey†; *V.C.H.*: Tempsford!; Harrold; Carlton; *G. H. Day*: Pavenham; *R. Townsend*: Odell*; *V. H. Chambers*!: roadside, Milton Ernest; *H. B. Souster*: railway banks, Souldrop, Felmersham; Southend sidings, Bedford.
Ivel: Sandy; *Abbot's annotated F.B.*: Meppershall; Southill, on railway bank!; *W.C. Herb.*: Rowney Warren, still there on roadside!; *J.McL., L.M.*: Sharpenhoe, in hedgerow on chalk!†; *D.M. Higgins in W.F.B.*: Harlington, on railway bank!*; *Miss Dickinson in W.F.B.*: Potton; *D. M. Higgins in Herb. Beds. Nat. Hist. Soc.*: Holwell; *J.E.L., J. Bot.*: disused chalk quarry, Sundon; railway banks, Flitwick, Henlow, Meppershall.
First record: Abbot, 1798.

ORNITHOPUS L.

Ornithopus perpusillus L. C.T.W. 49.18.1.   *Birdsfoot*
Sandy heaths and pastures   Native
Limited to the Lower Greensand, and locally abundant in all places where a small amount of turf has become established. Recorded for all neighbouring counties.
H.S. 22b, 25, 26, 46, 48.
Ouzel and Ivel: common in all likely situations.
First record: Abbot, 1795.

HIPPOCREPIS L.

Hippocrepis comosa L. C.T.W. 49.20.1.   *Horse-shoe Vetch*
Chalk hills   Native
Locally abundant on the Chalk, where it is found both on well-
established downland and less frequently on disturbed soils. The
problem of its absence from some of our chalk hills would well repay
study. Recorded for all neighbouring counties.
H.S. 59, 60, 62-63b, 65, 66a.
OUZEL, IVEL, LEA and COLNE: common over large areas of the
Chalk.
First record: 'on chalky hills everywhere', Hudson, *Flora Anglica
Ed. II*, 1778, 322.

ONOBRYCHIS Mill.

Onobrychis viciifolia Scop. C.T.W. 49.21.1.   *Sainfoin*
  Hedysarum onobrychis L., *O. sativa* Lam., *O. onobrychis* (L.) Karst.
Chalk hills, edges of cultivated fields, etc.   Native
There are two distinct forms of this species to be found in the county.
A native form, which is procumbent and has deep pink flowers,
is abundant on Knocking Hoe and appears sparingly on Galley
and Warden Hills. E. Milne-Redhead has grown this form at Kew
and has observed that it is constant in cultivation. The other is the
cultivated form, which is grown extensively as a fodder crop,
especially on the Chalk. It is a more robust form, with larger and
paler flowers. Escapes from cultivation are common, and readily
invade and become established on chalk hills, as is to be seen at
Totternhoe Knolls. It is interesting to note that its first record for
Britain was from Barton Hills by Gerard at a time when fodder
crops were not general, and it is likely that Gerard was observing
what was to him an unusual plant. It is of some significance that
Abbot also considered it to be rare. Recorded for all neighbouring
counties, and, as an escape from cultivation, for all botanical
districts of the county.
H.S. 59, 62, 67.
First record: 'upon Barton hill fower miles from Lewton in Bedfordshire on both sides of the hill'; *Gerard's Herbal*, 1597, 1064.

VICIA L.

Vicia sylvatica L. C.T.W. 49.22.6.   *Wood Vetch*
Woods   Native
Rare: recorded for all neighbouring counties except Hunts., but
with some doubt for Cambs.
OUSE: King's Wood; Sheerhatch Wood, where McLaren also found
it; *Plantae Bedford*.
OUZEL: Rushmere*.
IVEL: Eversholt; *F.B.*: Sharpenhoe Grove†*; *C.C.*
First record: Abbot, 1795.

V. cracca L. C.T.W. 49.22.4.   *Tufted Vetch*
Meadows, hedgerows, copses, etc.   Native
Common and well distributed: recorded for all districts and neighbouring counties.
H.S. 1a–2b, 9, 13, 21, 38, 39a, 56b, 57b, 60, 84a, 85.
First record: Abbot, 1798.

V. sepium L. C.T.W. 49.22.7.   .   *Bush Vetch*
Woodland rides, shady places, etc.   Native
Common: recorded for all districts and neighbouring counties.
H.S. 2b, 20a, 31a, 33, 44, 51a, 53a, 55b, 78, 79b, 80b, 83.
First record: Abbot, 1798.

V. sativa L. C.T.W. 49.22.9.   *Common Vetch*
Waste places, roadsides, etc.   Colonist
Common but somewhat variable. Recorded for all districts and neighbouring counties.
First record: Abbot, 1798.

V. angustifolia L. C.T.W. 49.22.10.   *Narrow-leaved Vetch*
Dry pastures   Native
Abundant on the Lower Greensand and frequent on light soils elsewhere. It is a very variable species and there are a number of records for var. *bobartii* (E. Forst.) Koch, a narrow leaved form, all from the Lower Greensand. Recorded for all districts except Cam, and for all neighbouring counties.
H.S. 22b, 25, 26, 48, 79a.
First record: Abbot, 1798.

V. lathyroides L. C.T.W. 49.22.11.   *Spring Vetch*
Pastures   Native
Rare, and limited to a few places on the Lower Greensand. G. C. Druce recorded var. *robusta* Druce from Ampthill in *B.E.C. 1920 Rep.* Recorded for Bucks. and Cambs.
OUZEL: Leighton Buzzard; *V.C.H.*
IVEL: Ampthill; Clophill; *Plantae Bedford.*: Maulden, locally abundant on a hill pasture!†; *F.B.*: wide roadside verge, Steppingley.
First record: Abbot, 1795; his stations are probably correct, but his herbarium specimen is *V. angustifolia*.

V. hirsuta (L.) Gray C.T.W. 49.22.1.   *Hairy Tare*
Pastures, roadsides, etc.   Native
Common: recorded for all districts and neighbouring counties.
H.S. 4, 25, 45a.
First record: Abbot, 1798.

V. tetrasperma (L.) Schreb. C.T.W. 49.22.2.   *Smooth Tare*
Pastures, roadsides, waste places, etc.   Native
Almost as common as *V. hirsuta* and growing in similar situations, but also found on disturbed soils. Recorded for all districts and neighbouring counties.
H.S. 15, 82.
First record: Abbot, 1798; but his herbarium specimen is *V. hirsuta*; more certainly, Herb. Miss Crouch, 1843.

V. tenuissima (Bieb.) Schinz & Thell. C.T.W. 49.22.3.
   V. gracilis Lois. non Soland. *Slender Tare*
Pastures, woodland rides, etc. Native
Rare, and often confused with robust forms of *V. tetrasperma*.
Recorded for all neighbouring counties.
Ouse: Clapham; *Newbould MS.*: Cardington†; *J.McL. in Bot. Rec.
Club 1884–6 Rep.*: Marston; *Ada Stimson in J.H. Notes*: Colesdon;
Roxton; *J.H. Notes*.
Ouzel: Shenley Hill; *A. Templeman in B.E.C. 1921 Rep.*
Ivel: woodland rides, Maulden*; hedgerow, Sundon.
Cam: hedgerow, Dunton.
Colne: Deadmansey Wood; *A. J. Wilmott*!
First record: W. W. Newbould in *Watson's Topographical Botany*,
1873.

LATHYRUS L.

Lathyrus sylvestris L. C.T.W. 49.23.6. *Wild Pea,*
Narrow-leaved Everlasting Pea
Woods, railway banks, etc. Native
An attractive species, frequent in many of our heavy boulder-clay
woods, and often making a fine show on railway embankments.
Abbot recorded it, as well as *L. latifolius*, and considered it common.
Recorded for all neighbouring counties.
Ouse: Colmworth; *W.H. 1875*: Sheerhatch Wood; *J.McL., L.M.*:
Twin Wood; *B.P.L.*: Exeter Wood; *J.McL. in Bot. Rec. Club 1884–6
Rep.*: Oakley Wood; *J.H. Notes*: Harrold; Odell; Turvey; *G. H.
Day*: Bromham; *B. B. West*: between Stevington and Stagsden;
H. B. Souster: Marston Thrift; Wilstead Wood; railway embankment,
Felmersham, where it is the dominant species for about 100 yards.
Ouzel: Aspley Wood†; *Herb. J.S.*
Ivel: Cainhoe; *W.C. Herb.*: Warden Wood; *J.McL.*: Southill;
Maulden; Clophill; *C.C.*: Flitwick Wood!*; *J.S. in J. Bot. 1889*:
Standalone Farm, Potton; between Shefford and Southill; both as
var. *platyphyllus* Retz; *J.E.L. in B.E.C. 1911 Rep.*: chalkpit, Sundon!;
C. C. Foss: spinney, Haynes; ditch, Flitwick.
Cam: hedgerow, Dunton; *H. B. Souster*: Cockayne Hatley Wood.
First record: Abbot, 1798.

L. latifolius L. C.T.W. p. 450 *Everlasting Pea*
Roadsides, railway banks, etc. Colonist
Rare: the past history of this species in the county was the subject of some con-
troversy. Abbot recorded it from Hawnes Wood and Bromham Spinney in *Plantae
Bedford.* and *F.B.*, and added Clapham in his annotated copy of *F.B. E. Bot.* tab.
1108, Dec. 1802, was drawn from a specimen sent by Abbot and represents the true
species, but his herbarium specimen labelled '*Lathyrus sylvestris latifolius*' is *L.
sylvestris* which still grows at Hawnes (Haynes) and near Bromham and Clapham.
L. latifolius grows as a naturalized species in all neighbouring counties.
Nene: Hinwick Lodge.
Ouse: gravel pit, Cople; roadside, Bromham.
Ouzel: roadside, Battlesden!*; *E.M.-R.*: roadside between Well Head and Dun-
stable Downs.
Ivel: Sundon; Silsoe; *H. B. Souster*: railway banks, Flitwick, Ampthill.
First certain record: E. Milne-Redhead and the author in *B.E.C. 1946–7 Rep.* 807.

R

L. tuberosus L. C.T.W. 49.23.5.   *Earth-nut Pea*
Pastures, roadside verges, etc.   Colonist
Rare: this interesting species was once limited in this country to Essex, but its range
is now extending.
OUSE: tunnel top, Ampthill, where it makes a fine show!*; *H. G. Oldfield.*
IVEL: rough pasture, Barton Hills, where I have looked for it in vain; *Herts. Nat. Hist. Soc.*
LEA: roadside near Luton and Dunstable Hospital, where it appeared until the
roadside verge was closely mown*; *F. L. Chesham.*
First record: F. L. Chesham and Herts. Nat. Hist. Soc. in *B.E.C. 1946–7 Rep.* 289:
it was discovered by each, independently, during the same afternoon.

L. pratensis L. C.T.W. 49.23.4.   *Meadow Vetchling*
Meadows, pastures, etc.   Native
Common and well distributed on a great variety of soils. Recorded
for all districts and neighbouring counties.
H.S. 1a, 1b, 10a, 10c, 21, 39a, 45b, 55b, 57b, 63a, 66b, 75.
First record: Abbot, 1798.

L. nissolia L. C.T.W. 49.23.2.   *Grass Vetchling, Grass Pea*
Grassy banks, roadsides, etc.   Native
This attractive species is frequent on the heavier boulder-clay soils
of the north and middle of the county. It appears also, more
sparingly, on gravels, the Chalk and Lower Greensand. It thrives
best against the competition of the taller grasses and herbs. It was
considered rare by Abbot and Saunders, which it probably was
until recently. *E. Bot.* tab. 112, June 1793, was drawn from a
specimen found at Ampthill by the Countess of Ossory and sent by
T. Vaux. The original bears the note 'from Short Wood near
Ampthill by favour of the Revd. Mr. Abbot' to which J. E. Smith
added 'is it from Stort Wood?'. Recorded for all neighbouring
counties.
H.S. 21, 25.
OUSE: too many stations to merit listing.
OUZEL: Leighton Buzzard; *E.M.-R.*
IVEL: too many stations to merit listing.
LEA: Dunstable; *S. A. Chambers in F.F.B.*: Skimpot; *W. D. Coales*:
Spittlesea; Selbourne Road, Luton; *H. B. Souster*: Warden Hills.
First record: *English Botany,* 1793.

L. aphaca L. C.T.W. 49.23.1.   *Yellow Vetchling*
Arable fields, etc.   Colonist
Rare and not established: recorded for all neighbouring counties.
OUSE: near Kempston; *J.McL., L.M.*: Bedford; *J.H. Notes.*
IVEL: Flitwick; *F. S. Lloyd in F.F.B.*: Biggleswade; *J.H. Notes*: Maulden; *B. B. West.*
LEA: Luton, garden weed, Old Bedford Road!†; *S.B.P.L.*: Warden Hills; *W.F.B.*:
rough pasture, south of Luton!*; *H. Cole*: arable field, north of Leagrave.
First record: J. McLaren, 1864.

L. montanus Bernh. C.T.W. 49.23.9.   *Bitter Vetch*
   Orobus tuberosus L.
Woods   Native
Very rare, and apparently limited to a very small area of the Lower
Greensand. Recorded for Herts., Bucks. and Northants.

OUZEL: Aspley Wood, where it was also found by W. Crouch; *Plantae Bedford.*: Birchmore; *F.B.*: King's Wood, sparingly by the side of one ditch!*; *S.B.P.L.*
First record: Abbot, 1795.

ROSACEAE

PRUNUS L.

Prunus spinosa L. C.T.W. 50.17.1.   *Sloe, Blackthorn*
Hedgerows, wood borders, etc.   Native
Common throughout the county. Thorn occurs frequently in place names, *e.g.* Chawston (Chaulestorne or Cealf's thorn in *Domesday Book*), Thorncote (Thornecote, 1206), a hamlet in Northill parish, and Thorn (Thorn, 1225), a hamlet in Houghton Regis parish—see *P.N.B.* Recorded for all districts and neighbouring counties.
H.S. 2a, 2b, 3b, 20b, 34b, 63b, 75.
First record: Abbot, 1798.

P. domestica L. C.T.W. 50.17.2.   *Wild Plum, Bullace*
Hedgerows   Denizen
Frequent. Wild plums occur, usually near houses, in most parts of the county. Subsp. **domestica** is without doubt the common form, but I have found subsp. **insititia** (L.) Poir. at Barton and Harlington, and it is probably awaiting record elsewhere. Abbot considered '*P. insititia*' to be rare, and gave Oakley and Renhold as stations, and *E. Bot.* tab. 84, Feb. 1801 was apparently drawn from material sent by him. *P. domestica sensu lato* is recorded for all districts except Nene, Kym and Cam, and for all neighbouring counties.
First record: '*P. insititia*', Abbot, 1798; '*P. domestica*', Abbot, in letter to J. E. Smith, 3 Sept. 1802.

P. avium (L.) L. C.T.W. 50.17.4.   *Gean, Wild Cherry*
Woods   Native
Frequent in woods on the Clay-with-Flints, where it makes fine trees and is undoubtedly native. Rare in the rest of the county, but often planted there. Recorded for all neighbouring counties.
H.S. 78, 81, 83.
OUSE: Oakley Hedges; *F.B.*: Biddenham; Sharnbrook; *W.H. 1875*: Harrold; *G. H. Day*: Woodcraft.
OUZEL: Aspley Wood!; *F.B.*: Lowe's Wood; between Tilsworth and Hockliffe; Wavendon Heath, v.c. 24 [Beds.].
IVEL: Warden Wood; Southill Wood; *J.McL. in J.H. Notes*: Exeter Wood; *M. L. Berrill in J.H. Notes*: Rowney Warren!; Shillington; *J.E.L., F.F.B.*: Keeper's Warren, *J.E.L. Diary*: Clophill!; *V. H. Chambers*: Folly Wood; Washer's Wood; Daintry Wood; Briar's Stocking; Chicksands Wood.
CAM: Wrestlingworth.
LEA and COLNE: too many stations to merit listing.
First record: Abbot, 1798.

P. cerasus L., C.T.W. 50.1.5, *Sour Cherry*, was considered common by Abbot, but his herbarium contains no specimen. Herb. McLaren, Luton Museum, contains two specimens labelled *P. cerasus* and *P. avium* but both are the latter. *Hillhouse List, 1875*, records it from Clapham, but again there is no specimen. I have seen it only by a roadside at Studham, v.c. 20 [Beds.], where H. B. Souster drew my attention to it. It was probably planted there. Recorded for all neighbouring counties except Hunts.

FILIPENDULA Mill.

Filipendula ulmaria (L.) Maxim. C.T.W. 50.2.2. *Meadow-sweet*
  Spiraea ulmaria L.
Marshes, riversides, ditches, etc.   Native
Common in all likely habitats and often dominating drying-out portions of marshland. Recorded for all districts except Colne, and all neighbouring counties.
H.S. 2a–3a, 6, 7, 10a, 10c, 13, 15, 18, 19a, 34a, 34b, 41, 42, 52–53b, 83, 84a.
First record: Abbot, 1798.

F. vulgaris Moench C.T.W. 50.2.1.   *Dropwort*
  Spiraea filipendula L., *F. hexapetala* Gilib.
Pastures   Native
Locally abundant on the chalk downs and appearing also in established pasture on basic clays and gravels throughout the county. Rarely seen in disturbed soils. Recorded for all districts except Colne, and all neighbouring counties.
H.S. 1a, 2a, 59, 60, 65–66b.
First record: Abbot, 1798.

RUBUS L.

A complex genus which received inadequate attention in the county until W. C. R. Watson visited it in 1946 and again in 1947 and 1948. In addition to useful field work he checked earlier records, incorporating the more reliable ones in his notes, and his paper in *Journ. Beds. Nat. Hist. Soc.* 1947 was the first reasonably complete account of the genus for the county. Brambles are widespread in the county but show great variability on the higher parts of the Lower Greensand and Clay-with-Flints. They appeared in many habitat studies but my visits were usually at times of the year when good material could not be collected. In some cases I was able to determine the plants in the field but have preferred to list them together here.
H.S. 1a–2a, 3b, 5, 18, 19b, 20b, 21, 27, 29–31a, 31b, 32–34a, 42, 43, 47, 49, 51, 52, 54, 55b, 63a, 63b, 69, 70b, 74, 77–81, 83.

[Subgen. *Idaeobatus* Focke]
Rubus idaeus L. C.T.W. 50.3.3.   *Raspberry*
Woods, heaths, etc.   Native
Frequent, especially on heaths on the Lower Greensand and Clay-with-Flints. Recorded for all districts, except Kym and Cam, and for all neighbouring counties.
H.S. 39a, 42.
First record: Abbot, 1798.

R. loganobaccus L. H. Bailey C.T.W. p. 463.   *Loganberry*
Heathy places   Garden escape
IVEL: Galley Hill, Sutton; *W.W.*!
LEA: Kidney Wood; *W.W.*!

[Subgen. *Rubus*]   *Blackberry, Bramble*
Bramble occurs as a place name in Bramingham (Bramblehangre
1240), a wood on a slope with brambles growing in it, see *P.N.B.*

[Sect. *Suberecti* P. J. Muell.]
R. nessensis W. Hall C.T.W. p. 466.
Rare: recorded also for Herts. and Northants.
OUZEL: King's Wood*; *W.W.*!
IVEL: between Southill Station and Warden; *E. M. Langley.*
First record: as *R. suberectus* Anders., A. Poulton in *W.H. 1876.*

R. scissus W. Wats.
Rare: recorded as *R. fissus* for Bucks. and Northants.
OUZEL: Heath and Reach*; *W.W.*!
First record: as *R. fissus* Lindl., *V.C.H.*, 1904.

R. plicatus Weihe & Nees C.T.W. p. 467.
OUZEL: Woburn; *G. C. Druce in Fl. of Bucks.*: near Baker's Wood*; Stockgrove;
W.W.!
First record: Druce, 1926.

R. sulcatus Vest. ex Tratt.
OUZEL: near Baker's Wood*; *W.W.*!

R. fissus Lindl.
Rare: recorded as *R. rogersii* E. F. Linton for Bucks.
OUZEL: Aspley Wood*; *W.W.*!

[Sect. *Silvatici* P. J. Muell.]
R. crespignyanus W. Wats.
OUZEL: near Woburn; *E. M. Langley in Herb. Kew, see W.W. in B.E.C. 1937 Rep.*

R. carpinifolius Weihe & Nees
Rare: recorded for Bucks. and Northants.
IVEL: abundant, Flitwick Moor*; Cooper's Hill; *W.W.*!
First record: Druce, *Fl. of Bucks.* 1926.

R. rhodanthus W. Wats.,
IVEL: Clophill; *W.W.*!

R. mercicus Bagn.
OUZEL: Aspley Wood, conf. W.W.*; *W. H. Mills*!
IVEL: Clophill*; *W.W.*!

R. nemoralis P. J. Muell.
Rare: recorded as *R. selmeri* Lindeb. for Bucks. and Northants.
IVEL: roadside, Clophill*; *W.W.*!

R. gratus Focke C.T.W. p. 467.
Rare: recorded for Bucks.
OUZEL: Aspley Heath, v.c. 24 [Beds.]*; *W.W.*!

R. lindleianus Ed. Lees C.T.W. p. 468.
Frequent: recorded for all neighbouring counties except Hunts.
OUZEL: King's Wood*; Stockgrove; Woburn; Aspley Wood*; *W.W.*!
IVEL: Sandy Heath*; Sutton Fen; Castle Hill, Sandy; Flitwick Moor; Cooper's
Hill; roadside, Clophill*; *W.W.*!
LEA: Spittlesea Wood; *W.W.*!
COLNE: Folly Wood; *W.W.*!
First record: W. W. Newbould in *W.H. 1876.*

R. horridisepalus (Sud.) W. Wats.
IVEL: roadside, Clophill*; *W.W.*!

R. laciniatus Willd. C.T.W. p. 470.
A frequently cultivated bramble which becomes readily naturalized.
OUSE: Clapham Park Wood!; *H. A. S. Key*: roadside, Bedford; *A. W. Guppy*.
IVEL: Warren Wood, Clophill*; *V. H. Chambers*.
LEA: Kidney Wood; *W.W.*!
First record: the author in *B.E.C. 1938 Rep.* 40.

R. macrophyllus Weihe & Nees C.T.W. p. 468.
Rare: early records by Saunders and E. M. Langley are to be doubted. Recorded
for Bucks., Herts. and Northants.
OUZEL: Woburn; *V.C.H.*: King's Wood*; *W.W.*!
LEA: Horsley's Wood*; *W.W.*!
First certain record: *V.C.H.*, 1904.

R. subinermoides Druce ex W. Wats.
Rare: recorded also for Northants.
COLNE: Whipsnade Common*; Deadmansey Wood; Elm Grove, v.c. 20 [Beds.];
W.W.!

R. amplificatus Ed. Lees
Rare: recorded also for Northants.
OUZEL: Heath and Reach; Woburn; Stockgrove; *W.W.*!
IVEL: Galley Hill, Sutton*; Cooper's Hill; *W.W.*!
COLNE: Pepperstock; *W.W.*!

R. libertianus Weihe (see W. Watson: *Watsonia I (1949)* 72)
OUZEL: hedgerows, Heath and Reach*; Woburn; *W.W.*!

R. pyramidalis Kalt. C.T.W. p. 469.
Frequent over much of the Lower Greensand: recorded for Bucks., Cambs. and
Northants.
OUZEL: Aspley Wood!; *G. C. Druce in J. Bot., 1897*: Woburn; *W. Moyle Rogers*:
King's Wood*; Heath and Reach; New Wavendon Heath, v.c. 24 [Beds.]; *W.W.*!
IVEL: Sandy Heath; *W.W.*!
First record: G. C. Druce, 1897.

R. schlechtendalii Weihe (see W. Watson: *Watsonia I (1949)* 73)
OUZEL: near Baker's Wood*; *W.W.*!
COLNE: Deadmansey Wood*; *W.W.*!

R. egregius Focke
OUZEL: Woburn Sands: *J. H. Riddelsdell in J. Bot. 1931*.
IVEL: roadside, Clophill*; *W.W.*!
First record: J. H. Riddelsdell, 1931.

R. incurvatus Bab. was recorded by E. M. Langley, but failing a specimen it should
be doubted. Recorded for Bucks.

R. cryptadenes Sud.
Rare: records of *R. argenteus* Weihe & Nees for Bucks. and Northants. may refer to
this.
LEA: near Luton Airport*; *W.W.*!

R. polyanthemos Lindeb. C.T.W. p. 469.
Frequent: recorded for Bucks., Cambs. and Northants.
OUSE: Tempsford; *W.W.*!
OUZEL: Aspley Wood!; *G. C. Druce in J. Bot. 1897*: Woburn Sands; *W. Moyle
Rogers*: Leighton Buzzard*; Heath and Reach; Wavendon Heath, v.c. 24 [Beds.];
W.W.!
IVEL: Sandy Heath; Sutton Fen; Biggleswade Common; Flitwick Moor; Cooper's
Hill; Potton; Clophill; *W.W.*!
COLNE: Whipsnade Zoo; *J. H. Riddelsdell in J. Bot. 1932*.
First record: G. C. Druce, 1897.

R. rhombifolius Weihe
Rare: recorded for Bucks. and Cambs.
IVEL: Cooper's Hill*; roadside, Clophill*; *W.W.*!

R. cardiophyllus Lef. & Muell. C.T.W. p. 469.
Frequent: recorded for Herts., Bucks. and Cambs.
OUSE: Tempsford; *W.W.*!
OUZEL: Woburn; *V.C.H.*: Woburn Sands; *W. Moyle Rogers*: Heath and Reach*;
Aspley Wood; *W.W.*!
IVEL: Maulden; *R. A. Pryor in Newbould MS.*: Sandy Heath; Cooper's Hill;
Rowney Warren; Clophill; *W.W.*!
COLNE: Pepperstock*; *W.W.*!
First record: R. A. Pryor, 1873.

[Sect. *Discolores* P. J. Muell.]
R. ulmifolius Schott f. C.T.W. p. 471.
   R. rusticanus Merc.
The common bramble of our hedgerows. It hybridizes easily with other brambles
producing both fertile and infertile forms. Recorded for all districts and neigh-
bouring counties.
First record: Abbot, 1798.

R. pubescens Weihe
Rare: Watson thinks that previous records for Beds., Herts., Bucks. and Northants.
were probably in error.
IVEL: roadside, Clophill*; *W.W.*!

R. procerus P. J. Muell.
A cultivated bramble which becomes naturalized.
COLNE: Studham Common*; *W.W.*!

R. falcatus Kalt.
Rare: records of *R. thyrsoideus* Wimm. for Bucks., Cambs. and Northants. may refer
to this.
OUSE: north of Sandy; *W.W.*!
IVEL: Sandy!*; *W. H. Mills*: Galley Hill, Sutton; Cooper's Hill; Potton; Clophill;
W.W.!

[Sect. *Vestiti* Focke]
R. macrostachys P. J. Muell.
OUZEL: King's Wood*; *W.W.*!
IVEL: roadside, Clophill; near Luton Airport*; *W.W.*!
LEA: Horsley's Wood; *W.W.*!
COLNE: Deadmansey Wood*; Ravensdell Wood, v.c. 20 [Beds.]; *W.W.*!

R. sciocharis (Sud.) W. Wats.
   R. sciaphilus Lange
A misunderstood species to which early Beds. records of *R. silvaticus* Weihe & Nees
and *R. salteri* Bab. should probably be referred. Recorded for Bucks. and
Northants.
OUZEL: Heath and Reach*; Woburn; Aspley Wood*; Woburn Park; Lowe's
Wood; Stockgrove; *W.W.*!
IVEL: Clophill; *W.W.*!
LEA and COLNE: Kidney Wood; *W.W.*!
First record: Druce, *Fl. of Bucks.*, 1926, unless a record of *R. silvaticus* from Woburn
by W. Moyle Rogers is to be referred to this.

R. vestitus Weihe C.T.W. p. 473.
One of our more common brambles to which most early records of *R. leucostachys*
auct. non Sm. should be referred. Recorded for all districts except Nene and Cam
and for Bucks., Cambs. and Northants.
First record: uncertain.

R. leucostachys Sm.
Rare: apparently limited to the Clay-with-Flints.
COLNE: Pepperstock; Whipsnade Common*; Studham Common; Ravensdell
Wood, v.c. 20 [Beds.]; *W.W.*!

R. adscitus Genev. C.T.W. p. 473.
OUZEL: Aspley Wood, det. W.W.*

R. criniger (E. F. Linton) Rog.
Rare: recorded for Bucks. and Cambs.
IVEL: Everton!; *W. H. Mills in Evans, Fl. of Cambs.*; Galley Hill, Sutton; Stratford;
W.W.!
COLNE: Studham Common*; *W.W.*!
First record: W. H. Mills, 1939.

[Sect. *Rotundifolii* W. Wats.]
R. hansenii Krause
   R. mucronatus auct.
Rare: recorded for Herts., Bucks. and Northants.
OUZEL: Woburn Sands; *J. H. Riddelsdell in J. Bot. 1931*: Aspley Woods*; Lowe's
Wood; *W.W.*!

A record of *R. rotundifolius* (Bab.) Blox., in *Journ. Beds. Nat. Hist. Soc. 1947*, was in
error.

[Sect. *Radulae* Focke]
R. radula Weihe
Widely distributed and more frequent in the north of the county, but not common.
Recorded for all districts and for Bucks., Cambs. and Northants.
First record: Woburn Sands, *E. F. Linton in J. Bot. 1893.*

R. discerptus P. J. Muell. C.T.W. p. 475.
   R. echinatus Rog.
Rare: recorded for Herts., Bucks. and Northants.
OUZEL: Woburn Sands; *W. Moyle Rogers*: King's Wood; Aspley Wood*; Leighton
Buzzard*; *W.W.*!
IVEL: Old Warden; *W.W.*!: Washer's Wood*.
LEA: Spittlesea Wood; Horsley's Wood*; *W.W.*!
COLNE: Stockwood; *E. M. Langley*: Pepperstock; Studham Common; Ravensdell
Wood, v.c. 20 [Beds.]; *W.W.*!
First record: W. Moyle Rogers.

R. rudis Weihe C.T.W. p. 476.
A frequent woodland bramble: recorded for Bucks. and Northants.
OUZEL: Woburn; *V.C.H.*: King's Wood*; Aspley Wood; New Wavendon Heath,
v.c. 24 [Beds.]; *W.W.*!: Baker's Wood.
LEA: Kidney Wood; Horsley's Wood*; *W.W.*!
COLNE: Studham Common; Folly Wood*; Pepperstock; *W.W.*!
First record: G. C. Druce, 1904.

R. echinatoides (Rog.) Druce C.T.W. 475.
Rare: recorded for Herts., Bucks. and Northants.
OUZEL: Heath and Reach*; *W.W.*!
COLNE: Whipsnade Common*; Folly Wood*; Ravensdell Wood, v.c. 20 [Beds.];
W.W.!

R. watsonii W. H. Mills
Frequent: recorded also for Herts., Bucks. and Hunts.
NENE: Hinwick, det. W.W.
OUSE: Clapham Park Wood, det. W.W.
OUZEL: King's Wood*; Aspley Wood; New Wavendon Heath, v.c. 24 [Beds.];
W.W.!
IVEL: Flitwick Moor; Cooper's Hill; Clophill; *W.W.*!
LEA: Kidney Wood; Spittlesea Wood; *W.W.*!
COLNE: Whipsnade Common*; Folly Wood; Pepperstock; Ravensdell Wood,
v.c. 20 [Beds.]; *W.W.*!: Zouche's Farm*.
First record: W. H. Mills in *Watsonia I (1949)* 135.

R. rhenanus P. J. Muell.
OUZEL: Aspley Wood*; *W.W.*!

[Sect. *Apiculati* Focke]
R. homalodontus Muell. & Wirtg.
COLNE: Deadmansey Wood*; *W.W.*!

R. foliosus Weihe
OUZEL: Aspley Wood*; *W.W.*!
LEA: Horsley's Wood*; Kidney Wood*; *W.W.*!

R. flexuosus Lef. & Muell. C.T.W. p. 476.
Frequent in woods; recorded for all neighbouring counties except Hunts.
OUZEL: Heath and Reach; *W.W.*!
LEA: Kidney Wood*; Spittlesea Wood*; Horsley's Wood; *W.W.*!
COLNE: Deadmansey Wood; Folly Wood; Ravensdell Wood, v.c. 20 [Beds.];
W.W.!

R. acutipetalus Lef. & Muell.
OUZEL: Aspley Wood*; *W.W.*!
LEA: Horsley's Wood*; *W.W.*!
COLNE: Folly Wood; *W.W.*!

R. fuscus Weihe
Rare: recorded for Herts. and Bucks.
IVEL: roadside, Clophill; *W.W.*!
LEA: Kidney Wood*; Horsley's Wood; *W.W.*!
COLNE: Pepperstock*; *W.W.*!

R. menkei Weihe
COLNE: Folly Wood*; *W.W.*!

R. adamsii Sud.
IVEL: outside Warden Wood, as *R. bloxamii* Ed. Lees; *E. M. Langley in J.H. Notes.*
LEA: Kidney Wood*; *W.W.*!
COLNE: Folly Wood; *W.W.*!
First record: E. M. Langley, *c.* 1916.

R. tereticaulis P. J. Muell.
OUZEL: Aspley Wood, det. W.W.; *C. Avery and J. E. Woodhead*!

R. pallidus Weihe C.T.W. p. 477.
OUZEL: Woburn; *W. Moyle Rogers.*
LEA: Kidney Wood, still there in 1948!*; *J.S. in J. Bot. 1889.*

R. insectifolius Lef. & Muell.
Rare: recorded for Cambs.
OUZEL: Aspley Wood; *W.W.*!
LEA: Kidney Wood; Horsley's Wood*; *W.W.*!
COLNE: Folly Wood; Pepperstock; *W.W.*!

R. scaber Weihe C.T.W. p. 477.
Rare: recorded for Bucks. and Northants.
OUZEL: King's Wood*; Aspley Wood*; Stockgrove; New Wavendon Heath, v.c. 24
[Beds.]; *W.W.*!
LEA: Horsley's Wood; *W.W.*!

R. microdontos Lef. & Muell.
LEA: Kidney Wood*; *W.W.*!

R. botryeros (Focke ex Rog.) Focke
Rare: recorded for Northants. as *R. cenomanensis.*
LEA: Horsley's Wood; Kidney Wood; *W.W.*!
COLNE: Whipsnade Common*; *W.W.*

R. luteistylus Sud.
OUZEL: Aspley Wood*; *W.W.*!

R. thrysiflorus Weihe
Rare: recorded for Bucks.
OUZEL: Woburn Sands, Beds. and Bucks., as *R. hirtus* subsp. *flaccidifolius* (P. J. Muell.); *W. Moyle Rogers, Handbook of British Rubi,* 89.

R. adornatiformis (Sud.) Bouvet
OUZEL: King's Wood; *W.W.*!

R. rufescens Lef. & Muell. C.T.W. p. 477.
Frequent: recorded for Herts., Bucks. and Cambs.
OUZEL: King's Wood*; Aspley Wood; *W.W.*!
LEA: Kidney Wood; Spittlesea Wood; Horsley's Wood*; *W.W.*!
COLNE: Pepperstock*; Whipsnade Common; Folly Wood*; Ravensdell Wood, v.c. 20 [Beds.]; *W.W.*!

R. melanoxylon Muell. & Wirtg.
LEA: Kidney Wood; *W.W.*!

R. heterobelus Sud.
IVEL: Washer's Wood, det. W.W.
LEA: Kidney Wood*; Horsley's Wood; *W.W.*!
COLNE: Deadmansey Wood*; Pepperstock; *W.W.*!

[Sect. *Grandifolii* Focke]
R. furvicolor Focke
LEA: Spittlesea Wood*; *W.W.*!

R. leightonii Ed. Lees ex Leight.
Rare: recorded for Bucks. and Hunts. as *R. anglicanus* Rog.
IVEL: Cooper's Hill*; *W.W.*!

R. angusticuspis Sud.
Rare: recorded as *R. obtruncatus* var. *angusticuspis* Sud. for Bucks. and Northants.
LEA: Spittlesea Wood*; *W.W.*!

R. rosaceus Weihe & Nees is recorded for Beds. on the authority of E. M. Langley by W. Moyle-Rogers in *J. Bot. 1915.* Recorded also for Herts., Bucks. and Northants., but Watson has not seen it in Beds. and discredits all records of *R. rosaceus* by Rogers.

[Sect. *Hystrices* Focke]
R. dasyphyllus Rog.
Rare: recorded for Cambs. and Northants.
OUZEL: Woburn; *V.C.H.*: Heath and Reach*; Stockgrove; *W.W.*!
IVEL: Flitwick Moor; Rowney Warren; *W.W.*!
COLNE: Whipsnade Zoo; *J. H. Riddelsdell in J. Bot. 1932*: Studham Common; Pepperstock; *W.W.*!
First record: G. C. Druce, 1904.

R. apricus var. **sparsipilus** W. Wats.
OUZEL: King's Wood*; *W.W.*!
LEA: Spittlesea Wood; Kidney Wood; *W.W.*!
COLNE: Folly Wood; Pepperstock; Ravensdell Wood, v.c. 20 [Beds.]; *W.W.*!

R. bavaricus Focke
IVEL: Daintry Wood*; *W.W.*!

R. fuscoater Weihe
Rare: recorded also for Bucks.
Clophill*; *W.W.*!
COLNE: Kidney Wood, det. W.W.; *W. H. Mills*!

R. adenolobus W. Wats. C.T.W. p. 480.
Rare: recorded as *R. koeleri* var. *cognatus* (N.E.Br.) Rog. for Bucks.
NENE: Wymington Scrub, det. W.W.*
LEA: Kidney Wood*; *W.W.*!

R. hylocharis W. Wats.
OUZEL: Heath and Reach; *W.W.*!

[Sect. *Glandulosi* P. J. Muell.]
R. angustifrons Sud.
LEA: Kidney Wood; *W.W.*!
COLNE: Deadmansey Wood; *C. Avery and J. E. Woodhead*!

[Sect. *Corylifolii* Focke]
R. balfourianus Blox. ex Bab. C.T.W. p. 482.
Rare: recorded for all neighbouring counties except Hunts.
OUZEL: Woburn Sands; *E. F. Linton in J. Bot. 1893.*
CAM: Cockayne Hatley, det. W.W.*
LEA: Chiltern Green Common, det. W.W.*
First record: E. F. Linton, 1893.

R. warrenii Sud.
Frequent: recorded for Herts. as *R. concinnus* Warren.
NENE: Wymington Scrub, det. W.W.*; Hinwick, det. W.W.
OUZEL: Woburn; Aspley Wood*; *W.W.*!
IVEL: Galley Hill, Sutton*; Northill; Flitwick; Cooper's Hill; Old Warden;
Rowney Warren; Clophill; *W.W.*!

R. sublustris Ed. Lees C.T.W. p. 482.
Frequent: recorded for Bucks., Cambs. and Northants.
OUZEL: Woburn Sands; Ridgmont; *W. Moyle-Rogers*: Heath and Reach; *W.W.*!
IVEL: Sandy; Sutton Fen; Flitwick Moor; Potton; *W.W.*!
CAM: Cockayne Hatley Wood; *W.W.*!
COLNE: Folly Wood; *W.W.*!
First record: W. Moyle-Rogers, *c.* 1911.

R. conjungens (Bab.) W. Wats. C.T.W. p. 483.
One of our more common brambles. Recorded for all districts except Kym and
for Cambs. and Northants.

R. tuberculatus Bab.
Rare: recorded for Cambs. and Northants.
NENE: Hinwick, det. W.W.*
OUZEL: King's Wood; Woburn*; *W.W.*!
LEA: between Luton and Dunstable*; Kidney Wood; *W.W.*!

R. britannicus Rog.
OUZEL: Leighton Buzzard; *W.W.*!

R. scabrosus P. J. Muell. C.T.W. p. 483.
Rare: recorded for Bucks., Cambs. and Northants.
LEA: Horsley's Wood*; *W.W.*!

R. myriacanthus Focke C.T.W. p. 483.
Rare: recorded for all neighbouring counties except Hunts.
COLNE: Folly Wood*; *W.W.*!

[Subgen. *Glaucobatus* Dum.]
R. caesius L. C.T.W. p. 484.   *Dewberry*
Ditches, etc.   Native
Common, and usually found on calcareous soils. Recorded for all
districts, except Lea and Colne, and for all neighbouring counties.
First record: Abbot, 1795.

Var. **arvalis** Reichb.
Ouzel: King's Wood; *W.W.*!
Ivel: Ampthill; *W.W.*!
Cam: Cockayne Hatley Wood; *W.W.*!
Lea: Chiltern Green Common; *W.W.*!
Colne: Folly Wood; Pepperstock; *W.W.*!

[Subgen. *Anoplobatus* Focke]
R. odoratus L.
Rare: a garden escape.
Ouse: near Lidlington; *Miss Robinson in B.E.C. 1911–4 Rep.*

GEUM L.

Geum urbanum L. C.T.W. 50.7.1.   *Herb Bennet, Wood Avens*
Woods, roadsides, waste places, etc.   Native
Common: recorded for all districts and neighbouring counties.
H.S. 3a, 3b, 5, 18, 19b, 20a, 32, 42, 52–55b, 69, 78, 81, 83.
First record: Abbot, 1798.

G. rivale L. C.T.W. 50.7.2.   *Water Avens*
Damp woods   Native
Very rare: recorded for all neighbouring counties except Hunts.
Ouse: Puttenham Wood, presumably Putnoe Wood, the station
given in *F.B.*, where it was found still growing in 1920 by A. W.
Guppy†*; *Plantae Bedford.*
Ouzel: Milton Bryan; *H. E. Pickering.*
Ivel: Fancott, where it appeared to be well established in a restricted
area in a small wood, but disappeared in 1942 when the wood was
cleared!*; *H. B. Sargent.*
First record: Abbot, 1795.

FRAGARIA L.

Fragaria vesca L. C.T.W. 50.6.1.   *Wild Strawberry*
Open rides in woods, rough disturbed ground, etc.   Native
Common especially on sandy and gravelly soils. Recorded for all
districts and neighbouring counties.
H.S. 2a–3b, 5, 18, 19b, 20a, 25, 26, 30a, 30b, 34a, 53b, 54, 68, 69,
77a, 77b, 79a, 81, 83.
First record: Abbot, 1798.

POTENTILLA L.

Potentilla anserina L. C.T.W. 50.4.5.   *Silverweed*
Roadsides, rough pastures, borders of marshes, etc.
Very common: a variable species, growing in a wide range of
habitats. The varieties *concolor* Wallr. and *discolor* Wallr., which
appear to be no more than habitat forms, were recorded in *J.E.L.*,
J. Bot. Recorded for all districts and neighbouring counties.
H.S. 2a, 3a, 6, 8, 10a, 10b, 13, 19a, 20a, 23, 51b, 53a, 55b, 56b,
57b, 61, 71, 73, 74b, 77a, 79a, 80a, 82, 84a, 85.
First record: Abbot, 1798.

P. argentea L. C.T.W. 50.4.6.   *Hoary Cinquefoil*
Disturbed soils in heathy and gravelly places   Native
Rare: all recent records are for the Lower Greensand. Records of
Saunders in *S.B.P.L.* from the border of Herts. and New Mill End
were probably in error. Recorded for all neighbouring counties
except Northants.
H.S. 25.
Ouse: Kempston gravel pits; *Plantae Bedford.*: Millbrook; *W.H.
1875.*
Ouzel: Aspley; *F.B.*
Ivel: Rowney Warren, apparently still there; *Plantae Bedford.*:
Ampthill; Shefford; *F.B.*: Southill†; *J.McL.*, *B.M.*: Flitwick;
Tingrith; *J.McL. in J.H. Notes*: Steppingley; Ridgmont†; Silsoe;
C.C.: Sandy; *J.E.L.*, *F.F.B.*: Maulden; *R. Townsend*: roadside, south
of Clophill*.
First record: Abbot, 1795.

P. reptans L. C.T.W. 50.4.14.   *Creeping Cinquefoil*
Roadsides, open pastures, etc.   Native
Common, and growing at its best in open habitats with limited
competition. Recorded for all districts and neighbouring counties.
H.S. 1a, 1b, 3a, 10a–10c, 13, 18, 19a, 20a, 21, 27, 34a, 45b, 52–53b,
54, 55b, 56b, 58b, 61, 68, 69, 73, 77a, 82, 85.
First record: Abbot, 1798.

P. anglica Laichard. C.T.W. 50.4.13.   *Trailing Tormentil*
  P. procumbens Sibth.
Heaths, heathy woods, etc.   Native
Not infrequent on the lighter soils, and probably more common than
the records given would indicate. Recorded for all neighbouring
counties.
H.S. 23, 30b, 79b.
Kym: Swineshead Wood.
Ouzel: Aspley Wood; *F.B.*: King's Wood*; Woburn Park.
Ivel: Ampthill; *F.B.*: Maulden; *J.McL.*, *L.M.*: Southill; *J.McL. in
J.H. Notes*: Cooper's Hill.
Lea: Kidney Wood.
Colne: Kidney Wood; Deadmansey Wood.
First record: Abbot, 1798.

P. erecta (L.) Räusch. C.T.W. 50.4.12.   *Common Tormentil*
  P. tormentilla Stokes
Heaths, dry heathy pastures, etc.   Native
Common in heathy places on the Lower Greensand and Clay-with-
Flints, and sparingly on gravelly soils. Recorded for all districts
except Cam, and for all neighbouring counties.
H.S. 23, 28, 30b, 35, 37, 38, 39b, 53b, 66b, 69, 74a, 79a, 80a.
First record: Abbot, 1798.

P. sterilis (L.) Garcke C.T.W. 50.4.3.   *Barren Strawberry*
   Fragaria sterilis L.
Dry banks, woodland rides, etc.   Native
Common and well distributed: recorded for all districts and neigh-
bouring counties.
H.S. 2a, 2b, 18, 31a, 34a, 34b, 51a, 53a, 54, 77a, 78, 80a, 80b, 81.
First record: Abbot, 1798.

P. palustris (L.) Scop. C.T.W. 50.4.2.   *Marsh Cinquefoil*
   Comarum palustre L.
Marshes   Native
Rare, and limited to a few marshes in the middle of the county.
Recorded for all neighbouring counties except Bucks.
H.S. 39a, 41.
OUZEL: Aspley; *Plantae Bedford.*
IVEL: Flitton Moor; *J.McL., L.M.*: Flitwick Moor, locally abundant
in a small area in the wettest parts!†*; *S.B.P.L.*: Westoning Moor,
abundant in a restricted area!; *C.C.*
First record: Abbot, 1795.

ALCHEMILLA L.

Alchemilla vestita (Buser) Raunk. C.T.W. 50.11.4. *Lady's-mantle*
   A. vulgaris auct.
Heathy woods and pastures   Native
Frequent on the Lower Greensand and Clay-with-Flints. *E. Bot.*
tab. 597, June 1799, was drawn from a specimen sent by Abbot from
Eversholt. The same plate was used for *Flora Bedfordiensis*. Abbot
considered that 'of all our natives this is the most elegant plant'.
Recorded for all neighbouring counties except Cambs.
OUSE: Pavenham; *J.H. Notes.*
OUZEL: Totternhoe; *W.F.B.*: Palmer's Shrubs*; King's Wood;
Potsgrove*.
IVEL: Eversholt; *Plantae Bedford.*: Sundon*; *S.B.P.L.*: Sharpenhoe;
W.F.B.: Fancott!; *P.T.*: Washer's Wood.
LEA: Bramingham; *S.B.P.L.*: Luton Hoo; *W.F.B.*: East Hyde*.
COLNE: Whipsnade, frequent on woodland verges!; *Plantae Bedford.*:
Pepperstock†*; *Herb. J.S.*: Caddington!; *V. H. Chambers*: woodland
verges, Studham.
First record: Abbot, 1795.

A. xanthochlora Rothm. C.T.W. 50.11.9.
   A. pratensis auct.
Pastures   Native
Rare: recorded also for Herts. and Bucks.
COLNE: plentiful in limited area on verge of drive, Luton Hoo*;
P.T. and the author in Watsonia I (1949) 42.

APHANES L.

Aphanes arvensis L. C.T.W. 50.12.1.   *Parsley Piert*
Alchemilla arvensis (L.) Scop., *A. aphanes* Leers
Arable fields, gravel pits, etc.   Colonist
Frequent: a species of neutral and calcareous soils, where it is seen
often as a cornfield weed. Distribution not yet fully known. Re-
corded for all neighbouring counties.
H.S. 17, 45a, 61, 71, 82.
Ouse: gravel pit, Willington*.
Ouzel: Totternhoe*.
Ivel: Markham Hills*; *B. Verdcourt*: railway cutting, Ampthill*;
Rowney Warren.
Lea: Chiltern Green*.
Colne: Studham; *E.M.-R. in Watsonia I (1949)* 166: arable fields,
Whipsnade*, Greencroft Barn, v.c. 20 [Beds.]*.
First record for the aggregate: Abbot, 1798.
First record for the segregate: E. Milne-Redhead, 1949.

A. microcarpa (Boiss. & Reut.) Rothm. C.T.W. 50.12.2.
Arable fields, heathy pastures, etc.   Native
Frequent: a species of acid soils, and recently separated from
A. arvensis by S. M. Walters, who has named all my material. It is
as common in natural habitats as in arable fields, and has every
appearance of being native. Recorded for Bucks. and Cambs., but
no doubt present in other neighbouring counties.
H.S. 22b, 23, 26, 27, 53a, 54, 77a.
Ouzel: heathy ground, Heath and Reach (see *Watsonia II (1951)*
43); rides, Aspley Wood*; King's Wood.
Ivel: roadside, Clophill*; Rowney Warren; Washer's Wood;
Daintry Wood.
Lea: rides, Horsley's Wood*.
Colne: Studham*.
First record: the author in *Watsonia II (1951)* 43.

AGRIMONIA L.

Agrimonia eupatoria L. C.T.W. 50.9.1.   *Common Agrimony*
Roadsides, wood borders, etc.   Native
Common, and well distributed: recorded for all districts and
neighbouring counties.
H.S. 2a, 2b, 10c, 19a, 21, 51b, 61, 68, 75, 80a, 80b.
First record: Abbot, 1798.

A. odorata (Gouan) Mill. C.T.W. 50.9.2.   *Sweet Agrimony*
Rough pastures, etc.   Native
Rare, and limited to light soils. Recorded for all neighbouring
counties.
Kym: Swineshead*; rough ground to north of Swineshead Wood,
v.c. 30 [Hunts.].
Ouzel: Stockgrove!*; *E.M.-R.*
First record: the author in *B.E.C. 1943-4 Rep.* 808.

ACAENA Mutis ex L.

Acaena anserinifolia (J. R. & G. Forst.) Druce C.T.W. 50.15.1.
Gravel pits, etc.   Colonist
An attractive species introduced with wool manure. It becomes readily established
and colonizes considerable patches.
OUSE: gravel pit, Wyboston!*; *C. C. Foss*: gravel pit, Cople; *P.T.*: gravel pit,
Eaton Socon.
IVEL: bank, by railway sidings, Langford; *J. E. Lousley*: arable field, Clifton*;
B. M. Vizard!
First record: C. C. Foss in *B.E.C. 1945 Rep.* 56.

POTERIUM L.

Poterium sanguisorba L. C.T.W. 50.14.1.   *Salad Burnet*
Pastures, railway banks, etc.   Native
Common on calcareous soils, and especially so on the Chalk.
Recorded for all districts and neighbouring counties.
H.S. 1b, 16, 59–61, 65–69.
First record: Abbot, 1798.

P. polygamum Waldst. & Kit. C.T.W. 50.14.2.   *Fodder Burnet*
Railway banks, edges of arable fields, etc.   Colonist
Rare: it is usually well established when it appears on railway banks, elsewhere it
is of only casual occurrence. Recorded for all neighbouring counties.
NENE: railway bank, Wymington.
KYM: arable field, Dean; *E.M.-R.*!
OUZEL: Tilsworth; *F. Rose.*
IVEL: Pegsdon; between Henlow station and Shefford; *J.E.L., F.F.B.*: railway
bank, Flitwick*.
LEA: Dunstable; Luton; *R. A. Pryor in J. Bot. 1876* 24: Limbury; *J.S. in Bot. 1889.*
First record: R. A. Pryor, 1876.

SANGUISORBA L.

Sanguisorba officinalis L. C.T.W. 50.13.1.   *Great Burnet*
Poterium officinale (L.) A. Gray
Pastures   Native
A feature of the drier portions of water meadows, especially by the
Ouse. It can, however, withstand comparatively dry conditions.
Absent from both extremely acid and calcareous pastures. Recorded
for all neighbouring counties.
H.S. 2a.
NENE: railway embankment, Wymington; meadow, Hinwick.
OUSE: Race Meadow; *Plantae Bedford.*: Fenlake; Cow Meadows;
F.B.: Hazelwood Lane, Cardington; *J.McL., L.M.*: Carlton; *W.H.
1875*: Oakley; Biddenham!; Bolnhurst; *A. W. Guppy*: claypits,
Stewartby; *H. B. Souster*: railway embankment, Souldrop; meadows,
Sharnbrook, Eaton Socon.
KYM: main drive, West Wood.
OUZEL: Totternhoe!†; *W.F.B.*: Eaton Bray; *C. C. Foss*: meadows,
Billington, Leighton Buzzard, Heath and Reach*, Potsgrove.
IVEL: Sharpenhoe; *W.F.B.*: meadow, Fancott.
First record: Abbot, 1795.

ROSA L.

A most difficult genus, for which many of the early records for the county must be doubted, as views on the limits of the species have changed. E. B. Bishop's paper *Notes on the Roses of Bedfordshire* (*B.E.C. 1938 Rep.*) was based on 300 gatherings made by H. Phillips and Enid McAlister Hall in some of the least promising rose territory of the county. Records from this paper are credited to Phillips (*H.P.*). Bishop, who followed Keller in his determinations, named further gatherings made mainly by V. H. Chambers and me (shown det. E.B.B.). R. Melville, who has with most British botanists followed Wolley-Dod's treatment, has since visited the county and named further gatherings (shown det. R.M.). It is possible that many of the names used are synonymous, but they are given here in the hope that they may assist work on the much-needed revision of the genus.

Rosa arvensis Huds. C.T.W. 50.16.1.   *Field Rose*
Hedgerows, wood borders, etc.   Native
Common: recorded for all districts except Nene and Ouzel, and for all neighbouring counties.
H.S. 2b, 3b, 32, 55b, 78–81, 83.
First record: Abbot, 1798.
   Var. **vulgaris** Ser.
IVEL: Barton; *H.P.*
LEA: Stopsley, det. E.B.B.*
COLNE: Kensworth, v.c. 20 [Beds.], det. E.B.B.*
   Var. **ovata** (Lejeune) Desv.
OUSE: Bromham; Colmworth; *H.P.*
   Var. **biserrata** Crép.
OUSE: Marston Mortaine; Colmworth; Felmersham; *H.P.*
IVEL: Barton; *H.P.*
   Var. **gallicoides** (Déségl.) Crép.
OUSE: Colmworth; *H.P.*

R. stylosa Desv. C.T.W. 50.16.4.
Hedgerows, etc.   Native
Rare: recorded for Bucks.
OUSE: Colmworth, as var. *congesta* (Rip.) R. Kell.; Felmersham, as var. *congesta*; *H.P.*
KYM: West Wood, as var. *systyla* (Bast.) Baker*; *N. Y. Sandwith.*
IVEL: Clophill, var. *systyla* and f. *fastigiata* (Bast.) Baker; *H.P.*
First record: unlocated, *Hillhouse List, 1876.*

R. canina L. C.T.W. 50.16.5.   *Dog Rose*
Hedgerows, scrub, etc.   Native
Common: probably the most variable species in the flora. Recorded for all districts and neighbouring counties.
H.S. 1b, 2a, 5, 19b, 20b, 21, 33–34b, 51, 52, 63b, 64, 66a, 67–69, 70b, 74a, 75, 76, 79, 83
First record: Abbot, 1798.

S

Var. **lutetiana** (Léman) Baker
OUSE: Bromham; Felmersham, f. *lasiostylis* Borb.; Colmworth; Felmersham; *H.P.*
OUZEL: Woburn Sands; *W. Moyle-Rogers.*
IVEL: Ridgmont; *W. Moyle-Rogers*: between Henlow and Shefford; *J.E.L., Diary*: Barton; *H.P.*
LEA: Luton†; *J.S., J. Bot. 1889*: Stopsley, f. *lasiostylis*, det. E.B.B.*
COLNE: Deadmansey Wood, f. *lasiostylis*, det. E.B.B.*
    Var. **sphaerica** (Gren.) Dum.
OUSE: Bromham; *H.P.*
IVEL: Barton; Clophill; *H.P.*
LEA: Limbury; *F.F.B.*: Dallow Lane, det. near this†; *Herb. J.S.*
    Var. **flexibilis** (Déségl.) Rouy
OUSE: Colmworth; *H.P.*
COLNE: Deadmansey Wood, det. E.B.B.*
    Var. **senticosa** (Ach.) Baker
OUSE: Cox's Pits; *J.H.*: Bromham; Marston Mortaine; Colmworth; *H.P.*
OUZEL: Aspley, f. *axyphylla* (Rip.) W.-Dod; Woburn, f. *mucronulata* (Déségl.) W.-Dod; det. E.B.B.*
IVEL: Barton; *H.P.*
    Var. **spuria** (Puget) W.-Dod
COLNE: Deadmansey Wood, f. *syntrichostyla* (Rip.) W.-Dod, det. E.B.B.*
    Var. **globularis** (Franch.) Dum.
OUSE: Wilden; *H.P.*
IVEL: Clophill, det. E.B.B.*
    Var. **ramosissima** Rau
OUZEL: Maiden Bower; *H.P.*
    Var. **squarrosa** Rau (var. *dumalis* sensu W.-Dod)
OUSE: Colmworth, f. *viridicata* (Puget) Rouy; Bromham, f. *viridicata*; *H.P.*
OUZEL: Woburn; *H.P.*
IVEL: Rowney Warren; *J.E.L., F.F.B.*: Barton; Clophill; Ampthill; *H.P.*
COLNE: Kensworth Common, v.c. 20 [Beds.], det. E.B.B.*
[Bishop refers to many gatherings of f. *cladoleia* (Rip.) W.-Dod but lists none]
    Var. **stenocarpa** (Déségl.) Rouy
OUSE: Felmersham; *H.P.*
OUZEL: Woburn, f. *conica* R. Kell; *H.P.*
IVEL: Barton; Ampthill; *H.P.*
    Var. **biserrata** (Mérat) Baker
COLNE: Deadmansey Wood, det. E.B.B.*
    Var. **carioti** (Chab.) Rouy
OUSE: Bromham; Clapham; *H.P.*
OUZEL: Maiden Bower; *H.P.*
IVEL: Ampthill; *H.P.*
    Var. **fraxinoides** H. Braun
KYM: West Wood; *N. Y. Sandwith.*
OUZEL: Aspley; *H.P.*
COLNE: Kensworth Common, v.c. 20 [Beds.]; Deadmansey Wood; det. E.B.B.*
    Var. **sylvularum** (Rip.) Rouy
[Bishop refers to several gatherings from various parts of the county]
OUSE: Bromham, f. *adscita* (Déségl.) Rouy; *H.P.*
OUZEL: Aspley Guise, f. *adscita*; *H.P.*
    Var. **andegavensis** (Bast.) Desp.
OUSE: Turvey; Wilden; Colmworth, f. *agraria* (Rip.) W.-Dod; *H.P.*
    Var. **verticillacantha** (Mérat) Baker
COLNE: Kensworth Common, det. E.B.B.*
    Var. **blondaeana** (Rip. ex Déségl.) Rouy
OUSE: between Wilden and Colmworth; *H.P.*

R. dumetorum Thuill. C.T.W. 50.16.5.
Hedgerows, etc.   Native
Frequent: often considered to be within the broad limits of *R. canina.*
Recorded for all districts except Cam, and for all neighbouring counties.

First record: J. Saunders in *J. Bot. 1889.*
  Var. **dumetorum**
OUSE: Marston Mortaine, f. *semiglabra* (Rip.) W.-Dod; Colmworth, f. *semiglabra*; *H.P.*
KYM: West Wood, f. *urbica* (Lem.) W.-Dod; *R.M.*
OUZEL: Woburn Sands, f. *urbica*; *W. Moyle-Rogers.*
IVEL: Cadwell; Stondon, f. *urbica*; *J.E.L., F.F.B.*
LEA: Limbury; *J.S., J. Bot. 1889.*
  Var. **ramealis** (Puget) W.-Dod
OUSE: Wilden; Felmersham; *H.P.*: riverside, Willington, det. R.M.*
IVEL: Flitwick; *H.P.*
  Var. **gabrielis** (F. Gér.) R. Kell.
IVEL: Flitwick Moor; *H.P.*
  Var. **platyphylla** (Rau) W.-Dod
COLNE: Kensworth, v.c. 20 [Beds.], det. E.B.B.*
  Var. **erecta** W.-Dod
NENE: Hinwick, det. E.B.B.*

R. dumalis Bechst.
  R. glauca Vill. ex Lois. non Pourr., *R. coriifolia* Fr. C.T.W. 50.16.6, *R. afzeliana* Fr. p.p.
Hedgerows, etc.   Native
Rare.
COLNE: Deadmansey Wood, det. E.B.B.*; *V. H. Chambers.*

R. obtusifolia Desv. C.T.W. 50.16.7.
Hedgerows, etc.   Native
Rare: recorded for Bucks.
  Var. **borreri** (Woods) W.-Dod
IVEL: Maulden; *H.P.*
  Var. **rothschildii** (Druce) W.-Dod
OUSE: Colmworth; *H.P.*
  Var. **tomentella** (Léman) Baker
OUSE: Marston Mortaine, f. *rectispina* R. Kell.; *H.P.*: Shelton; *H.P.*
[Bishop mentioned four gatherings of var. *capucinensis* R. Kell. from a small un-located radius, and referred a number of gatherings, with some uncertainty, to other varieties]

R. tomentosa Sm. C.T.W. 50.16.8.   *Downy Rose*
Hedgerows, etc.   Native
Rare: it is possible that records of *R. villosa* from Aspley, Bromham and Clapham Woods in *Abbot's annotated F.B.*, and from Hazlewood Lane in Herb. J. McLaren, British Museum, should be referred to this. Recorded for Bucks.
H.S. 80.
OUSE: Wilstead Wood, det. var. *typica* W.-Dod, R.M.*
KYM: West Wood, as var. *pseudocuspidata* (Crép.) Rouy!*; *R.M.*
OUZEL: near Aspley; *V.C.H.*
IVEL: south of Shefford; *J.E.L., J. Bot.*: Knocking Hoe, det. var. *typica*, R.M.*
COLNE: Deadmansey Wood, det. var. *dimorpha* (Bess.) Déségl., E.B.B., det. var. *typica*, R.M.*
First record: uncertain.

R. rubiginosa L. C.T.W. 50.16.11.   *Sweet Briar*
  R. eglanteria L. p.p.
Hedgerows, etc.   Native
Rare: but probably awaiting record in some stations. Recorded for
all neighbouring counties.
OUSE: Bromham, two gatherings determined by Bishop as 'a slight
approach to var. *echinocarpa* (Rip.) Gren.' and '? f. *horrida* Lange';
H.P.
LEA: near Daffodil Wood; hedge near New Mill End station†; *F.F.B.*
First record: Saunders.

R. micrantha Borrer ex Sm. C.T.W. 50.16.12.   *Sweet Briar*
Hedgerows, chalk scrub, etc.   Native
Rare: recorded for Bucks.
NENE: hedge near Podington; *T. B. Blow in Bot. Loc. Rec. Club 1875
Rep.*
IVEL: Barton; *H.P.*: foot of Barton Cutting; *R.M.*!
First record: T. B. Blow, 1875.

SORBUS L.

Sorbus aucuparia L. C.T.W. 50.22.1.   *Mountain Ash, Rowan*
  Pyrus aucuparia (L.) Ehrh.
Woods   ? Native
Frequent on the Lower Greensand, where it may be native; elsewhere it is almost
certainly planted. Abbot considered it to be rare. Recorded for all neighbouring
counties except Hunts.
H.S. 42.
OUZEL and IVEL: frequent in woods on the Lower Greensand.
COLNE: Deadmansey Wood!; *V. H. Chambers*: Long Wood, v.c. 20 [Beds.].
First record: Abbot, 1798.

S. aria (L.) Crantz C.T.W. 50.22.8.   *White Beam*
  Crataegus aria L., *Pyrus aria* (L.) Ehrh.
Woods, hedgerows, etc.   Native
Frequent and native in woods on the Clay-with-Flints in the Colne
district, but planted or bird-sown elsewhere. Recorded for all
neighbouring counties.
OUSE: between Fenlake and Goldington; *F.B.*
OUZEL: Aspley; *J.McL., L.M.*: Hockliffe.
IVEL: Warden; *V.C.H.*: Sandy; *J.E.L., J. Bot.*
LEA: near Chaul End!; *V. H. Chambers*: Leagrave; *H. B. Souster.*
COLNE: Dunstable Downs!; *H. B. Sargent*: Whipsnade Zoo; Blow's
Downs; Ravensdell Wood, v.c. 20 [Beds.]; Greencroft Barn, v.c. 20
[Beds.]*; Kensworth, v.c. 20 [Beds.]*.
First record: Abbot, 1798.

PYRUS L.

Pyrus communis L. C.T.W. 50.23.1.   *Wild Pear*
Woods, hedgerows, etc.   Of garden origin
Not infrequent, but in no place appearing to be native. Pears have long been
grown in the county and Peartree, at Stagsden, a place name now lost, appears as
early as 1347 (*P.N.B.*). Warden Pears took their name from Warden Abbey,
where they were cultivated. There are also many references in 16th century and
later literature to Warden pie. Recorded for all neighbouring counties.

Ouse: Thurleigh; *F.B.*: Lidlington; *W.C. Herb.*: Renhold; *J.McL., L.M.*: Cardington Wood; Exeter Wood; *J.McL. in J.H. Notes*: Marston; *W.H. 1875*: near Sheerhatch Wood; *J.E.L., J. Bot.*: Odell; *G. H. Day*: Ravensden; *B. B. West*.
Ouzel: hedgerow, between Chalgrave and Tebworth; *H. B. Souster*: hedgerow, the Litany*.
Ivel: Sundon; *S.B.P.L.*: Flitwick†; *J.S. in J. Bot. 1889*: Westoning; *G. D. Nicholls*: Potton Wood.
Lea: Leagrave; *H. B. Souster*.
First record: Abbot, 1798.

MALUS Mill.

Malus sylvestris Mill. C.T.W. 50.24.1. *Crab Apple*
  Pyrus malus L., *M. pumila* Mill.
Woods, hedgerows, etc. Native
Common. This comprises two subspecies: subsp. **sylvestris** (*M. acerba* Mérat) is considered to be native and I have observed it in all districts; subsp. **mitis** (Wallr.) Mansf. is said to be descended from garden stock and is almost as common. I have observed it in all districts except Nene. Appley (Appeleia *c.* 1150), in Chicksands parish, means an apple-tree clearing (*P.N.B.*). Recorded, for the aggregate, for all neighbouring counties.
First record: Abbot, 1798.

CRATAEGUS L.

Crataegus monogyna Jacq. C.T.W. 50.19.2. *Hawthorn*
Hedgerows, etc. ? Native
This is the common and frequently planted hawthorn of hedgerows in the county. It is usually the first shrub to appear in scrubland, especially so on the Chalk. Occasionally it develops into a tree, but is rarely more than 20 feet high. Hawnes (Hagenes in *Domesday Book*) probably owes its name to a spur of land on which a hawthorn stood (*P.N.B.*). Intermediates between this and the following species are common, and there are many forms of *C. monogyna* itself that have yet to be studied. Recorded for all districts and neighbouring counties.
H.S. 2b, 3b, 5, 18, 19b, 20b, 21, 38, 45b, 51, 52, 54, 55a, 59, 61, 63a, 63b, 65–69, 70b, 74a, 75, 76, 78–81, 83.
First record: Abbot, 1798.

C. oxyacanthoides Thuill. C.T.W. 50.19.1. *Midland Hawthorn*
  C. oxyacantha auct.
Woods Native
Less common than *C. monogyna*, but well distributed and seen at its best in woods on heavy clay soils. The ease with which it hybridizes with *C. monogyna* makes it impossible to give the distribution of the true species in detail. It appears in the pure, or reasonably pure, state in all districts and has been recorded for all neighbouring counties.
H.S. 32, 33, 34a, 34b.
First record: *Hillhouse List, 1876*.
First evidence: W. Crouch Herb., 1844, as *C. monogyna*.

CRASSULACEAE

SEDUM L.

Sedum telephium L. C.T.W. 51.1.2.   *Orpine, Livelong*
Sides of woods, roadsides, etc.   Colonist
Not infrequent and usually well established. Recorded for all neighbouring
counties except Hunts.
OUSE: Wyboston; *G. H. Day.*
OUZEL: Aspley Wood; *F.B.*: in derelict garden, King's Wood*; railway bank,
Stanbridgeford.
IVEL: Palmer's Wood; *J.McL., B.M.*: Maulden Green End, plentiful by roadside!;
M. L. Berrill in J.H. Notes: Clophill†; *Herb. J.S.*: Pedley Wood; *C.C.*: Flitwick;
W.F.B.: Potton; *H. B. Souster*.
LEA: New Mill End; *J.S. in Bot. Rec. Club 1880 Rep.*: Dane Street; *H. B. Souster*:
Birch Wood, Luton Hoo!; *P.T.*!: Horsley's Wood.
COLNE: Farley Bottom; *W.F.B.*: Woodside; *Pepperstock*†; *S.B.P.L.*: Caddington;
Lamb's Spinney, Studham; *H. B. Souster*: Deadmansey Wood*.
First record: Abbot, 1798.

S. acre L. C.T.W. 51.1.7.   *Wall-pepper, Biting Stonecrop*
Waste places, stonework on railways, etc.   Colonist
Frequent: considered common by Abbot, but rare by subsequent botanists. It has
increased in recent years. Most plentiful now on stonework by the railway, and
to be seen occasionally on disturbed soils on the Chalk. Recorded for all districts
and neighbouring counties.
H.S. 25.
First record: Abbot, 1798.

S. album L. C.T.W. 51.1.6.   *White Stonecrop*
Old walls, etc.   Wall denizen
Locally frequent on limestone walls in the north of the county,
rare elsewhere. Recorded for all neighbouring counties.
NENE: Hinwick*; railway bank, Wymington*.
OUSE: Stevington; Sharnbrook!; *F.B.*: Cotton End; *M. L. Berrill in
J.H. Notes*: Biddenham, on church wall!; *W.F.S. Diary*: Turvey;
H. B. Souster: Souldrop.
IVEL: Barton†; *S.B.P.L.*
First record: Abbot, 1798.

S. dasyphyllum L. C.T.W. 51.1.4.   *Thick-leaved Stonecrop*
Old walls   Wall denizen
Probably extinct: recorded for all neighbouring counties except Hunts.
OUSE: workhouse walls, Ford End, as *S. rupestre*; *J.McL., B.M.*
COLNE: Market Street; *Plantae Bedford*. [*Sedum minus circinato folio* was recorded on
the authority of Thomas Knowlton 'from the walls of Market Eit, near Market
Street', in Ray's *Synopsis Ed. III*—it is probable that Abbot picked up this record,
which may refer to v.c. 20].
First record: uncertain.

S. reflexum L. C.T.W. 51.1.10.
Old walls, etc.   Wall denizen
Considered common by Abbot, but rare now. I do not know Steparton† from
where McLaren collected it. Recorded for all neighbouring counties.
OUSE: Ford End; Oakley; Stevington; *Abbot's annotated F.B.*: Cardington; *J.McL.
in W.F.B.*
OUZEL: on chalk workings, Totternhoe; *E.M.-R.*
IVEL: Maulden; *W.C. Herb.*: old wall, Sundon*.
First record: Abbot, 1798.

SEMPERVIVUM L.

Sempervivum tectorum L. C.T.W. 51.2.1. *House Leek*
Cottage roofs, old walls, etc. Wall denizen
Not infrequent but rarely flowering. Recorded for all neighbouring counties
except Cambs.
OUSE: Lidlington; *L. W. Wilson.*
KYM: Keysoe; *B. B. West.*
OUZEL: Stanbridge!; *E.M.-R.*: Heath and Reach!; *P.T.*
IVEL: Cainhoe; *W.C. Herb.*: Flitwick!; *S.B.P.L.*: Harlington; *J.S., J. Bot. 1883*:
Shillington; Gravenhurst; *V. H. Chambers.*
LEA: Luton; *S.B.P.L.*
First record: Abbot, 1798.

SAXIFRAGACEAE

SAXIFRAGA L.

Saxifraga granulata L. C.T.W. 52.1.8. *Meadow Saxifrage*
Pastures Native
Frequent and appearing both on acid and alkaline soils. Recorded
for all neighbouring counties.
NENE: Hinwick; *W. Kitchin.*
OUSE: Lidlington; Bedford; *F.B.*: Cardington; *J.McL., B.M.*:
Millbrook; *W.H. 1875*: Roxton; *G. H. Day*: Cople; *R. Townsend*:
near rectory, Houghton Conquest!; *R. H. Goode.*
OUZEL: Rushmere; Shenley Hill.
IVEL: Cainhoe; *W.C. Herb.*: Alameda, Ampthill; *J.H.*: Steppingley;
C.C. in W.F.B.: Wake's End, Eversholt; *V. H. Chambers*: Southill
Park; Langford Bridge; *J.E.L., F.F.B.*: Ampthill Park; *C. C. Foss*:
Flitton, Flitwick Moor; Clophill; Potton; Maulden; Greenfield*.
LEA: Limbury; Marslets; *S.B.P.L.*: Luton Hoo; Blow's Downs!;
W.F.B.: Galley Hill.
First record: Abbot, 1798.

S. tridactylites L. C.T.W. 52.1.7. *Rue-leaved Saxifrage*
Old walls, dry soils, etc. Native
Uncommon, but well distributed: recorded for all neighbouring
counties.
NENE: old wall, Hinwick.
OUSE: Cardington; *J.McL., B.M.*: Moggerhanger; *W.F.S. Diary*:
old walls, Harrold, Biddenham; *G. H. Day*: Odell; *B. B. West*:
Oakley Church; Bromham Church; *A. W. Guppy*: Felmersham;
H. B. Souster: old walls, Milton Ernest, Bletsoe; *C. C. Foss*: Radwell
Bridge; Souldrop Church.
KYM: Swineshead Church; Knotting Church.
OUZEL: Leighton Buzzard Church.
IVEL: Ampthill; *J.S. in J. Bot. 1883*: Ridgmont; *W.C. Herb.*:
Sandy; *J.H. Notes*: Clophill; Rowney Warren, on sand!; *J.E.L.,
F.F.B.*: dry bank, Wilbury Hill; *R. Morse*: on glacial gravel, Ravens-
burgh Castle!; *H. B. Souster.*

Cam: old walls, Dunton, Eyeworth*.
Lea: Luton; *S.B.P.L.*: old wall, East Hyde*.
Colne: old wall, Studham.
First record: Abbot, 1798.

CHRYSOSPLENIUM L.

Chrysosplenium alternifolium L. C.T.W. 52.2.2.
Alternate-leaved Golden Saxifrage
Damp ditches   Native
Extinct: recorded for Hunts.
Ouse: Lidlington, where continued search has failed to reveal it
again; *W. C. Herb.* 1844.

C. oppositifolium L. C.T.W. 52.2.1. *Opposite-leaved Golden Saxifrage*
Wet ditches, marshy places, etc.   Native
Not infrequent by the sides of streams associated with the Lower
Greensand. Recorded for all neighbouring counties except Hunts.
H.S. 42.
Ouse: Lidlington, by a streamside!*; *W.H. 1875*: Holywell, Mill-
brook; *H. B. Souster*!
Ouzel: Aspley; *W.H. 1875*: New Wavendon Heath, v.c. 24 [Beds.]!;
P.T.
Ivel: 'ditch behind ye Church, Eversholt'!*; *Plantae Bedford.*:
Toddington†; *B.P.L.*: Flitwick Moor, scarce!; *M. and H. G.
Oldfield*: Maulden; *H. B. Souster*: plentiful in the lower parts of
Washer's Wood*; plentifully, Moors Plantation; ditches by the lake,
Tingrith Park.
First record: Abbot, 1795.

PARNASSIACEAE

PARNASSIA L.

Parnassia palustris L. C.T.W. 53.1.1.   *Grass of Parnassus*
Marshes, pastures, etc.   Native
Probably extinct. This species had an interesting distribution in the
county, as it grew not only in marshy places but also on the chalk
escarpment at Streatley. It grew here in Saunders's time with
Pinguicula vulgaris and *Carex pulicaris*, and I found it still there in
1917 but was unable to find either the butterwort or sedge. Simultaneously, and independently, Mrs. Boutwood who lived but a
short distance away, was also interested in its appearance there. We
each saw it every year until about 1925, when all the plants to be
found were uprooted by a misguided 'naturalist' in the hope that
he would be the last person to see it growing there. Recorded for
all neighbouring counties.
Ouse: Turvey; *Plantae Bedford.*: Stevington; *F.B.*

OUZEL: Totternhoe; *J.S., J. Bot. 1886*: it is interesting to note that Saunders did not include this station in later lists.

IVEL: 'at the bottome of Barton Hills'; *Parkinson, Theatrum Botanicum*, 429: Ampthill; *Plantae Bedford.*: Maulden Bogs; Gravenhurst Moor; *W.C. Herb.*: Markham Hills!†*; *J.S. in Bot. Rec. Club Rep. 1880.*

LEA: Limbury; Leagrave Marsh, where I have not seen it since 1919!†; *J.S. in Bot. Rec. Club Rep. 1880.*

First record: Parkinson, 1640.

GROSSULARIACEAE

RIBES L.

Ribes uva-crispa L. C.T.W. 56.1.5. *Gooseberry*
  R. grossularia L.
Woods, hedgerows, etc. Of garden origin
Rare: recorded for all neighbouring counties.
H.S. 5.
OUSE: Lidlington; *W.C. Herb.*: Cardington; *J.McL., B.M.*: Sheerhatch Wood; *J.E.L., F.F.B.*: Roxton; *G. H. Day*: Bromham; *A. W. Guppy*: Judge's Spinney.
KYM: railway sidings, Kimbolton, v.c. 30 [Hunts.]
OUZEL: Leighton Buzzard.
IVEL: Tripcock Wood; *S.B.P.L.*: Streatley, not infrequent in hedgerows!†*; *W.F.B.*: Toddington; Sundon; *H. B. Souster*: Greenfield; Barton, frequent in hedges remote from houses.
LEA: Luton; *V. H. Chambers*: Whitehill Wood; Arden Dell Wood.
COLNE: Caddington; *H. B. Souster*: derelict garden, Deadmansey Wood.
First record: W. Crouch, 1843.

R. nigrum L. C.T.W. 56.1.3. *Black Currant*
Sides of rivers Of garden origin
Rare: recorded for all neighbouring counties.
OUSE: 'squinancy berries—by the river at Blunham and elsewhere'; *Ray's Synopsis Ed. II*, 298; Fenlake; *J.McL., B.M.*: Turvey; Stevington; *B. B. West.*
OUZEL: riverside and sandpit, Leighton Buzzard.
IVEL: Flitwick, both by the river and on the Moor!*; *W.F.B.*: marsh at foot of Cooper's Hill; Ivel Navigation, Shefford.
LEA: Luton; *N. Spencer, Complete English Traveller*: Luton Hoo†; Leagrave; *J.S., J. Bot. 1889*: Limbury†; *Herb. J.S.*
First record: John Ray, 1690.

R. sylvestre Mert. & Koch C.T.W. 56.1.1. *Red Currant*
  R. rubrum L. nom. ambig.
Woods Of garden origin
Frequent, especially in woods in the marshy parts of the Lower Greensand. Recorded for all neighbouring counties.
H.S. 42, 43.
OUSE: Boughton Wood, Lidlington; *W.C. Herb.*: Turvey; Stevington, frequent in Woodcraft!; *B. B. West*: Holywell, Ridgmont; Pavenham.
OUZEL: near the lake, Battlesden; Rushmere.
IVEL: Southill Park; *J.McL., B.M.*: Flitwick Moor, especially common in Folly Wood!; *W.F.B.*: Cadwell; *J.E.L., Diary*: Maulden; Westoning Moor!; *H. B. Souster*: Toddington; Moors Plantation; marsh, Cooper's Hill.
LEA: Sundon; *Davis*: East Hyde; *A. M. Buck*: Luton Hoo†; *Herb. J.S.*: Copt Hall; *H. B. Souster.*
COLNE: Studham!†; *S.B.P.L.*: Folly Wood!; *H. B. Souster*: Kensworth Common; *L. J. Margetts*: Whipsnade Common*.
First record: W. Crouch, 1843.

DROSERACEAE

DROSERA L.

Drosera rotundifolia L. C.T.W. 57.1.1.   *Sundew*
Bogs   *Native*
Very rare: recorded for all neighbouring counties.
OUSE: Cleat Hill; *Mr. Roberts in V.C.H.*
IVEL: Ampthill Moor; *F.B.*: Maulden Common; *H. Brown*: Maulden Moor, probably the same station; *W.C. Herb.*: Flitwick Moor, where it increased in quantity after 1894 when the turf was removed to plant what is now Folly Wood; in the course of time it disappeared and continued search in recent years has failed to reveal it here, but in 1942 R. A. Palmer reported that he found one plant near Folly Wood†; *S.B.P.L.*
First record: Abbot, 1798.

D. anglica Huds. C.T.W. 57.1.2.   *Great Sundew*
Marshes   *Native*
Extinct: Abbot recorded both *D. anglica* and *D. longifolia* L. from Ampthill Bogs in *Plantae Bedford.* and *Flora Bedfordiensis*. The former contained a long note on the octogynous nature of the flowers of *D. anglica*, but Abbot added that he could see little difference between the two species except in the number of flowers in the spike. The original drawing of *D. anglica* for *E. Bot.* tab. 869, April 1801 has a note, 'among those sent by Rev. Mr. C. Abbot I found one as represented, none with 5 petals and 5 stamens'. Abbot's herbarium contains a correctly-named specimen of *D. anglica* but no specimen of his *D. longifolia*. It would appear that Abbot found *D. anglica*, but his record of *D. longifolia* (if he meant *D. intermedia* Hayne) is to be doubted. *D. anglica* has been recorded for Cambs. and Hunts. and *D. intermedia* for Bucks., Hunts. and Cambs.

LYTHRACEAE

LYTHRUM L.

Lythrum salicaria L. C.T.W. 59.1.1.   *Purple Loosestrife*
Riversides, marshy places, etc.   *Native*
Frequent over the greater part of the county, but not recorded for the Nene, Kym and Colne districts. Recorded for all neighbouring counties.
H.S. 6–8, 10c, 13, 14a, 15, 37, 39a, 53b.
First record: Abbot, 1798.

L. hyssopifolia L. C.T.W. 59.1.2.   *Grass Poly*
Grassy places, waste places, etc.   *? Native*
Possibly native in earlier times it now appears as a wool adventive only. *E. Bot.* tab. 292, Nov. 1795, was drawn from a specimen sent by Abbot. Recorded for all neighbouring counties.
OUSE: Oakley West Field; *Plantae Bedford.*: Goldington Green, where it was also found by McLaren; *F.B.*: 'Church Field'; *Abbot's annotated F.B.*
IVEL: railway siding, Flitwick*.
First record: Abbot, 1795.

PEPLIS L

Peplis portula L. C.T.W. 59.2.1.   *Water Purslane*
Ponds, wet rides in woods, etc.   *Native*
Rare: recorded for all neighbouring counties.

OUZEL: plentiful in cart ruts, King's Wood*; woodland rides, Stockgrove; ponds, Heath and Reach.
IVEL: Ampthill Moor, found also at Ampthill by C. Crouch*; *Plantae Bedford.*
LEA: Luton Hoo!†*; *J. Catt in J.S., J. Bot. 1889*: Chiltern Green, the pond in which it grew dried about 1942!†*; *W.F.B.*
COLNE: Studham Common; *W.F.B.*: pond, Kidney Wood*.
First record: Abbot, 1795.

THYMELAEACEAE

DAPHNE L.

Daphne laureola L. C.T.W. 60.1.2.   *Spurge Laurel*
Woods, hedgerows, etc.   Native
Frequent, especially on the Chalk and calcareous Boulder Clay.
Recorded for all districts and neighbouring counties.
H.S. 5, 33.
First record: Abbot, 1798.

ONAGRACEAE

CHAMAENERION Adans.

Chamaenerion angustifolium (L.) Scop. C.T.W. 62.3.1.
Epilobium angustifolium L.   *Rose-bay Willow-herb, Fireweed*
Cleared woodland, heaths, waste places, etc.   Colonist
Common: this attractive species was rare until comparatively recently. Abbot gave one station only, and Henry Brown's specimen came from Luton Park. W. Crouch collected a specimen from a garden and added, 'I have not found this plant in a wild state.' McLaren found it at Rowney Warren and Saunders gave only three stations for it. It is much more common now than it was 30 years ago, but I am unable to say when it began to increase in the county. Recorded for all districts and neighbouring counties.
H.S. 15, 22b, 25–27, 30a, 30b, 36, 38, 40, 46, 47, 49, 53a, 53b, 54, 64, 69, 70b, 77b, 79a–81.
First record: wood near Whipsnade, Abbot, 1795.

EPILOBIUM L.

All my gatherings of the more critical members of this genus have been identified by G. M. Ash, who also made one most profitable visit to the county in 1946.

Epilobium hirsutum L. C.T.W. 62.2.1. *Great Hairy Willow-herb,*
Codlins-and-Cream
Marshes, ditches, etc.   Native
Common: a white-flowered form has been observed at Wrestlingworth. Recorded for all districts and neighbouring counties.
H.S. 6, 7, 10c, 11, 13, 15, 16, 20a, 21, 34a, 39a, 43, 56a, 57a, 57b, 64, 84a–85.
First record: Abbot, 1798.

E. hirsutum × lamyi
OUSE: gravel pits, Eaton Socon*; *G. M. Ash*!
E. hirsutum × parviflorum
OUSE: gravel pits, Eaton Socon*; *G. M. Ash*!
E. hirsutum × tetragonum
OUSE: gravel pits, Eaton Socon*; *G. M. Ash*!

E. parviflorum Schreb. C.T.W. 62.2.2.
  E. villosum Curt.   *Small-flowered Hairy Willow-herb*
Marshes, wet ditches, etc.   Native
Common: recorded for all districts and neighbouring counties.
H.S. 9–10c, 14a, 16, 19a, 34a, 37, 51a, 52, 55b, 57a, 84a, 85.
First record: Abbot, 1798.
    E. parviflorum × roseum
KYM: Yelden*.
OUZEL: Salford; *G. C. Druce in B.E.C. 1936 Rep.*

E. tetragonum L.   *Square-stemmed Willow-herb*
  E. adnatum Griseb. C.T.W. 62.2.7.
Clearings in woods, gravel pits, etc.   Native
Well distributed, and probably more common than the records
indicate. Recorded for all neighbouring counties.
H.S. 16, 19a, 20a, 34a.
NENE: Wymington Scrub*.
OUSE: Bedford; Elstow; *Newbould MS.*: Putnoe Wood; *J.McL.*,
L.M.: gravel pit, Eaton Socon*; Wilstead Wood; Wootton Wood*;
Colesdon; gravel pit, Willington; Marston Thrift.
KYM: Melchbourne Park*.
OUZEL: Aspley Wood.
IVEL: Barton; clearing in wood, Fancott.
CAM: Dunton.
LEA: Stopsley.
COLNE: clearing, Deadmansey Wood.
First record: Abbot, 1798; but more certainly, *Newbould MS.*, 1873.

E. lamyi F. W. Schultz C.T.W. 62.2.8.
Woodland clearings, gravel pits, etc.   Native
Frequent: recorded for all neighbouring counties.
H.S. 16.
OUSE: gravel pits, Eaton Socon*; *G. M. Ash*!: gravel pit, Willington.
OUZEL: Aspley Wood; Heath and Reach.
IVEL: Warren Wood, Clophill*.
LEA: Arden Dell Wood.
COLNE: Deadmansey Wood*.
First record: the author in *B.E.C. 1943–4 Rep.* 808.
    E. lamyi × parviflorum
OUSE: gravel pits, Eaton Socon*; *G. M. Ash*!
    E. lamyi × tetragonum
OUSE: gravel pits, Eaton Socon*; *G. M. Ash*!

E. obscurum Schreb. C.T.W. 62.2.9.
Woodland clearings, etc.   Native
Frequent: recorded for all neighbouring counties.
H.S. 38, 43, 53a.

NENE: Wymington Scrub*.
OUSE: Elstow; *Newbould MS.*
OUZEL: Salford; *V.C.H.*: King's Wood; Rushmere*; Aspley Wood*;
Husborne Crawley*.
IVEL: Cooper's Hill*; Washer's Wood; Flitwick Moor.
LEA: Luton Hoo; *S.B.P.L.*: Limbury; *V.C.H.*
COLNE: Deadmansey Wood*; Whipsnade Common*; Ravensdell
Wood, v.c. 20 [Beds.]*.
First record: *Newbould MS.*, 1873.
   E. obscurum × palustre
IVEL: Flitwick Moor*.
   E. obscurum × roseum
OUZEL: Salford; *V.C.H.*

E. roseum Schreb. C.T.W. 26.2.5.   *Small-flowered Willow-herb*
Woodland clearings, damp places, etc.   Native
Infrequent: recorded for all neighbouring counties except Hunts.
OUSE: Candle Ford End; *Newbould MS.*
OUZEL: Salford; *V.C.H.*: Billington*; *V. H. Chambers*: Aspley
Wood*; Husborne Crawley*.
IVEL: Greenfield Mill!*; Harlington; Flitwick Moor; *V. H.*
Chambers: Clophill*.
LEA: Dunstable; *W. D. Coales*: Luton Hoo*.
First record: *Newbould MS.*, 1873.

E. lanceolatum Seb. & Mauri C.T.W. 62.2.4.
                                           Spear-leaved Willow-herb
Woodland clearings   Native
Rare: early records by C. Crouch were in error. Recorded for Bucks.
OUZEL: Aspley Wood!*; *V. H. Chambers in B.E.C. 1943–4 Rep. 808.*

E. adenocaulon Hausskn. C.T.W. 62.2.6.
Woodland clearings, gardens, etc.   Colonist
A recent introduction that has become one of our commonest willow-herbs.
Recorded for all neighbouring counties.
H.S. 16, 36, 44, 53a, 54.
NENE: Hinwick; Podington.
OUSE: gravel pits, Eaton Socon; *G. M. Ash*!; gravel pit, Willington.
KYM: railway sidings, Kimbolton, v.c. 30 [Hunts.]; *E.M.-R.*!
OUZEL: Aspley Wood*; King's Wood; Well Head; Rushmere; Heath and Reach.
IVEL: Fancott; *A. J. Wilmott*!: Tingrith; *E.M.-R.*: Cooper's Hill*; *V. H. Chambers*!:
Washer's Wood; Daintry Wood.
LEA: Kidney Wood.
COLNE: Deadmansey Wood*; Ravensdell Wood, v.c. 20 [Beds.]*.
First record: V. H. Chambers and the author in *B.E.C. 1943–4 Rep.* 724.
   E. adenocaulon × hirsutum
OUSE: gravel pits, Eaton Socon!; *N. Y. Sandwith.*
   E. adenocaulon × montanum
OUSE: Marston Thrift*.
OUZEL: King's Wood; *E.M.-R.*: Aspley Wood.
COLNE: Deadmansey Wood*.
   E. adenocaulon × obscurum
KYM: West Wood*.
IVEL: Washer's Wood (H.S. 53a)*.
   E. adenocaulon × parviflorum
OUSE: gravel pit, Eaton Socon*; *G. M. Ash*!: gravel pit, Willington*.

E. montanum L. C.T.W. 62.2.3.   *Broad-leaved Willow-herb*
Woods, shady places, etc.   Native
Common: recorded for all districts and neighbouring counties.
H.S. 3a, 18, 20a, 20b, 53a, 54, 55b, 69, 78, 79b, 83, [21].
First record: Abbot, 1798.
  E. montanum × obscurum
Ouse: Marston Thrift*.
Ouzel: King's Wood*; *J. P. M. Brenan*: Aspley Wood.
  E. montanum × parviflorum
Kym: Knotting.
Ouzel: Aspley Wood*.

E. palustre L. C.T.W. 62.2.10.   *Marsh Willow-herb*
Marshes   Native
Rare: recorded for all neighbouring counties.
H.S. 38, 39a, 41.
Ouse: Stevington Bogs; *F.B.*: Turvey; *W.H. 1875*: Bedford;
M. Brown in W.F.S. Diary.
Ouzel: Stockgrove*; *E.M.-R.*!
Ivel: Ampthill Bogs, where it was also found by W. Crouch; *F.B.*:
Flitwick, plentiful on the Moor!*; *J.McL., L.M.*: Harlington;
W.F.B.: Westoning Moor.
Lea: Limbury; *S.B.P.L.*
First record: Abbot, 1798.

OENOTHERA L.

Oenothera biennis L. C.T.W. 62.4.1.   *Evening Primrose*
Waste places, etc.   Colonist
Not infrequent, and usually well established. Recorded for all neighbouring
counties.
Ouzel: waste ground, Heath and Reach.
Ivel: rubbish dump, Sundon*; waste ground, Clophill.
Lea: railway sidings, Luton.

CIRCAEA L.

Circaea lutetiana L. C.T.W. 62.6.1.   *Enchanter's Nightshade*
Woods, etc.   Native
Common in shady places on moist soils throughout the county.
Recorded for all districts and neighbouring counties.
H.S. 3b, 18, 19b, 20a, 20b, 34b, 42, 43, 51a, 52, 53b, 55a, 55b, 83.
First record: Abbot, 1798.

HALORAGACEAE

MYRIOPHYLLUM L.

Myriophyllum verticillatum L. C.T.W. 63.1.1.
                                            Whorled Water-milfoil
Ponds, riversides, etc.   Native
Rare: a record by Saunders from Pepperstock in *S.B.P.L.* was in
error for *M. alterniflorum*. Recorded for all neighbouring counties.
H.S. 7.

Ouse: 'ditch below Bedford and Fenlake'; *Plantae Bedford.*: Fenlake; *J.McL.*, *L.M.*: Goldington†*; *J.S. in Bot. Rec. Club 1883 Rep.*: Oakley Bottom*; *C. C. Foss*!
IVEL: Ridgmont; *W.C. Herb.*: Cainhoe; Silsoe; *C.C.*: Park Wood, Old Warden; *J.E.L.*, *F.F.B.*
First record: Abbot, 1795.

M. spicatum L. C.T.W. 63.1.2.   *Spiked Water-milfoil*
Ponds, lakes, gravel pits, etc.   Native
Frequent: recorded for all neighbouring counties.
H.S. 6, 21.
Ouse: Ford End; Castle Mills; *F.B.*: Bedford!; *W.H. 1875*: Goldington†; *Herb. J.S.*: river, Pavenham; gravel pits, Eaton Socon*; brickpit, Brogborough; Felmersham.
Kym: lake, Melchbourne Park.
Ouzel: lake, Battlesden Park!*; *W.F.B.*: sandpit, Leighton Buzzard*; lakes, Woburn Park.
IVEL: gault pit, Cainhoe; *W.C. Herb.*: lake, Southill Park!; *J.McL.*, *L.M.*: between Shefford and Biggleswade; *J.S.*, *J. Bot. 1889*: Old Warden!; *J.E.L.*, *Diary.*
Lea: The Hyde; *R. A. Pryor in J. Bot. 1876*: Limbury†; *S.B.P.L.*
First record: Abbot, 1798.

M. alterniflorum DC. C.T.W. 63.1.3. *Alternate-flowered Water-milfoil*
Ponds   Native
Very rare or extinct: a record by Saunders from Limbury in *B.P.L.* was in error for *M. spicatum.* Recorded for all neighbouring counties.
Colne: Pepperstock†*, where I have looked for it in vain; *J.S. in J. Bot. 1889.*

HIPPURIDACEAE

HIPPURIS L.

Hippuris vulgaris L. C.T.W. 64.1.1.   *Mare's-tail*
Ponds, sides of lakes, etc.   Native
Rare: but usually dominating a considerable area when it appears. Recorded for all neighbouring counties.
Nene: pond, Hinwick.
Ouse: Bedford!; *Plantae Bedford.*: Turvey; *J.McL.*, *B.M.*: Harrold; *J.McL. in J.H. Notes*: Bromham!*; *B. B. West.*
IVEL: Shefford; *J.S.*, *J. Bot.*, *1889*: Stotfold Mill.
Lea: Luton Hoo†; *Davis.*
First record: Abbot, 1795.

CALLITRICHACEAE

CALLITRICHE L.

Callitriche stagnalis Scop. C.T.W. 65.1.1.   *Water Starwort*
Ponds, ditches, muddy paths in woods, etc.   Native
Common: although there are various records for *C. palustris*, *C.*

verna, C. vernalis and *C. obtusangula*, I have examined much herbarium and living material but cannot distinguish these among our broad-leaved water starworts. Recorded for all districts and neighbouring counties.

H.S. 7–9, 11, 14b, 34a, 53b, 84b.

First record: Abbot, 1798, as *C. verna*, common.

C. intermedia Hoffm. C.T.W. 65.1.5.

C. hamulata Kütz ex Koch, *C. autumnalis* auct.

Deep ponds, backwaters, etc.   Native

Common in backwaters and slow-flowing parts of the Ouse and frequent in suitable places elsewhere. Recorded for all neighbouring counties.

H.S. 7.

NENE: Hinwick*.

OUSE: common in backwaters of the river, gravel pits, etc.

KYM: pond, Riseley*.

OUZEL: pond, Heath and Reach.

IVEL: Clophill; *W.C. Herb.*

LEA: lake, Wardown Park*.

First record: Abbot, 1798.

LORANTHACEAE

VISCUM L.

Viscum album L. C.T.W. 61.1.1.   *Mistletoe*

Parasite on trees   Native

Uncommon and probably decreasing. Abbot considered it to be common and Saunders noted that it was frequent in the north and middle of the county. Charles Crouch observed that it grew on apple, hawthorn, maple, lime and black poplar. Recorded for all neighbouring counties except Hunts.

OUSE: Ravensden; Bromham; *W.H. 1875*: Odell*; Harrold; *G. H. Day*: Biddenham; *R. Townsend*: Thurleigh; *B. B. West*: Houghton Conquest Rectory!; *R. H. Goode.*

IVEL: near Pegsdon; *H. Brown*: Cainhoe; *W.C. Herb.*: Ampthill; *J.McL., B.M.*: Chicksands; *C.C.*: Silsoe Park, plentiful!; *W.F.B.*: between Silsoe and Clophill!; *H. B. Souster*: Maulden.

First record: Abbot, 1798.

Thesium humifusum DC. (SANTALACEAE), C.T.W. 67.1.1, *Bastard Toadflax*, was listed for the county in *Topographical Botany* on the authority of a specimen sent by Elizabeth Twining. The specimen is not in Watson's herbarium and I have no evidence of Elizabeth Twining's having visited the county. There is, however, a specimen in Herb. Watson collected by her at Royston, v.c. 20 Herts. It is possible that a slip of the pen caused Watson to list it for v.c. 30. Recorded for Herts., Bucks. and Cambs.

CORNACEAE

THELYCRANIA (Dum.) Fourr.

Thelycrania sanguinea (L.) Fourr.   *Dogwood*
 Cornus sanguinea L. C.T.W. 68.1.1.
Woods, hedgerows, etc.   Native
Common, especially so on calcareous soils. Recorded for all districts
and neighbouring counties.
H.S. 2a, 3b, 5, 18, 20b, 33, 34a, 34b, 51, 80.
First record: Abbot, 1798.

ARALIACEAE

HEDERA L.

Hedera helix L. C.T.W. 69.1.1.   *Ivy*
Woods, hedgerows, etc.   Native
Common: recorded for all districts and neighbouring counties.
H.S. 5, 32, 33, 51, 55a, 63a, 70a, 70b.
First record: Abbot, 1798.

UMBELLIFERAE

HYDROCOTYLE L.

Hydrocotyle vulgaris L. C.T.W. 70.1.1.   *Pennywort*
Marshes   Native
Rare: limited to marshy places on acid soils, and usually locally
abundant when it appears. Recorded for all neighbouring
counties.
H.S. 15, 35, 38, 40, 41.
OUZEL: Aspley; *F.B.*: Hockliffe, on hillside marsh!*; *E.M.-R.*: New
Wavendon Heath, v.c. 24 [Beds.].
IVEL: Ampthill; *F.B.*: Gravenhurst Moor; *W.C. Herb.*: Potton;
J.McL., *L.M.*: Flitwick Moor!†*; *S.B.P.L.*: Westoning Moor!;
W.F.B.: swamp between Sandy Warren and Biggleswade Common!;
J.E.L., *F.F.B.*: Flitton Moor!; *V. H. Chambers*: marsh, Long Lane,
Tingrith; *H. B. Souster*.
First record: Abbot, 1798.

SANICULA L.

Sanicula europaea L. C.T.W. 71.1.1.   *Wood Sanicle*
Woods   Native
Common: especially so in beech woods on the Chalk. Recorded for
all districts and neighbouring counties.
H.S. 5, 33, 69, 70a, 79b, 80b.
First record: Abbot, 1798.

T

CONIUM L.
Conium maculatum L. C.T.W. 71.13.1.   *Hemlock*
Damp places, waste ground, etc.   Native
Common in suitable habitats in the north and middle of the county.
It is dominant over the greater part of Sundon Rubbish Dump.
Recorded for all districts and neighbouring counties.
H.S. 47.
First record: Abbot, 1798.

SMYRNIUM L.
Smyrnium olusatrum L. C.T.W. 71.11.1.   *Alexanders*
Roadsides   Denizen
Rare, but usually long established. Recorded for all neighbouring
counties except Northants.
OUSE: Elstow!; Ravensden; *Plantae Bedford.*: Oakley; *F.B.*:
Wilshamstead†; *Herb. J.S.*: Bedford; *A. W. Guppy*: Stevington;
H. B. Souster: Milton Ernest; Radwell.
IVEL: Gravenhurst; *W.C. Herb.*: roadside, south of Silsoe!†*; *C.C.*:
Topler's Hill in some quantity*.
CAM: sparingly, Millow.
First record: Abbot, 1795.

BUPLEURUM L.
Bupleurum rotundifolium L. C.T.W. 71.14.2.   *Hare's-ear*
Arable fields, waste places, etc.   Colonist
Rare, but often appearing in the same field for a number of years.
Recorded for all neighbouring counties.
OUSE: Clapham; *J.McL., L.M.*: Bromham, *H. J. Heathcote in J.H.
Notes*: Bolnhurst; *I. Evans in J.H. Notes*: Bedford; *J.H. Notes.*
OUZEL: Woburn; *F.B.*
IVEL: Barton, where it was also found by Saunders†; *F.B.*: between
Pegsdon and Shillington; *Dame M. Tuke in J.E.L., F.F.B.*: Sundon
Rubbish Dump; *P.T.*!
CAM: Cockayne Hatley, where it has since been re-found by T.L.;
J.E.L., F.F.B.: Millowbury Farm, Dunton*; *H. and D. Meyer*!
LEA: Dunstable Road, ? Luton; *H. Brown*: Luton; *D. M. Higgins
in W.B.E.C. 1896–7 Rep.*: Maiden Common†; *Herb. J.S.*: waste
ground, Waller Street, Luton!; *H. Cole.*
First record: Abbot, 1798.

APIUM L.
Apium graveolens L. C.T.W. 71.16.1.   *Wild Celery*
Wet ditches   Native
Rare: recorded for all neighbouring counties.
OUSE: Medbury; Wilstead; Goldington; *F.B.*: Biddenham; *Abbot's
annotated F.B.*: ditches between Cotton End and Elstow; *J.McL.,
L.M.*: gravel pit, Wyboston; ditch, Elstow*.
IVEL: Shillington; *W.C. Herb.*: by the Mill, Barton†; *C.C.*: Campton;
J.E.L., F.F.B.: ditches, Flitwick Moor*.
First record: Abbot, 1798.

A. nodiflorum (L.) Reichb. f. C.T.W. 71.16.2.   *Fool's Water-cress*
  Sium nodiflorum L.
Marshes, sides of ponds, ditches, etc.   Native
Common, but variable. Small forms growing on mud are frequent,
and have been wrongly named *A. repens, e.g.* by McLaren. They
compare better with var. *longipedunculatum* F. W. Schultz, but merit
study. Recorded for all districts except Colne, and for all neighbour-
ing counties.
H.S. 7, 9, 10b, 10c, 13, 14a, 14b, 57a, 84b.
First record: Abbot, 1798.

A. inundatum (L.) Reichb. f. C.T.W. 71.16.4.   *Water Honewort*
  Sison inundatum L.
Ponds, ditches, etc.   Native
Rare: recorded for all neighbouring counties.
OUSE: stream, Tempsford, a welcome discovery of a species thought
to be extinct in the county!*; *I. J. Allison.*
IVEL: Ampthill; *Plantae Bedford.*: Clophill; *W.C. Herb.*
COLNE: Pepperstock†*; *J.S. in Bot. Rec. Club 1883 Rep.*
First record: Abbot, 1795.

CICUTA L.

Cicuta virosa L. C.T.W. 71.19.1.   *Cowbane*
Wet places   Native
Extinct: recorded for Herts., Bucks. and Cambs.
OUSE: Oakley Springs; *F.B.*

CARUM L.

Carum carvi L. C.T.W. 71.22.2.   *Caraway*
Meadows, waste places, etc.   Native
Rare: occurs as a native species in damp meadows, and as an
introduced species on railway banks and in waste places. *E. Bot.*
tab. 1503, Sept. 1805, was drawn from a specimen sent by Abbot
from Thurleigh, where he said it had grown for half-a-century.
Recorded for Herts., Bucks. and Northants.
OUSE: Thurleigh; *Plantae Bedford.*: Twin Wood; *B.P.L.*
OUZEL: meadow, Totternhoe, where it is plentiful and where the
seeds were collected by villagers until recently!†*; *W.F.B.*
IVEL: meadow, Gravenhurst, where it was also found by C. Crouch;
W.C. Herb.: railway bank, Flitwick†; *W.F.B.*: rubbish dump,
Sundon*.
First record: Abbot, 1795.

PETROSELINUM Hill

Petroselinum segetum (L.) Koch C.T.W. 71.17.2.
                                        Corn Parsley, Corn Caraway
  Sison segetum L., *Carum segetum* (L.) Benth. ex Hook. f.
Arable fields, banks, etc.   Colonist
Infrequent, but usually appearing regularly in the same fields.
Recorded for all neighbouring counties.
H.S. 86.

OUSE: Bedford; *Plantae Bedford.*: Goldington; Clapham; *F.B.*:
Cotton End; *M. L. Berrill in J.H. Notes*: roadside, Oakley; *E.M.-R.*!
OUZEL: Houghton Regis; *H. B. Sargent.*
IVEL: Gravenhurst; *W.C. Herb.*: Maulden; *M. L. Berrill in J.H.
Notes*: Barton Hill†; *W.F.B.*: Kitchen End; *C.C.*: arable field,
Wilbury!; *V.C.H.*: Shefford; *J.E.L., J. Bot.*: arable field, Streatley*.
LEA: railway bank, Leagrave!; *M. Brown*: garden weed, Luton!; *P.T.*:
arable fields, Limbury*, East Hyde, where it is locally abundant*.
First record: Abbot, 1795.

BUNIUM L.

Bunium bulbocastanum L. C.T.W. 71.23.1.   *Great Earth-nut*
   Carum bulbocastanum (L.) Koch
Chalky fields, etc.   ? Native
The most interesting Bedfordshire plant species. It is limited to a
comparatively small stretch of the Chalk from Cherry Hinton in
Cambridgeshire, where it was first discovered in Britain in 1835 by
W. H. Coleman, to just beyond Ivinghoe in Buckinghamshire. In
1840 Coleman found it in Bedfordshire and Hertfordshire, but it was
not until 1876 that its full distribution was studied by Pryor. He
found it to be abundant on the Chalk in South Bedfordshire, and
it is no less abundant now. It is most common in arable fields,
especially where cultivation has ceased, as has been too frequently
the case around Dunstable and Luton. It appears also on chalk
downland, but there is always the suspicion in these cases that the
soil has been disturbed at some time. It is very rare or absent on
well-established downland. In arable land it grows most frequently
in association with *Sinapis alba*, and on rough or broken downland
with *Anthyllis vulneraria*. The study of *Bunium* presents many problems. It occurs on the Continent, apparently growing there in
habitats similar to those in which it flourishes in Bedfordshire.
If at one time it had a wider British distribution it is difficult to
explain why it is now so limited. Agriculture on the North and South
Downs cannot have been very different from or more intensive
than it has been on the Chilterns. E. Milne-Redhead has drawn
my attention to the fact that its fruits ripen a few days before the
corn ripens; possibly this might not have been so further south. In
any case *Bunium*, when growing in arable land, depends only partly
on new plants being formed, as the tubers survive ploughing. An
examination of plants found in cornfields shows that a large proportion are in a reversed or side position. In arable reverting to pasture
it becomes the dominant species in the second year and remains so
for a number of years. My notes contain about fifty stations on the
Bedfordshire Chalk and Chalk Marl, where it also grows but less
abundantly. For convenience these have been mapped. Recorded
for the Ouzel, Ivel, Lea and Colne districts, and for Herts., Bucks.
and Cambs.
H.S. 66b, 67, 72, 73.
First record: W. H. Coleman in *E. Bot. Supplement*, Nov. 1841.

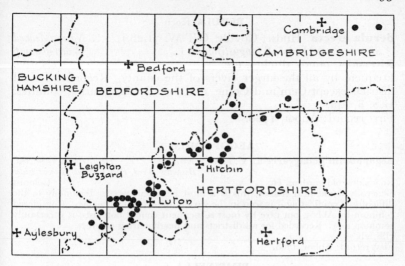

FIG. 20   DISTRIBUTION OF BUNIUM BULBOCASTANUM

SISON L.

Sison amomum L. C.T.W. 71.18.1.   *Stone Parsley*
Roadsides, rough pastures, etc.   Native
Common by roadsides in the north of the county, but less common
on the chalky soils of the south. Recorded for all districts and for all
neighbouring counties.
First record: Abbot, 1798.

FALCARIA Bernh.

Falcaria vulgaris Bernh. C.T.W. 71.21.1.
Prionitis falcaria (L.) Dum.
This has been known for many years in a field at Harrold* from whence it was
recorded in Hamson's Notes. The field was ploughed in 1942 and Miss Day, who
has taken a special interest in it, reports that the *Falcaria* still survives in the crop
and on the edge of the field. Its origin here is not known, but it is possible that it
may be a survival from the herb garden of the old Harrold Nunnery.

SIUM L.

Sium latifolium L. C.T.W. 71.27.1.   *Water-parsnip*
Riversides   Native
Common by the Ouse, especially above Bedford. Recorded for all
neighbouring counties except Herts.
H.S. 9.
OUSE: too many stations to merit listing.
IVEL: Stotfold, plentiful over a limited stretch of river.
First record: Abbot, 1798.

BERULA Koch

Berula erecta (Huds.) Coville C.T.W. 71.28.1.   *Narrow-leaved*
Sium erectum Huds., *S. angustifolium* L.   *Water-parsnip*
Riversides, ponds, ditches, etc.   Native
Frequent by all the larger rivers of the county. Recorded for all
districts except Cam and Colne, and for all neighbouring counties.
H.S. 8, 14a.
First record: Abbot, 1798.

AEGOPODIUM L.

Aegopodium podagraria L. C.T.W. 71.26.1.   *Herb Gerard, Goutweed,*
                                   Bishop's Weed, Pope's Weed, Ground Elder
Roadsides, old gardens, etc.   Colonist
Generally found near houses, this is a most persistent weed. Testimony to the
difficulty of eradicating it is seen in the variety of its common names. Considered
common by Abbot, but rare by most subsequent county botanists, it is certainly
common now. Recorded for all districts and neighbouring counties.
H.S. 2b.
First record: Abbot, 1798.

PIMPINELLA L.

Pimpinella saxifraga L. C.T.W. 71.25.1.   *Burnet-saxifrage*
Pastures, railway banks, chalk downs, etc.   Native
Common, especially on the Chalk. It is a very variable species and
pink-flowered forms are not uncommon. Var. **dissecta** Spreng. is
frequent. Recorded for all districts and neighbouring counties.
H.S. 1b, 10b, 45b, 59, 60, 62–63b, 65, 66a, 67, 74a, 75.
First record: Abbot, 1798.

P. major (L.) Huds. C.T.W. 71.25.2.   *Greater Burnet-saxifrage*
P. magna L.
Hedgerows, sides of woods, etc.   Native
Frequent, and thriving best on clay soils. It makes a very fine show
on the Clay-with-Flints around Caddington and Kensworth. It is
as variable as *P. saxifraga*, but the variants are less common with us.
The handsome var. **rubra** (Mérat) Fiori & Paoletti has been seen
in rough meadow-land at Souldrop*. I have not seen var. **bipin-
nata** (G. Beck) Burnat in the county, but there are good specimens
from Clapham Wood in Herb. J. McLaren, Luton Museum, and
from Bramingham† and Flitwick Wood† in Herb. Saunders.
Recorded for all districts and neighbouring counties.
H.S. 55b.
First record: 'in Bedfordshire woods', Ray's *Synopsis Ed. II*, 1696.

CONOPODIUM Koch

Conopodium majus (Gouan) Loret C.T.W. 71.24.1.
Bunium flexuosum Stokes, *C. denudatum* Koch   *Pignut, Earth-nut*
Woods, heaths, pastures, waste places, etc.   Native
Common and growing in a great variety of habitats. It is usually a
feature of well-manured waste places near farmhouses. Considered

rare by Abbot it is now recorded for all districts except Cam, and for all neighbouring counties.
H.S. 33, 38, 51a, 78, 83.
First record: Abbot, 1798.

CHAEROPHYLLUM L.

Chaerophyllum temulum L. C.T.W. 71.4.1.   *Rough Chervil*
C. temulentum L.
Roadsides, edges of woods, etc.   Native
Common: recorded for all districts and neighbouring counties.
H.S. 25, 27.
First record: Abbot, 1798.

SCANDIX L.

Scandix pecten-veneris L. C.T.W. 71.6.1.   *Shepherd's Needle,*
                                            Crow Needle (Batchelor)
Arable fields   Colonist
Common and well distributed. Recorded for all districts and neighbouring counties.
H.S. 4, 72, 73.
First record: Abbot, 1798.

ANTHRISCUS Pers.

Anthriscus caucalis Bieb.   *Bur Chervil*
Scandix anthriscus L., *A. neglecta* Boiss. & Reut. C.T.W. 71.5.1,
Cerefolium anthriscus (L.) Druce, *A. scandicina* Mansf.
Roadsides, dry open places, etc.   Native
Frequent on the Lower Greensand and also occurring sparingly on gravels. Recorded for all neighbouring counties.
OUSE: Bedford; *J.H. Notes*: Goldington; *B. B. West*: Elstow; *C. C. Foss*: Honey Hills; Willington.
OUZEL: Heath and Reach!†*; *S.B.P.L.*: Potsgrove.
IVEL: Clophill; *H. Brown*: Flitwick, locally abundant!; *W.F.B.*: Ampthill!; *V.C.H.*: Sandy!; Wilbury Hill; Rowney Warren; *J.E.L.*, *F.F.B.*: Maulden!; *J.E.L.*, *Diary*: Steppingley; Tingrith; *H. B. Souster*: Potton; rubbish dump, Sundon, where it appears regularly; disused gravel pit, Meppershall.
First record: Abbot, 1798.

A. sylvestris (L.) Hoffm. C.T.W. 71.5.2.   *Common Chervil, Keck,*
Chaerophyllum sylvestre L., *Cerefolium sylvestre* (L.) Bess. *Cow Parsley*
Roadsides, woods, etc.   Native
Common by roadsides. It grows less frequently in woods, but is the dominant species in elm woods at Tilsworth. Recorded for all districts and neighbouring counties.
H.S. 13, 32.
First record: Abbot, 1798.

SESELI L.

Seseli libanotis (L.) Koch C.T.W. 71.30.1.
Chalk hills   Native
Very rare: one of the more interesting Bedfordshire species. Apart
from its Bedfordshire station, where it grows in association with
Hypochoeris maculata, it is known only from Hertfordshire, Cambridgeshire and East Sussex. Since it was first found by Richard Morse at
Pegsdon it has increased considerably, especially since grazing
ceased on the hill, and at the time of writing it appears to be still
increasing. The colony shows some variability in the number of
flowering heads on the plants, and also in the form of the 'umbels',
of which flattened forms appear to be almost as common as convex
ones. Delicate pink-flowered forms also appear. There has been
much supposition as to the station for Hudson's record: 'inter St.
Albans et Stoney Stratford', in *Flora Anglica*, 1762. It would most
likely have been in the neighbourhood of Dunstable, but repeated
search has failed to locate it there.
H.S. 59.
IVEL: Pegsdon!*; *R. Morse in W.B.E.C. 1914–15 Rep.* 497.

OENANTHE L.

Oenanthe fistulosa L. C.T.W. 71.31.1.   *Water-dropwort*
Pond-sides, water-meadows, etc.   Native
Frequent: recorded for all neighbouring counties.
H.S. 9. 14b.
OUSE: Bromham; Goldington; Newnham; *Abbot's annotated F.B.*:
Pavenham; *B. B. West*: Harrold; *G. H. Day*: Bedford; *C. C. Foss*:
riverside, Radwell, Tempsford, Eaton Socon*; pond, Marston
Mortaine; Felmersham; Lidlington*.
OUZEL: Northall†; *W.F.B.*: ponds, Heath and Reach.
IVEL: Gravenhurst; *W.C. Herb.*: Flitwick†*; *S.B.P.L.*: Westoning;
W.F.B.: Harlington; *F.F.B.*: ponds, Long Lane, Tingrith*.
First record: Abbot, 1798.

O. silaifolia Bieb. C.T.W. 71.31.3.
   O. peucedanifolia auct.
Wet meadows   Native
Very rare or extinct: *E. Bot.* tab. 348, Sept. 1796, was drawn from
material sent by Abbot, whose herbarium contains a correctly-
named specimen. Recorded for all neighbouring counties except
Herts.
OUSE: Fenlake, Herb. J. McLaren, British Museum, contains a
specimen from Fenlake named *O. fistulosa* but determined as *O.
silaifolia* by A. J. Wilmott—a specimen in J.McL., L.M. named
O. pimpinelloides, is also this; *F.B.*
IVEL: Potton Marshes; *F.B.*
First record: Abbot, 1796.

O. lachenalii C. C. Gmel. C.T.W. 71.31.4. *Parsley Water-dropwort*
Wet meadows Native
Rare: it is interesting to note that a species usually found in salt marshes appears so far inland. Recorded for all neighbouring counties except Bucks.
H.S. 56b.
OUZEL: Cow Common*.
IVEL: meadow, Fancott*.
First record: *Hillhouse List, 1876*, with no details.

O. crocata L. C.T.W. 71.35.5. *Hemlock Water-dropwort*
Riversides ? Native
Extinct: recorded for all neighbouring counties except Cambs.
LEA: Luton Hoo Park, where I have looked for it in vain; *W.F.B.*
First record: *Hillhouse List, 1876*, with no details.

O. aquatica (L.) Poir. C.T.W. 71.31.6. *Fine-leaved Water-dropwort*
Phellandrium aquaticum L.
Abbot listed this as common and gave Bromham and Newnham as stations in his annotated copy of *F.B.* His herbarium contains a correctly-named specimen, which, however, may not have come from the county. Herb. McLaren, British Museum has an apparently correctly-named specimen from Fenlake. It is possible that the true *O. aquatica* may still occur in the county. Recorded for all neighbouring counties.

O. fluviatilis (Bab.) Colem. C.T.W. 71.31.7.
Rivers Native
Frequent in the larger rivers. Recorded for all neighbouring counties.
H.S. 7.
OUSE: R. Ouse as *O. phellandrium* Lam.; *J.McL., L.M.*: above Bedford; *R. A. Pryor in J. Bot. 1875*, 212: river, Tempsford, Oakley; backwater, Cardington Mill*.
OUZEL: near Heath and Reach†*; *S.B.P.L.*
IVEL: between Shefford and Biggleswade, it is frequent in the Ivel Navigation!†*; *J.S., J. Bot., 1889*: Sandy; *W.F.B.*: Southill*.
First record: R. A. Pryor, 1875.
First evidence: J. McLaren, 1864.

O. pimpinelloides L., C.T.W. 71.31.2, was recorded by Abbot in a letter to J. E. Smith, 3 Sept. 1802.

AETHUSA L.

Aethusa cynapium L. C.T.W. 71.32.1. *Fool's Parsley*
Arable fields, waste places, etc. Colonist
Common, and somewhat variable in the size of the individual plants. Recorded for all districts and neighbouring counties.
H.S. 4, 71, 72.
First record: Abbot, 1798.

FOENICULUM Mill.

Foeniculum vulgare Mill. C.T.W. 71.33.1. *Fennel*
Anethum foeniculum L., *F. foeniculum* (L.) Karst.
Waste places, railway banks, etc. Colonist
Frequent, and usually well established. Recorded for all neighbouring counties.
OUSE: Bedford; *L. W. Wilson*: Ravensden; *B. B. West*: Harrold; *G. H. Day*.
OUZEL: Leighton Buzzard; *W. D. Coales*.
IVEL: Eversholt; *F.B.*: Clophill; Shefford; *J.McL., B.M.*: Flitton; *J.McL. in J.H.
Notes*: Harlington, plentiful near station!; *S.B.P.L.*: Sandy!; *J.E.L., F.F.B.*:
Sundon Rubbish Dump*.
LEA: Luton, in quantity near Midland Road Station!†*; *S.B.P.L.*
First record: Abbot, 1798.

SILAUM Mill.

Silaum silaus (L.) Schinz & Thell. C.T.W. 71.34.1.
Sulphur Wort, Pepper Saxifrage
Peucedanum silaus L., *Silaus flavescens* Bernh.
Meadows, roadsides, railway banks, etc. Native
Common on basic clays, the Chalk Marl and Boulder Clay, but
absent on the Lower Greensand and Chalk. It is very rare in the
south of the county, from which it is only recorded at Bramingham†;
F.F.B. Recorded for all districts except Colne, and for all neigh-
bouring counties.
H.S. 1b, 21, 45b, 56b.
First record: Abbot, 1798.

ANGELICA L.

Angelica sylvestris L. C.T.W. 71.38.1. *Wild Angelica*
Marshes, wet meadows, woods, etc. Native
Common in the greater part of the county, but very rare in the
south, where suitable habitats do not exist. Recorded for all dis-
tricts and neighbouring counties.
H.S. 2a, 2b, 7, 10b, 10c, 13, 15, 34a, 38, 39a, 41, 43, 51a, 55b.
First record: Abbot, 1798.

PASTINACA L.

Pastinaca sativa L. C.T.W. 71.40.1. *Wild Parsnip*
P. sylvestris Mill., *Peucedanum sativum* (L.) Benth. ex Hook. f.
Rough pastures, roadsides, railway banks, etc. ? Native
Probably native on the Chalk, but it never occurs on well-established
downland. Otherwise it is common throughout the county as an
escape from cultivation. It would be difficult to single out the
stations in which it could be native. Recorded for all districts and
neighbouring counties.
H.S. 58b, 63a, 68.
First record: Abbot, 1798.

HERACLEUM L.

Heracleum sphondylium L. C.T.W. 71.41.1.
Hogweed, Cow Mumble, Cow Parsnip
Sides of woods, roadsides, etc. ? Native
Common: the var. **angustifolium** Huds., which was observed by
Abbot, is frequent and has been recorded for all districts except

Cam. The common var. **sphondylium** is recorded for all districts and neighbouring counties.
H.S. 1b–2b, 15, 18, 27, 32, 45b, 51a, 53a, 63a, 64, 70b, 76, 83, 85.
First record: Abbot, 1798.

A large *Heracleum* appears regularly as a garden escape on roadsides near Woburn and Ridgmont. The plants are cut down before they flower and it is impossible to state to which species they belong.

DAUCUS L.

Daucus carota L. C.T.W. 71.43.1.   *Wild Carrot*
Rough pastures, railway banks, waste places, etc.   ? Native
Common, especially on disturbed calcareous soils. It has every appearance of being native on the chalk downs but is a colonist elsewhere. Recorded for all districts and neighbouring counties.
H.S. 1a, 1b, 4, 21, 45a, 45b, 57a, 58b, 62, 63b, 64, 66a–67, 72, 73, 76.
First record: Abbot, 1798.

TORILIS Adans.

Torilis arvensis (Huds.) Link C.T.W. 71.8.2.   *Field Parsley,*
    Caucalis arvensis Huds.   *Spreading Hedge-parsley*
Arable fields   Colonist
Frequent, especially on calcareous soils, but by no means restricted to them. Recorded for all neighbouring counties.
H.S. 86.
NENE: Wymington.
OUSE: Stagsden; *C. Clarke in J.H. Notes*: Basmead; *C.R.O. List*: Great Barford; *E.M.-R.*
OUZEL: Houghton Regis; *S.B.P.L.*: Sewell; *E.M.-R.*: near Well Head; *H. B. Souster*: foot of Dunstable Downs; Totternhoe.
IVEL: Cainhoe; *W.C. Herb.*: Eversholt; *S.B.P.L.*: Pegsdon!; *J.E.L. Diary*: Flitwick*; Arlesey; Maulden.
CAM: Dunton.
LEA: Biscot; *S.B.P.L.*: Leagrave; Luton; Chiltern Green.
COLNE: Markyate, v.c. 30 [Herts.]; Studham.
First record: Abbot, 1798.

T. japonica (Houtt.) DC. C.T.W. 71.8.1.   *Upright Hedge-parsley*
    Caucalis anthriscus (L.) Huds.
Roadsides, wood borders, etc.   ? Native
Common: recorded for all districts and neighbouring counties.
H.S. 2a, 55b, 58b, 63a, 78.
First record: Abbot, 1798.

T. nodosa (L.) Gaertn. C.T.W. 71.8.3.   *Knotted Hedge-parsley*
    Caucalis nodosa (L.) Scop.
Field borders, arable fields, etc.   Colonist
Frequent, and like *T. arvensis* more so on calcareous soils. Recorded for all neighbouring counties.
H.S. 71.

Ouse: Milton Ernest: *J.H. Notes*: Basmead; *C.R.O. List*: Cardington; *E. M. Langley in J.H. Notes*: Lidlington; near Bromham Bridge; Elstow; *C. C. Foss*: Goldington!; *B. B. West*: Tempsford*; Biddenham.
Ouzel: Eaton Bray; *V.C.H.*
Ivel: Cainhoe; *W.C. Herb.*: Streatley!†; *S.B.P.L.*: Barton Hills!* as var. *pedunculata*; Wilbury Hill; Pegsdon!; *J.E.L., F.F.B.*: Flitton*.
Cam: Dunton.
Lea: chalk hill, near Luton (var. *pedunculata* (Rouy & Cam.) Druce, see *B.E.C. 1929 Rep.* 118); *H. Brown*: Leagrave; *S.B.P.L.*: Biscot; *H. B. Souster*: Luton; *V. H. Chambers*.
First record: Abbot, 1798.

CUCURBITACEAE
BRYONIA L.

Bryonia dioica Jacq. C.T.W. 72.1.1.   *White Bryony*
Hedgerows, scrub, etc.   Native
Common: female-flowered forms are the more frequent. Recorded for all districts and neighbouring counties.
H.S. 27, 29, 47, 52, 69.
First record: Abbot, 1798.

EUPHORBIACEAE
EUPHORBIA L.

Euphorbia platyphyllos L. C.T.W. 74.2.7.   *Broad-leaved Spurge*
Arable fields   Colonist
Locally common in two restricted areas. It is very variable in size. Recorded for all neighbouring counties.
H.S. 4.
Nene: common in arable fields in this district.
Ouse: Bedford; *F.B.*: Basmead; *C.R.O. List*: Clapham; *J.McL., L.M.*: Sharnbrook; *W.H. 1875*: Souldrop, in a number of fields in the north of the parish.
Ouzel: Lord's Hill*; *H. B. Souster*.
Ivel: A. J. Wilmott referred to 'nearest *E. corallioides* L.' a specimen named *E. platyphyllos* and collected by C. Crouch in Miss Green's garden, Eversholt, 1894, where it had been established 40 years*.
Cam: Cockayne Hatley!; *M. Brown and J.E.L. in J. Bot. 1931*: Dunton*: it appears regularly in both parishes.
First record: Abbot, 1798.

E. helioscopia L. C.T.W. 74.2.9.   *Sun Spurge,*
                    Milkweed (applied to all members of the genus)
Arable fields, waste places, etc.
Common and well distributed. Recorded for all districts and neighbouring counties.
First record: Abbot, 1798.

E. amygdaloides L. C.T.W. 74.2.17.   *Wood Spurge*
Woods   Native
Frequent in the heavier boulder-clay woods, but absent from large
areas of the county. Recorded for all neighbouring counties.
H.S. 30a, 34b.
OUSE: Clapham, still in Clapham Park Wood!; *F.B.*: Basmead;
C.R.O. List: Sheerhatch Wood; *J.E.L., F.F.B.*: Wilstead Wood;
Harrold Park Wood; Marston Thrift; King's Wood.
KYM: Worley's Wood.
OUZEL: Aspley, still in Aspley Wood!; *F.B.*: Heath and
Reach, plentiful in both King's Wood and Baker's Wood; *W.H.*
1875.
IVEL: Cainhoe Park Wood; *W.C. Herb.*: Warden Wood; *J.McL.,*
L.M.: Clophill Wood; *C.C.*: Flitwick Wood; *J.S. in J. Bot. 1883*:
Maulden Wood*; Northill.
COLNE: Pepperstock; *S.B.P.L.*
First record: Abbot, 1798.

E. esula L. C.T.W. 74.12.15.   *Leafy Spurge*
    E. virgata Waldst. & Kit. non Desf. C.T.W. 74.12.14.
Roadsides, waste places, etc.   Colonist
Rare, and usually well established. This species needs study, but is probably not
as variable as botanists have thought it to be. *E. virgata* f. *esulifolia* Thell. is inter-
mediate between what has passed as *E. esula* and *E. virgata* in this country. Re-
corded for all neighbouring counties except Hunts.
OUSE: railway, Willington, as *E. cyparissias*; *J.McL., B.M.*: Elstow, where it was
found also by L. W. Wilson*; *E. C. Day, 1903, in J.H. Notes*: riverside, Kempston!;
L. W. Wilson: roadside, Roxton!*; *H. B. Souster.*
OUZEL: roadside and gravel pits, Leighton Buzzard*.
IVEL: spinney to south of Biggleswade; Sundon Rubbish Dump.
LEA: Dallow Road!; *V. H. Chambers.*
COLNE: roadside, West Hyde!*; *W. D. Coales*: roadside near Kinsbourne Green,
E. virgata f. *esulifolia*.
First evidence: J. McLaren, c. 1880.
First record: J. Hamson, 1909; no name had been given to E. C. Day's earlier
discovery.

E. cyparissias L. C.T.W. 74.2.16.   *Cypress Spurge*
Waste places, railway banks, etc.   Colonist
Rare: *E. Bot.* tab. 840, Feb. 1801, was drawn from a specimen from Leete Wood
sent by Abbot. Recorded for all neighbouring counties except Hunts.
NENE: railway embankment, Wymington*.
OUZEL: Aspley Wood; *W.H. 1875*: Rushmere Lodge, Heath and Reach; *J.E.L.,*
F.F.B.
IVEL: Leete Wood; *C.A. in E. Bot. 1801*: Ampthill; *C.C.*
LEA: waste ground near Luton and Dunstable Hospital*; *H. B. Souster.*
First record: Abbot, 1801.

E. peplus L. C.T.W. 74.2.10.   *Petty Spurge*
Arable land, gardens, etc.   Colonist
Common: recorded for all districts and neighbouring counties.
H.S. 12.
First record: Abbot, 1798.

E. exigua L. C.T.W. 74.2.11.   *Dwarf Spurge*
Arable land   Colonist
Our commonest spurge. Recorded for all districts and neighbouring
counties.
H.S. 4, 71, 72, 82.
First record: Abbot, 1798.

E. lathyrus L. C.T.W. 74.2.2.   *Caper Spurge*
Roadsides, spinneys, etc.   Colonist
Rare, but usually well established. Recorded for all neighbouring counties.
Ouse: Cardington†; *J.McL. in Herb. J.S.*: Bedford; *J.H. Notes*: Bromham;
A. W. Guppy: Odell; Harrold; *G. H. Day*: roadside, Souldrop.
Ouzel: garden weed, Battlesden!*; *H. E. Pickering*.
Ivel: Maulden; *M. L. Berrill in J.H. Notes*: Stondon; *J. Pollard*.
Lea: plentifully in Arden Dell Wood; *H. B. Souster*: garden weed, Round Green.
First record: fir plantations, J. McLaren, Herb. Luton Museum, 1864.

MERCURIALIS L.

Mercurialis perennis L. C.T.W. 74.1.1.   *Dog's-mercury*
Woods   Native
Common, and often dominating large areas to the exclusion of all
other plants. Female plants are relatively scarce. Recorded for all
districts and neighbouring counties.
H.S. 2b–3b, 18, 31a–32, 33, 34b, 42, 51a, 53a–55b, 69, 78, 79b,
81, 83.
First record: Abbot, 1798.

M. annua L. C.T.W. 74.1.2.   *Annual Dog's-mercury*
Arable land, gardens, etc.   ? Native
Rare, but usually well established. Recorded for all neighbouring
counties.
Ouse: Ford End; *J.McL., B.M.*: Cardington†; *J.McL. in Bot. Rec.
Club 1884–6 Rep.*
Ivel: Southill; *J.E.L., J. Bot.*: Rowney Warren, in shady lane on
west side!*; *J.E.L., Diary*.
Lea: garden weed, Luton!*; *L. W. Wilson*.
First record: J. McLaren, 1864.

POLYGONACEAE

POLYGONUM L.

Polygonum aviculare L. C.T.W. 75.1.1.   *Knotgrass*
Arable fields, waste places, etc.   Colonist
Common, but one of the more variable species. Newbould, Pryor
and Saunders recorded many varieties which it would be of only
limited use to list. Recorded for all districts and neighbouring
counties.
H.S. 4, 12, 16, 17, 34a, 50, 63a, 71, 73, 82, 86.
First record: Abbot, 1798.

P. aequale Lindm. C.T.W. p. 693.
Arable fields, waste places, etc.   Colonist
The only one of the segregates of *P. aviculare* which I am able to recognize.
Recorded for all neighbouring counties.
OUZEL: Heath and Reach!*; *N. D. Simpson.*
IVEL: Flitwick*; Northill.
LEA: Farley Bottom*; *J.S.*

P. bistorta L. C.T.W. 75.1.5.   *Snake-root, Bistort*
Damp meadows   Native
Rare: recorded for all neighbouring counties.
OUSE: Thurleigh; *F.B.*: Harrowden; *J.McL., B.M.*
IVEL: Pennyfather's Moor*; Higham Gobion; *H. B. Souster.*
LEA: 'near Luton Hoo at a mill in the greatest profusion', it still
grows near the mill at East Hyde!*; *Plantae Bedford.*: meadow, now
Park Street Recreation Ground, Luton†; *W.F.B.*
First record: Abbot, 1795.

P. amphibium L. C.T.W. 75.1.6.   *Amphibious Bistort*
Ponds, lakes, slow-flowing rivers, etc.   Native
Common: terrestrial forms are frequent on waste ground. Recorded
for all districts except Colne, and all neighbouring counties.
H.S. 7, 11, 13, 58a.
First record: Abbot, 1798.

P. persicaria L. C.T.W. 75.1.7.   *Willow Weed, Persicaria*
Arable land, margins of ponds, etc.   Native
Common, but variable in the colour of the flowers and size of the
plants. Recorded for all districts and neighbouring counties.
H.S. 6, 12, 13, 34a, 82.
First record: Abbot, 1798.

P. lapathifolium L. C.T.W. 75.1.8.   *Pale Persicaria*
    P. scabrum Moench
Arable land, waste places, margins of ponds, etc.   Colonist
Frequent, especially on land reclaimed from bogs. Recorded for all
districts and neighbouring counties.
First record: Abbot, 1798.

P. nodosum Pers. C.T.W. 75.1.9.
    P. maculatum (Gray) Bab., *P. petecticale* (Stokes) Druce
Arable land, wet places, waste places, etc.   Native
Frequent, and variable. The form *salicifolia* C. E. Moss, usually
found in waste places, deserves study. Recorded for all neighbouring
counties.
OUSE: Great Barford; *N. Y. Sandwith.*
OUZEL: Woburn Park, f. *salicifolia*.
IVEL: Campton; Rowney Warren; *J.E.L., B.E.C.*: ditches, Biggles-
wade*; *A. J. Wilmott and E.M.-R.*!
LEA: River Lea near Luton; *R. A. Pryor in J. Bot. 1876*: Luton Hoo;
rubbish dump, Luton, f. *salicifolia*.
COLNE: arable field, Studham; *E.M.-R.*!
First record: R. A. Pryor, 1876.

P. hydropiper L. C.T.W. 75.1.10.　　　　　　　*Water-pepper*
Muddy and marshy places　　　　　　　　　　　　　Native
A species usually of acid soils. Recorded for all neighbouring counties.
H.S. 6, 13, 35, 53b.
OUSE: frequent on muddy shores of the river.
OUZEL and IVEL: too many stations to merit listing.
First record: Abbot, 1798.

P. mite Schrank C.T.W. 75.1.11.
Wet places　　　　　　　　　　　　　　　　　　Native
Rare: recorded for all neighbouring counties.
OUSE: Goldington; *B. Verdcourt in B.E.C. 1944-5 Rep.* 810: Tempsford*.

P. mite × persicaria
OUSE: Fenlake*; *E.M.-R.*!

P. minus Huds., C.T.W. 75.1.12, is of doubtful occurrence. Abbot recorded it from Elstow and Cardington in *Flora Bedfordiensis*, and added Goldington in his annotated copy, but his herbarium did not contain a specimen. Saunders recorded it from Flitwick in *S.B.P.L.* and *J. Bot. 1883* but not in subsequent lists. His herbarium specimen is *P. hydropiper*. Recorded for all neighbouring counties.

P. convolvulus L. C.T.W. 75.1.13.　　　　　*Black Bindweed*
Arable fields, waste places, etc.　　　　　　　　Colonist
Common: recorded for all districts and neighbouring counties.
H.S. 4, 12, 13, 16, 17, 44, 71.
First record: Abbot, 1798.

RUMEX L.

R. acetosella L. C.T.W. 75.4.3.　　　　　　*Sheep's Sorrel*
Heaths, heathy pastures, etc.　　　　　　　　　　Native
Common on acid soils, but absent from considerable areas. Recorded for all neighbouring counties.
H.S. 22a, 23-29, 38, 44, 46, 48, 74a, 76, 77a.
OUSE: rough ground, Marston Thrift.
KYM: railway sidings, Kimbolton Station, v.c. 30 [Hunts.].
OUZEL and IVEL: too many stations to merit listing.
LEA: Limbury†; *Herb. J.S.*: Biscot; *H. B. Souster*: Chiltern Green.
COLNE: Whipsnade Common!; Studham Common!; *V. H. Chambers*: near Ravensdell Wood, v.c. 20 [Beds.].
First record: Abbot, 1798.

R. tenuifolius (Wallr.) Löve C.T.W. 75.4.2.
Heaths　　　　　　　　　　　　　　　　　　　Native
On more acid soils than the preceding, and awaiting further record.
H.S. 22b, 27.
OUZEL: Heath and Reach*.
IVEL: Rowney Warren*.

R. acetosa L. C.T.W. 75.4.4. *Sorrel, Alum* (Westoning)
Pastures, meadows, etc. Native
Common: a species of more varied habitats than *R. acetosella*.
Recorded for all districts and neighbouring counties.
H.S. 1a, 15, 16, 23, 28, 29, 37, 39b, 41, 45a, 47, 74b, 75, 82, 85.
First record: Abbot, 1798.

R. hydrolapathum Huds. C.T.W. 75.4.6. *Great Water-dock*
Riversides Native
Frequent by the larger rivers. Recorded for all neighbouring
counties.
H.S. 7, 8, 14a.
OUSE: common by the river.
OUZEL: Leighton Buzzard, as *R. aquaticus* L.!*; *J.S. in J. Bot.
1883*: Heath and Reach!; *J.S. in J. Bot. 1889*: Hockliffe; Billington.
IVEL: too many stations to merit listing.
LEA: New Mill End!†; *F.F.B.*: Luton Hoo.
First record: Abbot, 1798.

R. crispus L. C.T.W. 75.4.13. *Curled Dock*
Rough pastures, roadsides, waste places, etc. ? Native
Our commonest dock. Recorded for all districts and all neighbouring counties.
H.S. 4, 6–9, 11, 13, 16, 19a, 20a, 21, 29, 34a, 34b, 53a, 57b, 58b,
64, 71, 73, 82, 84a, 85.
First record: Abbot, 1798.
Var. **arvensis** Hardy
KYM: Melchbourne*, f. *unicallosus* (Peterm.) Lousley
IVEL: Arlesey*, f. *unicallosus*
R. crispus × **sanguineus**
OUSE: gravel pit, Eaton Socon*; *J. E. Lousley in Watsonia II (1951)* 52.
R. crispus × **obtusifolius**
OUSE: Bedford; *Newbould MS.*: Great Barford*; *N. Y. Sandwith*.
IVEL: Rowney Warren; *J.E.L., F.F.B.*: Southill Park*; *E.M.-R.*!

R. obtusifolius L. C.T.W. 75.4.14. *Broad-leaved Dock*
Rough pastures, roadsides, waste places, etc. ? Native
Common: recorded for all districts and neighbouring counties.
H.S. 6, 14b, 37, 44, 51a, 52, 57a, 57b, 77a, 83, 84a.
First record: Abbot, 1798.
Subsp. **arvensis** (Fr.) Danser
IVEL: Flitwick*; Southill*.
R. obtusifolius × **pulcher**
IVEL: Chicksands Lodge; Rowney Warren; *J.E.L., F.F.B.*
R. obtusifolius × **sanguineus**
OUZEL: Sheep Lane*; *E.M.-R.*!

U

R. pulcher L. C.T.W. 75.4.15.   *Fiddle Dock*
Rough pastures   ? Native
Locally abundant on sandy and gravelly pastures in the east of the
county. Considered common by Abbot. Recorded for all neigh-
bouring counties.
OUSE: Cardington†*; *J.McL.*, *B.M.*: Eastcotts; Cople; Oakley;
Eaton Socon; Bromham.
IVEL: Cainhoe; *W.C. Herb.*: Clophill†; Silsoe; Pulloxhill; *C.C.*:
Henlow!*; Campton; Rowney Warren; Warden Abbey; *J.E.L.*,
F.F.B.: Northill; Southill; Potton; Shillington.
LEA: Luton Hoo; *W.F.B.*
First record: Abbot, 1798.
    Subsp. **divaricatus** (L.) Murb.
IVEL: railway siding, Arlesey*; *J. E. Lousley!*

R. sanguineus L. C.T.W. 75.4.16.   *Red-veined Dock, Wood Dock*
    R. condylodes Bieb.
Woodland rides, hedgerows, etc.   Native
Common: a species essentially of shade conditions. Recorded for
all districts and neighbouring counties.
H.S. 3a, 5, 7, 8, 10b, 10c, 18, 19a, 20a, 32–34b, 42, 43, 52, 55b, 83.
First record: Abbot, 1798.

R. conglomeratus Murr. C.T.W. 75.4.17. *Clustered or Sharp Dock*
Riversides, meadows, etc.   Native
Common: a species of open habitats. Not recorded by Abbot,
although his herbarium has a specimen named *R. acutus*, presumably
Smith's species, but not included in *Flora Bedfordiensis*. Recorded
for all districts and neighbouring counties.
H.S. 6, 9, 11, 13, 14b, 56a, 57a, 84a, 84b.
First record: T. B. Blow in *Bot. Rec. Club 1875 Rep.*

R. palustris Sm. C.T.W. 75.4.19.   *Marsh Dock*
    R. limosus auct.
Pondsides   Native
Rare: recorded for all neighbouring counties except Bucks.
OUSE: Goldington, as *R. maritimus*; *J.McL.*, *L.M.*
IVEL: pond, Warden Abbey!†*; *J.E.L.*, *J. Bot.*
First record: J. E. Little, 1919.
First evidence: J. McLaren, 1864.

R. maritimus L. C.T.W. 75.4.20.   *Golden Dock*
    R. aureus Mill.
Pond and riversides   Native
Rare: *E. Bot.* tab. 725, April 1800, was drawn from material sent
by Abbot. Recorded for all neighbouring counties.
OUSE: Goldington, where McLaren found *R. palustris*; *Plantae
Bedford.*: Putnoe; *E. M. Langley in J.H. Notes*: Moor End, Felmer-
sham*.
LEA: Luton Hoo!*; New Mill End!†*; *S.B.P.L.*
First record: Abbot, 1795.

URTICACEAE
URTICA L.

Urtica dioica L. C.T.W. 76.3.2.   *Stinging-nettle*
Waste places, arable land, woods, etc.   Colonist
Common, and usually an indication of excess of nitrogenous matter
in the soil; consequently growing plentifully around farm buildings
and rabbit warrens and in woods near farms. Recorded for all
districts and neighbouring counties.
H.S. 6, 7, 9, 13, 18, 22b, 25, 29, 31a–31c, 34a, 41–43, 47, 52–55b,
57a, 81, 84b, 85.
First record: Abbot, 1798.

U. urens L. C.T.W. 76.3.1.   *Small Stinging-nettle*
Arable land   Colonist
Common in the market-gardening areas on rich soils, less frequently
elsewhere. Recorded for all districts and neighbouring counties.
H.S. 17, 44, 47, 50.
First record: Abbot, 1798.

PARIETARIA L.

Parietaria diffusa Mert. & Koch C.T.W. 76.1.1.   *Wall Pellitory*
    P. ramiflora auct., *P. officinalis* auct.
Old walls   Wall denizen
Frequent, especially on church walls, except in the extreme south
of the county. I know it only on Dunstable Priory Church wall in
the Lea district, and it is not known in the Colne district. Recorded
for all neighbouring counties.
First record: Podington, O. St. John Cooper in *Collections towards the
history and antiquities of Bedfordshire*, 1783.

CANNABACEAE
HUMULUS L.

Humulus lupulus L. C.T.W. 77.1.1.   *Hop*
Hedgerows   Of garden origin
Frequent, especially near towns and villages. Recorded for all districts and
neighbouring counties.
First record: Abbot, 1798.

ULMACEAE
ULMUS L.

I have received considerable assistance with the elms from R.
Melville. Apart from J. E. Little's records I have rejected all others
except those made by or determined by Melville. Elms are variable
and hybridize freely, which adds to the complexity of the genus.
H.S. (for elms generally) 5, 20b, 32, 34b, 52.

Ulmus glabra Huds. C.T.W. 78.1.1.   *Wych Elm*
  U. montana Stokes
Roadsides, wood borders, etc.   Native
Not infrequent. Stanbridge Wood, an unusual example of an
elm wood, is entirely *U. glabra*. Recorded for all neighbouring
counties.
NENE: Podington*.
KYM: Dean*.
OUZEL: Aspley; *J.E.L.*, *F.F.B.*: Woburn*; Stanbridge Wood;
Blackgrove Wood.
IVEL: Sandy; Campton; *J.E.L.*, *F.F.B.*
First record: Abbot, 1798.

   U. glabra × **plotii**
OUSE: Felmersham, known as the Felmersham Elm; *G. H. Day*.
KYM: Dean*.
OUZEL: Tebworth; Tilsworth*; *R. Melville*!
IVEL: Harlington*; *R. Melville*!
COLNE: Kensworth, v.c. 20 [Beds.], *ad glabra**; *R. Melville*!

U. carpinifolia Gleditsch C.T.W. 78.1.5.   *Smooth Elm*
  U. nitens Moench
Roadsides, wood borders, etc.   Native
Probably frequent in the north of the county. Recorded for all
neighbouring counties.
OUSE: Milton Ernest*; *E.M.-R.*!
KYM: Melchbourne*.
IVEL: Chicksands; *J.E.L.*, *B.E.C.*: Warden Abbey; *J.E.L.*, *Diary*.
First record: J. E. Little, 1936.

   U. carpinifolia × **glabra**
NENE: Podington, *ad plotii**.
OUSE: Clapham Park*.
KYM: Melchbourne*.
IVEL: Old Warden; *J.E.L.*, *Diary*.
CAM: Eyeworth*.

   U. carpinifolia × **plotii**
OUSE: Great Barford; *R. Melville*: Bletsoe*; *E.M.-R.*
KYM: Knotting*; *E.M.-R.*!
OUZEL: Tilsworth*; *R. Melville*!
IVEL: Harlington*; *R. Melville*!

   U. carpinifolia × **cornubiensis**
LEA: Luton Hoo*; *E.M.-R.*!
COLNE: Caddington*; *R. Melville*!

   U. carpinifolia × **coritana**
OUSE: Milton Ernest*; *E.M.-R.*!

U. × **hollandica** Mill. C.T.W. p. 719.   *Dutch Elm*
  U. major Sm.   Probably planted
Rare: one of the named forms of *U. carpinifolia* × *glabra*. Recorded for Herts.,
Bucks. and Northants.
OUZEL: Tebworth*; *R. Melville*!

U. plotii Druce C.T.W. 78.1.6.   *Plot's Elm*
   U. minor auct.
Roadsides, wood borders, etc.   Native
Frequent, and well distributed.   Recorded for all neighbouring
counties.
KYM: Dean*; Shelton*.
IVEL: Stondon; *J.E.L. in B.E.C. 1925 Rep.* 894: Pegsdon*; *R. Melville*.
LEA: Luton Hoo*; *E.M.-R.*!
COLNE: frequent near the church, Kensworth, v.c. 20 [Beds.]*;
E.M.-R.
First record: J. E. Little, 1925.

U. procera Salisb. C.T.W. 78.1.2.   *English Elm*
   U. anglica Druce, *U. campestris* auct., *U. sativa* auct.
Roadsides, etc.   Native
The common elm of our roadsides and field boundaries. Recorded
for all neighbouring counties.
OUZEL: Tilsworth*; Tebworth*; Hockliffe*; *R. Melville*!
IVEL: Harlington*; *R. Melville*!
LEA: Slaughter's Wood*.
COLNE: Kensworth, v.c. 20 [Beds.]; *R. Melville*.
First record: uncertain.

U. cornubiensis Weston   *Cornish Elm*
   U. stricta (Ait.) Lindl. C.T.W. 78.1.3, *U. minor* auct.
Parklands   Planted
Rare: recorded for all neighbouring counties except Cambs.
KYM: Swineshead*.

U. coritana Melville C.T.W. 78.1.4.   *Coritanian Elm*
Roadsides, wood borders, etc.   Native
Probably frequent in the north of the county.
NENE: Farndish, var. *angustifolia* Melville*.
KYM: West Wood*; Knotting; Melchbourne*.
OUZEL: Eggington*; *E.M.-R.*!
IVEL: Topler's Hill; *R. Melville*.
   U. coritana × glabra
CAM: Wrestlingworth*.

BETULACEAE

BETULA L.

Betula pendula Roth   *Silver Birch*
   B. alba L. p.p., *B. verrucosa* Ehrh. C.T.W. 81.1.1.
Woods   Native
Frequent, especially on cleared woodland on the Lower Greensand
and Clay-with-Flints; but it appears throughout the county except
on the Chalk and Boulder Clay. I have seen the restricted species
in all districts except the Cam, and it is recorded for all neighbouring
counties. Var. *dalecolica* (L.) Schneider grows in the ponds at New
Wavendon Heath, v.c. 24 [Beds.]*, where it was probably planted.

Birch is of some antiquity in the county: Little Barford appears as
Bereford, the ford by the birch trees, in *Domesday Book* and Birch-
more appears as Birchemore as early as 1227 (*P.N.B.*).
H.S. 3b, 27, 35, 53b, 79, 80.
First record: for the aggregate, Abbot, 1798.
 B. pendula × **pubescens**
Appears in all places where both parents grow in some quantity.
H.S. 22.
OUZEL: Heath and Reach; New Wavendon Heath, v.c. 24 [Beds.].
IVEL: Sutton Fen.

B. pubescens Ehrh. C.T.W. 81.1.2.   *Birch*
Woods, marshy places, etc.   Native
Less common than *B. pendula*, but well distributed. It is the common
species of marshy places on the Lower Greensand.
H.S. 22, 23, 30a, 35, 36, 40, 42, 43, 49, 77.
NENE: Wymington Scrubs!; *E.M.-R.*
OUSE: Jackdaw Hill*.
KYM: Melchbourne Park.
OUZEL: Aspley Wood; King's Wood; New Wavendon Heath, v.c.
24 [Beds.]; Heath and Reach.
IVEL: Washer's Wood; Folly Wood; Flitwick Moor; Sutton Fen;
Cooper's Hill; Moors Plantation.
LEA: Kidney Wood.
COLNE: Dell Wood, Studham; *E.M.-R.*
First record: the author in *B.E.C. 1943-4 Rep.* 811.

ALNUS Mill.

Alnus glutinosa (L.) Gaertn. C.T.W. 81.2.1.   *Alder*
  Betula alnus L., *A. rotundifolia* Stokes, *A. alnus* (L.) Britton
Riversides, marshy woods, etc.   Native
Frequent: woods formed in the marshy places associated with the
Lower Greensand are usually almost entirely alder. Elsewhere it
grows regularly by streams and rivers. Recorded for all neighbour-
ing counties.
H.S. 38, 39a, 40, 85.
NENE: Hinwick; *W. Kitchin.*
OUSE: Fenlake; *F.B.*: Milton Ernest; Harrold; *J.H.*: Bromham;
A. W. Guppy: Odell; *G. H. Day*: Ampthill Park; *H. B. Souster*!:
Kempston; Blunham.
KYM: Melchbourne Park.
OUZEL: Battlesden Park.
IVEL: Clophill; *W.C. Herb.*: Flitwick!*; *S.B.P.L.*: Sandy; *J.E.L.*,
F.F.B.: Southill; *B. B. West*: Eversholt; *V. H. Chambers*: Fancott;
Westoning Moor; Washer's Wood, some fine trees in dampest part
of wood, felled 1950.
LEA: Luton, a few trees in New Bedford Road!†*; *Herb. J.S.*: East
Hyde; *V. H. Chambers*: Leagrave Marsh.
First record: Abbot, 1798.

CORYLACEAE
CARPINUS L.
Carpinus betulus L. C.T.W. 82.1.1. *Hornbeam*
Woods, hedgerows, etc. Native
Frequent as a native tree in woods on the Clay-with-Flints, and appearing as a planted tree in all parts of the county. Abbot considered it to be rare. The records given below are for the native species. Recorded for all districts and neighbouring counties.
H.S. 5, 33, 70, 79.
LEA: frequent in hedgerows between Luton Airport and Chiltern Green.
COLNE: Deadmansey Wood*.
First record: Abbot, in letter to J. E. Smith, 3 Sept. 1802.

CORYLUS L.
Corylus avellana L. C.T.W. 82.2.1. *Hazel, Cob-nut*
Woods, hedgerows, etc. Native
Common throughout the county. Hazlewood, in Millbrook parish, a place-name now lost, appears as Hesilwood as early as 1330 (*P.N.B.*). Recorded for all districts and neighbouring counties.
H.S. 3b, 5, 18, 19b, 20b, 30a, 32–34b, 43, 51, 53a, 54, 63a, 75, 77, 78, 80, 81, 83.
First record: Abbot, 1798.

FAGACEAE
QUERCUS L.
Quercus robur L. C.T.W. 83.3.3. *Common Oak*
  Q. pedunculata Ehrh. ex Hoffm.
Woods, etc. Native
Common, especially in woods on the Boulder Clay. Oakley appears as Accleya, an oak clearing, in 1060, Ruxox, in Flitwick parish, as Rokeshoc (Roke's Oak) in 1174–7 and Eggington as Ekendon, an oak-grown hill, in 1195 (*P.N.B.*). Recorded for all districts and neighbouring counties.
H.S. 2a, 5, 15, 19b, 20b, 24, 28, 30a, 32, 33, 34a, 34b, 35, 39b, 40, 42, 43, 46, 47, 49, 52, 53a, 55a, 74a, 78–80.
First record: Abbot, 1798.

Q. petraea (Mattuschka) Liebl. C.T.W. 83.3.4. *Sessile Oak*
  Q. sessiliflora Salisb.
Woods Native
Rare, and apparently limited to woods on the Lower Greensand. Recorded for all neighbouring counties.
OUSE: Salford; *V.C.H.*
OUZEL: near Heath and Reach, it is frequent in King's Wood!*;
V.C.H.: New Wavendon Heath, v.c. 24 [Beds.]!*; *V. H. Chambers.*
IVEL: Ampthill; *V.C.H.*
LEA: planted, Luton Hoo; *S.B.P.L.*
First record: *Hillhouse List, 1875.*

Q. cerris L. C.T.W. 83.3.1.   *Turkey Oak*
Parks, hedgerows, etc.   Planted
Frequent: recorded for all neighbouring counties except Hunts.
NENE: Hinwick; Podington; Wymington, edge of Great Hayes Wood.
OUSE: Bromham Park; *A. W. Guppy*: Millbrook; *L. J. Tremayne*: Souldrop; edge of
Wootton Wood.
KYM: Melchbourne Park.
OUZEL: Little Billington!*; *V. H. Chambers*: Aspley Wood*; Battlesden.
IVEL: Silsoe Park, as '*Q. sessilis*'; *W.C. Herb.*: Sutton; *J.E.L., Diary*: Potton; Clop-
hill; Washer's Wood.
LEA: Pope's Meadow, Luton.
First evidence: W. Crouch, 1844.

CASTANEA Mill.

Castanea sativa Mill. C.T.W. 83.2.1.   *Sweet Chestnut*
  Fagus castanea L., *C. castanea* (L.) Karst.
Woods, parks, etc.   Planted
Frequent on the Lower Greensand, where it has been planted more extensively
than in other parts of the county. Abbot considered it to be rare. Recorded for all
neighbouring counties except Hunts.
H.S. 26, 31, 49, 77.
OUSE: roadside, Kempston.
KYM: Worley's Wood.
OUZEL: Aspley Wood!*; *W.H. 1875*: Battlesden Park; Sheep Lane; Rushmere;
Woburn Park.
IVEL: Silsoe, frequent in woods in neighbourhood!; *F.B.*: Cainhoe; *W.C. Herb.*:
Rowney Warren!; *J.McL., B.M.*: Millbrook!; Sandy!; *W.H. 1875*: Clophill,
very common in this parish!; *J.E.L., J. Bot.*: Sutton; *J.E.L., Diary*: Flitwick
Park.
LEA: Horsley's Wood.
COLNE: Badger Dell Wood; Clements End; *H. B. Souster*: Caddington.
First record: Abbot, 1798.

FAGUS L.

Fagus sylvatica L. C.T.W. 83.1.1.   *Beech*
Woods, etc.   Native
Common, and native on the Chalk where it forms a few fine woods,
as in the upper part of Leete Wood and at Maulden Firs. Judge's
Spinney is a fine example of a native beech wood away from the
Chalk. Beech occurs in isolated trees, probably planted, throughout
the county. In parks the copper beech is frequent, and var. *laciniata*
Vignet is planted at Luton Hoo. Recorded for all districts and
neighbouring counties.
H.S. 5, 31, 59, 65, 70, 79–81.
First record: Abbot, 1798.

SALICACEAE

SALIX L.

This genus presents difficulties, as the species belonging to it hybridize
freely. Willows and osiers are also frequently planted, the planted
forms themselves often being artificially produced hybrids; it is
often difficult in the field to distinguish these planted trees from the
native ones. Abbot was interested in willows at a time when
botanists were beginning to become aware of their complexity. A

long list of osiers was made by T. Orlebar Marsh, which would no doubt be interesting if it were decipherable, and if it were possible to identify the forms he mentions. I have had considerable assistance in naming my specimens from R. Melville and R. D. Meikle.

Place names are often derived from willows and sallows: Willey Hundred (Wilge Hundred in *Domesday Book*) means a hundred which had its meeting-place by a willow; Salph End, in the parish of Renhold, is Salcho (a spur of land with willows) and Salford is Saleford (a ford by the willows) in *Domesday Book*; but Willington (Welitone in *Domesday Book*) could refer to a farm by the willows or to Wila's farm (*P.N.B.*).

Salix fragilis L. C.T.W. 84.2.3.   *Crack Willow*
Riversides, etc.   Native
Frequent: recorded for all neighbouring counties.
H.S. 6.
NENE: Wymington.   ·
OUSE: Longholme; *J.H. Notes*: riverside, Felmersham.
OUZEL: The Litany; *E.M.-R.*
IVEL: Warden Great Wood; *J.McL., L.M.*: between Shefford and Biggleswade; *J.S. in J. Bot. 1889*: Barton Springs†; *Herb. J.S.*: large tree, College Farm, Pegsdon; *J.E.L. in J. Bot. 1922*: Flitwick Moor; *E.M.-R.*: Washer's Wood*; Clophill*; Westoning Moor*.
LEA: Limbury†; *Herb. J.S.*
COLNE: Whipsnade Zoo!; *V. H. Chambers*.
First record: Abbot, 1798.

S. alba L. C.T.W. 84.2.2.   *White Willow*
Riversides   Native
Common in suitable habitats. Recorded for all neighbouring counties.
H.S. 11.
NENE: Wymington.
OUSE: Longholme; *J.H. Notes*: Ford End†; *Herb. J.S.*: Felmersham.
KYM: Knotting; Swineshead; Melchbourne.
OUZEL: Salford Mill*.
IVEL: Cainhoe; *W.C. Herb.*: Arlesey; Shefford; Cadwell; Barton!*; *J.E.L., F.F.B.*: Campton; Henlow; Warden Abbey; between Biggleswade Common and Sandy Warren; *J.E.L., Diary*: Flitwick Moor; *A.J. Wilmott!*: Girtford Bridge*; *E.M.-R.!*: Greenfield Mill*; *V. H. Chambers*.
CAM: Dunton; Wrestlingworth*.
LEA: Limbury; Biscot; *J.S. in J. Bot. 1889*: Marslets†; *Herb. J.S.*
First record: Abbot, 1798.
      Var. **caerulea** (Sm.) Sm. C.T.W. p. 753.   *Cricket-bat Willow*
IVEL: Radwell; *J.E.L., F.F.B.*
LEA: Limbury*.
      S. alba × triandra
IVEL: Cadwell Bridge, as *S. alba* var. *vitellina × triandra*, planted; *J.E.L. in J. Bot. 1922*, 79.

S. × **russelliana** Sm. (*S. alba* × *fragilis*) C.T.W. p. 754.   *Bedford Willow*
This recalls the long correspondence between Abbot and J. E. Smith as the result
of which Abbot won his point and the willow was named in honour of the 6th Duke
of Bedford.
H.S. 38.
NENE: Podington, *ad fragilis**.
OUSE: Cranfield*.
KYM: Tilbrook, v.c. 30 [Hunts.]*.
OUZEL: Woburn; *C.A.*: Tilsworth, *ad alba**; The Litany*; *E.M.-R.*!: Blackgrove
Wood, *ad fragilis**; Salford Mill.
IVEL: Greenfield Mill*; Clophill*; *V. H. Chambers*!: Girtford Bridge*; *E.M.-R.*!:
Broom*; Flitwick Moor*.
CAM: Wrestlingworth, *ad alba**.

S. triandra L. C.T.W. 84.2.4.   *Almond Willow*
Riversides, osier beds, etc.   Planted
Frequent, but often planted. It was considered rare by Abbot. Recorded for all
districts except Nene, Cam and Colne, and for all neighbouring counties.
First record: Abbot, 1798.

S. purpurea L. C.T.W. 84.2.5.   *Purple Osier*
Riversides   Native
Rare: recorded for all neighbouring counties.
H.S. 13.
NENE: Podington*.
OUSE: Thurleigh; *F.B.*: Fenlake Meadows; *J.McL.*, *L.M.*: Radwell;
V. H. Chambers: Tempsford*; Willington*.
OUZEL: unlocalized; *V.C.H.*
IVEL: between Shefford and Biggleswade!*; *J.S. in J. Bot. 1889*:
Campton; Cadwell; *J.E.L.*, *F.F.B.*
LEA: Biscot; Limbury!†*; *S.B.P.L.*: Bedford Road, near Luton!;
Herb. J.S.
COLNE: Whipsnade Zoo, possibly planted*.
First record: Abbot, 1798.

S. × **rubra** Huds. (*S. purpurea* × *viminalis*) C.T.W. p. 758
Recorded for all neighbouring counties except Hunts.   ? Planted
OUSE: Bedford; *Botanist's Guide*: Turvey, var. *purpureoides* Gren. & Godr.; *W. R.
Linton in B.E.C. 1887 Rep.*
KYM: Knotting*; *E.M.-R.*!
OUZEL: Battlesden Lake*.
LEA: Limbury, *ad viminalis**; Biscot*.

S. viminalis L. C.T.W. 84.2.7.   *Common Osier*
Riversides, osier beds, etc.   ? Planted
Frequent, but often planted. Recorded for all neighbouring counties.
H.S. 64.
OUSE: Radwell; *V. H. Chambers*.
KYM: Knotting.
OUZEL: Palmer's Shrubs; Milton Bryan.
IVEL: Clophill; *W.C. Herb.*: between Shefford and Biggleswade; *J.S.
in J. Bot. 1889*: Ickwell; Arlesey; Cadwell; *J.E.L.*, *F.F.B.*: between
Shefford and Southill; *J.E.L.*, *Diary*: Flitwick Moor; *R. Morse*.
CAM: Brook Farm, Wrestlingworth*.
LEA: Leagrave Marsh†; Stopsley†; *Herb. J.S.*
First record: Abbot, 1798.

S. caprea L. C.T.W. 84.2.8. *Goat Willow, Great Sallow*
Woods, hedgerows, etc. Native
Common: recorded for all districts and neighbouring counties.
H.S. 2b, 19b, 34a, 58b, 64, 79.
First record: Abbot, 1798.

S. × smithiana Willd. (*S. caprea × viminalis*)
    S. sericans Tausch
OUSE: Stagsden; *W. R. Linton in W.B.E.C. 1888–9 Rep.*: Carlton Moor End*;
E.M.-R.
OUZEL: Thorn*; Houghton Regis*; *V. H. Chambers.*
IVEL: Cadwell; Meppershall; Stondon; Sandy!; *J.E.L., F.F.B.*: Flitwick Moor*;
Harlington; *R. Melville*: Barton*.
LEA: Limbury!†*; *J.S. in J. Bot. 1889*: Bramingham Shott; *F.F.B.*: Dunstable*;
E.M.-R.
COLNE: Caddington*; *R. Melville*!

S. calodendron Wimm.
    S. dasyclados auct. C.T.W. p. 761.
OUSE: St. Neot's Bridge, 1888!*; pond at Turvey, 1890; both specimens in Herb.
B.M.; *E. F. Linton* (see *R. D. Meikle in Watsonia II (1952) 246*).

S. aurita L. C.T.W. 84.2.11. *Eared Willow*
Woods Native
Rare: recorded for all neighbouring counties.
H.S. 38.
OUZEL: King's Wood!*; *R. Melville.*
IVEL: Warden; *F.B.*: Flitwick!*; *V.C.H.*
First record: Abbot, 1798.
    S. aurita × caprea (*S. × capreola* A. Kerner ex Anderss.)
OUZEL: King's Wood, *ad aurita*.
LEA: Luton, as *S. cinerea*, det. this, J.E.L.†; *Herb. J.S.*

S. atrocinerea Brot. C.T.W. 84.2.9. *Common Sallow*
Woods, etc. Native
Common: recorded for all districts and neighbouring counties.
H.S. 3b, 38, 39a.
First record: Abbot, as *S. cinerea*, in letter to J. E. Smith, 3 Feb. 1802.
    S. atrocinerea × aurita
    S. lutescens auct., *S. multinervis* auct.
OUZEL: King's Wood; *R. Melville*: Battlesden.*
IVEL: Pegsdon; Flitwick Moor, where it was also found by R. Melville; *J.E.L.,*
F.F.B.: between Biggleswade Common and Sandy Warren; *J.E.L., Diary.*
CAM: Brook Farm, Wrestlingworth*.
LEA: Limbury*.
    S. atrocinerea × caprea
    S. reichardtii auct.
OUSE: Souldrop*.
OUZEL: King's Wood; *R. Melville.*

S. cinerea L. C.T.W. 84.2.10.
Marshy places Native
A species of Eastern England. Earlier records refer probably to
S. atrocinerea. Recorded for Cambs.
H.S. 42, 43.

Kym: West Wood; *R. Melville in B.E.C. 1946–7 Rep.* 222.
Ivel: Flitwick Moor; *R. Melville*: marsh, foot of Cooper's Hill*;
Moors Plantation*.
Cam: Brook Farm, Wrestlingworth.
First record: R. Melville, 1947.

S. repens L. C.T.W. 84.2.14.   *Creeping or Dwarf Willow*
Marshes   Native
Extinct: recorded for all neighbouring counties except Northants.
Ivel: Ampthill; *F.B.*: Maulden Moor; *W.C. Herb.*: Flitwick Moor†;
W.F.B.
First record: Abbot, 1798.

POPULUS L.

Populus canescens (Ait.) Sm. C.T.W. 84.1.2.   *Grey Poplar*
Woods, hedgerows, etc.   Planted
Frequent: recorded for all neighbouring counties.
Nene: Hinwick; Great Hayes Wood.
Kym: Melchbourne.
Ouzel: Thorn!*; *V. H. Chambers*: Houghton Regis!*; Heath and Reach!*;
E.M.-R.: between Hockliffe and Tilsworth*.
Ivel: Shillington; *W.F.B.*: Westoning!*; *C.C.*: Old Warden Park; *J.E.L. in
J. Bot. 1916*: Southill Park!*; *E.M.-R.*!
Lea: Luton Hoo!; *W.F.B.*
Colne: Caddington†; *J.S. in J. Bot. 1883.*
First record: *Hillhouse List, 1875.*

P. tremula L. C.T.W. 84.1.3.   *Aspen*
Woods, etc.   Native
Frequent in woods on the Lower Greensand, as at King's Wood,
Heath and Reach, where it is undoubtedly native, but it is also
planted throughout the county to such an extent that it is difficult
to distinguish planted from native trees. Aspen occurs in place
names: Aspley (Asplea in 969) and Apsley End (Aspele in *Domesday
Book*) both mean an aspen-tree clearing (*P.N.B.*). Var. *villosa*
Lang. is recorded from Salford Wood in *V.C.H.* Recorded for all
districts except Lea, and for all neighbouring counties.
H.S. 19b, 20b, 30a, 34a, 34b.
First record: Abbot, 1798.

P. nigra L. C.T.W. 84.1.4.   *Black Poplar*
Riversides   Native
Frequent by all the larger rivers, where the trees seem to be usually
var. **betulifolia** (Pursh) Torr. Considered rare by Abbot. Recorded
for all neighbouring counties.
H.S. 45b.
Kym: near Dean.
Ouse, Ouzel and Ivel: too many stations to merit listing.
Lea: Luton Hoo: *S.B.P.L.*
First record: Abbot, 1798.
        Var. **plantierensis** (Simon-Louis) C. K. Schneider
Roadsides, etc.   Planted
More common than the records indicate.
Nene: Farndish*.
Ivel: meadow near Flitwick Mill*.

P. italica (Du Roi) Moench C.T.W. 84.2.5.   *Lombardy Poplar*
Roadsides, etc.   Planted
Superficially like *P. nigra* var. *plantierensis* and more common than the records
indicate. Recorded for Herts., Bucks. and Cambs.
IVEL: Campton; *J.E.L., F.F.B.*: roadside, Silsoe*.

P. × canadensis Moench (*P. deltoidea* Marsh. × *nigra*) C.T.W. 84.1.6.
Var. **serotina** (Hartig) Rehd.   *Black Italian-poplar*
Roadsides, etc.   Planted
A frequently planted tree, known only in the male form. Recorded for Herts.,
Bucks. and Cambs.
NENE: Wymington.
OUZEL: Shenley Hill*; Bidwell; *E.M.-R.*
IVEL: above Cadwell Bridge; *J.E.L. in B.E.C. 1912 Rep.*: Sutton; Ickleford;
J.E.L., F.F.B.: Fancott*.
LEA: old Sewage Works, Luton*; *E.M.-R.*!
First record: J. E. Little, 1912.
Var. **marilandica** (Poir.) Rehd.
Less common than the above. Recorded for Herts.
OUZEL: Shenley Hill*; *E.M.-R.*
LEA: old Sewage Works, Luton*; *E.M.-R.*!

P. gileadensis Rouleau C.T.W. 84.1.7.   *Balm of Gilead*
P. tacamahacca auct.
Roadsides, etc.   Planted
Rare: recorded for Bucks.
OUZEL: Leighton Buzzard*; *E.M.-R.*!: Milton Bryan*.
IVEL: Cadwell; Stotfold; *J.E.L., F.F.B.*
LEA: Luton Hoo!*; *E.M.-R.*
COLNE: Kensworth, v.c. 20 [Beds.]*; *R. Melville*!: Whipsnade Zoo*.

P. alba L. C.T.W. 84.1.1.   *White Poplar*
Roadsides, etc.   Planted
Frequent: J. E. Little noted that he had observed female forms only. Abbot
considered it rare. Recorded for all neighbouring counties.
NENE: Hinwick.
OUSE: Fenlake; Stevington!; *F.B.*: Basmead; *C.R.O. List*: Cardington; *J.McL. in
J.H. Notes*: Bromham; *B. B. West.*
OUZEL: Leighton Buzzard; *E.M.-R.*!
IVEL: Cainhoe; *W.C. Herb.*: Shillington; *S.B.P.L.*: Cadwell Bridge; *J.E.L. in
J. Bot. 1916*: Barton Springs!†*; *Herb. J.S.*: College Farm, Pegsdon*; *E.M.-R.*!
LEA: New Bedford Road, north of Luton.
First record: Abbot, 1798.

VACCINIACEAE

VACCINIUM L.

Vaccinium myrtillus L. C.T.W. 85.13.2.   *Huckleberry, Bilberry*
Heaths, heathy woods, etc.   Native
Limited to a small area of the Lower Greensand, where it is locally
abundant. It has no doubt become more scarce in recent years
with the extension of afforestation. Recorded for Herts. and Bucks.
OUZEL: Aspley Wood, where it still survives in a few rides!*;
Plantae Bedford.: Heath and Reach†; *B.P.L.*: New Wavendon
Heath, plentifully in the open rides, v.c. 24 [Beds.]*.
First record: Abbot, 1795.

V. oxycoccus L.   *Cranberry*
 Oxycoccus quadripetalus Gilib., *O. palustris* Pers. C.T.W. 85.14.1.
Bogs   Native
Extinct, as it is also in Herts. and Cambs.
IVEL: Ampthill Moor; *Plantae Bedford.*: Potton Marshes; *F.B.*
First record: Abbot, 1795.

ERICACEAE

CALLUNA Salisb.

Calluna vulgaris (L.) Hull C.T.W. 85.11.1.   *Ling, Heather*
 Erica vulgaris L.
Heathy places   Native
Frequent on heaths and in heathy places on the Lower Greensand
and Clay-with-Flints, usually in some quantity when it appears,
but absent from the greater part of the county. Recorded for all
neighbouring counties.
H.S. 22a, 23, 24, 27, 35, 47, 66b, 74a.
OUZEL: Aspley, frequent in the rides in Aspley Wood!; *Abbot's
annotated F.B.*: Woburn, sparingly in parts of Woburn Park!;
W.H. 1875: plentiful on the heath at Heath and Reach; heathy
rides, King's Wood*; abundant, New Wavendon Heath, v.c. 24
[Beds.].
IVEL: Sandy, plentifully on Sandy Warren, etc.!; Ampthill, abun-
dant on Cooper's Hill!; Everton; *Abbot's annotated F.B.*: Maulden,
sparingly in various parts of the parish!; *W.C. Herb.*: Rowney
Warren, locally abundant!; *J.McL., L.M.*: Potton, 1855*; *J. M.
Vipian*: Flitwick†; *Herb. J.S.*
LEA: Chiltern Green Common, barely surviving!; *P.T.*: Galley
Hill, sparingly in clay pocket*.
COLNE: Whipsnade Heath, plentiful!; *V. H. Chambers*: Deadmansey
Wood; *P.T.*: Studham Common; sparingly at West Hyde.
First record: Abbot, 1798.

ERICA L.

Erica cinerea L. C.T.W. 85.12.4.   *Bell-heather*
Heaths   Native
Extinct: recorded for all neighbouring counties, but doubtfully for
Hunts.
IVEL: Potton; *J.McL., in J.H. Notes* (? error).
COLNE: Pepperstock, where it was still found by Saunders†*;
H. Brown: heath near Markyate; *Miss Crouch in W.C. Herb.* (it is
possible that these two stations are one and may have been in v.c. 20
Herts).
First record: H. Brown, 1838.

E. tetralix L. C.T.W. 85.12.1.   *Cross-leaved Heath*
Heaths   Native
Extinct: recorded for all neighbouring counties.
OUZEL: Woburn; *W.H. 1875.*
IVEL: Ampthill Bogs; Potton Marshes, where it was still found by
McLaren; *F.B.*: Maulden Bogs; *W.C. Herb.*: Flitwick; *J.McL. in
J.H. Notes.* ·
First record: Abbot, 1798.

PYROLACEAE

PYROLA L.

Pyrola minor L. C.T.W. 86.1.1.   *Common Wintergreen*
Heathy woods   Native
Very rare or extinct: recorded for Herts. and Bucks.
OUZEL: Aspley, *S.B.P.L.*
COLNE: woods near Whipsnade and Pepperstock; *Plantae Bedford.*:
Saunders recorded it from Pepperstock, but his station was in v.c.
20 Herts.
First record: ·Abbot, 1795.

MONOTROPACEAE

MONOTROPA L.

Monotropa hypopitys L. C.T.W. 87.1.1.   *Yellow Bird's-nest*
  Hypopitys hypopitys (L.) Small
Woods   Native
Rare, and apparently limited to beech woods. I have not been able
to work out the distribution of the two segregate species of Clapham,
Tutin and Warburg's Flora. Recorded for all neighbouring
counties except Hunts.
OUSE: Basmead; *C.R.O. List*: Turvey; *H. L. Munby in Herb. B.M.*:
Colworth; *W.H. 1875*: Twin Wood; *W. G. Nash in J.H. Notes*:
Judge's Spinney!; *H. B. Souster.*
OUZEL: wood below Whipsnade Downs, *M. hypopitys* of *C.T.W.*!;
W. D. Coales.
IVEL: Streatley, in woods at top of Markham Hills, *M. hypopitys* of
C.T.W.!*; *S.B.P.L.*: Tingley Wood Plantation; *A. Long in J.E.L.*,
F.F.B.: wood by old Bedford road, Barton; *F. L. Chesham.*
LEA: woods near Luton; *Plantae Bedford.*: Luton Hoo; *H. Brown*:
New Mill End Road, I have not seen it here since 1945!*; *Davis*:
wood near Galley Hill, *i.e.* Maulden Firs; *F.F.B.*
COLNE: copse near Isle of Wight Farm, v.c. 20 [Beds.]; *Fl. Herts.*
First record: 'in Bedfordshire woods', Hudson, *Flora Anglica Ed. II*,
1778, 175.

PRIMULACEAE

HOTTONIA L.

Hottonia palustris L. C.T.W. 90.2.1.   *Water Violet*
Ditches, etc.   Native
Very rare: recorded for all neighbouring counties.
OUSE: Eaton Socon; near the Mill, where it was also found by
McLaren; *Plantae Bedford.*: Cardington; *J.McL. in J.H. Notes*:
Tempsford, well established in a stream with *Apium nodiflorum* and
Baldellia ranunculoides, an interesting discovery of a species believed
to be extinct in the county!*; *Margaret Dawes*.
IVEL: Potton Marshes; *F.B.*: Potton; *D. M. Higgins in J. Bot. 1883*:
Sandy, possibly the same station†; *D. M. Higgins in S.B.P.I.*:
Biggleswade; *H. Read in J.H. Notes*.
First record: Abbot, 1795.

PRIMULA L.

Primula elatior (L.) Hill C.T.W. 91.1.4.   *Oxlip*
Woods   Native
One of the more interesting Bedfordshire plant species, it is limited
in Britain to two areas. The larger is on the border of Essex and
Suffolk, and the smaller on the border of Cambridgeshire and
Huntingdonshire. The solitary Bedfordshire station is on the edge
of the latter. Two theories have been put forward to account for its
limited distribution: one, that it is limited by geographical factors,
is neither supported nor disproved by evidence from the county.
There are boulder-clay woods in the county near by which appear
in every other detail like Potton Wood but have no doubt greater
rainfall and differing ranges of temperature. The other theory
suggests that the oxlip had once a wider distribution and that the
primrose, a comparative newcomer, has restricted its range by
hybridization. This is supported, to a small degree, by evidence
from the county, as the oxlip is limited to a corner of Potton Wood
and is associated with the hybrid, whereas in Cockayne Hatley
Wood there are only primroses and the hybrid. There are many
records of oxlip for the county but all, except the one given below,
are in error. Recorded for Hunts. and Cambs.
CAM: Potton Wood!*; *H. and D. Meyer in B.E.C. 1934 Rep.* 832.
    P. elatior × vulgaris
CAM: Potton Wood!; *H. and D. Meyer in B.E.C. 1934 Rep. 832:* Cockayne
Hatley Wood*; *H. and D. Meyer!*

P. vulgaris Huds. C.T.W. 90.1.5.   *Primrose*
    P. acaulis (L.) Hill
Woods   Native
Common throughout the county especially in boulder-clay woods,
but in danger of diminishing through the bad coppicing of woods in
recent years. The so-called variety *caulescens* Koch, probably a

back-cross with *P. veris*, was recorded in *S.B.P.L.* from Sundon, Streatley† and Harlington, and by McLaren from Kempston Wood. I have seen it at Whipsnade and in Potton Wood. Recorded for all districts except Nene, where it appears only as a garden escape, and all neighbouring counties.

H.S. 2b, 18, 19b, 20b, 23, 30a, 31c, 32, 34a, 34b, 51a, 52–55a, 80b, 81.

First record: Abbot, 1798.

P. veris L. C.T.W. 90.1.3.   *Cowslip, Paigle*
   P. officinalis (L.) Hill
Pastures   Native
Common in meadows on clay soils and on chalk downland, but absent or very rare on acid soils. It is rarely seen in places where the soil has been disturbed. Recorded for all districts and neighbouring counties.

H.S. 1b, 23, 56b, 59, 60, 62, 65.

First record: Abbot, 1798.

P. veris × vulgaris C.T.W. p. 803.   *Hoaxlip*
   P. variabilis Goupil non Bast.
Not uncommon, and likely to occur in any place where both parents are present. It has been much confused with *P. elatior*, the earlier records of which must be referred to this. At its best it makes a very handsome plant.
H.S. 2b, 52.
Ouse:Marston; *W.H. 1875*: White's Wood; Twin Wood; *A. W. Guppy*: Sheerhatch Wood*; *L. W. Wilson*: Galsey Wood; *C. C. Foss*: Stevington; railway bank, Souldrop.
Kym: West Wood.
Ouzel: Blackgrove Wood.
Ivel: Cainhoe; *W.C. Herb.*: Barton; *D. M. Higgins in W.B.E.C. 1900–1 Rep.*: Stanfordbury Farm, Shefford; *J.E.L., J. Bot.*: Westoning; *G. D. Nicholls*: Washer's Wood.
Cam: Cockayne Hatley Wood!; *J.E.L. and M. Brown in B.E.C. 1934 Rep.*
Lea: George Wood†*; Bramingham Wood†; *Herb. J.S.*: Horsley's Wood*; *P.T.*
Colne: near Long Wood, v.c. 20 [Beds.].
First record: Abbot, as *P. elatior*, common, 1798.

LYSIMACHIA L.

Lysimachia vulgaris L. C.T.W. 90.4.3.   *Yellow Loosestrife*
Riversides   Native
A feature of the reedy shores of the Ouse, and apparently not extending to any considerable degree along its tributaries. Recorded for all neighbouring counties.

H.S. 7.

Ouse: common by the river.

Ouzel: by ornamental pond, Heath and Reach, probably planted!; *W. D. Coales.*

Ivel: Flitwick Moor; *M. Oldfield.*

First record: Abbot, 1798.

L. nummularia L. C.T.W. 90.4.2.   *Creeping Jenny, Herb Tuppence*
Damp and shady places   Native
Frequent in woodland rides and marshy places. It often does not
flower and apparently never fruits. Recorded for all districts and
neighbouring counties.
H.S. 3a, 10b, 10c, 18, 20a, 23, 34a, 34b, 52–53b.
First record: Abbot, 1798.

L. nemorum L. C.T.W. 90.4.1.   *Wood* or *Yellow Pimpernel*
Woods   Native
Limited to woods on the Lower Greensand and Clay-with-Flints,
where it is often a feature of the rides. Recorded for all neighbour-
ing counties.
H.S. 23, 53a, 53b, 77a, 77b, 79b.
OUZEL: Aspley Wood!; *Plantae Bedford.*: King's Wood; Palmer's
Shrubs.
IVEL: Thrift Wood; *H. B. Souster*: Washer's Wood.
LEA: Chiltern Green Wood!†; *S.B.P.L.*: Kidney Wood.
COLNE: Deadmansey Wood*; Oldhill Wood; Kidney Wood*;
Ravensdell Wood, v.c. 20 [Beds.].
First record: Abbot, 1795.

ANAGALLIS L.

Anagallis tenella (L.) Murr. C.T.W. 90.7.1.   *Bog Pimpernel*
Bogs, marshes, etc.   Native
Very rare: recorded for all neighbouring counties.
H.S. 10a.
OUSE: Stevington Moor, the re-discovery of this, in or near Abbot's
old station, by Miss Day in 1943, is one of the more important recent
botanical finds in the county—it grows plentifully in one small
marsh (H.S. 10a), but is absent from the similar marshes close by!*;
Plantae Bedford.
OUZEL: Totternhoe Mead, where it was apparently plentiful until
part of the mead was drained†; Heath and Reach†*; *S.B.P.L.*
IVEL: Ampthill Moor; *Plantae Bedford.*: Maulden Bogs; *W.C. Herb.*:
Flitwick; *J.McL., B.M.*
First record: Abbot, 1795.

A. arvensis L. C.T.W. 90.7.2.   *Scarlet Pimpernel*
Arable fields, gardens, waste places, etc.   Colonist
Common: colour forms are frequently observed. Recorded for all
districts and neighbouring counties.
H.S. 4, 11, 19a, 20a, 53b, 54, 71–73, 82, 86.
First record: Abbot, 1795.

A. foemina Mill.   *Blue Pimpernel*
   A. caerulea Schreb. non L., *A. arvensis* subsp. *foemina* (Mill.)
Schinz & Thell. C.T.W. 90.7.2.
Arable fields, gardens, etc.   Colonist
Rare: the herbarium specimens cited below are correctly named,
but in cases where records are not supported by specimens there

may have been confusion with blue-flowered forms of *A. arvensis*. Recorded for Herts., Bucks. and Northants., and doubtfully for Cambs.

OUSE: Oakley West Field; *Plantae Bedford.*: Cardington; *J.McL.*, *J. Bot. 1884*: Wootton; *S. H. Adams in J.H. Notes*: garden weed, Bedford; *J.H. Notes*: Box End; *R. Townsend*: regular in its appearance at Harrold!*; *G. H. Day*.

OUZEL: arable field, Heath and Reach!; *J. P. M. Brenan*.

IVEL: Barton Hill; *F.B.*: Pulloxhill†; Flitton; Kitchen End*; *C.C.*

LEA: garden weed, Luton*.

First record: Abbot, 1795.

CENTUNCULUS L.

Centunculus minimus L. C.T.W. 90.8.1. *Chaffweed*
Heaths   Native
Probably extinct, as it is apparently also in Herts., Bucks. and Cambs. *E. Bot.* tab. 531, Dec. 1798, was drawn from a specimen sent by Abbot.

IVEL: Ampthill Moor; *F.B.*

SAMOLUS L.

Samolus valerandi L. C.T.W. 90.10.1. *Brook Weed*
Marshes, riversides, etc.   Native
Not uncommon by the Ouse in places liable to flooding, rare elsewhere. Recorded for all neighbouring counties except Bucks.

OUSE: banks of the river above Bedford; *Plantae Bedford.*: Goldington Brook; *J.McL.*, *L.M.*: Newnham Bridge; *J.H. Notes*: Bedford†; *Herb. J.S.*: Bromham Bridge!*; *W.F.B.*: Roxton; *G. H. Day*: Biddenham; *W. D. Coales*: Oakley Bottom; *C. C. Foss*!: Harrold; *C. C. Foss and G. H. Day*: on wall at side of weir, Tempsford; *P.T.*!: lock, Great Barford; *H. B. Souster*!

KYM: ditch, Pertenhall.

OUZEL: marsh on hillside, Hockliffe!*; *E.M.-R.*

IVEL: Gravenhurst; *W.C. Herb.*

First record: Abbot, 1795.

OLEACEAE

FRAXINUS L.

Fraxinus excelsior L. C.T.W. 92.1.1. *Ash*
Woods, roadsides, etc.   Native
This rivals the oak in being the commonest Bedfordshire tree. It is strange that it does not occur in any place names. Recorded for all districts and neighbouring counties.

H.S. 3b, 5, 18, 20b, 30a, 32–34b, 43, 45b, 51, 53a, 53b, 55a, 69, 70, 80, 81, 83, 85.

First record: Abbot, 1798.

LIGUSTRUM L.

Ligustrum vulgare L. C.T.W. 92.3.1.   *Privet*
Woods, hedgerows, etc.   Native
Common, and native on the Chalk, where it occurs in clearances in
beech woods and in hedgerows. It appears less frequently in woods
on the Boulder Clay throughout the county, but it may have been
planted in some cases. Recorded for all districts and neighbouring
counties.
H.S. 2b, 3b, 5, 20b, 32, 34a, 34b, 51, 70a, 70b.
First record: Abbot, 1798.

APOCYNACEAE

VINCA L.

Vinca major L. C.T.W. 93.1.2.   *Greater Periwinkle*
Roadsides, copses, etc.   Denizen
Not uncommon, and usually well established. The original drawing for *E. Bot.*
tab. 514, Nov. 1798, is marked 'Abbot'. Recorded for all neighbouring counties.
NENE: roadside, Farndish.
OUSE: Ravensden; Clapham; *F.B.*: Stagsden; *W.H. 1875*: Roxton; *G. H. Day*:
Odell Wood!; Bedford; *B. B. West.*
KYM: copse, Nether Dean!; *M. Dalton.*
OUZEL: Aspley Wood; *J.McL., L.M.*: sandpit, Leighton Buzzard!; *W. D. Coales.*
IVEL: Clophill; *W.C. Herb.*: Clifton; Arlesey; Southill Park; *J.E.L., J. Bot.*:
hedgerow, Higham Gobion.
LEA: railway cutting, south of Luton*.
First record: Abbot, 1798.

V. minor L. C.T.W. 93.1.1.   *Lesser Periwinkle*
Roadsides, copses, etc.   Denizen
As frequent and as well established as the preceding. Recorded for all neighbour-
ing counties.
NENE: roadside, Hinwick.
OUSE: Hill Plantation; *J.McL., L.M.*: Stagsden; *A. W. Guppy*: Kempston; *B. F.*
Haylock; Chawston; Odell; *B. B. West.*
KYM: Dean Grange; *Miss Dalton*: Little Staughton churchyard.
OUZEL: Aspley; *F.B.*
IVEL: Potton; *F.B.*: Chicksands Dell; *W.C. Herb.*: Maulden Wood!*; *M. L. Berrill*
in J.H. Notes: Ridgmont; *C.C.*: Cadwell; *J.E.L., F.F.B.*: Tingrith; *H. B. Souster.*
LEA: Limbury†; New Mill End; *S.B.P.L.*
COLNE: Whipsnade; *W. D. Coales*: roadside, West Hyde; *L. J. Margetts*; roadside,
Markyate, v.c. 30 [Herts.]; roadside, Kensworth, v.c. 20 [Beds.].
First record: Abbot, 1798.

GENTIANACEAE

BLACKSTONIA Huds.

Blackstonia perfoliata (L.) Huds. C.T.W. 94.4.1.   *Yellow-wort*
  Chlora perfoliata (L.) L.   Native
Frequent in places where the soil has been previously disturbed on
the Lower Chalk, Gault, Oxford Clay and Oolite. It is not eaten
by rabbits and often makes a fine show in the vicinity of rabbit
warrens. Like so many species now frequent on disturbed soils it
was considered rare by Abbot. A most attractive species, it remains

fresh in water longer than any other wild plant I know. Recorded for all neighbouring counties.

H.S. 68.

NENE: railway embankment, Wymington!; *R. R. B. Orlebar.*

OUSE: Bromham Grange Farm; *Plantae Bedford.*: Milton Hill; *F.B.*: Stevington; Ravensden; Thurleigh; Oakley; *Abbot's annotated F.B.*: Basmead; *C.R.O. List*: Hammer Hill; *J.McL., L.M.*: Sharnbrook; *W.H. 1875*: Clapham; *J.H. Notes*: Harrold; railway embankment, Souldrop!; *G. H. Day*: Colesdon; *A. W. Guppy.* OUZEL: sandpit, Leighton Buzzard, an unusual habitat with us.

IVEL: Cainhoe Park Wood; *W.C. Herb.*: Sundon, in chalk scrub!; *Davis*: Warden Wood; *J.McL., B.M.*: Streatley, in chalk scrub!*; *F.F.B.*: Barton, in chalk scrub!†; *Herb. J.S.*: Pegsdon, in scrub and over-grazed pasture!; Shefford; *J.E.L., F.F.B.*: claypits, north of Sandy station, Henlow, Arlesey.

CAM:-scrub, Dunton.

First record: Abbot, 1795.

CENTAURIUM Hill

Centaurium minus Moench C.T.W. 94.3.3. *Common Centaury*
  C. umbellatum Gilib., *Chironia centaurium* (L.) Curt., *Erythraea centaurium* (L.) Borkh., *Centaurium centaurium* (L.) Druce
Woodland rides, pastures, etc. Native
Frequent on light soils, but by no means limited to them. White-flowered forms are seen occasionally. Recorded for all districts and neighbouring counties.

H.S. 3a, 11, 20a, 23, 34a, 53b, 54, 58b, 77a, 79a, 80a.

First record: Abbot, 1798.

C. pulchellum (Sw.) Druce C.T.W. 94.3.1.
  Erythraea pulchella (Sw.) Fr.
Woodland rides, etc. Native
Rare, but possibly overlooked. Recorded for all neighbouring counties except Bucks.

H.S. 34a.

NENE: Great Hayes Wood.

OUSE: Marston Thrift!*; *V. H. Chambers.*

KYM: West Wood*.

IVEL: disused brickfield, Sandy.

CAM: Cockayne Hatley Wood*.

First record: G. C. Druce, *Fl. of Bucks.*, 1926.

GENTIANELLA Moench

Gentianella amarella (L.) H. Sm. C.T.W. 94.6.4. *Felwort,*
  Gentiana amarella L. *Autumnal Gentian*
Pastures Native
Frequent on the Chalk, where it appears both in well-established turf and in places where the soil has been previously disturbed. It

also occurs occasionally in pastures on the Boulder Clay. A white-flowered form has been seen at Pegsdon. Recorded for all neighbouring counties except Hunts.

H.S. 59, 60, 62, 65, 66a, 67, 68.

NENE: Knapwell Hill; *W. Kitchin.*

OUSE: Turvey; Bromham; *F.B.*: Thurleigh; *Abbot's annotated F.B.*: Basmead; *C.R.O. List*: Ravensden; *J.McL.*, *L.M.*: Marston; *A. Stimson in J.H. Notes*: Stagsden; *R. Townsend*: Wilden; *G. H. Day.*

OUZEL: Dunstable Downs!; *J.McL.*, *B.M.*: Totternhoe Knolls; clay pocket, King's Wood.

IVEL: 'on the Barton Hills upon a waste chalky ground'!; *Parkinson, Theatrum Botanicum,* 407: Pegsdon!; *J.E.L.*, *F.F.B.*: Sundon Hills; scrub, Streatley*; Sharpenhoe Hills.

LEA: Leagrave Marsh†; Warden Hills!†; *Herb. J.S.*: the Lynches; *H. B. Souster*: in scrub and on hill slope, Galley Hill.

COLNE: Dunstable East Hill, v.c. 20 [Beds.]; *Fl. Herts.*: Kensworth Common; near Ravensdell Wood, v.c. 20 [Beds.].

First record: Parkinson, 1640.

G. anglica (Pugsl.) E. F. Warb. C.T.W. 94.6.6.

Gentiana anglica Pugsl., *G. lingulata* var. *praecox* (Townsend) Wettst.

Chalk hills   Native

Rare, or overlooked. Recorded also for Bucks.

OUZEL: Dunstable Downs; *G. C. Druce in B.E.C. 1923 Rep.*

IVEL: Pegsdon, det. H. W. Pugsley; *J.E.L.*, *F.F.B.*

First record: G. C. Druce, 1923.

G. germanica (Willd.) E. F. Warb. C.T.W. 94.6.3.

Gentiana germanica Willd.

Pastures, old chalk pits, etc.   Native

Rare, and always appearing with us in places where the soil has been disturbed at some time. It is a handsome species, with intermediates with *G. amarella* that merit some further study. Recorded also for Herts. and Bucks.

OUSE: Ravensden Hill; *E. M. Langley in J.H. Notes*, a doubtful record.

IVEL: Harlington; *D. M. Higgins in W.B.E.C. 1894–5 Rep.*: chalk pit, Chalton Cross; *W.F.B.*: locally plentiful in a few old chalk pits near Sundon*.

COLNE: rough pasture near Greencroft Barn, v.c. 20 [Beds.]*.

First record: D. M. Higgins, 1895.

G. campestris (L.) H. Sm. C.T.W. 94.6.1. (*Gentiana campestris* L.). There is doubt about the appearance of this in the county. Abbot recorded it in *Plantae Bedfordiensis* from Barton Hills and Dunstable, but his herbarium contains no specimen. It is possibly a repetition of Parkinson's record, which refers to *G. amarella*. It is recorded for all neighbouring counties except Cambs., but in all cases with some doubt.

MENYANTHACEAE

MENYANTHES L.

Menyanthes trifoliata L. C.T.W. 95.1.1. *Bogbean, Buckbean*
Marshes, wet meadows, etc. Native
Rare, and growing only in places for the greater part of the year
under water. Recorded for all neighbouring counties.
H.S. 39a, 41.
Ouse: Paradise, near Bedford; Biddenham South Field, where I am
told it still grows; Kempston Church, still apparently growing near
here; *F.B.*: Felmersham, *R. Lucas*.
Ouzel: pondside, Heath and Reach, possibly planted.
Ivel: Ampthill Moor; *F.B.*: Gravenhurst; *W.C. Herb.*: Flitton,
locally abundant on the Moor!; *S.B.P.L.*: Ridgmont; *C.C. in W.F.B.*:
Westoning Moor!; *G. D. Nicholls*: by the lake, Hawnes Park,
possibly planted; *C. C. Foss*: locally abundant in the wettest parts of
Flitwick Moor*.
Lea: Limbury†; Leagrave Common, where the leaves were to be
seen until 1948—with the gradual drying out of the marsh it had
not flowered for many years; *W.F.B.*
First record: Abbot, 1798.

NYMPHOIDES Hill

Nymphoides peltata (S. G. Gmel.) Kuntze C.T.W. 95.2.1.
N. nymphoides (L.) Druce *Fringed Waterlily*
Ponds, etc. Native
Rare: recorded for all neighbouring counties except Herts.
Ouzel: Leighton Buzzard; *H. Brown*: Stanbridgeford, in a pond,
since dried†; *S. A. Chambers in F.F.B.*
First record: H. Brown, 1839.

BORAGINACEAE

CYNOGLOSSUM L.

Cynoglossum officinale L. C.T.W. 97.1.1. *Hound's-tongue*
Woods, rough pastures, etc. Native
Frequent on heavy clay soils in the Ouse district, rare elsewhere.
A white-flowered form has been seen at Wrestlingworth. Recorded
for all neighbouring counties.
Ouse: Bromham; Biddenham; Elstow; *F.B.*: Willington; *J.McL.*,
L.M.: Cox's Pits; Wilshamstead!†; *J.McL.*, *B.M.*: Renhold;
Pavenham!; *W.H. 1875*: Mowsbury; Danish camp, Howbury;
J.H. Notes: Cleat Hill; *V.C.H.*: Great Barford; Oakley; Felmersham;
G. H. Day: Radwell; *K. Hayes in B.E.C. 1934 Rep.*: Tempsford;
A. W. Guppy: Kempston Wood End; *R. Townsend*: Cardington*;
L. W. Wilson: Woodcraft; Wootton Wood.
Ivel: Beadlow; *Miss Harradine in J.H. Notes*: lane by Tingley Wood;
H.Cole: chalkpits and rough pastures, Pegsdon, apparently increasing.

CAM: Eyeworth; plentiful in rough pasture, Brook Farm, Wrestling-worth.
LEA: near Biscot; *Davis.*
First record: Abbot, 1798.

C. germanicum Jacq. C.T.W. 95.1.2.   *Green Hound's-tongue*
C. montanum auct.
Rough pastures, etc.   Native
Rare: records from Wilstead, in *B.P.L.*, and Cleat Hill, in *J.H. Notes,* are to be doubted, as there are no specimens to support them. Recorded for all neighbouring counties except Herts. and Cambs.
OUSE: Wootton Wood, where I have been unable to re-find it*; *the author with B. B. West, in B.E.C. 1943–4 Rep.* 738.
OUZEL: between Leighton Buzzard and Woburn, probably in Beds.: *A. Wallis in Fl. of Bucks.,* 233.
First record: A. Wallis, 1926.

SYMPHYTUM L.

Symphytum officinale L. C.T.W. 97.4.1.   *Comfrey*
Riversides, streamsides, ditches, etc.   Native
Frequent by the Ivel and its streams, rare elsewhere. Abbot recorded two species, *S. officinale* and *S. patens* Sibth., both of which, it is interesting to note, he considered rare. *S. patens* is now considered to be a red-flowered variety of *S. officinale,* and J. E. Little made a number of records of it as such. Red-flowered forms are very common with us; indeed I have not seen in Bedfordshire the cream-flowered forms which are so much a feature of the Hunting-donshire fens. It is possible that pure *S. officinale* is absent with us and that all that passes for it is the result of hybridization with *S. peregrinum.* The records given below refer to plants which appear to agree with the description of *S. officinale.* Recorded for all neighbour-ing counties.
OUSE: Bedford: *Abbot's annotated F.B.*: ditches, Holcot Wood.
KYM: ditches, Dean.
OUZEL: Bidwell!; *V. H. Chambers*: Husborne Crawley.
IVEL: Chicksands, as *S. patens*; *W.C. Herb.*: Greenfield!†; Barton!*; *J.S. in Bot. Rec. Club 1881–2 Rep.*: Sharpenhoe!; Flitwick!; *J.S.,* in *J. Bot. 1883*; Biggleswade; *J.H. Notes*: Clifton!; Campton!; north of Radwell; *J.E.L., F.F.B.*: Clophill!; Flitton!; *V. H. Chambers*: Harlington; Ampthill; Shefford.
LEA: Luton Hoo!; Leagrave; *W.F.B.*: East Hyde.
COLNE: Caddington.
First record: Abbot, 1798.

S. peregrinum Ledeb. C.T.W. 97.4.3.   *Blue Comfrey*
Roadsides, waste places, ditches, etc.   Colonist
This is the common comfrey of the county. Its history and the nature of its increase would well repay study. Recorded for all districts except Colne, and for all neighbouring counties.
First evidence: by the Ivel, Shefford, as *S. patens*; *Herb. McLaren, Luton Museum,* 1864.

S. tuberosum L. *Tuberous Comfrey*
Woods, etc. Denizen
Rare: in addition to the record given below it occurs in churchyards as an obvious
escape from cultivation. Recorded for all neighbouring counties except Bucks.
LEA: near the waterfall in the park, near Luton, it is still abundant in a few woods
and in waste places at Luton Hoo!†*; *Susan Foster in Herb. Watson*, 1841.

PENTAGLOTTIS Tausch

Pentaglottis sempervirens (L.) Tausch C.T.W. 97.7.1. *Alkanet*
Anchusa sempervirens L.
Waste places, etc. Colonist
Rare, but usually well established. Recorded for all neighbouring counties.
OUZEL: Aspley; *W.H. 1875*: Aspley Heath, v.c. 24 [Beds.]; *V. H. Chambers*:
plentiful at Garside's Pits, Leighton Buzzard*.
IVEL: Cainhoe; *W.C. Herb.*: Tingrith; *J.E.L., F.F.B.*: Galley Hill, Sutton!;
H. B. Souster.
CAM: churchyard, Wrestlingworth*.
LEA: waste ground, Luton; *W. N. Thorpe.*
First record: W. Crouch, 1845.

LYCOPSIS L.

Lycopsis arvensis L. C.T.W. 97.8.1. *Field Bugloss*
Rough pastures, waste places, sand and gravel pits Colonist
Common on the Lower Greensand and frequent on gravels.
Recorded for all neighbouring counties.
H.S. 48.
OUSE: Elstow; Newnham; Bedford; *F.B.*: arable fields, Oakley;
H. B. Souster!
OUZEL: too many stations on the Lower Greensand to merit listing.
IVEL: common on the Lower Greensand and on glacial gravels in
the neighbourhood of Fancott, Toddington and Barton.
LEA: near Biscot; *Davis.*
First record: Abbot, 1798.

MYOSOTIS L.

Myosotis scorpioides L. *Water Forget-me-not*
M. palustris (L.) Hill C.T.W. 97.10.1.
Wet places, damp woodland rides, etc. Native
Common, and well distributed throughout the county. A record of
the closely allied *M. secunda* A. Murr. (*M. repens* auct.) by the author
in *B.E.C. 1943–4 Rep.* was in error but an unlocalized specimen in
Herb. McLaren, British Museum is this. Recorded for all districts
and neighbouring counties.
H.S. 6–9, 10b, 13, 14b, 37, 43, 57a, 57b, 84a–85, [27].
First record: Abbot, 1798.

M. cespitosa K. F. Schultz C.T.W. 97.10.4.
Wet and marshy places Native
Less widely distributed than *M. scorpioides*, but often replacing it on
calcareous soils. Recorded for all districts except Colne, and for all
neighbouring counties.
H.S. 37.
First record: *Newbould MS.*, 1873.

M. sylvatica Hoffm., C.T.W. 97.10.7., is of doubtful occurrence in the county. It was listed from Basmead (*C.R.O. List*) and in the *Hillhouse List, 1876*, with no details. There are no specimens to support Saunders's records in *Field Flowers of Bedfordshire* from Tilsworth and Limbury. It has appeared as a garden outcast at Clapham Park Wood*, and as such has been recorded from Herts., Bucks. and Northants. Recorded, apparently as the native species, from Cambs.

M. arvensis (L.) Hill C.T.W. 97.10.8.   *Common Forget-me-not*
Arable fields, woodland rides, waste places, etc.   Colonist
Common: robust specimens have been named var. *sylvestris* Schlecht. and have been mistaken for *M. sylvatica*, and dwarfed specimens mistaken for *M. hispida*. Recorded for all districts and neighbouring counties.
H.S. 3a, 12, 19a, 20a, 23, 27, 31a, 31b, 34a, 43, 45b, 52–54, 55b, 61, 69, 71, 72, 77b, 81–83, 86.
First record: Abbot, 1798.

M. hispida Schlecht. C.T.W. 97.10.10.   *Early Forget-me-not*
  M. collina auct.
Dry pastures, heathy places, old walls, etc.   Native
Frequent on the Lower Greensand, rare elsewhere. Recorded for all neighbouring counties.
H.S. 22b, 47.
Ouse: Cardington*; *J.McL. in J. Bot. 1884*: railway bank, Souldrop; *E.M.-R.*!: gravel pit, Willington.
Ouzel and Ivel: too many stations to merit listing.
Cam: Cockayne Hatley Wood; *J.E.L., Diary*.
Lea: Leagrave; Luton Hoo!; *F.F.B.*
Colne: near Half Moon, south of Dunstable, as var. *mittenii* H. C. Wats.; *Fl. Herts.*: Whipsnade Common!*; *V. H. Chambers*.
First record: J. McLaren, Herb. Luton Museum, 1864.

M. discolor Pers. C.T.W. 97.10.9.   *Variegated Forget-me-not*
  M. versicolor Sm.
Heaths, arable fields, old walls, etc.   Native
More common than *M. hispida*, but like it in being limited to light and well-drained soils. Recorded for all neighbouring counties.
H.S. 22b, 45a, 77a.
Nene: locally plentiful on railway embankment, Wymington*.
Ouse: Cardington†; *J.McL. in Herb. J.S.*: Marston Thrift, in more heathy parts of the wood; railway embankment between Souldrop and Sharnbrook.
Ouzel and Ivel: too many stations especially on the Lower Greensand to merit listing.
Lea: near Luton; *J.S. in J. Bot. 1883*: Horsley's Wood.
Colne: rough ground near Deadmansey Wood; *E.M.-R.*!: arable field, Kensworth; arable field near Ravensdell Wood, v.c. 20 [Beds.].
First record: Clophill, W. Crouch, 1845.

LITHOSPERMUM L.

Lithospermum officinale L. C.T.W. 97.11.1. *Common Gromwell*
Woodland verges, hedgerows, rough pastures, etc. Native
Common on basic clays and calcareous soils in the north of the
county, infrequent elsewhere. It was considered rare by Abbot.
Recorded for all neighbouring counties.
H.S. 1a.
NENE: common in this district.
OUSE: frequent especially to the north of the river.
KYM: Knotting!*; *L. W. Wilson*: Keysoe!; Newton Gorse; *C. C. Foss.*
OUZEL: Listed in *V.C.H.* but I have not seen it.
IVEL: Cainhoe; *W.C. Herb.*: Hammer Hill; *B.P.L.*: Warden Wood;
J.McL. in J.H. Notes: Pegsdon Hills!; *F. S. Lloyd in F.F.B.*: Sheer-
hatch Wood; *J.E.L., J. Bot.*: Hawnes Park.
CAM: Dunton; *H. B. Souster.*
LEA: Moat Island, Limbury; *W.F.B.* [other records by Saunders in
S.B.P.L. are probably in error, as the same stations are given for
L. arvense in *W.F.B.*].
First record: Abbot, 1798.

L. arvense L. C.T.W. 97.11.3. *Corn Gromwell*
Arable fields   Colonist
Common, especially on calcareous soils. Recorded for all districts
and neighbouring counties.
H.S. 4, 12, 72, 86.
First record: Abbot, 1798.

ECHIUM L.

Echium vulgare L. C.T.W. 97.13.1. *Viper's Bugloss*
Rough pastures   Colonist
Not infrequent, and with us growing on a great variety of soils.
Recorded for all neighbouring counties.
OUSE: Clapham; Stevington; *Abbot's annotated F.B.*: Basmead;
C.R.O. List: Cardington Mill; *J.McL. in J.H. Notes*: railway
siding, Elstow; *A. W. Guppy.*
KYM: Worley's Wood; *C. C. Foss!*
IVEL: near Pegsdon; *H. Brown*: Cainhoe; *W.C. Herb.*: near Barton;
Davis: Potton; *J.McL., L.M.*: Southill; *J.McL., B.M.*: Sandy;
Millbrook; *W.H. 1875*: Flitwick Mill; *M. L. Berrill in J.H. Notes*:
Clophill; Steppingley; *C.C.*: Barton Hills, plentiful on glacial
gravel!; old chalk quarries, Sundon.
LEA: railway bank, south of Luton, where it is plentiful!*; New
Mill End†; *W.F.B.*
First record: Abbot, 1798.

CONVOLVULACEAE

CALYSTEGIA R.Br.

Calystegia sylvestris (Willd.) Roem. & Schult. C.T.W. 98.2.2.
Volvulus inflatus Druce p.p.
Hedgerows, waste places, etc.   Colonist
Almost as common in the county as *C. sepium*. Recorded for all districts and neigh-
bouring counties.

C. sepium (L.) R.Br. C.T.W. 98.2.1.   *Great Bindweed*
Convolvulus sepium L., *Volvulus sepium* (L.) Junger
Hedgerows, etc.   Native
Common: pink-flowered forms are not infrequent. Recorded for all
districts and neighbouring counties.
H.S. 7, 10c, 13, 42.
First record: Abbot, 1798.

A record of *C. soldanella* (L.) R.Br. from Cardington; *J.McL. in J.H. Notes* was
obviously in error.

CONVOLVULUS L.

Convolvulus arvensis L. C.T.W. 98.1.1.   *Bindweed, Wheatbine*
Arable fields, roadsides, waste places, etc.   Colonist
Common, and a pest on arable land. Recorded for all districts
and neighbouring counties.
H.S. 1b, 4, 12, 17, 44, 48, 61, 64, 71–73, 86.
First record: Abbot, 1798.

CUSCUTA L.

Cuscuta epilinum Weihe C.T.W. p. 850.   *Flax Dodder*
A parasite on flax   Casual
Very rare: recorded for all neighbouring counties except Hunts.
OUSE: Park Lane; *J.McL., B.M.*
IVEL: Stondon; *R. Long in J.E.L., F.F.B.*
First record: J. McLaren, c. 1880.

C. europaea L. C.T.W. 98.3.1.   *Great Dodder*
Rough ground usually near rivers   Native
With us a parasite on *Urtica dioica*, but spreading to adjoining
plants. It is rare, but usually reappears in the same place for a
number of years. Considered common by Abbot. Recorded for all
neighbouring counties.
OUSE: Odell; *B. B. West*: riverside, Harrold Bridge*, Milton
Ernest, Stevington.
OUZEL: Totternhoe; *S.B.P.L.*: Leighton Buzzard!; *B. Verdcourt.*
IVEL: Flitwick; *R. H. Webb, see R. A. Pryor in J. Bot. 1881*: Cainhoe;
W.C. Herb.: Kitchen End*; Hillfoot, Pulloxhill†*; *C.C.*: Clophill;
F. L. Chesham.
LEA: Dumb Hills, a doubtful record; *S.B.P.L.*
First record: Abbot, 1798.

CUSCUTA

C. epithymum (L.) Murr. C.T.W. 98.3.2. *Lesser* or *Common Dodder,*
Hair Weed, Beggar Weed (Batchelor)
Heaths, pastures, etc. Native
A parasite, and found in Bedfordshire mainly on *Calluna, Thymus*
pulegioides and various species of clover. Rare, notwithstanding the
large number of records. Recorded for all neighbouring counties.
Ouse: Basmead; *R. A. Pryor in Newbould MS.*: Salford; *V.C.H.*: on
clover, Harrold*; *G. H. Day*: on clover, near Judge's Spinney!*;
R. Lucas.
Ouzel: Heath and Reach, it still grows on *Calluna* on Rushmere
Heath!*; *J.S. in J. Bot. 1889*: Leighton; *V.C.H.*: Totternhoe;
F.F.B.: Woburn Park, on *Calluna*; New Wavendon Heath, v.c. 24
[Beds.], on *Calluna.*
Ivel: Ampthill Warren; Barton Hill, where it still grows on
Thymus!; *F.B.*: Rowney Warren†; *J.McL., L.M.*: Southill; *J.S. in*
J. Bot. 1889: Pegsdon; *J. Pollard*: Harlington; *F.F.B.*
Lea: Dunstable; *Journ. Roy. Agric. Soc.* 1873: Someries; *B.P.L.*: near
Luton; Biscot†; *J.S. in J. Bot. 1889*: railway, Luton†; *Herb. J.S.*:
Stopsley Common*; *J.S.*: Luton; *D. M. Higgins in W.B.E.C. 1905*
Rep.: Warden Hills; Limbury; *F.F.B.*
First record: Abbot, 1798.

SOLANACEAE

SOLANUM L.

Solanum dulcamara L. C.T.W. 99.4.1. *Woody Nightshade,*
Bittersweet
Hedgerows, waste places, woods especially near water, etc. Native
Common, and widely distributed in varied habitats. Recorded for
all districts and neighbouring counties.
H.S. 6, 13, 39a, 40, 43, 64, 74b, 83, 84b, 85.
First record: Abbot, 1798.

S. nigrum L. C.T.W. 99.4.2. *Black Nightshade*
Arable fields Colonist
Common on the richer soils, especially on the alluvium in the Ouse
and Ivel valleys. The well-marked var. **atriplicifolium** (Desf.)
Dunal, is found frequently and is probably a wool adventive. Re-
corded for all districts except Kym, and for all neighbouring counties.
H.S. 13, 17, 34a, 50.
First record: Abbot, 1798.

ATROPA L.

Atropa bella-donna L. C.T.W. 99.2.1. *Deadly Nightshade*
Hedgerows, old gardens, waste places, etc. ? Native
Rare, and in none of its recent stations appearing to be native,
though it is possible that it was native in Abbot's station. Recorded
for all neighbouring counties.

NENE: old garden, Hinwick House*.

OUSE: Bromham; *J.McL.*, *B.M.*: Marston; *G. E. Brown in C. Crouch's annotated copy of W.F.B.*

IVEL: Shefford; *W.C. Herb.*: Ampthill station, where it was well established until 1950 when it was cut down!†; *F.F.B.*: Vicarage garden, Harlington; *C.C.*: brickworks, Arlesey; *J.E.L.*, *J. Bot.*: roadside, Broom; *I. M. Allison*: Stotfold; *K. B. West*: tomato fields, Meppershall*; *shown to me by a W.I. member*: railway sidings; Flitwick.

LEA: woods, Luton Hoo†; *S.B.P.L.*: yard, Red Lion Hotel, Dunstable!; *C. C. Foss.*

COLNE: near Whipsnade; *F.B.*

First record: Abbot, 1798.

DATURA L.

Datura stramonium L. C.T.W. 99.5.1.   *Thorn-apple*
Arable fields, etc.   Casual
Frequent. As this occurs so regularly in fields where shoddy has been applied there can be no doubt that in addition to being a normal casual it is also a wool adventive. In these conditions a purple-flowered form (*D. tatula* L.) is not infrequent. It is never established, and there seems to be little point in listing its many stations. Recorded for all districts except Kym and Colne, and for all neighbouring counties.
First record: Goldington Fields, *Herb. J. McLaren, Luton Museum*, 1864.

HYOSCYAMUS L.

Hyoscyamus niger L. C.T.W. 99.3.1.   *Henbane,*
                                                 Alice Water (Harrold, etc.)
Arable fields, disused sand and gravel pits, etc.   Colonist
Not infrequent, and often well established. It was considered common by Abbot. Recorded for all neighbouring counties.

NENE: Podington; well established in old gravel pits; *R. R. B. Orlebar.*

OUSE: St. Cuthbert's; *Abbot's annotated F.B.*: cornfields, Biddenham; *J.McL.*, *L.M.*: Cardington Mill; *J.McL.*, *B.M.*: Cotton End; Stagsden; *B.P.L.*: Bromham Mill; *E. M. Langley in J.H. Notes*: Elstow Abbey; *A. W. Guppy*: Honey Hills; *R. Townsend*: Howbury Park; *J. Wooding*: Castle Hill, Eaton Socon; *C. C. Foss*: Roxton; *G. H. Day*: Eastcotts; *Miss Tabb*: Milton Ernest; *B. B. West*: Blunham; *E.M.-R.*: garden weed, Houghton Conquest.

OUZEL: old sandpit, Leighton Buzzard.

IVEL: Maulden; *M. L. Berrill in J.H. Notes*: Biggleswade, at one time cultivated here for medicinal purposes; *J.H. Notes*: Brook Green Farm, Barton; *G. Horsler*: weed in small holding, Flitwick.

LEA: Barton Road; *Davis*: Limbury†; *J.S. in J. Bot. 1883*: south of Luton; *W.F.B.*: Arden Dell Wood; *P.T.*: Dunstable; *H. B. Souster*: waste ground, Round Green; Chiltern Green.

COLNE: road between Markyate and Dunstable, v.c. 20 [Beds.]; *Fl. Herts.*: waste ground near Whipsnade Heath!; *S. A. Chambers in F.F.B.*: Aley Green; *L. J. Margetts.*

First record: Abbot, 1798.

SCROPHULARIACEAE
VERBASCUM L.

Verbascum thapsus L. C.T.W. 100.1.1. *Great Mullein, Aaron's Rod*
Roadsides, rough pastures, waste places, etc. Colonist
Frequent on lighter soils, and usually well established, but in many cases an obvious garden escape. It is recorded for all districts, with too many stations to merit listing, and for all neighbouring counties.
H.S. 25.
First record: Abbot, 1798.

V. nigrum L. C.T.W. 100.1.4. *Dark Mullein*
Roadsides, waste places, railway banks, etc. ? Native
More frequent on the Chalk than *V. thapsus*, and elsewhere growing sparingly on well drained soils. Little noted a cream-flowered variety at Southill (*J. Bot. 1919*). It was considered rare by Abbot. Recorded for all neighbouring counties.
H.S. 69.
OUSE: Harrold; *G. H. Day*: railway sidings, Bedford.
OUZEL: Aspley; *F.B.*: Woburn; *Abbot's annotated F.B.*
IVEL: Sandy; Warden; *F.B.*: Clophill; *W.C. Herb.*: Stanford; *J.McL. in J.H. Notes*: Shefford; *A. W. Guppy*: Barton, plentiful in disused chalk pits, etc.!; Southill; *J.E.L., J. Bot.*
CAM: Cockayne Hatley; *T.L.*
LEA: abundant, Luton; *J. Anderson in New Bot. Guide II*, 1837: Dallow Lane!†; *Herb. J.S.*: gravel pit, Limbury*; railway between Chaul End and Skimpot, plentifully; roadside near Warden Hills; frequent as a weed of waste places at Round Green and Stopsley.
COLNE: south of Dunstable, where it has since been found by L. J. Margetts, v.c. 20 [Beds.]; *Fl. Herts.*
First record: Abbot, 1798.

LINARIA Mill.

Linaria vulgaris Mill. C.T.W. 100.3.4. *Toadflax*
Antirrhinum linaria L., *L. linaria* (L.) Karst.
Old chalk pits, railway sidings, etc. Colonist
Locally abundant on the Chalk, and on large railway sidings, but absent from considerable areas in the North Bedfordshire Uplands and Lower Greensand. Recorded for all neighbouring counties.
H.S. 1b, 15, 38, 63b, 66b, 71–73.
NENE: railway bank, Wymington.
OUSE: Fenlake; St. Leonard's; *Abbot's annotated F.B.*: Harrold; *G. H. Day*: Bedford, plentiful on waste ground near the railway!; *A. W. Guppy*: roadside, Felmersham.
OUZEL: rare, except on the Chalk, and the railway; roadside, Battlesden.
IVEL: plentiful on the Chalk: Clifton; Arlesey; *J.E.L., F.F.B.*: Shefford; *H. B. Souster*: Biggleswade Common; Flitwick Moor.
CAM: roadside, Wrestlingworth.
LEA: very common in this district.
COLNE: common on the Chalk: roadside, Studham.
First record: Abbot, 1798.

L. repens (L.) Mill. C.T.W. 100.3.3. *Pale Toadflax*
Old chalk pits, railway ballast, etc. Colonist
Not infrequent in chalk pits, and a marked feature of railway ballast in the south of the county. Recorded for all neighbouring counties.
OUSE: on Park Road, ? Cardington; *J.McL., L.M.*
OUZEL: railway, Sewell; *E.M.-R.*
IVEL: arable field, Lower Sundon; *H. B. Souster*: railway, Sundon, Harlington, Ampthill.
LEA: Dunstable Road, Luton; *J. Foster in Herb. Watson, 1841, also H. Brown, 1841*: St. Ann's Hill; *Davis*: Skimpot, where it is still plentiful*; *R. A. Pryor in J. Bot. 1876*: chalk pit, foot of Dunstable Downs!; *V. H. Chambers*: railway, Dallow Road, Luton L.N.E. Station, Dunstable, Leagrave, East Hyde.
First record: J. Foster or H. Brown, 1841.

L. repens × vulgaris
IVEL: railway between Sundon and Harlington, with both parents*; *P.T.*!
LEA: railway, Luton!; *R. A. Pryor in J. Bot. 1876.*

CHAENORHINUM (DC.) Reichb.

Chaenorhinum minus (L.) Lange C.T.W. 100.4.1. *Small Toadflax*
Antirrhinum minus L., *Linaria minor* (L.) Desf.
Railway ballast, arable fields, etc.   Colonist
Common everywhere on the railway, usually growing between the
metals, elsewhere appearing occasionally as a weed of arable land
on calcareous soils. The records given below exclude those for the
railway. It was considered rare by Abbot. Recorded for all
neighbouring counties.
H.S. 45a, 72.
NENE: arable field, Hinwick.
OUSE: Oakley West Field; *Plantae Bedford.*: between Bedford and
Harrowden; *J.McL., L.M.*: Willington; *E. M. Langley in J.H. Notes*:
gravel pit, Eaton Socon*.
KYM: arable field near Pertenhall; *C. C. Foss.*
OUZEL: arable field, Totternhoe.
IVEL: Barton Hills!; *Plantae Bedford.*: Cainhoe; *W.C. Herb.*: Shefford
Hill; *J.McL., L.M.*: Stotfold; *J.E.L., F.F.B.*: arable field, Wingfield;
V. H. Chambers: arable field, Pegsdon.
CAM: arable field, Dunton.
LEA: Leagrave!; *S.B.P.L.*: Lynches; *H. B. Souster*: arable field,
Galley Hill!; *V. H. Chambers*: arable fields, Luton*.
COLNE: arable field, Pepperstock; chalk quarry, Mount Pleasant.
First record: Abbot, 1795.

KICKXIA Dum.

Kickxia spuria (L.) Dum. C.T.W. 100.5.1.   *Fluellen*
Antirrhinum spurium L., *Linaria spuria* (L.) Mill.
Arable fields   Colonist
Well distributed throughout the county, and more common on light
soils than elsewhere. Considered rare by Abbot. Recorded for all
districts and neighbouring counties.
H.S. 4, 71.
First record: Abbot, 1798.

K. elatine (L.) Dum. C.T.W. 100.5.2.   *Sharp-leaved Fluellen*
Antirrhinum elatine L., *Linaria elatine* (L.) Mill.
Arable fields   Colonist
Less common than *K. spuria*, but often found growing in association
with it. It was also considered rare by Abbot. Recorded for all
neighbouring counties.
NENE: Podington.
OUSE: Bromham; Clapham; *F.B.*: Basmead; *C.R.O. List*: Cotton
End; *M. L. Berrill in J.H. Notes*: Roxton; *G. H. Day*: Ravensden;
B. B. West: Souldrop; Stagsden.

KYM: Keysoe; *B. B. West*: Knotting.
OUZEL: top of Lord's Hill!; *W. D. Coales*: Houghton Regis!*;
E.M.-R.: Totternhoe.
IVEL: Cainhoe; *W.C. Herb.*: Ampthill; *J.H. Notes*: Shefford;
Pegsdon!*; *J.E.L.*, *F.F.B.*: Wingfield; Barton!; *V. H. Chambers*:
Streatley; *H. B. Souster*: Toddington; Northill; Silsoe.
CAM: Dunton.
COLNE: Chaul End; *H. B. Souster*.
First record: Abbot, 1798.

CYMBALARIA Hill

Cymbalaria muralis Gaertn., Mey. & Scherb. C.T.W. 100.6.1.
Ivy-leaved Toadflax
Antirrhinum cymbalaria L., *Linaria cymbalaria* (L.) Mill.
Old walls   Wall denizen
Frequent on old walls and similar habitats throughout the county. Considered
rare by Abbot. Recorded for all districts except Cam, and for all neighbouring
counties.
First record: Abbot, 1798.

ANTIRRHINUM L.

Antirrhinum orontium L. C.T.W. 100.2.1.   *Lesser Snapdragon*
Arable fields   Colonist
Rare: recorded for all neighbouring counties except Hunts.
OUZEL: Leighton Buzzard Heath; *J. Anderson* in *New Bot. Guide, II*:
sandy field, Heath and Reach, possibly the same station* (H.S. 44).
First record: J. Anderson, 1837.

SCROPHULARIA L.

Scrophularia nodosa L. C.T.W. 100.7.1.   *Figwort*
Damp woods, hedgerows, etc.   Native
Common and widely distributed: recorded for all districts and
neighbouring counties.
H.S. 19a–20a, 31a, 34a, 52, 53b, 54, 77a, 78, 79b, 83.
First record: Abbot, 1798.

S. aquatica L. C.T.W. 100.7.2.   *Water Figwort, Water Betony*
Riversides, wet ditches, etc.   Native
Common in all suitable habitats. J. E. Little recorded plants as var.
pubescens Bréb. from Cadwell in *B.E.C. 1923 Rep.* Recorded for all
districts and neighbouring counties.
H.S. 6, 7, 10b, 10c, 13, 14a, 16, 19a, 42, 43, 57a, 85.
First record: Abbot, 1798.

MIMULUS L.

Mimulus guttatus DC. C.T.W. 100.8.1.   *Monkey-flower*
    M. langsdorffii Donn ex Greene
Streamsides   Denizen
Rare: recorded for Herts., Bucks. and Northants.
IVEL: ditch by footpath between Ampthill and Maulden, 1931, *C.C.*: streamside,
Stotfold, where it has probably escaped from an ornamental garden; *H. and D.
Meyer*.
LEA: riverside, East Hyde, where it was established for about forty years, but
became extinct about 1936 when Luton sewage works were planned!*; *W.F.B.*
First record: Saunders, 1897.

Y

LIMOSELLA L.

Limosella aquatica L. C.T.W. 100.9.1.   *Mudwort*
Shores of lakes, ponds, etc.   Native
Rare: *E. Bot.* tab. 357, Nov. 1796, was drawn from a specimen sent by Abbot. Recorded for all neighbouring counties.
Ouse: Goldington Green, it was still there in McLaren's time; *Plantae Bedford.*: Fenlake†*; *J.McL.*, *B.M.*: Cardington, probably the same station†; *J.McL. in Herb. J.S.*
Ouzel: plentiful by shore of Lower Drakelow Pond*.
First record: Abbot, 1795.

DIGITALIS L.

Digitalis purpurea L. C.T.W. 100.12.1.   *Foxglove*
Cleared woodland, rough pastures, etc.   Native
Frequent, and undoubtedly native in open woods on the Lower Greensand and Clay-with-Flints; a garden escape elsewhere. White-flowered forms are not unusual. It was considered rare by Abbot. Recorded for all neighbouring counties.
H.S. 28.
Ouse: Ampthill Park; *H. B. Souster*!
Ouzel: common on the Lower Greensand in this district.
Ivel: Maulden!; *F.B.*: Clophill!; *J.McL. in J.H. Notes*: Ampthill; *F.F.B.*: Flitwick Moor!; *R. Morse in J.E.L. Diary*: White Wood, v.c. 30 [Hunts.]; *H. B. Souster.*
Lea: Luton Hoo, planted!†; *F.F.B.*: Horsley's Wood; *H. B. Souster*: waste ground, Luton*.
Colne: Deadmansey Wood; Greencroft Barn, v.c. 20 [Beds.].
First record: Abbot, 1795.

VERONICA L.

Veronica officinalis L. C.T.W. 100.13.5.   *Common Speedwell*
Heathy woods, heaths, etc.   Native
Common in woods on the Lower Greensand and Clay-with-Flints, but absent in large areas of the county. Recorded for all neighbouring counties.
H.S. 3b, 19a, 19b, 23, 25, 28, 30a, 53b, 54, 77a, 79a–80a, 81.
Nene: Great Hayes Wood; Wymington Scrubs.
Ouse: Exeter Wood; *J.McL.*, *B.M.*: Cotton End; *M. L. Berrill in J.H. Notes*: Hanger Wood; *C. C. Foss*: Harrold; *A. W. Guppy*: Marston Thrift; rough pasture, Colesdon*; Sheerhatch Wood; King's Wood.
Kym: West Wood; Swineshead Wood; Galsey Wood; Dean.
Ouzel and Ivel: common in woods and on heathy pastures on the Lower Greensand, absent or rare on the Oxford Clay and Gault.
Lea: Kidney Wood; Horsley's Wood.
Colne: frequent in woods on the Clay-with-Flints.
First record: Abbot, 1798.

V. chamaedrys L. C.T.W. 100.13.7.   *Germander Speedwell*
Woodland rides, shady places, heaths, etc.   Native
Common throughout the county: recorded for all districts and
neighbouring counties.
H.S. 18, 22b, 28, 34a, 34b, 51b–53a, 61, 63a, 69, 78–80a, 81, 83.
First record: Abbot, 1798.

V. montana L. C.T.W. 100.13.6.   *Wood Speedwell*
Woods   Native
Not infrequent in woods on light soils, especially on the Clay-with-
Flints. Considered rare by Abbot, and apparently not known to
Hamson and Saunders. Recorded for all neighbouring counties,
but with some doubt for Northants.
OUZEL: Aspley Wood; *V. H. Chambers*: wood, Battlesden Park.
IVEL: 'ditch behind church, Eversholt'; *Plantae Bedford*.
LEA: Arden Dell Wood*; *V. H. Chambers*!: frequent in woods in
Luton Hoo Park; George's Wood, plentifully.
COLNE: Folly Wood; *E.M.-R.*!: Elm Grove Wood, v.c. 20 [Beds.]*;
Ravensdell Wood, v.c. 20 [Beds.]; Kidney Wood.
First record: Abbot, 1795.

V. scutellata L. C.T.W. 100.13.4.   *Marsh Speedwell*
Marshes   Native
Rare: recorded for all neighbouring counties.
OUSE: Fenlake; Stevington; *F.B.*: Basmead; *C.R.O. List*: Marston
Wood End; *V. H. Chambers*: Bromham; *B. B. West*: plentiful in
marsh at Tempsford*.
OUZEL: Northall (? Beds.)†; *S.B.P.L.*: sparingly in marsh near
Heath and Reach*.
IVEL: Ampthill; *Plantae Bedford.*: Gravenhurst; *W.C. Herb.*: Flitwick,
plentiful on the Moor*; *S.B.P.L.*
LEA: Limbury†; *S.B.P.L.*
First record: Abbot, 1795.

V. beccabunga L. C.T.W. 100.13.1.   *Brooklime*
Ponds, streams, ditches, etc.   Native
Common in all likely places in the county. Recorded for all districts
and neighbouring counties.
H.S. 6, 8, 10a–10c, 14b, 56a, 57a.
First record: Abbot, 1798.

V. anagallis-aquatica L. C.T.W. 100.13.2.   *Water Speedwell*
Streams, ponds, ditches, etc.   Native
More frequent than the records given below would indicate; its
full distribution in the county needs to be worked out. Recorded
for all neighbouring counties.
OUZEL: Eaton Bray; *G. C. Druce in B.E.C. 1927 Rep.* 313.
IVEL: Gravenhurst; *W.C. Herb.*: Fancott*.
LEA: Marslets†; *Herb. J.S.*: Leagrave Marsh*.

First record: for the aggregate, Abbot, 1798; for the restricted species, G. C. Druce, 1927.
First evidence: W. Crouch, 1844.

V. anagallis-aquatica × **catenata**
IVEL: Ivel Navigation, Shefford*, det. J. H. Burnett; *E.M.-R.*!

V. catenata Pennell C.T.W. 100.13.3.   *Water Speedwell*
V. aquatica Bernh. non Gray
Marshes, pondsides, etc.   Native
Probably more common than the preceding species, and certainly more widely distributed. It appears to grow more readily in marshy places, while *V. anagallis-aquatica* is more common in water. Recorded for all neighbouring counties.
H.S. 8, 14b.
NENE: marsh, Hinwick*.
OUSE: Cople; *J. P. M. Brenan*: Great Barford; *E.M.-R.*: water meadows, Eaton Socon*, Felmersham, Sharnbrook.
KYM: pondside, Pertenhall*; marsh, Melchbourne*.
OUZEL: Eaton Bray; *G. C. Druce in Fl. Bucks.*
IVEL: Biggleswade; Warden Abbey!; *J.E.L.*, *J. Bot.*: pondside, Pulloxhill*.
CAM: pondside, Cockayne Hatley*.
First record: J. E. Little, 1919.

V. serpyllifolia L. C.T.W. 100.13.12.   *Thyme-leaved Speedwell*
Woodland drives, dry pastures, old walls, etc.   Native
Common, and variable according to its habitat. Recorded for all districts and neighbouring counties.
H.S. 11, 20a, 26, 34b, 53a, 54, 55b, 61, 77a, 80a, 82.
First record: Abbot, 1798.

V. arvensis L. C.T.W. 100.13.14.   *Wall Speedwell*
Arable fields, dry pastures, old walls, etc.   Native
Generally common, and with a wide distribution in varied habitats. It appears as a native species in closely-cropped pastures, but is no doubt a colonist elsewhere. Recorded for all districts except Kym, and for all neighbouring counties.
H.S. 22a, 22b, 27, 28, 56b, 82.
First record: Abbot, 1798.

V. persica Poir. C.T.W. 100.13.20.   *Buxbaum's Speedwell*
V. tournefortii C. C. Gmel. p.p.
Arable land   Colonist
Very common: recorded for all districts and all neighbouring counties.
H.S. 16, 17, 50, 72, 86.
First record: T. B. Blow in *Bot. Rec. Club 1876 Rep.*
First evidence: Eversholt, *J. McLaren in Herb. Luton Museum*, 1864.

V. polita Fr. C.T.W. 100.13.21.   *Grey Speedwell*
V. didyma auct.
Arable fields, old walls, etc.   Colonist
Less common than *V. persica*, but more common than *V. agrestis*. Plants were recorded as var. *thellungiana* Lehm. from Edworth by J. E. Little in *B.E.C. 1928 Rep.* Recorded for all districts and neighbouring counties.
H.S. 1a, 12, 17, 44, 71.
First record: R. A. Pryor in *J. Bot. 1876*, 22.

V. agrestis L. C.T.W. 100.13.22.   *Field Speedwell*
Arable fields, gardens, etc.   Colonist
The least common of our field speedwells. Most of the early records refer to either
V. persica or *V. polita*, and Abbot's specimen is the latter. Recorded for all neigh-
bouring counties.
NENE: arable field, Hinwick*.
KYM: Melchbourne*.
OUZEL: Aspley Guise; *J.E.L., Diary*: Pinfold Farm*.
IVEL: arable fields, Greenfield, Sandy*.
CAM: arable field, Dunton.
LEA: Luton†; *M. Brown in B.E.C. 1927 Rep.*: garden weed, Stopsley*.
First certain record: *Hillhouse List, 1876*; but there is an unlocalized specimen,
correctly-named, in *Herb. J. McLaren, Luton Museum*, 1864.

V. hederifolia L. C.T.W. 100.13.19.   *Ivy-leaved Speedwell*,
                                        Winter Weed (Batchelor)
Arable fields, dry banks, etc.   Colonist
Common, and well distributed, but thriving best on light soils.
Recorded for all districts and neighbouring counties.
H.S. 12.
First record: Abbot, 1798.

EUPHRASIA L.

Euphrasia nemorosa (Pers.) Mart. C.T.W. 100.17.13.   *Eyebright*
Woodland drives, heathy pastures, chalk downs, etc.   Native
This is apparently the common eyebright of the county. H. W.
Pugsley named specimens collected by me at Colesdon and Deadmansey Wood as var. *collina* Pugsl., and material sent by J. E. Little
from Deacon Hill as var. *ciliata* Drabble. Recorded for all districts
and neighbouring counties.
H.S. 2a, 59, 60, 62, 66a, 67, 68, 75, 80a.
First record: for the aggregate (*E. officinalis* L.), Abbot, 1798; for the
restricted species, D. M. Higgins in *Watson Bot. Exchange Club
1899–1900 Rep.*

E. pseudokerneri Pugsl. C.T.W. 100.17.16.
Chalk downs
Locally common, growing both in established turf and in chalk
scrub. Recorded for Herts., Bucks. and Cambs.
H.S. 59.
OUZEL: Totternhoe; *D. M. Higgins in W.B.E.C. 1899–1900 Rep.*
IVEL: Deacon Hill!; *J.E.L. in W.B.E.C. 1922 Rep.*: Knocking Hoe!*;
J.E.L., Diary: chalk downs, Streatley.
LEA: Blow's Downs, east side; *L. J. Margetts*.
COLNE: Whipsnade Zoo.
First record: D. M. Higgins, 1900.

ODONTITES Zinn

Odontites verna (Bellardi) Dum. C.T.W. 100.18.1.   *Red Eyebright*,
                                                 Red Bartsia
    Euphrasia odontites L., *Bartsia odontites* (L.) Huds., *O. rubra* Gray
Rough pastures, arable fields, etc.   ? Native

Common: much of the Bedfordshire material has been referred to subsp. *serotina* (Wettst.) E. F. Warb., but no serious work has yet been done on the distribution of the subspecies of this in the county. Recorded for all districts and neighbouring counties.
H.S. 2a, 4, 19a, 80a, 82.
First record: Abbot, 1798.

PEDICULARIS L.

Pedicularis palustris L. C.T.W. 100.14.1.   *Lousewort*
Marshes   Native
Very rare or extinct: it is interesting to note that Abbot originally considered it to be common. Recorded for all neighbouring counties.
OUSE: Oakley; Clapham; Stevington; *Abbot's annotated F.B.*
OUZEL: Totternhoe Mead†; Heath and Reach; *S.B.P.L.*
IVEL: Flitton Marshes; *J.McL., L.M.*: Flitwick Marsh where it appeared on the wettest parts of the Moor until about 1926!†*; *J.McL., B.M.*: Harlington; *W.F.B.*: Westoning; *F.F.B.*
LEA: records from New Mill End and Limbury in *W.F.B.* are probably in error.
First record: Abbot, 1798.

P. sylvatica L. C.T.W. 100.14.2.   *Red-rattle*
Damp pastures, bogs on acid soils, etc.   Native
Rare, and probably decreasing. Recorded for all neighbouring counties.
OUZEL: Aspley; *J.McL., L.M.*: Heath and Reach!; *S.B.P.L.*: Milton Bryan; *H. E. Pickering.*
IVEL: Ampthill; *F.B.*: Clophill Heath; *W.C. Herb.*: Eversholt, where it was also found by H. E. Pickering*; Flitwick; *J.McL., B.M.*: marshy slopes by Washer's Wood*.
LEA: Someries†; *S.B.P.L.*
First record: Abbot, 1798.

RHINANTHUS L.

Rhinanthus crista-galli L.   *Yellow-rattle*
    R. minor Ehrh. non L. C.T.W. 100.15.2.
Pastures, roadside verges, etc.   Native
Frequent, and well distributed in the county, but seen at its best in grassland on calcareous soils. Recorded for all districts and neighbouring counties.
H.S. 41, 72.
First record: Abbot, 1798.

Records of *R. major* Ehrh. by Saunders from Barton in *S.B.P.L.* and Harlington in *W.F.B.* are in error as the specimens are *R. crista-galli*. *R. stenophyllus* (Schur) Druce is listed for Bedfordshire in *Fl. of Bucks.*, but I can find no other evidence of its appearance in the county.

MELAMPYRUM L.

Melampyrum cristatum L.　C.T.W. 100.16.1.　*Crested Cow-wheat*
Woodland rides, roadsides, etc.　　　　　　　　　　　　Native
One of the more interesting Bedfordshire plant species, as it reaches
its western limit, as a native species, in the county.　Although
Abbot considered it common, no subsequent county botanists did,
and it is certainly not common now.　Recorded for all neighbouring
counties, but doubtfully so for Bucks.
OUSE: 'it is no less plentiful than there about Blunham'; *Ray's
Synopsis Ed. II* 1696, 164; Clapham; *Abbot's annotated F.B.*: Twin
Woods†; *J.McL.*, *L.M.*: Basmead; *C.R.O. List*: Little Wood,
Roxton; *G. H. Day*: Eaton Socon; *E.M.-R.*: roadside, Honeydon!*;
roadside, Souldrop!; *A. W. Guppy*.
IVEL: Polton Wood, ? Potton Wood; *W. Christy in Herb. Watson*,
1829: Deadman's Oak; *J.McL.*, *B.M.*
CAM: near orchards, Cockayne Hatley!; *T.L.*: Cockayne Hatley
Wood.
First record: John Ray, 1696.

M. arvense L.　C.T.W. 100.16.2.　　　　　　　*Field Cow-wheat*
Edges of arable fields　　　　　　　　　　　　　　　Colonist
Rare, but well established in its two main Bedfordshire stations,
sporadic elsewhere.　Like *M. cristatum* it is essentially an East Anglian
species, and reaches its western limit in the county.　Recorded also
for Herts. and Northants.
OUSE: Stagsden, plentiful in one field!; *D. M. Higgins in W.B.E.C.
1904–5 Rep.*: Bromham, one plant only; *A. Wooding*.
OUZEL: foot of Lord's Hill, Bidwell, in some quantity!*; *W.D. Coales*.
IVEL: scarce in one field, Higham Gobion!; *L. C. Chambers*.
LEA: one plant, Streatley!; *L. C. Chambers*.
First record: D. M. Higgins, 1905.

M. pratense L.　C.T.W. 100.16.3.　　　　　*Common Cow-wheat*
Woods, shady places, etc.　　　　　　　　　　　　　Native
Very rare, and limited to woods on light and heathy soils.　It is a
species I have not seen in the county, although I have seen it in
plenty in adjoining parts of Hertfordshire.　Recorded for all neigh-
bouring counties.
OUSE: Exeter Wood; *J.McL.*, *B.M.*: Millbrook; *W.H. 1875*.
OUZEL: Aspley Wood, where it was also found by W. Crouch and
McLaren; *F.B.*: Woburn, specimen in Herb. Beds. Nat. Hist. Soc.;
W.H. 1875.
IVEL: Warden Wood; *J.McL.*, *L.M.*
LEA: Luton Hoo; Dumb Hill Wood; *F.F.B.*
COLNE: Pepperstock†; *F.F.B.*: Whipsnade; *H. B. Sargent*.
First record: Abbot, 1798.

M. sylvaticum L., C.T.W. 100.16.4, was recorded, as var. *laurifolium* Beauv., from
Woburn by G. C. Druce in *B.E.C. 1917 Rep*. It is possible that Druce was mistaken. Recorded also for Bucks.

OROBANCHACEAE

OROBANCHE L.

Orobanche rapum-genistae Thuill. C.T.W. 101.2.3.

Greater Broomrape

Heathy places   Native

With us a parasite on broom, and limited to a small area of the county. Recorded for all neighbouring counties except Hunts.

IVEL: Castle Hill, Clophill; *W.C. Herb.*: Rowney Warren, as *O. elatior*, where it still appears frequently!*; *J.McL., L.M.*: Chicksands Wood; *J.H. Notes*: Flitwick†; *Herb. J.S.*: between Rowney Warren and Haynes; *J.E.L., Diary*.

First record: for the aggregate, Abbot, who in common with other early botanists recorded only *O. major* auct.; for the restricted species, J. Hamson, 1924.

First evidence: W. Crouch, 1844.

O. elatior Sutton C.T.W. 101.2.6.   *Tall Broomrape*
   O. major L. p.p.

Rough pastures, roadsides, etc.   Native

A parasite on *Centaurea scabiosa*, and more common and widely distributed than *O. rapum-genistae*. It is more frequent than the records indicate, as I have ignored records by Abbot from Oakley, Aspley and Ampthill, by C. Crouch from Flitwick and Ridgmont and by Saunders from Maulden, because of doubt to which species they should be referred. Recorded for all neighbouring counties.

OUSE: Bromham, it is frequent in this parish!; *J.H. Notes*: Harrold*; *G. H. Day*.

IVEL: plentiful in the neighbourhood of Wilbury Hill*.

First record: uncertain.

O. minor Sm. C.T.W. 101.2.8.   *Lesser Broomrape*
   O. apiculata Wallr.

Arable fields, roadsides, gravel pits, etc.   Native

Frequent, but uncertain in its appearance. It is usually a parasite on clovers and allied species and is often abundant in a clover field, but it has also been found on *Picris echioides*, *Crepis vesicaria* and other unrelated species. The size of the individual plants is very variable, and large plants have been mistaken for *O. elatior*. Recorded for all neighbouring counties.

OUSE: Bushmead*; *G. A. Battcock*: gravel pit, Eaton Socon, on various host plants*.

OUZEL: Houghton Regis; *E.M.-R.*: Dunstable; *W. D. Coales*.

IVEL: Sharpenhoe, as *O. major*, on clover; *S.B.P.L.*: Ampthill; Flitwick; Harlington†; *J.S., J. Bot. 1889*: Ridgmont; Pulloxhill*; Henlow; *C.C.*: Holwell; *J.E.L., F.F.B.*: Streatley; *W. D. Coales*: Clophill*; *B. Verdcourt*: Sundon*.

LEA: Stopsley Common; *W.F.B.*: Dunstable; *C. C. Foss*: Chiltern Green, in quantity; *H. Cole*: Chalton Cross.
COLNE: Bury Farm, Caddington; *L. J. Margetts*: roadside, Caddington.
First record: *Hillhouse List, 1876.*

LATHRAEA L.

Lathraea squamaria L. C.T.W. 101.1.1.   *Toothwort*
Woods, woodland verges, etc.   Native
Parasite on the roots of hazel, and with us apparently limited to woods on light soils. Recorded for all neighbouring counties except Hunts.
OUSE: Milton Hill; *J.McL., B.M.*: Oakley Hill, probably the same station; *J.H. Notes.*
IVEL: Sundon; *S.B.P.L.*: Pedley Wood; *C.C.*
LEA: Luton Hoo, plentiful both in woods and grassland!†; East Hyde; Stopsley!; *S.B.P.L.*
COLNE: between Whipsnade and Studham!; *Plantae Bedford.*: copse, near Zouche's Farm, v.c. 20 [Beds.]; Little John's Wood, Caddington; *L. J. Margetts*: Heath Wood*; Badger Dell Wood*.
First record: Abbot, 1795.

LENTIBULARIACEAE

UTRICULARIA L.

Utricularia vulgaris L. C.T.W. 102.2.1.   *Bladderwort*
Ditches, backwaters, gravel pits, etc.   Native
Very rare, but probably lingering in a vegetative state in some stations. Recorded for all neighbouring counties.
H.S. 11.
OUSE: Bromham Ditches; *Plantae Bedford.*: Bedford†; *A. Ransome see J.S., J. Bot. 1889*: gravel pit, Felmersham, where it was found by the author in company with M. W. Cornish in a vegetative state in 1949 and in flower by C. C. Foss and the author in 1951*.
IVEL: Everton; *Plantae Bedford.*
First record: Abbot, 1795.

U. minor L. C.T.W. 102.2.4.   *Lesser Bladderwort*
Pools, ditches, etc.   Native
Extinct: recorded for all neighbouring counties, but doubtfully for Bucks.
OUSE: Ampthill; Potton Marshes; *F.B.*

PINGUICULA L.

Pinguicula vulgaris L. C.T.W. 102.1.3.   *Butterwort*
Marshes   Native
Probably extinct: a species with a most unusual distribution in the county. It appeared in three heathy bogs where it might have been expected, and on two chalk hills, no doubt in small damp pockets on ridges of the Totternhoe Stone. Recorded for all neighbouring counties.

OUZEL: Totternhoe Mead, where it became extinct owing to drainage begun about 1888—in 1907 (see *Trans. Herts. Nat. Hist. Soc. XIII*, lix) *Parnassia palustris* and *Anagallis tenella* which grew with it were already extinct, but *P. vulgaris* was still surviving on the side of a ditch†; *W. G. Smith in J. Bot. 1886.*
IVEL: Ampthill Moor; *Plantae Bedford.*: Maulden Bogs; *W.C. Herb.*: Pegsdon Hills (i.e. Knocking Hoe) where it was observed annually for or by J. E. Little until 1921, when the drought of that year caused it to become extinct; *J. Pollard in J. Bot. 1875*: Markham (i.e. Streatley) Hills, where it grew in association with *Parnassia palustris* and *Carex pulicaris*—Mrs. Boutwood and I independently found *Parnassia* there in 1917 and observed it annually until 1926, but neither of us saw either the butterwort or sedge; *C.C.*
First record: Abbot, 1795.

VERBENACEAE

VERBENA L.

Verbena officinalis L. C.T.W. 104.1.1.   *Vervain*
Roadsides, dry banks, etc.   Colonist
Frequent, but rarely found far from houses. It is usually well established. Abbot considered it to be common. Recorded for all neighbouring counties.
H.S. 57a.
NENE: '*Verbena majus flore coeruleo* is reckoned a specific for many diseases' [Podington]; *O. St. John Cooper, Collections towards the history and antiquities of Bedfordshire.*
OUSE: Newnham; Biddenham; Ford End; *F.B.*: Cardington; *J.McL., B.M.*: Fenlake; Bromham; *W.H. 1875*: Goldington; Wilden; *J.H. in V.C.H.*: Cotton End; *M. L. Berrill in J.H. Notes*: Elstow; *J.H. Notes*: Odell; *B. B. West*: Wyboston; *H. B. Souster*: Tempsford.
OUZEL: Totternhoe where it grows both in the north end of the village and at Well Head!*; *S.B.P.L.*
IVEL: Meppershall; *W.C. Herb.*: Silsoe Park; *J. Pollard*: near Harlington; *W. Durant!*: Sutton; Biggleswade.
LEA: near St. Ann's Hill; *Davis*: Limbury!; *S.B.P.L.*: Dallow Farm†; *Herb. J.S.*: New Mill End!; *H. B. Souster.*
COLNE: Markyate, v.c. 30 [Herts.].
First record: O. St. John Cooper, 1783.

LABIATAE

MENTHA L.

I am indebted to R. A. Graham, who has named all my material and made many useful suggestions regarding the identity of early records.

Mentha arvensis L. C.T.W. 105.1.3.   *Corn Mint*
Woodland rides, arable fields, etc.   Colonist
Common: the most widely distributed of our mints, but very variable. In woodland rides it is a lank and straggling plant, but in arable fields it is usually dwarfed and compact. Recorded for all districts and neighbouring counties.
H.S. 23, 72.
First record: Abbot, 1798.

Var. **densifoliata** Briq.
OUZEL: New Wavendon Heath, v.c. 24 [Beds.]*
IVEL: Barton*; Northill*.
COLNE: Deadmansey Wood*.

M. × **verticillata** L. (*M. aquatica* × *arvensis*) C.T.W. p. 934.
M. sativa L.
Riversides, marshy places, etc.   Native
Occasional: recorded for all neighbouring counties.
OUSE: Goldington; *J.McL. in J. Bot. 1884*: Cardington; *J.McL.*
KYM: Melchbourne*.
OUZEL: Rushmere; Battlesden*; Heath and Reach, with var.
rivalis Sole*; Woburn, with var. *paludosa* (Sole) Dr.*
IVEL: Clophill Mills; *Abbot's annotated F.B.*: Chalton*.
First record: Abbot in letter to J. E. Smith, 3 Sept. 1802.

M. × smithiana R. Grah., C.T.W. p. 935 (*M. rubra* Sm. non Mill.) was listed
for v.c. 30 in *Comital Flora*, but I can find no basis for its inclusion. Recorded for all
neighbouring counties.

M. **aquatica** L. C.T.W. 105.1.4.   *Water Mint*
Riversides, marshes, etc.   Native
Common in all suitable habitats. Recorded for all districts except
Colne and for all neighbouring counties.
H.S. 3a, 6–8, 10a, 10b, 13, 14a, 14b, 34a, 39a, 43, 56b, 84a–85.
First record: Abbot, 1798.
Var. **lobeliana** G. Beck.
IVEL: Arlesey*; *E.M.-R.*: Henlow*.
OUZEL: Cow Common.

M. **longifolia** (L.) Huds. C.T.W. 105.1.6.   *Horse Mint*
M. sylvestris L.
Roadsides, waste places, etc.   Colonist
Rare: recorded for all neighbouring counties except Hunts.
OUSE: Watkin's Paddock; *Abbot's annotated F.B.*
IVEL: Silsoe; Barton; *Abbot's annotated F.B.*: in the orchard at Toddington par-
sonage; *E. Forster in Bot. Guide*: Clophill!; *W.C.*: Flitwick†; *S.B.P.L.*
LEA: Dunstable, Town Station, var. *horridula* Briq.*; *W. D. Coales*: Bedford Road,
north of Luton, var. *horridula**; Vauxhall Works, var. *horridula**.
First record: Abbot, in letter to J. E. Smith, 3 Sept. 1802.

M. rotundifolia (L.) Huds., C.T.W. 105.1.7, is of doubtful occurrence in the
county. It was recorded from Silsoe, 1799, by Abbot in his annotated copy of
Flora Bedfordiensis and in a letter to J. E. Smith, 3 Sept. 1802. There is another
doubtful record from Basmead in the *C.R.O. List*. Recorded for all neighbouring
counties except Hunts.

LYCOPUS L.

Lycopus europaeus L. C.T.W. 105.2.1.   *Gipsy-wort*
Riversides, marshy places, etc.   Native
Common by all the larger rivers in the county. Recorded for all
districts except Nene and Colne, and for all neighbouring counties.
H.S. 7, 9, 13, 14a.
First record: Abbot, 1798.

ORIGANUM L.

Origanum vulgare L. C.T.W. 105.3.1.   *Marjoram*
Chalk scrub, roadsides, etc.   Native
Common on disturbed soils on the Chalk, rare elsewhere on cal-
careous soils. Recorded for all neighbouring counties.
H.S. 68.
NENE: railway bank, Wymington.
OUSE: Colesdon; Cleat Hill; Ravensden; *A. W. Guppy.*
KYM: Riseley; *A. W. Guppy*: The Grange, Dean, ? garden escape.
OUZEL: common in suitable places on the Chalk.
IVEL: Eversholt Wood; *J.McL., L.M.*: common in suitable places
on the Chalk.
LEA and COLNE: common on the Chalk.
First record: Streatley, Barton; Abbot, 1795.

THYMUS L.

Thymus pulegioides L. C.T.W. 105.4.1.   *Thyme*
    T. ovatus Mill., *T. chamaedrys* Fr., *T. serpyllum* auct.
Chalk hills, heaths, etc.   Native
The common thyme of the county, and to be found in all likely
places. Recorded for all neighbouring counties.
H.S. 1a, 27, 56b, 59–62, 63b, 65, 66a, 67, 68, 75.
NENE: railway cutting, Wymington*.
OUSE: Exeter Wood; *M. L. Berrill in J.H. Notes*: Carlton Hall Farm;
Packhorse Way, Chellington; *G. H. Day*: Tempsford*; *P.T.*!:
railway, Souldrop.
OUZEL: common on the Chalk and Chalk Marl.
IVEL: common on the Chalk, less common on the Lower Greensand;
Rowney Warren.
LEA: common on the Chalk: Chiltern Green Common*.
COLNE: common on the Chalk and on heaths on the Clay-with-
Flints.
First record: Abbot, 1798.

T. drucei Ronn.   *Thyme*
Chalk hills   Native
Rare: I am indebted to C. D. Pigott, who not only confirmed that
all my gatherings of thyme were *T. pulegioides*, but visited the county
to satisfy himself, and us, that this second species also appears.
Recorded for Herts. and Bucks.
COLNE: near Whipsnade*; *C. D. Pigott.*

CALAMINTHA Mill.

Calamintha ascendens Jord. C.T.W. 105.7.2. *Common Calamint*
    Satureja ascendens (Jord.) Maly, *Melissa calamintha* auct.
Banks by roadsides, etc.   Colonist
Occasional: recorded for all neighbouring counties except Hunts.

OUSE: Cople, where it was found by McLaren and still grows!*;
Bedford; *Plantae Bedford.*: Lidlington; *W.C. Herb.*: Biddenham, as
C. nepeta; *J.McL.*, *L.M.*: Wootton; *W.H. 1875*: Cardington; *J.McL.
in J. Bot. 1884.*
KYM: roadside, Yelden.
OUZEL: Woburn†; *W.H. 1875*: Battlesden*.
IVEL: between Shefford and Biggleswade; *J.S. in J. Bot. 1889*:
Silsoe, well established!*; *C.C.*: between Cadwell Bridge and
Wilbury; *J.E.L.*, *J. Bot.*: Hills End, Eversholt; *H. B. Souster*:
roadside, Clophill!; *B. Verdcourt*.
LEA: Maiden Common; *W.F.B.*
First record: Abbot, 1795.

C. nepeta (L.) Savi (*Satureja nepeta* (L.) Scheele), C.T.W. 105.7.3, was recorded
from Ravensden, 'named by Kew', by Edgar Evans in *Hamson's Notes*; but failing
more definite evidence it cannot be included. Recorded for Bucks., Cambs. and
Hunts.

ACINOS Mill.

Acinos arvensis (Lam.) Dandy C.T.W. 105.8.1. *Basil-thyme*
Thymus acinos L., *Calamintha acinos* (L.) Clairv., *Satureja acinos* (L.)
Scheele
Disturbed soils on the Chalk, railways, etc. Colonist
Frequent around rabbit warrens and on the edges of quarries on the
Chalk, and as a railway weed growing only on ballast by the metals.
Considered rare by Abbot, who gave Barton Hills and Aspley, the
latter, it should be noted, on the Greensand, where I have not found
it. Recorded for all neighbouring counties.
H.S. 61, 71.
NENE: railway, Wymington*.
OUSE: frequent on the railway.
OUZEL and IVEL: frequent on the Chalk and in the Ivel district on
the railway.
LEA: frequent on the Chalk and common on the railway.
COLNE: frequent on the Chalk.
First record: Abbot, 1795.

CLINOPODIUM L.

Clinopodium vulgare L. C.T.W. 105.9.1. *Wild Basil,*
Hedge Calamint
Roadsides, scrub, railway banks, etc. Colonist
Common, and especially so on disturbed land on calcareous soils.
A white-flowered form has been seen at East Hyde. Recorded for
all districts and neighbouring counties.
H.S. 1a–2a, 66b, 68, 69, 79a.
First record: Abbot, 1798.

SALVIA L.

Salvia pratensis L. C.T.W. 105.11.2.   *Meadow Sage*
Waste places, etc.   ? Native
Rare: recorded for all neighbouring counties except Hunts.
OUSE: Biddenham; Ford End; probably one station and credited to T. O. Marsh
in *Flora Bedfordiensis*; *Plantae Bedford.*
LEA: Stopsley Common†; *E. R. Hunt in Herb. J.S.*
First record: ? Abbot, 1795.

S. horminoides Pourr. C.T.W. 105.11.3.   *Wild Clary*
   S. verbenaca auct.
Roadsides, etc.   Denizen
Well established on sandy and gravelly soils in various parts of the county. Con-
sidered common by Abbot. Recorded for all neighbouring counties.
OUSE: St. Cuthbert's; Clapham; Elstow; *F.B.*: Cardington; *J.McL., B.M.*:
Biddenham!; Bromham; *J.H. Notes*: Harrold; *G. H. Day*: Queen's Park, Bedford!*;
L. W. Wilson.
OUZEL: Totternhoe; *F.F.B.*
IVEL: Maulden; *W.C. Herb.*: Toddington†; *S.B.P.L.*: near Sandy!; *A. B. Sampson*:
Silsoe; Shillington, frequent to north of the village!*; *C.C.*: Potton, plentiful near
the church!; Southill; Henlow!; *J.E.L., J. Bot.*: Sampshill; *Westoning W.I.*
CAM: Wrestlingworth.
First record: Abbot, 1798.

PRUNELLA L.

Prunella vulgaris L. C.T.W. 105.13.1.   *Self-heal*
Pastures, woodland rides, etc.   Native
Common in a great variety of habitats and on various soils. White-
flowered forms are not uncommon. Recorded for all districts and
neighbouring counties.
H.S. 2a, 3a, 4, 7, 10b–11, 19a–20a, 21, 34a, 34b, 53a–54, 55b,
56b, 57a, 60, 61, 67, 69, 76, 77a, 79a, 80a–83.
First record: Abbot, 1798.

STACHYS L.

Stachys officinalis (L.) Trev. C.T.W. 105.14.1.   *Betony*
   Betonica officinalis L.
Woodland rides, heathy pastures, etc.   Native
Frequent, and well distributed in the county, but generally absent
on calcareous soils. Recorded for all districts and neighbouring
counties.
H.S. 53a, 74a.
First record: Abbot, 1798.

S. arvensis (L.) L. C.T.W. 105.14.3.   *Field Woundwort*
Arable fields   Colonist
Rare: recorded for all neighbouring counties.
IVEL: Barton Hills, but I cannot find a specimen to support the
record; *B.P.L.*: arable field, Silsoe!*; *H. B. Souster.*
LEA: near Dunstable Downs; *Abbot, in letter to J. E. Smith, 3 Sept.
1802.*
First record: Abbot, 1802.

S. palustris L. C.T.W. 105.14.6. *Marsh Woundwort*
Marshy places, riversides, etc. Native
Common in all suitable habitats, especially by the Ouse and Ivel.
Recorded for all neighbouring counties.
H.S. 6, 8, 11, 13, 15.
OUSE: too many stations to merit listing.
OUZEL: Woburn; *S.B.P.L.*: Leighton Buzzard; Heath and Reach.
IVEL: too many stations to merit listing.
LEA: persisting as a garden weed at Stopsley, where it was intro-
duced mistakenly for a mint.
First record: Abbot, 1798.

S. sylvatica L. C.T.W. 105.14.7. *Hedge Woundwort*
Hedgebanks, wood borders, etc. Colonist
Common: recorded for all districts and neighbouring counties.
H.S. 1b, 2a, 32, 78, 83.
First record: Abbot, 1798.

S. germanica L. C.T.W. 105.14.4.
Pastures, etc. ? Colonist
Extinct: recorded for Bucks. and Northants. *E. Bot.* tab. 829, Jan. 1801, was
drawn from a specimen sent from a 'hill two miles from Bedford' by J. Hemsted.
OUSE: Clapham Hill; *Abbot's annotated F.B.*
LEA: near Luton; *Abbot's annotated F.B.*
First record: J. Hemsted, 1801.

BALLOTA L.

Ballota nigra L. C.T.W. 105.15.1. *Black Horehound, Stinking Henbit*
Hedgebanks, roadsides, waste places, etc. Colonist
Common: the ordinary Bedfordshire and British form is not var.
ruderalis (Sw.) Koch, which is the true *B. nigra* L. Recorded for
all districts and neighbouring counties.
First record: Abbot, 1798.

GALEOBDOLON Adans.

Galeobdolon luteum Huds. C.T.W. 105.16.1. *Weasel Snout,*
Lamium galeobdolon (L.) L. *Yellow Archangel, Yellow Dead-nettle*
Woods, woodland rides, etc. Native
Frequent, especially in woods on light soils. Fasciated forms are
frequent near Caddington. Recorded for all districts except Nene,
and for all neighbouring counties.
H.S. 78, 81.
First record: Abbot, 1795.

LAMIUM L.

Lamium amplexicaule L. C.T.W. 105.17.1. *Henbit*
Arable fields Colonist
Common, especially on the lighter soils. Cleistogamic forms are
frequent. Recorded for all districts except Colne, and for all
neighbouring counties.
H.S. 17, 25, 44, 45a, 50.
First record: Abbot, 1798.

L. hybridum Vill. C.T.W. 105.17.3.   *Cut-leaved Dead-nettle*
Arable fields, etc.   Colonist
Occasional: recorded for all neighbouring counties.
H.S. 44.
NENE: fields, Wymington.
OUSE: Roxton; *G. H. Day*: fields, Souldrop.
OUZEL: plentiful in disused sandpit, Leighton Buzzard*; Totternhoe*.
IVEL: Maulden!; *M. L. Berrill in J.H. Notes*: Southill; Clophill; *J.E.L., J. Bot.*:
Tingrith*; Westoning; Flitwick; Beeston.
LEA: garden weed, Luton†; *J.S. in J. Bot. 1889*: railway bank, Sundon Park.
COLNE: Kensworth*.
First record: *Hillhouse List, 1876*.

L. purpureum L. C.T.W. 105.17.4.   *Red Dead-nettle*
Arable fields, waste places, etc.   Colonist
Common: white-flowered forms have been seen at Felmersham and
Ampthill. Recorded for all districts and neighbouring counties.
H.S. 12, 16.
First record: Abbot, 1798.

L. album L. C.T.W. 105.17.5.   *White Dead-nettle*
Roadsides, hedgebanks, waste places, etc.   Colonist
Common: recorded for all districts and neighbouring counties.
First record: Abbot, 1798.

GALEOPSIS L.

Galeopsis angustifolia Ehrh. ex Hoffm. C.T.W. 105.19.1.
   G. ladanum auct.   *Narrow-leaved Hemp-nettle*
Arable fields   Colonist
Frequent in arable fields on calcareous soils, with us appearing in
two distinct forms.

(a) with adpressed hairs on the calyx tube, and narrow leaves
    tapering towards each end.
NENE: Podington*; railway bank, Wymington*.
IVEL: Cainhoe Park Farm; *W.C. Herb.*: Flitwick†; *Herb. J.S.*:
Noon Hill.
CAM: Dunton*.
LEA: Luton*; Leagrave*.

(b) with patent hairs on the calyx tube, and broad leaves rounded
    at the base.
H.S. 71, 72.
OUZEL: Totternhoe*.
IVEL: Barton; near Ravensburgh Castle.
LEA: Dallow Lane; *Herb. J.S.*: Warden Hills*.
First record: for the aggregate, Abbot, 1798.

G. tetrahit L. C.T.W. 105.19.4.   *Common Hemp-nettle*
Rough pastures, arable land, woods, etc.   Native
Common, and well distributed, growing in a great variety of
habitats. White-flowered forms are not infrequent. Recorded for
all districts and neighbouring counties.
H.S. 15, 43, 71, 83.
First record: Abbot, 1798.

G. speciosa Mill. C.T.W. 105.19.6. *Large-flowered Hemp-nettle*
Arable land Colonist
Rare: recorded for all neighbouring counties.
OUSE: cornfield, Elstow; *C. C. Foss.*
IVEL: between Hawnes and Chicksands Priory; *Abbot in letter to J. E. Smith, 3 Sept. 1802:* Flitwick Marsh, where it appears regularly in fields reclaimed from the marsh!†*; *S.B.P.L.*
First record: Abbot, 1802.

NEPETA L.

Nepeta cataria L. C.T.W. 105.20.1. *Cat-mint*
Hedgebanks, roadsides, waste places, etc. Colonist
Frequent, and usually well established. Abbot considered it to be common. Recorded for all neighbouring counties.
NENE: Sharnbrook Summit; *V. H. Chambers.*
OUSE: Biddenham; *Abbot's annotated F.B.*: Medbury; *J.McL., B.M.*: Stagsden; *J. Ekins in J.H. Notes*: Bromham; *B. B. West*: near the church, Carlton; *C. C. Foss*: Pavenham; *H. B. Souster!*
OUZEL: Totternhoe!; *W.F.B.*: Stanbridgeford; *F.F.B.*: Eaton Bray; *W. D. Coales*: Potsgrove!; *E.M.-R.*
IVEL: Cainhoe; *W.C. Herb.*: Rowney Warren; *J.McL., L.M.*: Maulden; *M. L. Berrill in J.H. Notes*: Steppingley; Westoning; Sundon; *C.C.*: Southill; *J.E.L., F.F.B.*: Stanfordbury; *J.E.L., Diary*: Rowney Warren; *V. H. Chambers*: Ampthill; *B. Verdcourt*: Sundon Rubbish Dump.
LEA: Leagrave!*; Limbury; Streatley†; *S.B.P.L.*: Dunstable; *E.M.-R.*: Great Bramingham; *H. B. Souster.*
COLNE: between Markyate and Dunstable; *Fl. Herts.*: Caddington; v.c. 20 [Herts.]!*; *W. D. Coales.*
First record: Abbot, 1798.

GLECHOMA L.

Glechoma hederacea L. C.T.W. 105.21.1. *Ground Ivy*
Nepeta hederacea (L.) Trev.
Woodland rides, rough pastures, etc. Native
Common, especially on heavy and wet clays, but growing in a wide range of habitats. Recorded for all neighbouring counties.
H.S. 2b, 3a, 6, 13, 18, 19a, 20a, 20b, 23, 27, 31a–32, 34a, 34b, 38, 42, 43, 51a, 52, 53a–54, 55b, 77a, 78, 79a, 80a, 81, 83.
First record: Abbot, 1798.

MARRUBIUM L.

Marrubium vulgare L. C.T.W. 105.22.1. *White Horehound*
Waste places, arable fields, etc. Colonist
Rare: a well-established plant in many stations, but more recently introduced as a wool adventive. Recorded for all neighbouring counties.
OUSE: Elstow; Cople; *F.B.*: Cardington; *J.McL.*: Biddenham; *R. Townsend in J.H. Notes*: Bedford; *A. W. Guppy*: gravel pits, Eaton Socon*, Wyboston.
OUZEL: roadside, Leighton Buzzard!*; *W. D. Coales.*
IVEL: Everton; *F.B.*: Deepdale; *J.McL., L.M.*: Sundon†; *S.B.P.L.*: railway sidings, Flitwick, Henlow, Biggleswade, etc.; arable fields, Beeston*, Maulden, etc.
LEA: waste ground near Luton; *W.F.B.*: water tower, Luton; *C. C. Foss*: Midland Road Station, Luton; *W.W.!*
First record: Abbot, 1798.

SCUTELLARIA L.

Scutellaria galericulata L. C.T.W. 105.23.1. *Skull-cap*
Marshes, riversides, etc. Native
Frequent in suitable habitats, especially in association with the Ouse and Ivel. Recorded for all neighbouring counties.

H.S. 15, 39a, 42, 43.
OUSE: too many stations to merit listing.
OUZEL: Battlesden; *S.B.P.L.*: Leighton Buzzard; meadows, Heath and Reach; marshes, Woburn.
IVEL: too many stations to merit listing.
LEA: Luton Hoo; *A. W. Franks in Herb. Epsom Coll.*: New Mill End!; *S.B.P.L.*
First record: Abbot, 1798.

TEUCRIUM L.

Teucrium scorodonia L. C.T.W. 105.24.4.   *Wood Sage*
Open woodland, heathy places, etc.   Native
Locally abundant on the Lower Greensand: recorded for all neighbouring counties.
H.S. 23, 27, 30a, 30b, 31b.
OUZEL and IVEL: too many stations to merit listing.
First record: Abbot, 1798.

AJUGA L.

Ajuga chamaepitys (L.) Schreb. C.T.W. 105.25.1.   *Ground-pine*
   Teucrium chamaepitys L.
Arable fields, rough downland, etc.   Native
Limited to a very restricted area of the Chalk, and always on soils previously disturbed. A species of Eastern England it is recorded for Herts. and Cambs., and doubtfully for Northants.
IVEL: near Pegsdon; *H. Brown*: Barton Hills, 1841, where it grows on the Chalk and Glacial Gravel!†*; *I. Brown in Herb. Watson*: Streatley†; *S.B.P.L.*: arable field near Tingley Wood; *C. Swain.*
LEA: Luton Downs; *J. Sibley in F.B.*: foot of Warden Hill; *F.F.B.*
First record: Abbot, 1798.

A. reptans L. C.T.W. 105.25.2.   *Bugle*
Woodland rides, wet meadows, etc.   Native
Common, especially in wet situations. Recorded for all districts and neighbouring counties.
H.S. 2b, 3b, 18, 19b, 20a, 23, 30a–31c, 33–34b, 37, 43, 51a, 52–55a, 77a, 77b, 79a, 80a, 81, 83.
First record: Abbot, 1798.

PLANTAGINACEAE

PLANTAGO L.

Plantago major L. C.T.W. 106.1.1.   *Great Plantain*
Waste places, arable land, etc.   Colonist
Common, especially on previously disturbed soils, in open habitats. Recorded for all districts and neighbouring counties.
H.S. 3a, 6, 8, 16, 19a, 20a, 27, 34a, 34b, 53a, 54, 55b, 56b, 64, 71, 80a, 82.
First record: Abbot, 1798.

P. media L. C.T.W. 106.1.2.   *Hoary Plantain*
Pastures   Native
Common, especially on calcareous soils. Recorded for all districts
and neighbouring counties.
H.S. 1b, 8, 13, 59, 62, 65–67, 75.
First record: Abbot, 1798.

P. lanceolata L. C.T.W. 106.1.3.   *Ribwort Plaintain*
Pastures, waste places, arable land, etc.   Colonist
Common, and probably the most widely distributed species in the
county, as it readily invades most native habitats. A dwarfed form
with compact roundish heads is common on the chalk hills.
Recorded for all districts and neighbouring counties.
H.S. 1a–2a, 8–10b, 11, 13, 20a, 21, 25–27, 44–45b, 48, 51b, 56b–57b,
58b, 59, 61–69, 74b–76, 80a, 82.
First record: Abbot, 1798.

P. coronopus L. C.T.W. 106.1.5.   *Buck's-horn Plantain*
Heathy places   Native
Frequent on the Lower Greensand: recorded for all neighbouring
counties except Hunts.
OUZEL: Aspley; *F.B.*: Heath and Reach, in the neighbourhood of
which it is plentiful!*; *S.B.P.L.*
IVEL: Ampthill, frequent on Cooper's Hill!†*; *F.B.*: Sandy;
Abbot's annotated F.B.: Clophill!†; *H. Brown*: Maulden; *W.C. Herb.*:
Rowney Warren; *J.McL., B.M.*: Flitwick; *S.B.P.L.*: near Sutton
Fen.
First record: Abbot, 1798.

CAMPANULACEAE
JASIONE L.
Jasione montana L. C.T.W. 107.5.1.   *Sheep's-bit*
Sandy heaths, etc.   Native
Limited to the Lower Greensand, where it is locally frequent. Large-
flowered forms, *cf.* var. *major* Mert. & Koch, appear in a sandpit
near Flitwick station. Recorded for all neighbouring counties
except Hunts.
OUZEL: Aspley, where it is frequent!; *Plantae Bedford.*: frequent on
greensand exposures near Heath and Reach*.
IVEL: Potton; Clophill!; *Plantae Bedford.*: Maulden!; *W.C. Herb.*:
Everton Heath; *J.McL., L.M.*: Flitwick!†*; *J.McL., L.M.*: Millbrook; *F.F.B.*: heathy ground, Sandy; old sandpit, Flitton.
First record: Abbot, 1795.

CAMPANULA L.
Campanula glomerata L. C.T.W. 107.2.5.   *Clustered Bellflower*
Chalk hills   Native
Common on most of our chalk hills, and seen at its best in places
where the soil has at some time been disturbed. It is recorded from

Aspley Guise in *Hillhouse List 1875*, probably in error. Recorded for all neighbouring counties except Hunts.
H.S. 59, 60, 62, 65, 68.
OUZEL, IVEL, LEA and COLNE: common in all likely places on the Chalk.
First record: Abbot, 1798.

C. latifolia L. C.T.W. 107.2.1.   *Great Bellflower*
Woods   Native
Not infrequent in woods on light soils with a shallow clay covering. *E. Bot.* tab. 302, Feb. 1796, was drawn from a Bedfordshire specimen sent at Abbot's request by Mr. Sibley of Market Street. Recorded from Herts., Bucks. and Northants., and as a garden escape from Northants.
H.S. 18.
OUSE: Twin Wood; *A. W. Guppy*: Stevington; *J. Wooding*: Holcot Wood.
OUZEL: Leighton Buzzard, in a wood near the canal; *J. Anderson in New Bot. Guide, II*, 1837: Aspley; *V.C.H.*: King's Wood; Palmer's Shrubs.
IVEL: Eversholt; *Plantae Bedford.*: between Hawnes and Maulden; *Abbot's annotated F.B.*: Cainhoe Park Wood; *W.C. Herb.*: Rowney Warren; *J.McL., B.M.*: Barton Leete Wood; *S.B.P.L.*: Maulden Wood; *M. L. Berrill in J.H. Notes*: Chicksands Wood; Washer's Wood*.
LEA: Luton Park; *H. Brown.*
COLNE: Whipsnade; *Plantae Bedford.*: Studham Bottom; *F.B.*: near Zouche's Farm, v.c. 20 [Beds.]; *Fl. Herts.*: Farley Bottom†; *W.F.B.*: Cradle Spinney, Caddington; *L. J. Margetts*: spinney, Farley Estate; *H. B. Souster*: wood on west slope of Blow's Downs, v.c. 20 [Beds.].
First record: Abbot, 1795.

C. trachelium L. C.T.W. 107.2.2.   *Nettle-leaved Bellflower*
Woods   Native
More frequent and more widely distributed than the preceding, growing more generally on the heavier clay soils. Recorded for all neighbouring counties.
H.S. 69.
OUSE: Moggerhanger; Thurleigh; Clapham; *Abbot's annotated F.B.*: Basmead; *C.R.O. List*: Sharnbrook; *W.H. 1875*: Oakley; *B.P.L.*: Hammer Hill; *M. L. Berrill in J.H. Notes*: Brown's Wood!; *C. C. Foss*: Stagsden; *A. W. Guppy*: Ravensden; *B. B. West*: Stevington; Sheerhatch Wood; Roxton*.
KYM: Riseley; *G. H. Day*: Hay Wood; *M. Dalton.*
IVEL: Warden Wood; *J.McL., B.M.*: Streatley; Barton, sparingly in Leete Wood!; *S.B.P.L.*: Potton Wood.
CAM: Cockayne Hatley Wood!; *J.E.L., F.F.B.*
First record: Abbot, 1798.

C. rapunculoides L. C.T.W. 107.2.3.   *Creeping Bellflower*
Hedgebanks, edges of arable fields, etc.   Colonist
Uncommon, but usually well established. Recorded for all neighbouring counties.
NENE: near Hinwick House.
OUSE: Bedford; *J.H. Notes*.
OUZEL: Dunstable Downs; *W. D. Coales*.
IVEL: Cainhoe, as *C. trachelium*; *W.C. Herb.*: Clophill; *C.C.*: Ampthill; *D. M. Higgins in W.B.E.C. 1901–2 Rep.*: allotments, Flitwick.
LEA: near Luton; *J. Foster in Herb. Watson*: Luton Hoo; *B.P.L.*: railway bank, south of Luton†; *D. M. Higgins in W.B.E.C. 1896–7 Rep.*
COLNE: persisting for some years on edges of arable fields between Long Wood and Ravensdell Wood, v.c. 20 [Beds.]*.
First record: J. Foster, 1841.

C. rotundifolia L. C.T.W. 107.2.6.   *Harebell*
Pastures   Native
Especially common on the chalk hills and pastures on the Lower Greensand. It occurs less frequently on heaths on the Clay-with-Flints, but it may occur anywhere where the competition of taller plants does not exclude it. Recorded for all districts except Kym and Cam, and for all neighbouring counties.
H.S. 22a, 28, 59, 60, 62, 63b, 65, 66b, 68, 69, 74a, 75.
First record: Abbot, 1798.

LEGOUSIA Durande

Legousia hybrida (L.) Delarb.   *Venus's Looking-glass, Corn Bellflower*
   Campanula hybrida L., *Specularia hybrida* (L.) A. DC. C.T.W. 107.3.1.
Arable fields   Colonist
Frequent in arable fields on the Chalk, Oolite and Boulder Clay; rare or absent on the Lower Greensand and Clay-with-Flints. Recorded for all neighbouring counties.
H.S. 71, 72.
NENE: plentiful in a number of fields at Podington and Wymington.
OUSE: Biddenham; Bedford; Elstow; *Abbot's annotated F.B.*: Basmead; *C.R.O. List*: Cardington; *E. M. Langley in J.H. Notes*: Oakley!; Stevington; *B. B. West*: near Hanger Wood, *E. Proctor*.
OUZEL: Totternhoe*.
IVEL: Ridgmont; *W.C. Herb.*: Streatley!; Barton!; *S.B.P.L.*: Arlesey!; Knocking Hoe; *J.E.L., F.F.B.*: Sundon; Fancott; Higham Gobion.
LEA: Biscot†; *Herb. J.S.*: Chaul End*; *J.S.*: Galley Hill!; *V. H. Chambers*: Dunstable; *W. D. Coales*: Chiltern Green.
COLNE: below Dunstable East Hill, *i.e.*, Blow's Downs; *Fl. Herts*.
First record: Abbot, 1798.

RUBIACEAE
GALIUM L.

Galium mollugo L. C.T.W. 109.3.3.   *Great Hedge-bedstraw*
Roadsides, rough pastures, railway banks, etc.   Native
Common, but somewhat variable: var. *insubricum* Gaudin was

recorded from Clapham by Hamson. Recorded for all districts and neighbouring counties.
H.S. 1a, 2a, 3a, 63a, 63b, 65, 66b–68.
First record: Abbot, 1798.

G. mollugo × verum
OUSE: roadside, Moggerhanger*.
OUZEL: foot of Dunstable Downs; *B. Verdcourt.*
IVEL: rough pasture, Barton Hills*.

G. erectum Huds. C.T.W. 109.3.4.   *Erect Hedge-bedstraw*
Pastures   Colonist
Rare, and probably introduced in recent years with grass seed. Most of the old records are to be doubted. Abbot's herbarium specimen is a form of *G. mollugo.*
Recorded for all neighbouring counties.
OUSE: Ford End; *F.B.*: Cardington; *J.McL.*: Colworth; *W.H. 1875.*
OUZEL: Woburn Park; *J. Pollard.*
IVEL: Flitwick Marsh, but the specimen is not satisfactory†; *S.B.P.L.*: Westoning; *F.F.B.*: Wilbury Hill; *J.E.L. in B.E.C. 1930 Rep.*: Sundon Hill*.
LEA: woods near Dunstable; *F.B.*: plentiful in rough pasture, Luton Hoo*.
First record: uncertain.

G. saxatile L.   *Heath Bedstraw*
   G. hercynicum Weigel C.T.W. 109.3.6, *G. procumbens* Stokes
Heaths, etc.   Native
Common in heathy places on the Lower Greensand and Clay-with-Flints. Recorded for all neighbouring counties.
H.S. 22a–24, 27, 28, 35, 37, 39b, 46, 47, 49, 74a, 79a.
OUZEL and IVEL: too many stations to merit listing.
LEA: Warden Hill, in clay pocket!; *F.F.B.*: Chiltern Green Common.
COLNE: common in heathy places.
First record: Abbot, 1798.

G. pumilum Murr. C.T.W. 109.3.7.   *Slender Bedstraw*
   G. sylvestre Poll. non Scop., *G. pusillum* auct.
Pastures   Colonist
Rare, probably introduced with grass seed. Abbot's record from Shefford is to be doubted as his herbarium specimen is a weak lateral shoot of *G. mollugo.* Recorded for Herts., Bucks. and Northants.
COLNE: in limited quantity, Blow's Downs, v.c. 20 [Beds.]!*; *E.M.-R. in B.E.C. 1943–4 Rep.* 728.

G. palustre L. C.T.W. 109.3.8.   *Marsh Bedstraw*
Marshes, etc.   Native
Common, but variable. Var. *elongatum* (C. Presl) Rouy & Fouc. was recorded from Flitwick Moor; *V.C.H.*, and from Warden Abbey and Biggleswade; *J.E.L., J.Bot.*, and var. *witheringii* Sm. from the south of Luton in *W.F.B.* Recorded for all districts and neighbouring counties.
H.S. 3a, 6, 9, 13, 14a, 15, 19a, 20a, 23, 34a, 34b, 37–39a, 41, 56a, 74b, 84a, 85.
First record: Abbot, 1798.

G. uliginosum L. C.T.W. 109.3.10. *Bog Bedstraw*
Marshy places Native
Frequent. Recorded for all neighbouring counties.
H.S. 10a, 37–39a, 41, 85.
OUSE: Stevington!; *F.B.*: water meadow, Eaton Socon.
OUZEL: Heath and Reach†*; *S.B.P.L.*: Horsemoor Farm; Cow Common.
IVEL: Potton; Ampthill!; *F.B.*: Clophill; Gravenhurst; *W.C. Herb.*: Flitwick Moor!; *W.F.B.*: Cadwell; *J.E.L.*, *F.F.B.*: Westoning Moor; near Washer's Wood.
LEA: Marslets†; *Herb. J.S.*: Leagrave Marsh*.
First record: Abbot, 1798.

G. verum L. C.T.W. 109.3.5. *Lady's Bedstraw*
Roadsides, pastures, railway banks, etc. Native
A common species with a wide range of habitats. Recorded for all districts and neighbouring counties.
H.S. 1a–2a, 21, 28, 45b, 56b, 59–62, 65, 66a–67, 74a, 75.
First record: Abbot, 1798.

G. tricorne Stokes C.T.W. 109.3.11. *Rough Corn-bedstraw*,
G. spurium auct. *Pin Burs* (Batchelor)
Arable fields Colonist
Frequent on calcareous soils, especially on the Chalk and Oolite.
Miss Kitchin reports that lacemakers in the Podington area used to collect its fruits to cover the heads of pins to protect their fingers.
Recorded for all neighbouring counties.
H.S. 72.
NENE: plentiful in a number of fields near both Podington and Wymington.
OUSE: Twin Wood; *J.McL.*, *L.M.*: Hillfoot; *J.McL.*, *B.M.*: Cranfield; *A. E. Ellis*: Harrold; *G. H. Day*: Thurleigh!; *E.M.-R.*: Stagsden; *C. C. Foss*: Souldrop.
KYM: Knotting; near Kimbolton Station, v.c. 30 [Hunts.].
OUZEL: Totternhoe!; *W.F.B.*: Houghton Regis; *E.M.-R.*: Tilsworth.
IVEL: Streatley!*; Harlington; *B.P.L.*: Pegsdon!; *W.F.B.*: Mead Hook; *C.C.*: Holwell; Stanfordbury; *J.E.L.*, *F.F.B.*: Tebworth
V. H. Chambers: Barton; Arlesey; railway sidings, Henlow.
CAM: Dunton.
LEA: Eaton Green; *W.F.B.*: between Luton and Dunstable; *F.F.B.*: Dallow Rd., Luton†; *Herb. J.S.*
COLNE: Kensworth, v.c. 20 [Beds.].
First record: Abbot, 1798.

G. aparine L. C.T.W. 109.3.12. *Goosegrass, Cleavers, Scratchgrass*
Hedgerows, waste places, arable fields, etc. Colonist
Common: recorded for all districts and neighbouring counties.
H.S. 4, 5, 7, 12, 13, 18, 20a, 32, 34a, 41–43, 51b–53a, 54, 55b, 71, 78, 82, 83, 85.
First record: Abbot, 1798.

G. cruciata (L.) Scop. C.T.W. 109.3.1.   *Crosswort*
Valantia cruciata L.
Heaths, roadside verges, etc.   Native
Considered common by Abbot, but it is certainly not so now.
Frequent on light and heathy soils, it is absent from considerable
areas of the county. Recorded for all neighbouring counties.
OUSE: Wilstead; *J.McL.*, *L.M.*: Lidlington; *C.C.*
OUZEL: Little Billington; *S. A. Chambers in F.F.B.*: Clipstone; *V. H.
Chambers*: Heath and Reach; Leighton Buzzard*; Stockgrove.
IVEL: near Pegsdon; *H. Brown*: Cainhoe; *W.C. Herb.*: Clophill;
J.E.L., *J. Bot.*: Barton; *W.F.B.*
LEA: Luton Hoo!†; *S.B.P.L.*: railway, south of Luton; *D. M.
Higgins.*
COLNE: Studham Common; near Long Wood, v.c. 20 [Beds.].
First record: Abbot, 1798.

ASPERULA

Asperula odorata L. C.T.W. 109.2.1.   *Sweet Woodruff*
Heathy woods   Native
Common in woods on the Clay-with-Flints, frequent in woods on the
Lower Greensand. Recorded for all neighbouring counties, but
doubtfully for Hunts.
H.S. 78, 80b, 81.
OUSE: Great Barford; *G. H. Day.*
OUZEL: Aspley; King's Wood!; *Plantae Bedford.*: Palmer's Shrubs.
IVEL: Cainhoe Park Wood; *W.C. Herb.*: Warden Wood; *J.McL.*,
L.M.: Eversholt; *J.McL.*, *B.M.*: Streatley; *H. B. Sargent*: Washer's
Wood.
LEA and COLNE: common in almost every wood on the Clay-with-
Flints.
First record: Abbot, 1795.

A. cynanchica L. C.T.W. 109.2.3.   *Squinancy Wort*
Chalk hills   Native
Common on the Chalk, where it appears both in well-established
downland and in places where the soil has been disturbed. Recorded
for all neighbouring counties except Hunts.
H.S. 59, 60, 62, 63b, 65, 66a, 67.
OUZEL, IVEL, LEA and COLNE: locally abundant in all likely
habitats.
First record: Abbot, 1795.

SHERARDIA L.

Sherardia arvensis L. C.T.W. 109.1.1.   *Field Madder*
Arable fields   Colonist
Common and well distributed, but considered rare by Abbot.
Recorded for all districts and neighbouring counties.
H.S. 61, 71, 86.
First record: Abbot, 1798.

CAPRIFOLIACEAE

SAMBUCUS L.

Sambucus nigra L. C.T.W. 110.1.2.   *Elder*
Woods, hedgerows, etc.   Native
Common, and well distributed. It grows particularly plentifully
on disturbed ground and in the neighbourhood of rabbit warrens
on the Chalk. Var. **nigra** is recorded for all districts and neigh-
bouring counties.
H.S. 18, 27, 31c, 34b, 42, 47, 51, 54, 69, 70a, 74a, 81, 83.
First record: Abbot, 1798.

Var. **aurea** Sweet
LEA: Leagrave; *D. P. Young.*
Var. **laciniata** L.
OUZEL: near Plough Inn, Eaton Bray; *H. B. Souster.*
IVEL: garden escape, Barton; *F.F.B.*
LEA: garden escape, Luton; *F.F.B.*

S. ebulus L. C.T.W. 110.1.1.   *Danewort*
Hedgerows   Denizen
Rare: a species remarkable for its long persistence. There is a
legend about this, as about *Anemone pulsatilla*, that it grows where
Danish blood was spilt. Abbot, no doubt in error, listed it as
common. Recorded for all neighbouring counties.
OUSE: Kempston, where McLaren still found it; Dropshort, near
which it still grows on the road to Stagsden!; *Plantae Bedford.*:
Stagsden, no doubt the latter station; *W.H. 1875.*
IVEL: 'the nether field, behind Barton Church, is called Dunstall or
Danestall field, where they say the Danes had an overthrowe in
Battel, and in that field doth yet grow Danesweed or Danesbloud,
soe called, they believe, because the Danes bloud there spilt did
bring up this weed which here will not be destroyed'—I am
indebted to I. J. O'Dell, not only for this source, but for searching
every field until he found the plant still growing there!; *F. Taverner*:
Sundon; *Davis.*
LEA: Limbury, where it is still plentiful in a limited area!†*;
S.B.P.L.: Luton Hoo; *W.F.B.*
COLNE: hedge, Farley Bottom; *L. J. Margetts.*
First record: Francis Taverner, *The history and antiquities of Hexton,
B.M. Add. MSS. 6223, c. 1640.*

VIBURNUM L.

Viburnum opulus L. C.T.W. 110.2.2.   *Guelder Rose*
Woods, hedgerows, etc.   Native
Frequent and widely distributed. Recorded for all districts and
neighbouring counties.
H.S. 30a, 79, 80.
First record: Abbot, 1798.

V. lantana L. C.T.W. 110.2.1.   *Wayfaring Tree*
Hedgerows, rough downland, etc.   Native
Common, especially on calcareous soils, and seen at its best in chalk
scrub and hedgerows. Recorded for all districts and neighbouring
counties.
H.S. 3b, 5, 33, 53b, 66a, 68–70a.
First record: Abbot, 1798.

LONICERA L.

Lonicera periclymenum L. C.T.W. 110.5.2.   *Honeysuckle*
Woods, hedgerows, etc.   Native
Common: recorded for all districts and neighbouring counties.
H.S. 2b, 3b, 19b, 27, 30a, 32–34b, 40, 43, 49, 51, 53a, 53b, 78,
79b–81, 83. .
First record: Abbot, 1798.

ADOXACEAE
ADOXA L.

Adoxa moschatellina L. C.T.W. 111.1.1.   *Moschatel*
Woods   Native
Frequent in woods on the Lower Greensand and Clay-with-Flints,
rare elsewhere. Recorded for all neighbouring counties.
H.S. 83.
OUSE: Renhold, reported to be still there by L. W. Wilson; *Plantae
Bedford.*: Clapham; *F.B.*: Lidlington; *C.C.*: foot of Cleat Hill;
Muriel Hamson in J.H. Notes.
OUZEL: Aspley, where it was still found by W. Crouch and J.
McLaren; *Plantae Bedford.*: near Fox and Hounds, Potsgrove; King's
Wood.
IVEL: Sundon; *S.B.P.L.*: Streatley!†; *B.P.L.*: Noon Hill; *H. B.
Souster*: Leete Wood; Washer's Wood.
LEA: Dunstable; *W. G. Smith in J. Bot. 1876*: Limbury; *S.B.P.L.*:
Bramingham, plentiful in George Wood!†*; *F.F.B.*: Icknield Way,
near Lilley Hoo; *J.E.L.*, *B.E.C.*
COLNE: wood, north of Zouche's Farm, v.c. 20 [Beds.]; *D. Jenks in Fl.
of Herts.*: Caddington!; Whipsnade!; *F.F.B.*: Badger Dell Wood!;
V. H. Chambers: plentiful in woods near Kensworth, v.c. 20 [Beds.].
First record: Abbot, 1795.

VALERIANACEAE
VALERIANA L.

Valeriana officinalis L. C.T.W. 112.2.1.   *Valerian*
   V. sambucifolia Mikan f.
Woods, rough pastures, etc.   Native
Common, and well distributed in the greater part of the county,
but rare or absent in the south. It is a very variable species, and
has long attracted the attention of botanists. A number of records

are for *V. sambucifolia*, which more recent botanists discard. It is obviously a species that demands still more study. Recorded for all districts except Lea and Colne, and for all neighbouring counties.
H.S. 1b–3b, 13, 38, 43, 53b.
First record: Abbot, 1798.

V. dioica L. C.T.W. 112.2.3. *Marsh Valerian*
Marshes *Native*
Frequent, occurring in all likely places in the county. Recorded for all neighbouring counties.
H.S. 10a, 14b, 38, 39a.
Ouse: Caldwell; Bromham; *F.B.*: Harrowden; *J.McL., B.M.*: Honeydon; *A. W. Guppy*: Stevington Marshes; water meadow, Eaton Socon.
Ouzel: Totternhoe!; Leighton Buzzard; Stanbridgeford; *F.F.B.*: near Palmer's Shrubs; Battlesden; Heath and Reach.
Ivel: Gravenhurst Moor; *W.C. Herb.*: Southill Lake; *J.McL., L.M.*: Westoning Moor!; Flitwick Moor!; *F.F.B.*: Fancott.
Lea: New Mill End!*; Luton Hoo; *F.F.B.*: Limbury Marsh†*; *Herb. J.S.*: Leagrave Common.
First record: Abbot, 1798.

VALERIANELLA Mill.

Valerianella locusta (L.) Betcke C.T.W. 112.1.1. *Lamb's Lettuce*,
Valeriana locusta L., *Valerianella olitoria* (L.) Poll. *Corn-salad*
Arable fields, old walls, dry banks, etc. *Colonist*
Common, and well distributed. It was considered common by Abbot, but his herbarium specimen is *V. dentata*, which he did not list. A form with very large fruits has been found on Sundon Rubbish Dump*. Recorded for all districts except Kym, and for all neighbouring counties.
First record: Abbot, 1798.

V. dentata (L.) Poll. C.T.W. 112.1.5. *Sharp-fruited Corn-salad*
Arable fields, railway banks, etc. *Colonist*
Common, especially on calcareous soils. Forms with hairy fruits are not infrequent. Recorded for all districts except Kym, and for all neighbouring counties.
H.S. 71, 72.
First record: W. Crouch Herbarium, 1845.

V. rimosa Bast. C.T.W. 112.1.3.
Arable fields *Colonist*
Rare: recorded for all neighbouring counties.
Ivel: plentiful in arable fields to the east of Barton!*; *W. D. Coales in Watsonia II* (1951) 44.

V. carinata Lois.
Old walls *Colonist*
Rare: recorded for Herts., Bucks. and Northants.
Ivel: wall, Tingrith Church, 1953*; *H. Cole*.

DIPSACACEAE

DIPSACUS L.

Dipsacus fullonum L. C.T.W. 113.1.1.   *Teasel*
D. sylvestris Huds., *D. fullonum* subsp. *sylvestris* (Huds.) Clapham
Riversides, waste places, etc.   Native
Common, especially on clay soils: recorded for all districts and
neighbouring counties.
H.S. 2a, 7, 13, 20a, 34a, 52, 58b.
First record: Abbot, 1798.

D. sativus (L.) Scholler
  D. fullonum L. p.p.
Rough pastures   Colonist
Rare: recorded for Herts. and Bucks.
LEA: Bradger's Hill, with intermediates with *D. fullonum* and appearing for a num-
ber of years!*; *P.T.*

D. pilosus L. C.T.W. 113.1.2.   *Small Teasel*
Damp woods, riversides, etc.   Native
Rare: recorded for all neighbouring counties.
OUSE: Goldington; Medbury; *Plantae Bedford.*: Kempston; *F.B.*:
Lidlington; *W.C. Herb.*: Park Lane, Cardington; *J.McL.*, *L.M.*:
Colesdon; Roxton!; *G. H. Day*: Stagsden; side of King's Wood*.
IVEL: about Toddington and Clophill; *E. Forster in Bot. Guide*:
Warden Wood; *J.McL.*, *L.M.*: Steppingley; Westoning; Beckerings-
park*; *C.C.*: near Stondon; *J. Pollard*: Beadlow; *Miss Harradine in
J.H. Notes*: Flitwick Moor!*; *F. Seymour Lloyd in W.F.B.*: Shefford;
Maulden; *H. B. Souster*: Chicksands Great Wood*; Rowney
Warren; Clophill.
First record: Abbot, 1795.

SCABIOSA L.

Scabiosa columbaria L. C.T.W. 113.3.1.   *Small Scabious*
Pastures   Native
Common on the Chalk, where it grows both on established down-
land and on disturbed soils. It occurs also, but less frequently, on
oolitic exposures in the north of the county. It grows regularly on
the roadside verge south of Clophill, from which neighbourhood
W. Crouch obtained his specimen, apparently on the Lower Green-
sand. A white-flowered form has been seen at Pegsdon. Recorded
for all neighbouring counties.
H.S. 10b, 59, 60, 63a, 63b, 65, 66a, 67.
OUSE: Cardington; *J.McL. in J.H. Notes*: Cleat Hill, *J.McL.*, *B.M.*:
Bromham; *J.H. Notes*: Felmersham; Chellington; Turvey; *G. H.
Day*: slopes above river between Pavenham and Stevington.
OUZEL, IVEL, LEA and COLNE: common on the Chalk.
First record: Abbot, 1798.

SUCCISA Haller

Succisa pratensis Moench C.T.W. 113.4.1.   *Devil's-bit*
  Scabiosa succisa L.
Damp woods, pastures, etc.   Native
A species with an unusual range of habitats in the county. Locally
abundant in the rides of heavy clay woods, it occurs also on some
chalk downland and in meadowland on the Chalk Marl. It is very
variable in the size of the plants and in its differing habitats, and
white-flowered forms are not infrequent. Recorded for all districts
except Colne, and for all neighbouring counties.
H.S. 2a, 10a, 10b, 19a, 19b, 30a, 30b, 38, 56b, 59–61, 66b.
First record: Abbot, 1798.

KNAUTIA L.

Knautia arvensis (L.) Coult. C.T.W. 113.2.1.   *Field Scabious,*
  Scabiosa arvensis L.   *Pincushion*
Arable fields, rough pastures, waste places, etc.   Colonist
Common in most of the county but unaccountably rare in the north-
east. A white-flowered form has been observed at Arlesey. Re-
corded for all districts except Kym, and for all neighbouring
counties.
H.S. 61, 63a, 64, 71–73, 76, 86.
First record: Abbot, 1798.

COMPOSITAE

EUPATORIUM L.

Eupatorium cannabinum L. C.T.W. 114.22.1.   *Hemp Agrimony*
Riversides, ditches, damp woods, etc.   Native
Common by the Ouse and its main tributaries, rare elsewhere.
It is an attractive species, and often dominates considerable
stretches of a riverside. Recorded for all neighbouring counties.
H.S. 6, 10c.
NENE: Wymington Scrubs, in open woodland!; *V. H. Chambers.*
OUSE, OUZEL and IVEL: too many stations to merit listing.
First record: Abbot, 1798.

SOLIDAGO L.

Solidago virgaurea L. C.T.W. 114.17.1.   *Golden-rod*
Heaths, heathy woods, etc.   Native
Rare, and limited to the Lower Greensand. Recorded for all
neighbouring counties except Hunts.
OUZEL: Aspley, in open heathland!†; *Plantae Bedford.*: Birchmore;
Potton Wood; *F.B.*: *Woburn*; *P.T.*: locally abundant at Heath and
Reach*; sparingly, King's Wood.
IVEL: Clophill, on open greensand!; *J.McL. in J.H. Notes*: Maulden;
M. L. Berrill in J.H. Notes.
First record: Abbot, 1798.

BELLIS L.

Bellis perennis L. C.T.W. 114.21.1.   *Daisy*
Meadows, garden lawns, etc.   Native
Common, but limited in its habitats and so less ubiquitous than it
might appear to be. A rayless form appears regularly at Greencroft
Barn, v.c. 20 [Beds.] and on Whipsnade Heath. Recorded for all
districts and neighbouring counties.
H.S. 1a, 3a, 8, 9, 10a, 10b, 14b, 16, 19a, 20a, 27, 28, 45b, 56b, 58b,
62, 68, 75, 76, 80a.
First record: Abbot, 1798.

ERIGERON L.

Erigeron acer L. C.T.W. 114.20.1.   *Blue Fleabane*
Pastures, waste places, etc.   Colonist
Frequent on clay soils where the ground has at some time been
disturbed. It readily colonizes brick pits, both on the Gault and
Oxford Clay. It is found more sparingly on the Chalk, and is not
infrequent on ballast by the railway. It was considered rare by
Abbot. Recorded for all districts except Kym, and for all neighbour-
ing counties.
H.S. 45a, 45b, 58b, 64.
First record: Abbot, 1795.

E. canadensis L. C.T.W. 114.20.4.   *Canadian Fleabane*
Waste places, railway sidings, roadsides, etc.   Colonist
A comparatively recent introduction, but now common, especially on the Lower
Greensand. Recorded for all districts except Kym and Cam, and for all neigh-
bouring counties.
H.S. 13, 16, 22b, 27, 44, 58b.
First record: J. McLaren in *J. Bot. 1884*, 250.

FILAGO L.

Filago germanica (L.) L. C.T.W. 114.13.1.   *Cudweed*
Heaths, disturbed soils, etc.   Native
Frequent on the Lower Greensand and on river and glacial gravels.
Recorded for all neighbouring counties.
H.S. 25, 27, 44, 45a.
OUSE: Basmead Manor; *C.R.O. List*: Oakley; Bromham!; Milton
Ernest; *B. B. West*: Harrold; Roxton; *G. H. Day*: gravel pits,
Eaton Socon, Willington.
OUZEL: sandpits, Leighton Buzzard; plentiful in neighbourhood of
Heath and Reach*.
IVEL: Cainhoe; *W.C. Herb.*: Arlesey; Southill; Barton Hills, on
glacial gravel!; *J.E.L., F.F.B.*: roadside, Maulden, Clophill;
Rowney Warren*; railway bank, Ampthill.
LEA: Limbury†; *Herb. J.S.*
COLNE: Pepperstock†; *Herb. J.S.*: Whipsnade Heath; *V. H.
Chambers*.
First record: Abbot, 1798.

F. spathulata C. Presl C.T.W. 114.13.3.
Sandy places
Rare, or overlooked. I have tried unsuccessfully to distinguish this and the
following species from the preceding. Recorded for Herts., Bucks. and Cambs.
OUZEL: Birchmore; *W.H. 1875*, spec. in Herb. Birmingham Univ.
IVEL: Flitwick†; *B.P.L.*
First record: W. Hillhouse, 1875.

F. apiculata G.E.Sm. C.T.W. 114.13.2.
Sandy places
Rare, or overlooked. Recorded for Herts., Bucks. and Cambs., and doubtfully for
Northants.
OUZEL: Aspley; *W.H. 1875*, spec. in Herb. Birmingham Univ.: Woburn;
W. Moyle-Rogers in F.F.B.: Heath and Reach, etc.; *V.C.H.*
IVEL: Shefford; *T. B. Blow in Bot. Rec. Club Rep. 1875*: Flitwick†*; *S.B.P.L.*:
Maulden; *J.E.L., J. Bot.*
First record: Abbot, 1798.

F. minima (Sm.) Pers. C.T.W. 114.13.5.   *Slender Cudweed*
    F. montana L. p.p.
Heaths, sandy places, etc.   Native
Frequent on and, unlike *F. germanica*, limited to the Lower Greensand. As it is unable to withstand the competition of taller species
its period of existence on previously disturbed soils is limited. A
record in the name of M. Brown from Caddington in *W.F.S. Diary*
was probably in error. Considered rare by Abbot. Recorded for
all neighbouring counties.
H.S. 22a, 22b, 25, 27, 46, 48.
OUZEL and IVEL: too many records to merit listing.
First record: Abbot, 1795.

Antennaria dioica (L.) Gaertn. C.T.W. 114.16.1.
                              Mountain Cat's-ear, Cat's-foot
Chalk hills   Native
Very rare and apparently limited to well-established downland.
I have looked for it in vain in all the recorded stations. Recorded
for all neighbouring counties except Bucks.
OUZEL: Dunstable Downs; *F.F.B.*
IVEL: Barton Hills, where it was growing at least until 1936 when it
was last seen by Richard Morse†; *Herb. R. H. Webb*: Sharpenhoe
Clappers, 1926*; *H. A. J. Martin*: Streatley Hills, possibly same
station, c. 1926; *M. E. Boutwood.*
First record: R. H. Webb, c. 1840, see A. R. Pryor, *J. Bot. 1876.*

GNAPHALIUM L.
Gnaphalium uliginosum L. C.T.W. 114.14.4.   *Marsh Cudweed*
Pondsides, arable fields, waste places, etc.   Native
A common species with a wide range of habitats, but usually those of
a damp and open nature. More common on acid soils than else-
where. Recorded for all districts except Cam, and for all neighbour-
ing counties.
H.S. 13, 16.
First record: Abbot, 1798.

G. sylvaticum L. C.T.W. 114.14.1.   *Wood Cudweed*
  G. rectum Sm.
Heathy woods, heaths, etc.   Native
Frequent in woods on the Lower Greensand and Clay-with-Flints.
E. Bot. tab. 124, Aug. 1793, was drawn from a specimen lent by
Abbot. Recorded for all neighbouring counties.
H.S. 23.
OUSE: Lidlington; *C.C.*
OUZEL: Aspley, still frequent in Aspley Wood!*; *Plantae Bedford.*:
Heath and Reach, frequent in King's Wood!; *F.F.B.*
IVEL: Clophill!; Southill; *F.B.*: Everton Wood; *Abbot's annotated
F.B.*: Steppingley; *W.C. Herb.*: Keeper's Warren; *J.McL.*, *L.M.*:
Blackham Firs, Ridgmont; *C.C.*: Portobello Farm, Sutton; *J.E.L.*,
Diary: Sandy Heath; *J.E.L.*, *J. Bot.*: foot of Cooper's Hill.
LEA: Chiltern Green Wood, frequent in Horsley's Wood, no doubt
intended!*; *W.F.B.*
COLNE: plentiful in Deadmansey Wood and Byslip Wood.
First record: Abbot, 1793.

INULA L.

Inula helenium L. C.T.W. 114.11.1.   *Elecampane*
Hedgerows, rough pastures, etc.   Colonist
Rare. This attractive species, recorded so early for the county, is
still with us. It is possibly now of garden origin, as it is rarely far
from houses. Recorded for all neighbouring counties.
OUSE: Ravensden, found there still by L. W. Wilson in 1936*;
Stevington; *F.B.*: Cox's Pits; *J.McL. in J.H. Notes*: Turvey; *W. W.
Mason in B.E.C. 1918 Rep.*: Thurleigh Lane; *A. Ransom*: Cleat Hill;
W. Durant.
KYM: hedgerow near Swineshead!; *G. H. Day.*
OUZEL: 'it groweth plentifully in the fieldes on the left hande as you
go from Dunstable to Puddle Hill'; *Gerard's Herbal*, 1062.
IVEL: Pulloxhill; Higham Gobion; *C.C.*
First record: Gerard, 1597.

Inula conyza DC. C.T.W. 114.11.4.   *Ploughman's Spikenard*
  Conyza squarrosa L., *Inula squarrosa* (L.) Bernh. non L.
Rough pastures   Native
Frequent on sandy and gravelly soils throughout the county. Re-
corded for all neighbouring counties.
H.S. 25, 58b.
OUSE: Bromham; Medbury; *Abbot's annotated F.B.*: Wilstead;
Limbersey Lane; *M. L. Berrill in J.H. Notes*: Park Lane; *J.McL.*,
B.M.: Chawston; *G. H. Day*: Eaton Socon, plentifully on Castle
Hill.
KYM: Dean; *B. B. West.*
OUZEL: Aspley Wood†; *Herb. J.S.*: roadside, King's Wood!;
E.M.-R.

IVEL: Ampthill; *Abbot's annotated F.B.*: Clophill!; *W.C. Herb.*: Ireland; *J.McL.*, *L.M.*: Barton, on glacial gravel!*; Flitwick; *W.F.B.*: rough pasture, Warden Abbey; roadside between Stondon and Henlow; claypit, Arlesey.
LEA: near the Lodge; *Davis*: Dallow Road†; *S.B.P.L.*: railway siding, Dunstable.
First record: Abbot, 1798.

PULICARIA Gaertn.

Pulicaria dysenterica (L.) Bernh. C.T.W. 114.12.1.   *Fleabane*
  Inula dysenterica L.
Marshy places, roadsides, etc.   Native
Common in the north and middle of the county, especially on clay soils. Recorded for all districts except Lea and Colne, and for all neighbouring counties.
H.S. 7, 10c, 21, 57a, 57b, 58b.
First record: Abbot, 1798.

P. vulgaris Gaertn. C.T.W. 114.12.2.   *Small Fleabane*
  P. pulicaria (L.) Karst.
Marshy places   Native
Apparently extinct. Recorded for Herts. and Cambs., and doubtfully for Northants.
OUSE: Goldington Green, where McLaren still found it in 1864; *Plantae Bedford.*: Ravensden; *F.B.*
First record: Abbot, 1795.

BIDENS L.

Bidens cernua L. C.T.W. 114.3.1.   *Nodding Bur-marigold*
Marshes, riversides, etc.   Native
Rare: it is probable that Abbot's record of *B. minimus* refers to var. *minima* (Huds.) Boiss. Recorded for all neighbouring counties.
OUSE: Clapham; *F.B.*: Goldington; *Abbot's annotated F.B.*: Kempston Pits, as *B. minimus*; *F.B.*: Renhold; *W.H. 1875*: Harrold; *G. H. Day*.
OUZEL: Woburn; *W.H. 1875*.
IVEL: Clophill; *F.B.*: Flitton; *J.McL.*, *L.M.*: Flitwick; *S.B.P.L.*: Ampthill; *F.F.B.*: Biggleswade; *J.E.L.*, *J. Bot.*: Tingrith; *M. Brown in W.F.S.*, *Diary*.
LEA: Luton Hoo!; *B.P.L.*: New Mill End, where it has become very scarce since the opening of Luton Sewage Works!†*; *Herb. J.S.*
First record: Abbot, 1798.

B. tripartita L. C.T.W. 114.3.2.   *Three-cleft Bur-marigold*
Riversides, etc.   Native
Much more common than the preceding, and found frequently by pools in sand, clay and gravel pits, as well as by rivers and streams. Recorded for all districts except Nene, Cam and Colne, and for all neighbouring counties.
H.S. 13.
First record: Abbot, 1798.

AA

GALINSOGA Ruiz & Pav.

Galinsoga parviflora Cav. C.T.W. 114.4.1.   *Kew Weed, Gallant Soldier*
Arable fields, waste places, etc.   Colonist
A comparatively recent introduction, which is now not uncommon on sandy
soils, and reappears in the same field for a number of years. Recorded for Herts.
and Hunts.
H.S. 50.
IVEL: Galley Hill; *J.E.L. in B.E.C. 1932 Rep.*: Potton!*; *H. Gilbert-Carter*: waste
ground, Ampthill!; *C. C. Foss*: Sundon Rubbish Dump*; *E.M.-R.*!: frequent near
Potton, Sandy and Everton; arable field, Old Warden.
LEA: garden weed, Stockwood!; *J. Cowley.*
First record: J. E. Little, 1932.

ACHILLEA L.

Achillea millefolium L. C.T.W. 114.24.1.   *Milfoil, Yarrow*
Pastures, roadsides, waste places, etc.   ? Native
Very common: pink-flowered forms are frequent. Recorded for all
districts and neighbouring counties.
H.S. 1a–2a, 13, 25, 27–29, 45b, 48, 50, 51b, 56b, 57a, 58b, 63a, 63b,
66a–67, 70a, 74a, 75.
First record: Abbot, 1798.

A. ptarmica L. C.T.W. 114.24.2.   *Sneezewort*
Riversides, etc.   Native
Common by the Ouse, but also occurring sparingly in woods
throughout the county and as a casual in waste places. Recorded
for all neighbouring counties.
H.S. 7.
NENE: Great Hayes Wood.
OUSE: common by the river.
KYM: West Wood.
OUZEL: ditch by Aspley Wood; *H. B. Souster*: King's Wood.
IVEL: Tingrith Park; *Mrs. Twidell in J. Bot. 1889*: Harlington†;
Mrs. Twidell in Herb. J.S.: ditch, Attwood's Hill, Toddington;
H. B. Souster: Maulden Wood.
LEA: station platform, Leagrave; railway siding, Luton*.
First record: Abbot, 1798.

ANTHEMIS L.

Anthemis cotula L. C.T.W. 114.23.2.   *Stinking Mayweed*
Arable fields, waste places, etc.   Colonist
Common: recorded for all districts except Lea, and for all neigh-
bouring counties.
H.S. 11, 16.
First record: Abbot, 1798.

A. arvensis L. C.T.W. 114.23.3.   *Corn Chamomile*
Arable fields, railway banks, etc.   Colonist
A rare species with us, and often confused with the preceding. It
should be noted that it flowers early and is not to be seen when the
other mayweeds and chamomiles are in flower.

OUSE: Oakley West Field; *F.B.*: Lidlington; *C.C.*

IVEL: Ridgmont; *C.C.*: Flitwick, where it appears regularly on a railway bank!†*; *S.B.P.L.*: Pegsdon; Holwell, v.c. 30 [Herts.]; *J.E.L.*, *J. Bot.*
First record: Abbot, 1798.

A. nobilis L., C.T.W. 114.23.4, is of doubtful occurrence in the county. *Herb. J.McL.*, *B.M.* contains a wrongly-named specimen labelled 'railway bank', and *Herb. Saunders* another wrongly-named specimen from Leagrave. Saunders recorded it from 'Ravensbury Hill' in *B.P.L.*, but not in subsequent lists, and from Limbury in *S.B.P.L.* and Butterfield's Green in *W.F.B.*, but there are no specimens to support these records. There is a specimen from Segenhoe Manor in Herb. Pollard, but it is doubtful if this was of a wild plant, as Crouch, who had lived at Segenhoe and knew Pollard well, would almost certainly have recorded it. Recorded for all neighbouring counties except Hunts.

CHRYSANTHEMUM L.

Chrysanthemum segetum L. C.T.W. 114.28.1. *Corn-marigold,*
Goolds (Batchelor)
Arable fields, gravel pits, railway sidings, etc. Colonist
Frequent on sandy and gravelly soils, and usually regular in its appearance. Considered common by Abbot and Hamson, it is certainly not so now. Recorded for all neighbouring counties.
OUSE: Bedford; *A. B. Sampson*: Cardington; *L. J. Tremayne*: Stewartby; Bromham, frequent in the gravel pits!; *A. W. Guppy*: Harrold; Roxton; *G. H. Day*.
OUZEL: Tebworth; *H. B. Souster*: Woburn*; Potsgrove.
IVEL: Cainhoe; *W.C. Herb.*: Flitwick; *S.B.P.L.*: Haynes!; Sutton; Sandy Warren; *J.E.L.*, *F.F.B.*: Ampthill; Potton; Shefford; railway sidings, Harlington, Flitwick; arable field, Maulden.
LEA: Limbury; *S.B.P.L.*: railway siding, Luton.
COLNE: cornfield, Caddington; *L. J. Margetts*.
First record: Abbot, 1798.

C. leucanthemum L. C.T.W. 114.28.2. *Dog Daisy, Ox-eye Daisy,*
Moon Daisy
Pastures, chalk hills, railway banks, etc. Native
Common, and often making a fine show in meadows. Recorded for all districts and all neighbouring counties.
H.S. 1a–2a, 21, 45a, 45b, 56b, 59, 60, 63a, 64, 76.
First record: Abbot, 1798.

MATRICARIA L.

Matricaria maritima L. subsp. **inodora** (L.) Clapham
C.T.W. 114.27.1. *Scentless Mayweed*
Chrysanthemum inodorum (L.) L.
Arable land, waste places, etc. Colonist
Common: a form lacking disc florets has been seen at East Hyde*.
Recorded for all districts and neighbouring counties.
H.S. 16, 17, 34a, 50, 58b, 64, 82, 86.
First record: Abbot, 1798.

M. recutita L.   *Wild Chamomile*
   M. chamomilla L. p.p. C.T.W. 114.27.2.
Arable fields, waste places, etc.   Colonist
Less common than the preceding, but well distributed in the
county. It is more frequent on light soils than elsewhere. Recorded
for all districts and neighbouring counties.
H.S. 4.
First record: Abbot, 1798.

M. matricarioides (Less.) Porter C.T.W. 114.27.3.   *Rayless Mayweed*
   M. suaveolens (Pursh) Buchen. non L.
Roadsides, farm tracks, etc.   Colonist
A comparatively recent introduction, but now common throughout the county.
Recorded for all districts and neighbouring counties.
H.S. 27.
First record: J. E. Little in *J. Bot. 1917*, 49.

TANACETUM L.

Tanacetum vulgare L. C.T.W. 114.29.1.   *Tansy*
Rough pastures, roadsides, etc.   Colonist
Frequent on the Lower Greensand, where it often colonizes grassy banks and
makes a fine show. Abbot considered it rare. Recorded for all neighbouring counties.
OUSE: Bromham Grange; *F.B.*: Bedford Friars; *Abbot's annotated F.B.*: near Little
Barford Power Station; *H. B. Souster!*: railway sidings, Bedford.
OUZEL and IVEL: too many stations to merit listing.
LEA: Leagrave!; East Hyde; *B. Verdcourt*: railway sidings, Luton.
COLNE: near the Packhorse, Kensworth!; *W. D. Coales*.
First record: Abbot, 1798.

ARTEMISIA L.

Artemisia absinthium L. C.T.W. 114.31.3.   *Wormwood*
Roadsides, rough pastures, etc.   Of garden origin
Frequent, usually found near to houses, and well established. Abbot considered
it to be common. Recorded for all neighbouring counties.
OUSE: Cardington*; *J.McL. in W.F.B.*: Elstow; *J.McL., L.M.*: Brogborough;
C.C.: Harrold; *G. H. Day*: Stagsden; refuse tip, Bedford; railway sidings, Bedford.
KYM: roadside, Tilbrook; v.c. 30 [Hunts.].
OUZEL: Leighton Buzzard; *W.F.S. Diary*: roadside, Heath and Reach*.
IVEL: Biggleswade!; *I. Brown in Herb. Watson*: Cainhoe; *W.C. Herb.*: Sandy;
J.McL., B.M.: Southill; *J.E.L., J. Bot.*: Potton.
LEA: Luton; *W.F.S. Diary*.
First record: Abbot, 1798.

A. vulgaris L. C.T.W. 114.31.1.   *Mugwort*
Waste places, roadsides, etc.   Native
Common throughout the county: recorded for all districts and
neighbouring counties.
H.S. 13, 50, 63a.
First record: Abbot, 1798.

TUSSILAGO L.

Tussilago farfara L. C.T.W. 114.9.1.   *Coltsfoot*
Disturbed soils   Native
Common especially on clay soils. It readily colonizes clay, chalk
and sand pits. Recorded for all districts and neighbouring counties.
H.S. 11, 16, 21, 34b, 45a, 45b, 58b, 63b, 64, 69, 71, 72.
First record: Abbot, 1798.

PETASITES Mill.

Petasites hybridus (L.) Gaertn., Mey. & Scherb.
C.T.W. 114.10.1.   *Butterbur*
Tussilago petasites L., *P. ovatus* Hill, *P. petasites* (L.) Karst.
Wet meadows, riversides, etc.   Native
Frequent, and usually locally abundant where it appears. Staminate-
flowered plants are the more general in the county but Miss Day
succeeded in finding pistilate-flowered plants at Felmersham in
1949. Recorded for all neighbouring counties.
H.S. 6.
NENE: Farndish; *W. Kitchin*: Hinwick.
OUSE: Oakley; Cardington; Cople; *F.B.*: near Castle Mill;
J.McL., L.M.: Pavenham!; Felmersham!; *J.H. Notes*: Chellington;
G. H. Day: Biddenham; Bromham; *A. W. Guppy*: Stevington*;
Blunham.
KYM: Dean Grange; *M. Dalton.*
IVEL: Clophill; *W.C. Herb.*: Eversholt; *J.McL., B.M.*: Flitwick,
plentiful near the Mill!*; *J.H. Notes*: Greenfield Mill!; *J. E.
Cooper*: between Biggleswade and Shefford; *J.S., J. Bot. 1889*:
Arlesey; Cadwell; *J.E.L., J. Bot.*: Campton!; *J.E.L., F.F.B.*:
Chicksands; *V. H. Chambers*: Maulden; *H. B. Souster*: Stotfold
Mill.
LEA: Marslets, now extinct!†; *Herb. J.S.*
First record: Abbot, 1798.

SENECIO L.

Senecio aquaticus Hill C.T.W. 114.7.2.   *Marsh Ragwort*
Wet meadows, marshes, etc.   Native
Common in the Ouse water meadows, sparingly elsewhere. Re-
corded for all neighbouring counties.
H.S. 8, 13, 14b, 41.
OUSE: common in water meadows by the river.
OUZEL: damp meadows, Heath and Reach, Leighton Buzzard.
IVEL: Clophill!; *W.C. Herb.*: Biggleswade!; Shefford; *J.E.L.,
F.F.B.*: damp meadow, Upper Gravenhurst!; *V. H. Chambers*:
Southill; Pennyfather's Moor; Westoning Moor.
LEA: Leagrave, at one time common on the Marsh!†; *J. S. Herb.*:
damp meadow, East Hyde.
First record: Abbot, 1798.

S. jacobaea L. C.T.W. 114.7.1.   *Ragwort*
Dry places, waste ground, etc.   Native
Common, but of uneven distribution. It is not eaten by rabbits,
and thrives in the neighbourhood of rabbit warrens and on heathland heavily grazed by rabbits. Recorded for all districts and
neighbouring counties.
H.S. 2a, 16, 17, 20a, 22b, 23, 26–28, 46, 48, 50, 51b, 53a, 54, 61,
64, 76, 80a.
First record: Abbot, 1798.

S. erucifolius L. C.T.W. 114.7.4.   *Hoary Ragwort*
Rough pastures, railway banks, etc.   Native
Common on calcareous soils whether of clay or chalk, but absent
or very rare on acid soils where it is entirely replaced by *S. jacobaea*.
It appears, but with uncertainty, on gravels. Considered common
by Abbot and rare by Saunders, it is certainly common now. Re-
corded for all districts and neighbouring counties.
H.S. 1a, 1b, 3a, 21, 25, 45a, 45b, 51b, 58b, 59, 62, 63a, 64, 68, 75,
79a.
First record: Abbot, 1798.

S. squalidus L. C.T.W. 114.7.5.   *Oxford Ragwort*
Railway sidings, waste places, etc.   Colonist
A recent introduction which has increased rapidly: it appears most frequently by
the railway, by which means it no doubt entered the county. Recorded for all
districts except Nene and Cam, and for all neighbouring counties.
First record: the author in *B.E.C. 1943–4 Rep.*

S. viscosus L. C.T.W. 114.7.7.   *Sticky Groundsel*
Waste places, roadsides, etc.   Colonist
Frequent and increasing. It was reported by Abbot to J. E. Smith in a letter dated
2 Nov. 1804, but was not recorded by any later botanists. A specimen was shown
to me by L. W. Wilson in 1936, and I have since seen it in all districts except Cam.
Recorded for all neighbouring counties.
First record: Abbot, 1804.

S. sylvaticus L. C.T.W. 114.7.6.   *Heath* or *Wood Groundsel*
Heathy places   Native
Frequent on heaths and commons on the Lower Greensand and
Clay-with-Flints. Recorded for all neighbouring counties.
H.S. 25, 29, 47.
OUSE: Bedford; *J. Pollard*: Clapham; *J.H. Notes*.
OUZEL: Aspley!; *Abbot's annotated F.B.*: Heath and Reach*.
IVEL: Sandy!; *Abbot's annotated F.B.*: Rowney Warren!; *J.McL.*,
B.M.: Maulden; *J. Pollard*: Flitwick, frequent on drier parts of the
Moor!†*; *Herb. J.S.*: Tingrith; *H. B. Souster*: Clophill!; *P.T.*:
White Wood, v.c. 30 [Hunts.]; Beckeringspark Moor; Bunker's
Hill.
COLNE: Whipsnade; *H. B. Souster*: Studham Common.
First record: Abbot, 1798.

S. vulgaris L. C.T.W. 114.7.8.   *Groundsel*
Arable land, waste places, etc.   Colonist
Common: recorded for all districts and neighbouring counties.
H.S. 12, 16, 17, 25, 27, 34a, 45a, 47, 50, 51b, 63a, 71, 73, 82.
First record: Abbot, 1798.
       Var. **hibernicus** Syme
    Var. *radiatus* auct.
A well marked variety, and found with us only on railway ballast.
OUSE: Oakley*.
OUZEL: Leighton Buzzard.
IVEL: Harlington*; rubbish dump, Sundon.

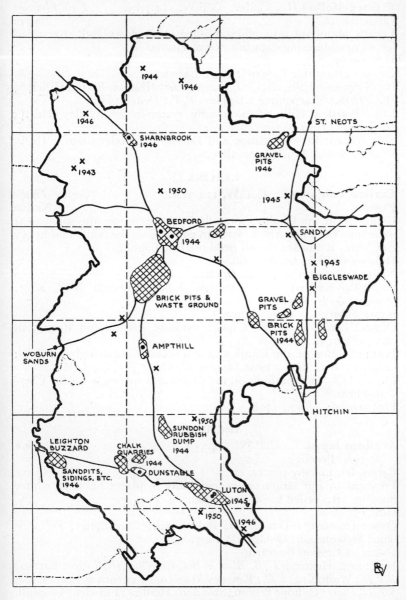

FIG. 21   DISTRIBUTION OF SENECIO SQUALIDUS

Oxford Ragwort was first observed in the county in 1943 but may have been present earlier. It is now abundant on waste ground throughout the county, especially near the railway.

S. integrifolius (L.) Clairv. C.T.W. 114.7.13.   *Field Fleawort*
Chalk downs   Native
Locally abundant on well-established chalk downland. Recorded
for all neighbouring counties except Hunts.
H.S. 59, 60, 65.
OUZEL: Dunstable Downs!; *V. H. Chambers.*
IVEL: Pegsdon Hills, 1841!; *I. Brown in Herb. Watson*: Barton Hills!†*;
W.C. Herb.: Sharpenhoe Clappers; *P.T.*: Deacon Hill.
LEA: Warden Hills, it is equally common on Galley Hill!†;
H. Brown c. 1838.
First record: between Silsoe and Luton, S. Goodenough in Herb.
Forster, British Museum, *c.* 1810.

CARLINA L.

Carlina vulgaris L. C.T.W. 114.32.1.   *Carline Thistle*
Pastures   Native
Common on the Chalk, especially in places where the soil has at
some time been disturbed. Rare in pastures on the oolite and heavy
clay soils. Recorded for all neighbouring counties.
H.S. 59–62, 63b, 65, 66a, 68.
NENE: Podington; *R. R. B. Orlebar.*
OUSE: Ravensden; *Abbot's annotated F.B.*: Renhold; *W.H. 1875*:
Dungee; *G. H. Day.*
KYM: Tilbrook, v.c. 30 [Hunts.].
OUZEL: common on the Chalk: roadside, Heath and Reach, in
clay pocket; *E.M.-R.*!
IVEL: common on the Chalk and also occurring sparingly on clays:
Cainhoe; *W.C. Herb.*: near Deadman's Oak; *J.McL., L.M.*
LEA and COLNE: common on the Chalk and sparingly on the Clay-
with-Flints.
First record: Abbot, 1798.

ARCTIUM L.

Arctium lappa L. C.T.W. 114.33.1.   *Great Burdock*
   A. majus Bernh.
Riversides, marshy places, etc.   Native
Frequent by our larger rivers, especially in places liable to winter
flooding. Recorded for all neighbouring counties.
H.S. 13.
OUSE: Harrold*; Tempsford; Biddenham; Cardington; Felmers-
ham; Sharnbrook; Oakley; Willington.
OUZEL: Leighton Buzzard.
IVEL: near Hexton; *T. B. Blow in Bot. Rec. Club 1876 Rep.*: Barton;
B.P.L.: Westoning; *C.C.*: Rowney Warren, by stream at lower end!;
J.E.L., Diary: Clifton; Broom; Stondon; Henlow; Flitwick; Clophill.
CAM: Mob's Hole.
LEA: Limbury†; *S.B.P.L.*
First record: Abbot, 1798, for the aggregate; Blow, 1876, for the
segregate.

A. nemorosum Lejeune   *Wood Burdock*
   A. vulgare A. H. Evans p.p. C.T.W. 114.33.2.
Streamsides, edges of woods, etc.   Native
Frequent, and generally growing in drier situations than *A. lappa*.
A revision has recently been made of the genus, and the segregates of
this species in the county need now to be studied. The records
given below, except that from Eaton Socon, are for *A. nemorosum* in
its broad limits. Recorded for all neighbouring counties.
H.S. 52.
NENE: Wymington; *C. C. Foss*!
OUSE: meadow, Eaton Socon*; *J. P. M. Brenan*: meadow, Bromham.
KYM: Worley's Wood.
OUZEL: Woburn; Blackgrove Wood.
IVEL: Southill Wood; *W.F.B.*: Pegsdon; *J.E.L.*, *J. Bot.*
CAM: streamside, Dunton.
COLNE: Deadmansey Wood.
First record: Saunders, 1897.

A. minus (Hill) Bernh. C.T.W. 114.33.3.   *Burdock, Clot Bur*
Roadsides, woods, waste places, etc.   Native
This is the common burdock in the county. It grows on all soils and
appears to be absent only on the greensand heaths and established
downland. Recorded for all districts and neighbouring counties.
H.S. 3a, 5, 13, 20a, 27, 31a, 34a, 34b, 42, 53a, 54, 55b, 69, 80b, 83.
First record: T. B. Blow in *Bot. Rec. Club 1876 Rep.*

CARDUUS L.

Carduus nutans L. C.T.W. 114.34.3.   *Musk Thistle,*
                                     Nodding Thistle
Rough pastures   Native
Common, especially on the Chalk, it occurs also on clays and
gravels, and more frequently where the soil has at some time been
disturbed. White-flowered forms have been observed. Recorded
for all districts and neighbouring counties.
H.S. 61.
First record: Abbot, 1798.

C. tenuiflorus Curt. C.T.W. 114.34.1.   *Slender Thistle*
Railway sidings, arable fields, gravel pits, etc.   Wool adventive
Rare, but in some cases established for a number of years. A record by Saunders in
J. Bot. 1883 is in error. It would have been interesting to know more of Abbot's
record. Recorded for Herts., Hunts. and Northants.
OUSE: Cox's Pits; *F.B.*: gravel pits, Eaton Socon, well established for some years*.
IVEL: railway sidings, Flitwick, Biggleswade; arable fields, Beeston, Maulden.
First record: Abbot, 1798.

C. crispus L. C.T.W. 114.34.4.   *Welted Thistle*
   C. acanthoides L. C.T.W. 114.34.5, *C. polyacanthos* Curt. non Lam.
Roadsides, rough pastures, waste places, etc.   Native
Common, although it is surprising to note that Hamson wrote that
he had not seen it. There has been much confusion with this

species. Druce considered that the common Welted Thistle is
C. acanthoides and that *C. crispus* is a rare alien. Clapham, Tutin and
Warburg hold that the reverse is the case. It is difficult to know to
what the records refer of *C. acanthoides* by Little from Southill in
J. Bot. 1919 and from Arlesey in his Diary. Recorded for all districts
and neighbouring counties.
H.S. 4, 52, 57a, 57b, 64, 71.
First record: Abbot, 1798.

CIRSIUM Mill.

Cirsium eriophorum (L.) Scop. C.T.W. 114.35.1. *Woolly Thistle*
  Carduus eriophorus L., *Cnicus eriophorus* (L.) Roth
Rough pastures   Native
Common on the heavy clay soils of the north of the county, and
frequent around Sundon where it grows on Boulder Clay and the
Chalk. One of our more attractive species, it is recorded for all
neighbouring counties.
NENE, OUSE and KYM: too many stations to merit listing.
IVEL: Meppershall; *W.C. Herb.*: Ireland; *J.McL.*, *L.M.*: Sundon!†;
S.B.P.L.: Higham Gobion; *C.C.*: Houghton Regis!; *V. H. Chambers*:
Harlington!; *W.F.B.*
CAM: Wrestlingworth.
LEA: Bramingham!; Round Green; *W.F.B.*
First record: Abbot, 1798.

C. vulgare (Savi) Ten. C.T.W. 114.35.2.   *Spear Thistle*
  Carduus lanceolatus L., *Cirsium lanceolatum* (L.) Scop. non Hill,
Cnicus lanceolatus (L.) Willd.
Rough pastures, waste ground, etc.   Colonist
Our commonest thistle. Thistle occurs as a place name in Thistly-
grounds, at Dunton, which appears as Blakethistle as early as 1256
(*P.N.B.*). The ground here is very rough, but *Picris echioides* is
more common than any of the thistles. Recorded for all districts
and neighbouring counties.
H.S. 5, 7, 9, 11, 13, 22b, 27, 28, 45a, 45b, 51b, 52, 53a, 55b, 56b,
58b, 61, 64, 66b, 75, 82, 84a, 85.
First record: Abbot, 1798.

C. acaulon (L.) Scop. C.T.W. 114.35.6. *Dwarf* or *Stemless Thistle*
  Carduus acaulos L., *Cnicus acaulos* (L.) Willd.
Pastures   Native
Common on the chalk hills and oolitic exposures, but occurring also
on the Boulder Clay and Greensand. The form *caulescens* Reichb.,
with stems of variable height, is common. Recorded for all districts
and neighbouring counties.
H.S. 1a, 1b, 28, 56b, 59–62, 65–68, 73, 74a, 75.
First record: Abbot, 1798.

C. arvense (L.) Scop. C.T.W. 114.35.4. *Field* or *Creeping Thistle*
  Serratula arvensis L., *Cnicus arvensis* (L.) Roth
Arable fields, etc. Colonist
Common on arable land, but readily invading natural habitats,
where it fails to become established. It is variable, and forms with
no or few spines (*cf.* var. *mite* Wimm. & Grab.) are frequent. Recorded for all districts and neighbouring counties.
H.S. 1a, 2a, 4, 7–13, 15–17, 20a, 21, 23, 27–29, 34a, 44, 45b, 51b,
53a–54, 56b, 58b, 60, 61, 63a, 64, 66b, 69, 71, 72, 75, 79, 80a–82,
84a, 85.
First record: Abbot, 1798.

C. palustre (L.) Scop. C.T.W. 114.35.3. *Marsh Thistle*
  Carduus palustris L., *Cnicus palustris* (L.) Willd.
Marshes, damp woods, etc. Native
Common in all likely stations in the county: white-flowered forms are
not infrequent. Recorded for all districts and neighbouring counties.
H.S. 3a, 10a–10c, 15, 19a, 23, 34b, 37–39a, 40–43, 51a, 52–54,
55b–57b, 77a, 78, 79a, 80a, 83, 84a, 85.
First record: Abbot, 1798.

ONOPORDUM L.

Onopordum acanthium L. C.T.W. 114.37.1. *Scotch Thistle*
Roadsides, waste places, etc. Garden escape
Rare, but well established in a few places usually near houses. Recorded for all
neighbouring counties.
OUSE: St. Cuthbert's; *Abbot's annotated F.B.*: Ford End; *J.McL. in J.H. Notes*:
Biddenham; *J.McL., L.M.*: Cotton End; *M. L. Berrill in J.H. Notes*: Bromham;
Wyboston; *A. W. Guppy*: roadside, Eaton Socon.
IVEL: Cainhoe; *W.C. Herb.*: Maulden; *M. L. Berrill in J.H. Notes*: Flitwick;
W.F.B.: Sandy, near the station!; Wilbury Hill!; *J.E.L., F.F.B.*: Stanford; *H. B.
Souster*: Harlington; Haynes; Shefford; chalk tip, Barton Cutting*; Sutton.
LEA: near Luton; *W.F.B.*
COLNE: Watling Street, south of Dunstable, v.c. 20 [Beds.]; *Fl. Herts.*
First record: Abbot, 1798.

SILYBUM Adans.

Silybum marianum (L.) Gaertn. C.T.W. 114.36.1. *Milk Thistle*
  Carduus marianus L., *Mariana mariana* (L.) Hill
Waste places, arable fields, etc. Casual
Rare, and only occasionally established. In recent years it has appeared as a wool
adventive. Abbot listed it as common in *Flora Bedfordiensis*, but added two stations
in his annotated copy. Recorded for all neighbouring counties.
OUSE: Bedford, where it was also found by L. W. Wilson*; *Abbot's annotated F.B.*:
Willington; *J.McL., L.M.*: Ford End; Cardington; *J.McL. in J.H. Notes*.
IVEL: Sandy; *Abbot's annotated F.B.*: Ampthill Park; *B.P.L.*: Flitwick!; *W.F.B.*:
Southill Station; *J.E.L., F.F.B.*: arable field, Maulden; *H. B. Souster!*: rubbish
dump, Sundon; arable field, Beeston*.
LEA: south of Luton; *W.F.B.*
First record: Abbot, 1798.

SERRATULA L.

Serratula tinctoria L. C.T.W. 114.40.1. *Saw-wort*
Woods, rough pastures, etc. Native
Rare and apparently limited to light soils. It was considered
common by Abbot. Recorded for all neighbouring counties.

H.S. 30a.
NENE: tunnel baulk, Wymington.
OUSE: Thurleigh Lane; *R. Townsend*: Ravensden Wood End;
A. W. Guppy: Cleat Hill; *W. Durant*.
OUZEL: Aspley Wood†; *J.McL.*, *L.M.*: Woburn; *W.H. 1875*:
Salford; *V.C.H.*: Milton Bryan; *H. E. Pickering*: rides, King's
Wood*.
IVEL: Ridgmont; *W.C. Herb.*: Tingrith; *H. E. Pickering*: rough
pastures, Sundon*.
First record: Abbot, 1798.

CENTAUREA L.

Centaurea scabiosa L. C.T.W. 114.39.1.   *Greater Knapweed*
Rough pastures, roadsides, etc.   Native
Common: white-flowered forms are frequently seen. Recorded for
all districts and neighbouring counties.
H.S. 59, 61, 62, 63b–66a, 67, 68, 72, 73, 86.
First record: Abbot, 1798.

C. cyanus L. C.T.W. 114.39.2.   *Cornflower, Bluebottle*
Cornfields, waste places, roadsides, etc.   ? Colonist
Considered by Abbot and other early botanists to be a common cornfield weed.
It is now only a plant of waste places, and probably a garden escape. Recorded
for all neighbouring counties.
OUSE: Ridgmont; *Herb. Miss Crouch*: Cotton End; *M. L. Berrill in J.H. Notes*:
Clapham; *A. B. Sampson in Herb. Kew*: Ravensden; *B. B. West*: Kempston;
gravel pits, Cople, Eaton Socon.
OUZEL: Woburn; *W.H. 1875*: Totternhoe; *H. B. Sargent*: Leighton Buzzard:
B. Verdcourt.
IVEL: Steppingley; *W.C. Herb.*: Maulden; *M. L. Berrill in J.H. Notes*: Chalton
Cross; *V. H. Chambers*: rubbish dump, Sundon; north of Potton.
LEA: Biscot; Limbury†*; *S.B.P.L.*: Skimpot; *S. A. Chambers in W.F.B.*: Braming-
ham; *V. H. Chambers*: arable field, Warden Hills*; rubbish dump, Luton.
First record: Abbot, 1798.

C. nigra L. C.T.W. 114.39.5.   *Lesser Knapweed, Hardheads*
Pastures, roadsides, etc.   Native
A common but very variable species the forms of which in the
county need more study. The common form on the Chalk and on
calcareous soils elsewhere in the county is subsp. **nemoralis** (Jord.)
Gugl. Recorded for all districts and neighbouring counties.
H.S. 1a–2a, 10b, 10c, 13, 45a, 45b, 51b, 56b, 58b–60, 62–63b,
65–68, 74a, 74b, 76.
First record: Abbot, 1798.

C. calcitrapa L. C.T.W. 114.39.7.   *Star-thistle*
Waste places, arable land, etc.   Casual
Rare: recorded for all neighbouring counties.
OUSE: Biddenham; Eaton Socon; *F.B.*: Park Lane; *J.McL.*, *B.M.*: Cardington
Mill; *E. M. Langley in J.H. Notes*.
IVEL: Wilbury Hill; *J.E.L. in B.E.C. 1913 Rep.*: Flitwick!*; *M. Holdsworth*:
railway siding, Flitwick, an unusual form*.
First record: Abbot, 1798.

C. solstitialis L. C.T.W. 114.39.8. *Yellow Star-thistle, St. Barnaby's Thistle*
Waste places, arable land, etc. Casual
Rare, but introduced recently as a wool adventive. Recorded for all neighbouring counties.
OUSE: Goldington; *J.McL. in J. Bot. 1884*: in clover, Cotton End, *M. L. Berrill in J.McL., B.M.*: Austin Canons; *P. Wyatt in J.H. Notes*: Cardington Mill†; *Herb. J.S.*
OUZEL: Maiden Bower; *S. A. Chambers in W.F.B.*
IVEL: Clophill; *W.C. Herb.*: Wilbury Hill; *J.E.L., J. Bot.*: Steppingley; *C.C.*: near Rowney Warren; *J.E.L., Diary*: railway siding, Flitwick*; arable fields, Maulden, Old Warden.
LEA: Biscot†*; *W.F.B.*: near Bramingham*; *L. C. Chambers.*
First record: W. Crouch, 1845.

ARNOSERIS Gaertn.

Arnoseris minima (L.) Schweigg. & Koerte C.T.W. 114.43.1.
Hyoseris minima L. *Lamb's Succory*
Arable fields Colonist
Rare, and limited to the Lower Greensand. It apparently appears regularly in the neighbourhood of Sandy, where I have looked for it in vain. Recorded for all neighbouring counties except Hunts.
E. Bot. tab. 95, March 1793, was drawn from a specimen sent by Abbot.
OUZEL: Aspley; *E. Bot.*: Leighton Buzzard; *V.C.H.*
IVEL: Ampthill; *E. Bot.*: Deepdale†; *J.McL., L.M.*: Potton; *J.McL., B.M.*: Sandy; Maulden; Clophill; *C.C.*: Portobello Farm, Sutton; *J.E.L., F.F.B.*
First record: Abbot, 1793.

LAPSANA L.

Lapsana communis L. C.T.W. 114.42.1. *Nipplewort*
Roadsides, waste places, etc. Colonist
Common: recorded for all districts and neighbouring counties.
H.S. 4, 31a–32, 34a, 55b, 63a, 72.
First record: Abbot, 1798.

PICRIS L.

Picris echioides L. C.T.W. 114.46.1. *Bristly Ox-tongue,*
Pod Thistle
Rough pastures, waste places, roadsides, etc. Colonist
Frequent on the basic clays, rare or absent on the Lower Greensand and Clay-with-Flints. Recorded for all neighbouring counties.
H.S. 3a, 11, 21, 58b.
NENE: Wymington; Hinwick; Great Hayes Wood.
OUSE: too many stations on the Oxford Clay to merit listing.
KYM: Knotting; Swineshead.
IVEL: frequent on the Gault and Chalk Marl.
CAM: Thistly-grounds Farm, Dunton, especially abundant.
LEA: Bramingham; *S.B.P.L.*
COLNE: garden weed, Stockwood!; *J. Cowley*: below Blow's Downs; *L. J. Margetts.*
First record: Abbot, 1798.

P. hieracioides L. C.T.W. 114.46.2.   *Rough or Hawkweed Ox-tongue*
Roadsides, waste places, etc.   Colonist
More common and widely distributed than the preceding, but
mainly limited to calcareous soils. Recorded for all districts and
neighbouring counties.
H.S. 1a, 58b, 63a, 63b.
First record: Abbot, 1795.

CREPIS L.

Crepis biennis L. C.T.W. 114.54.5.   *Rough Hawk's-beard*
Rough pastures, etc.   Colonist
Rare. This has been so much confused with *C. vesicaria* that
Saunders's two records in *F.F.B.*, both supported by inmature
specimens, must be doubted. It is to be regretted that McLaren
left no specimen. Recorded for all neighbouring counties.
OUZEL: Aspley Heath; *J.McL. in J.H. Notes*: roadside near Leighton
Buzzard; *E.M.-R.*!
IVEL: [Pulloxhill†; *F.F.B.*]: rubbish dump, Sundon; roadside,
Henlow*.
CAM: fieldside, Dunton*.
LEA: [Leagrave†; *F.F.B.*].
COLNE: roadside, near Greencroft Barn, v.c. 20 [Beds.], well estab-
lished*; *E.M.-R.*!
First record: uncertain.

C. capillaris (L.) Wallr. C.T.W. 114.54.6.   *Smooth Hawk's-beard*
   C. tectorum auct.
Rough pastures, roadsides, waste places, etc.   ? Native
Common, but very variable. Little recorded var. *diffusa* DC. from
Galley Hill, Sutton, in *J. Bot. 1919*, and made a number of manu-
script notes on var. *anglica* Druce & Thell. Recorded for all districts
and neighbouring counties.
H.S. 1a, 1b, 11, 16, 22b, 23, 25, 27, 28, 34a, 44, 45a, 46, 48, 50, 53a,
58b, 61, 63a, 64, 68, 69, 75, 77a, 80b.
First record: Abbot, 1798.

C. vesicaria L. subsp. **taraxacifolia** (Thuill.) Thell.
   C. taraxacifolia Thuill. C.T.W. 114.54.2.   *Beaked Hawk's-beard*
Rough pastures, roadsides, etc.   Colonist
A comparatively recent introduction, but now very common,
especially on calcareous soils. Saunders at first mistook it for *C.
nicaeensis* Balb., and as such recorded it in *J. Bot. 1890* and in
subsequent lists. The record of *C. foetida* L. in *B.P.L.*, but not in
later lists, may also refer to this. Recorded for all districts and all
neighbouring counties.
H.S. 17, 29, 58b, 64.
First record: D. M. Higgins in *W.B.E.C. 1894–5 Rep.*, although
C. Crouch and Saunders collected specimens as early as 1890.

HYPOCHOERIS L.

Hypochoeris maculata L. C.T.W. 114.44.3. *Spotted Cat's-ear*
Chalk hills Native
Very rare. This was J. E. Little's most noteworthy discovery in the
county, and the more creditable in that he identified it on vegetative
characters only. In his day the hill on which it grows was well
grazed with sheep and only the non-flowering rosettes were to be
seen. Since sheep were taken off the hill about 1930 and it has been
grazed only occasionally by cattle and by rabbits it has been allowed
to flower and has much increased. With *Seseli libanotis*, with which
it grows, it is supposed to spread into Hertfordshire only a few yards
away, but I have not seen it on the other side of the boundary.
It occurs in Cambridgeshire and has been recorded doubtfully for
Northants. Otherwise it is confined in Britain to five very limited
areas on the Chalk or Limestone.
H.S. 59.
IVEL: Pegsdon!*; *J.E.L. in B.E.C. 1913 Rep.* 389.

H. radicata L. C.T.W. 114.44.1. *Long-rooted Cat's-ear*
Rough pastures, heaths, waste places, etc. Native
Very common.
Recorded for all districts and neighbouring counties.
H.S. 7, 11, 14b, 21, 22b, 24–27, 29, 37, 38, 46, 48, 76, 79a.
First record: Abbot, 1798.

H. glabra L. C.T.W. 114.44.2. *Smooth Cat's-ear*
Sandy heaths Native
Rare. This seems to grow best in places where the soil has at some
time been disturbed. *E. Bot.* tab. 575, April 1798, was drawn from
a specimen sent by Abbot. Recorded for all neighbouring counties
except Hunts.
IVEL: Ampthill Warren; Sandy Warren†; *Plantae Bedford.*: Potton;
Sutton; *J.E.L., Diary*: in disused sandpit, Bunker's Hill*; in rough
ground and on railway bank, Deepdale.
First record: Abbot, 1795.

LEONTODON L.

Leontodon hispidus L. C.T.W. 114.45.2. *Rough Hawkbit*
Pastures, roadsides, etc. Native
Common, especially on calcareous soils and particularly on the
Chalk. Recorded for all districts and neighbouring counties.
H.S. 1a–2a, 11, 21, 45a, 45b, 59, 60, 62–66a, 67–69, 75, 77a.
First record: Abbot, 1798.

L. autumnalis L. C.T.W. 114.45.1. *Autumnal Hawkbit*
Pastures, rough ground, etc. Native
A common but variable species: the well-marked var. **pratensis**
Koch has been observed in the water meadows, Eaton Socon*;

E.M.-R.! and on Flitwick Moor; *E.M.-R.*! Recorded for all districts
and neighbouring counties.
H.S. 2a, 8, 28, 44, 58b, 69.
First record: Abbot, 1798.

L. taraxacoides (Vill.) Mérat   *Hairy Hawkbit*
   L. leysseri G. Beck C.T.W. 114.45.3, *L. hirtus* auct., *L. nudicaulis*
auct.
Pastures, heaths, etc.   Native
Frequent in pastures throughout the county: 'var. *lasiolaena* Bisch.'
was recorded by J. E. Little from Barton Hills in *J. Bot. 1919*.
Recorded for all districts and neighbouring counties.
H.S. 20a, 56b, 64, 68, 69.
First record: Abbot, 1798.

                        TARAXACUM Web.

Taraxacum officinale Web. C.T.W. 114.55.1.   *Dandelion*
   Leontodon taraxacum L., *T. vulgare* Schrank, *T. taraxacum* (L.) Karst.
Pastures, arable land, etc.   Native
Common, but very variable. For the many microspecies recorded
for the county the reader is referred to *B.E.C. Reports 1928*, 619,
620, 629; *1931*, 563, 565; *1937*, 430. Few contemporary botanists
will view these with any seriousness. Recorded for all districts and
neighbouring counties.
H.S. 1a, 5, 8, 9–10c, 14b, 15, 18, 19a, 22b, 26–28, 32–34a, 38, 45a,
51b–53a, 55b, 56b, 58b, 63a–65, 70a, 76, [13, 62, 69, 75].
First record: Abbot, 1798.

T. laevigatum (Willd.) DC. C.T.W. 114.55.4.   *Lesser Dandelion*
   T. erythrospermum Andrz. ex Bess.
Dry pastures   Native
Frequent on the Lower Greensand, and probably not uncommon on
the Chalk. Recorded for all neighbouring counties.
H.S. 25, 45a, 48.
OUZEL: Heath and Reach; *V.C.H.*
IVEL: Ampthill; *V.C.H.*: Tingrith; *J.E.L.*, *F.F.B.*: Streatley†;
Herb. J.S.: Rowney Warren; Maulden.
LEA: Dunstable East Hill, *i.e.* Blow's Downs; *Fl. Herts*.
First record: Saunders, 1897.

T. palustre (Lyons) DC.   *Marsh Dandelion*
   T. paludosum (Scop.) Schlecht. C.T.W. 114.55.2.
Marshes, damp meadows, etc.   Native
Rare: the most distinct of our dandelions. It shows great variability
in each of the two stations in which I have seen it. Recorded for all
neighbouring counties except Hunts.
H.S. 56b.
OUZEL: Cow Common*; *A. J. Wilmott*.
IVEL: Southill; *J.E.L.*, *F.F.B.*: Cainhoe*; *H. B. Souster*!
First record: J. E. Little, 1936.

LACTUCA L.

Lactuca virosa L. C.T.W. 114.49.3.
Waste places, etc.   Colonist
Rare, and usually well established. It has been frequently recorded, but often in
error for *L. serriola*. I have preferred to list below the more certain records. Re-
corded for all neighbouring counties.
OUSE: Marston, 1898; *G. E. Brown*: plentiful in gravel pits, Cople.
IVEL: 'North Road, two-and-a-half miles from Baldock by the turning to Calde-
cote'!*; *W. H. Coleman in Fl. Herts.* 1841: rubbish dump, Shefford!; *J.E.L. in
B.E.C. 1934 Rep.*: roadside between Stotfold and Henlow; waste ground, Three
Counties Station; gravel pit, Meppershall.
LEA: Streatley!; waste ground, Luton; *P.T.*
COLNE: Caddington; *L. J. Margetts*.
First record: W. H. Coleman, 1841.

L. serriola L. C.T.W. 114.49.2.   *Prickly Lettuce*
Waste places, etc.   Colonist
A species which increased greatly between 1935 and 1945, but which now appears
to be diminishing. It was especially common on disturbed calcareous soils, but
by no means limited to them. It is difficult to know whether a record of *L. scariola*
by Abbot, in a letter to J. E. Smith, 3 Feb. 1802, refers to this or the preceding
species. Recorded for all districts except Colne, and all neighbouring counties.
H.S. 7, 16.
First certain record: J. E. Little in *W.B.E.C. 1932 Rep.*

MYCELIS Cass.

Mycelis muralis (L.) Reichb. C.T.W. 114.50.1.   *Wall Lettuce*
  Lactuca muralis (L.) Gaertn.
Old walls, woods, etc.   ? *Native*
Rare: it appears in the county on walls, but more frequently in
open woods on the Chalk. Recorded for all neighbouring counties.
H.S. 69.
IVEL: wood, Sharpenhoe Clappers; *E.M.-R.*: Flitwick Manor;
sparingly in Leete Wood.
LEA: Dallow Road, Luton; Luton Hoo, plentiful on wall on east
side of the park!†*; *S.B.P.L.*: Maulden Firs.
COLNE: roadside by Long Wood; v.c. 20 [Beds.].
First record: Saunders, 1885.

SONCHUS L.

Sonchus arvensis L. C.T.W. 114.51.2.   *Corn Sowthistle*
Arable land, waste places, etc.   Colonist
Common: a form with glabrous involucral bracts, *cf.* var. *glabrescens*
Guenth., Grab. & Wimm., has been found by P. Taylor on a hilly
pasture at Sundon. Recorded for all districts and neighbouring
counties.
H.S. 58b, 63a, 64, 71, 86.
First record: Abbot, 1798.

S. asper (L.) Hill C.T.W. 114.51.4.   *Spiny Sowthistle*
Waste places, arable land, etc.   Colonist
Common: recorded for all districts and neighbouring counties.
H.S. 9, 11–13, 16, 25, 44, 51b, 57b, 64, 86.
First record: G. C. Druce in *Bot. Rec. Club 1884–6 Rep.*, but Abbot's
S. oleraceus is represented by two specimens, one of which is *S. asper*.

S. oleraceus L. C.T.W. 114.51.3.   *Sowthistle*
Waste places, arable fields, etc.   Colonist
As common as *S. asper* but more variable. Records were made of
var. *triangularis* Wallr. from near Woburn Sands by G. C. Druce in
B.E.C. 1920 Rep., and var. *laceras* Wallr. by H. Phillips from Wootton
Pillinge in *B.E.C. 1933 Rep.* Recorded for all districts and neigh-
bouring counties.
H.S. 6, 12, 17, 34a, 50, 51b, 53a, 63a.
First record: Abbot, 1798.

TRAGOPOGON

Tragopogon pratensis L. C.T.W.114.47.1.   *Goat's Beard,*
                                                    John-go-to-bed-at-noon
Pastures, waste places, etc.   Native or Colonist
Common, and very variable. Subsp. **minor** (Mill.) Rouy, is
common throughout the county, with every appearance of being
native; but subsp. **pratensis** is found on railway sidings, in waste
places, etc., and appears to be at best only a colonist. Forms with
long florets appear also on roadsides, farm tracks, etc., but these are
probably not the true subsp. *pratensis*. Recorded for all districts and
neighbouring counties.
H.S. 1a, 25, 62, 65, 66b, 67, 76 (all subsp. *minor*).
First record: Abbot, 1798.

HIERACIUM L.

Most of my Hawkweed material was seen by H. W. Pugsley before
the publication of his *Prodromus*. P. D. Sell and C. West have since
examined it and further gatherings, and I am grateful to them for
their determinations.

Hieracium pellucidum Laest. C.T.W. 1153.
Railway embankments   Colonist
NENE: Wymington*.

H. exotericum Jord.
Railway embankments   Colonist
Rare: recorded also for Herts. and Cambs.
H.S. 1a.
NENE: Wymington*.
LEA: East Hyde, f. *grandidens* (Dahlst.) Pugsl.*
        Var. **sublepistoides** (Zahn) Pugsl.
NENE: Wymington*.
OUSE: Moor End, Felmersham*; Oakley.

H. vulgatum (Fr.) Almq.
There are many old records for this and for *H. murorum*, but the one given below is
only one which can be listed with certainty.
IVEL: Ampthill; *A. E. Ellis in Pugsley's Prodromus.*
A plant collected on the railway at Wymington* was referred to under *H. lepidulum*
Stenström in Pugsley's *Prodromus*, but Sell and West think that this is doubtful,
although the plant belongs to the section *Eu-Vulgata* Series *genuina*, as does *H.
lepidulum.*
Gatherings of two different plants from Podington* and Clophill* both belong to
the Section *Vulgata Eu-Vulgata*, Series *Sciaphila*, but cannot be placed by Sell and
West with any known British species.

H. anglorum (Ley.) Pugsl. C.T.W. p. 1148.

Heaths, etc. Native

Rare: recorded also for Herts.

OUZEL: side of Aspley Wood*; P. D. Sell and C. West!

IVEL: Clayshill (? Clophill); *J.McL. in Pugsley's Prodromus.*

H. lachenalii C. C. Gmel. C.T.W. p. 1148.

H. sciaphilum auct.

Heathy places, dry banks, etc. Native

Frequent, and probably native on the Lower Greensand and Clay-with-Flints, but a colonist on railway banks, etc. Recorded for Herts. and Cambs.

NENE: Wymington; *H. K. Airy Shaw in Herb. Kew.*

OUZEL: Leighton Buzzard; *G. C. Druce in J. Bot. 1891*: between Woburn and Woburn Sands!; *C. W. Muirhead*: railway bank, Sewell!*; *E.M.-R.*: roadside near King's Wood*; *P. D. Sell and C. West*!

IVEL: railway bank, Holwell; *J.E.L., B.E.C.*: railway, south of Harlington*.

LEA: roadside, Pepperstock*.

First record: Druce, 1891.

H. tridentatum Fr. C.T.W. 1142.

Heathy places Native

IVEL: near Shefford; *Newbould MS.*: Ampthill; *E. Forster in Pugsley's Prodromus.*

First record: doubtful.

H. umbellatum L. C.T.W. 1142.

Heaths, heathy woods, etc. Native

Not infrequent on the Lower Greensand. Recorded for all neighbouring counties, but doubtfully for Cambs.

OUZEL: Aspley; *F.B.*: Woburn; *Abbot's annotated F.B.*: plentiful at Heath and Reach, with var. *coronopifolium* Bernh.*.

IVEL: Clophill*; *J.McL., L.M.*: Ampthill; *J. Brown in Pugsley's Prodromus*: Maulden, with var. *coronopifolium*!*; *C.C.*

LEA: Chiltern Green; *W.F.B.*

First record: Abbot, 1798.

H. bladonii Pugsl. C.T.W. p 1138.

Sides of woods, heaths, etc. Native

Apart from *H. pilosella* this is our commonest hawkweed. Sell and West find that the characters given in the *Prodromus* are insufficient to distinguish this from *H. perpropinquum* (Zahn) Pugsl., and records for the latter are included here. There are many early records for *H. boreale* auct. and *H. sabaudum* auct., which I have excluded as they could refer to this or the two following species. Recorded for all neighbouring counties.

OUSE: Woodcraft*.

OUZEL: Aspley Wood*; King's Wood*; specimens from both these stations were also determined *H. perpropinquum* by Pugsley.

IVEL: Clophill!*; *the author in Pugsley's Prodromus.*
COLNE: Deadmansey Wood*.
First record: for the aggregate, Aspley, as *H. sabaudum*, Abbot,
1795.

H. rigens Jord.
Sides of woods, heaths, etc.   Native
IVEL: side of Clophill Warren Wood*.

H. vagum Jord. C.T.W. p. 1139.
Railway banks   Colonist
Not uncommon.
NENE: railway bank, Wymington*.
IVEL: railway banks, Harlington*, Southill*, Westoning*.
LEA: railway bank, Leagrave*.

H. pilosella L. C.T.W. p. 1164.   *Mouse-ear Hawkweed*
Rough pastures, etc.   Native
Common, especially on rough ground on all soils. Recorded for all
districts and all neighbouring counties.
H.S. 1a, 22b, 23, 28, 29, 45a, 45b, 47, 48, 60–62, 65, 66a, 67–69,
74a, 75, 76, 79a.
First record: Abbot, 1798.

Var. **concinnatum** F. J. Hanb.
IVEL: Rowney Warren; *P. D. Sell.*

ANGIOSPERMAE: MONOCOTYLEDONES

ALISMATACEAE

BALDELLIA Parl.

Baldellia ranunculoides (L.) Parl. C.T.W. 115.1.1.
Lesser Water-plantain
Alisma ranunculoides L., *Echinodorus ranunculoides* (L.) Engelm.
Marshy places   Native
Rare: recorded for all neighbouring counties.
OUSE: Marston, on the authority of Miss Stevenson (no doubt Ada
Stimson, to whom it is credited, without station, in *Hillhouse List,
1875*); *V.C.H.*: pond, Tempsford, with *Hottonia palustris* and *Apium
inundatum*!*; *I. J. Allison*: gravel pit, Wyboston*.
IVEL: Ampthill Bogs; *F.B.*
First record: Abbot, 1798.

ALISMA L.

Alisma plantago-aquatica L. C.T.W. 115.3.1.   *Water-plantain*
Pondsides, streams, ditches, etc.   Native
Common throughout the county, but never abundant: a handsome
species. Recorded for all districts and neighbouring counties.
H.S. 7–9, 11, 21, 58a.
First record: Abbot, 1798.

A. lanceolatum With. C.T.W. 115.3.2.

Pondsides, streams, ditches, etc.   Native

Rare, but no doubt overlooked and awaiting record from a number of stations. Recorded for all neighbouring counties.

OUSE: Stevington*; *B. Verdcourt in Watsonia I (1949)* 56.

IVEL: Flitwick Marsh; but there is no specimen to support the record; *S.B.P.L.*

First record: uncertain.

SAGITTARIA L.

Sagittaria sagittifolia L. C.T.W. 115.5.1.   *Arrow-head*

Riversides, ditches, etc.   Native

Frequent by our larger rivers: recorded for all neighbouring counties.

OUSE: Bedford!†; *Herb. J.S.*: Pavenham; Oakley*; Roxton; Eaton Socon.

OUZEL: Leighton Buzzard; *S.B.P.L.*: Heath and Reach.

IVEL: Shillington; *W.F.B.*: Clifton; Shefford!; Biggleswade; *J.E.L., F.F.B.*: Girtford Bridge; *H. B. Souster.*

LEA: Luton Hoo; *W.F.B.*

First record: Abbot, 1798, common.

BUTOMACEAE

BUTOMUS L.

Butomus umbellatus L. C.T.W. 116.1.1.   *Flowering Rush*

Riversides, ditches, etc.   Native

One of our more attractive plant species. Frequent by the Ouse and Ivel, it is, however, nowhere common. Abbot listed it as rare in *Flora Bedfordiensis* but amended this to *ubique* in his annotated copy. Recorded for all neighbouring counties.

H.S. 14a.

NENE: Hinwick Lodge, possibly planted.

OUSE: Kempston; *W.H. 1875*: Cardington Brook; *J.McL., L.M.*: Harrold; *G. H. Day*: Tempsford Bridge; *H. B. Souster*: Great Barford; *C. C. Foss*: Bromham*; Stevington; Pavenham; Oakley; Felmersham; Eaton Socon.

KYM: Melchbourne Park; Tilbrook, v.c. 30 [Hunts.].

OUZEL: Leighton Buzzard; *B. Verdcourt.*

IVEL: Gravenhurst; *W.C. Herb.*: Shillington; Campton; *C.C.*: between Shefford and Biggleswade!; *J.S., J. Bot. 1889*: Shefford; Biggleswade.

CAM: Wrestlingworth.

LEA: New Mill End†; *S.B.P.L.*

First record: Abbot, 1798.

HYDROCHARITACEAE

ELODEA Michx.

Elodea canadensis Michx. C.T.W. 117.3.1.   *Canadian Waterweed*
Rivers, pools in sandpits, gravel pits, etc.   Denizen
Appearing in all likely places. Recorded for all neighbouring counties.
H.S. 7, 11.
OUSE: common in the river and gravel pits, brickpits, etc.
OUZEL: common in the lakes at Woburn Park and Stockgrove
and in sandpits.
IVEL: Chicksands; Shefford, common in the Ivel Navigation!;
Southill!; *J.E.L.*, *F.F.B.*: lake, Flitwick Manor.
LEA: brickpit, Stopsley; lake, Wardown.
First record: brooks near the river, J. McLaren, Herb. Luton
Museum, 1864.

LAGAROSIPHON Harv.

Lagarosiphon major (Ridl.) Moss
Ponds   Colonist
This is frequently used in aquaria, which may account for its origin in the disused
chalk pit at Arlesey*, where it was first found in a wild state in Britain by E. Milne-
Redhead and the author in 1944, and where it is well established and spreading to
neighbouring pits.

HYDROCHARIS L.

Hydrocharis morsus-ranae L. C.T.W. 117.1.1.   *Frog-bit*
Slow-running rivers, backwaters, etc.   Native
Rare or extinct: it still grows a few yards over the county boundary
at St. Neots. Recorded for all neighbouring counties.
OUSE: Fenlake Ditches, it was still there in McLaren's time;
Plantae Bedford.: Castle Mills; *F.B.*: Goldington†; *W.F.B.*
IVEL: ditches near Biggleswade on the way to Northill; *Plantae Bedford.*
First record: Abbot, 1795.

JUNCAGINACEAE

TRIGLOCHIN L.

Triglochin palustris L. C.T.W. 119.1.1.   *Marsh Arrow-grass*
Marshes, wet meadows, etc.   Native
Rare: but usually growing in some abundance when it appears.
Recorded for all neighbouring counties.
H.S. 10a, 56a.
OUSE: Hassock Meadow; *F.B.*: King's Mead Marsh; *J.McL.*, *L.M.*:
Odell; *G. H. Day*: water meadow, Great Barford; marsh, Stevington.
OUZEL: marsh near Houghton Regis!; *E.M.-R.*: Cow Common;
Totternhoe Mead.
IVEL: Ampthill Bogs; *F.B.*: Gravenhurst, in meadow!*; Westoning;
C.C.
LEA: Leagrave Marsh, where I have not seen it since 1942!*; near
Bramingham Shott†; *S.B.P.L.*
First record: Abbot, 1798.

POTAMOGETONACEAE

POTAMOGETON L.

I have had considerable assistance from J. E. Dandy and G. Taylor
with this genus. J. E. Dandy has also made visits to the county to
study the distribution of the various species.

Potamogeton natans L. C.T.W. 122.1.1.   *Broad-leaved Pondweed*
Ponds, rivers, etc.   Native
Common and well distributed. Recorded for all districts and
neighbouring counties.
H.S. 11, 13, 21, 58a.
First record: Abbot, 1798.

P. polygonifolius Pourr. C.T.W. 122.1.2.   *Bog Pondweed*
  P. oblongus Viv.
Heathy pools   Native
Probably extinct, and there are few stations left in which it could
grow. A specimen in Herb. Saunders from Pepperstock is *P. natans*.
Recorded for all neighbouring counties.
IVEL: Maulden Moor, as *P. natans*; *W.C. Herb.*: Flitwick Moor†;
J.S. in Bot. Rec. Club 1884–6 Rep.
First record: Saunders, 1886.
First evidence: W. Crouch, *c.* 1845.

P. lucens L. C.T.W. 122.1.5.   *Shining Pondweed*
Rivers   Native
Limited to the Ouse and Ivel, where it is not uncommon. Recorded
for all neighbouring counties.
OUSE: Bedford!†; *J.S. in J. Bot. 1889*: Fenlake!*; *J.McL., B.M.*:
Eaton Socon; Pavenham.
IVEL: Shefford, plentiful in Ivel Navigation!†*; *J.S. in J. Bot. 1889*:
Chicksands; *W.F.B.*
First record: Abbot, 1798.

P. heterophyllus Schreb. is recorded, with no station, in *Hillhouse List, 1876,* but without a specimen to support the record it is impossible to say what species was intended. *P. heterophyllus* is a synonym of *P. gramineus* L.

P. alpinus Balb. C.T.W. 122.1.10.   *Reddish Pondweed*
Ponds, gravel pits, etc.   Native
Rare: recorded for all neighbouring counties except Hunts.
OUSE: gravel pit, Cople*.
IVEL: Clophill, as *P. perfoliatus*; *W.C. Herb.*: gravel pit, Henlow*.
First evidence: W. Crouch, 1845.

P. praelongus Wulf. C.T.W. 122.1.11.   *Long-stalked Pondweed*
Rivers   Native
Rare, and apparently limited to the Ouse at and below Bedford.
Recorded for all neighbouring counties except Herts.

Ouse: just above Bedford; *R. A. Pryor in J. Bot. 1875*, 212: plentiful in the Ouse between Bedford and Cardington Mill*; *J. E. Dandy*!: Eaton Socon.
First record: R. A. Pryor, 1875.

P. perfoliatus L. C.T.W. 122.1.12.   *Perfoliate Pondweed*
Rivers, ponds, etc.   Native
Frequent in the Ouse and Ivel, rare elsewhere. Recorded for all neighbouring counties.
Ouse: Bedford!†; Newnham; *F.B.*: Fenlake*; Bromham; Pavenham*.
Ivel: Shefford, plentiful in the Ivel Navigation!*; *J.E.L.*, *J. Bot.*: Clifton; *J.E.L.*, *F.F.B.*: Biggleswade*; Southill*; *J. E. Dandy*!
Lea: Limbury†; *S.B.P.L.*
First record: Abbot, 1798.

P. friesii Rupr. C.T.W. 122.1.14.   *Flat-stalked Pondweed*
    P. mucronatus Schrad. ex Sond., *P. compressus* auct.
Ponds, rivers, etc.   Native
Rare: recorded for all neighbouring counties except Herts.
Ouse: Fenlake, where it was also found by J. E. Dandy in 1944!*; *F.B.*
Ouzel: Lower Drakelow Pond, where it is abundant!*; *A. B. Jackson.*
Ivel: Biggleswade†; *J.S. in J. Bot. 1886.*
First record: Abbot, 1798.

P. pusillus L. C.T.W. 122.1.16.
    P. panormitanus Biv.
Ponds, rivers, etc.   Native
Frequent: it is probable that early records should in most cases be referred to *P. berchtoldi*, which is much more common. Recorded for all neighbouring counties.
Ouse: Bedford*; *J. E. Dandy*!: gravel pit, Eaton Socon*; *E.M.-R.*!
Kym: lake, Melchbourne Park.
Ouzel: lake, Woburn Park*; *J. E. Dandy*: sandpit, Leighton Buzzard*; pond, Heath and Reach.
Ivel: Southill Lake; *J.E.L. in W.B.E.C. 1913 Rep.*
First certain record: J. E. Little, 1913.

P. berchtoldi Fieb. C.T.W. 122.1.18.   *Small Pondweed*
    P. pusillus auct.
Ponds, rivers, etc.   Native
More common than the preceding: recorded for all neighbouring counties.
Ouse: Elstow, as *P. pusillus*; *J.McL.*, *L.M.*: gravel pit, Felmersham*.
Ouzel: Heath and Reach, as *P. pusillus*!†; *Herb. J.S.*: lake, Battlesden*; *J. E. Dandy*!: pond, Rushmere*.
Ivel: Gravenhurst, as *P. pusillus*†; *Herb. J.S.*: ditches, Flitwick Moor*.
Lea: New Mill End, as *P. pusillus*; *Herb. J.S.*
First evidence: J. McLaren, 1864.

P. trichoides Cham. & Schlecht. C.T.W. 122.1.19.
Hair-like Pondweed
Ponds   Native
Rare: recorded for Cambs. and Hunts.
IVEL: Southill Lake, 1930, mixed with *P. pusillus* and named *P. panormitanus*, see Dandy and Taylor in *J. Bot. 1938*, 166; *J.E.L. in Herb. Brit. Mus. and Herb. Kew*.

P. crispus L. C.T.W. 122.1.22.   *Curled Pondweed*
Ponds, rivers, etc.
Probably our most common pondweed: recorded for all districts and neighbouring counties.
H.S. 21.
First record: Abbot, 1798.

P. pectinatus L. C.T.W. 122.1.24.   *Fennel-leaved Pondweed*
Ponds, rivers, etc.   Native
Frequent, and well distributed in the county, but considered rare by both Abbot and Saunders. Recorded for all neighbouring counties.
H.S. 7.
OUSE: Fenlake; *F.B.*: Cardington; *J.McL.*, *B.M.*: Bedford!†*; *Herb. J.S.*: Pavenham*; Tempsford; Great Barford*; Blunham Mill.
KYM: Melchbourne Lake*.
OUZEL: Battlesden Lake; lake, Woburn Park*; *J. E. Dandy*!: sandpit, Leighton Buzzard.
IVEL: Shefford!†*; *W.F.B.*: Southill Lake!; *J.E.L.*, *J. Bot.*: claypit, Arlesey*; lake, Ickwell Bury.
LEA: Luton Hoo!; *W.F.B.*: river, East Hyde.
First record: Abbot, 1798.

P. densus L. C.T.W. 122.1.25.   *Opposite-leaved Pondweed*
Ponds, ditches, etc.   Native
Rare: recorded for all neighbouring counties.
OUSE: Ford End; *F.B.*: Fenlake; *J.McL.*, *L.M.*: ponds, Tempsford.
OUZEL: Cow Common, in some quantity*.
IVEL: Shillington; *W.C. Herb.*: near Flitwick Mill!*; *M. L. Berrill in J.H. Notes*: Biggleswade Common; Cadwell; *J.E.L.*, *F.F.B.*: pond, Arlesey.
LEA: Limbury; Luton; *S.B.P.L.*: Luton Hoo Lake; *Herb. J.S.*
First record: Abbot, 1798.

ZANNICHELLIACEAE
ZANNICHELLIA L.
Zannichellia palustris L. C.T.W. 124.1.1.   *Horned Pondweed*
Ponds, etc.   Native
Common: recorded for all districts except Colne, and for all neighbouring counties.
H.S. 11, 58a.
First record: *Hillhouse List, 1876*.

LILIACEAE

NARTHECIUM Huds.

Narthecium ossifragum (L.) Huds. C.T.W. 127.2.1.
Anthericum ossifragum L.   *Bog Asphodel*
Bogs   Native
Extinct: recorded for Bucks., Hunts. and Cambs.
OUZEL: bogs near Woburn; *Plantae Bedford.*
IVEL: Ampthill Bogs; *F.B.*
First record: Abbot, 1795.

CONVALLARIA L.

Convallaria majalis L. C.T.W. 127.4.1.   *Lily-of-the-Valley*
Woods   Native
Rare: one of the more attractive of our species. According to
Magna Britannia (1787) the London markets were generally sup-
plied from the Bedfordshire woods. How well it was known in the
eighteenth century is shown in extracts from the *Blecheley Diary of
William Cole*: 3 June 1766, 'Tansley went to Aspley Wood for some
Lillies of the Valley'; 7 June 1767, 'Tansley went, as usual, to get me
some Lillies of the Valley'. Abbot figured it in *Flora Bedfordiensis*,
which led Sowerby to use other than Bedfordshire material for
English Botany. Recorded for all neighbouring counties except
Hunts.
H.S. 23.
OUZEL: Aspley Wood (see above), where it was also found by Abbot,
W. Crouch, McLaren and Saunders†; *W. Cole*: Birchmuth (? Birch-
more) Wood; *Plantae Bedford*: King's Wood, where it grows sparingly
in one part of the wood under oak and more plentifully in a heathy
part in association with bracken!*; *J.S. in Bot. Rec. Club, 1880 Rep.*
First record: William Cole, 1766.

POLYGONATUM Mill.

Polygonatum multiflorum (L.) All. C.T.W. 127.5.3.   *Solomon's Seal*
   Convallaria multiflora L.
Woods, shady places, etc.   ? Native
Rare: probably native in Abbot's day, but doubtfully so now. *E. Bot.* tab. 279, Oct.
1795, was apparently drawn from a specimen sent by Abbot from Thurleigh.
Recorded for all neighbouring counties.
OUSE: Thurleigh; *Plantae Bedford.*: between Chellington and Felmersham*;
G. H. Day.
OUZEL: Aspley Wood; *W.C. Herb.*
IVEL: Potton Wood; *Plantae Bedford.*: Flitwick Woods; *J.McL., L.M.*: Tingrith;
J.McL., B.M.: copse near Ickwell, remote from houses*; *H. B. Souster*: Maulden,
well established on verge of wood*.
LEA: Luton Hoo†; *S.B.P.L.*
First record: Abbot, 1795.

Maianthemum bifolium (L.) Schmidt (*Unifolium bifolium* (L.) Greene), *May
Lily*, C.T.W. 127.6.1, was said by Saunders, in *Field Flowers of Bedfordshire*, to be
represented in the Kew Herbarium by a specimen collected 'under fir trees at
Aspley Wood'. I have been unable to find this specimen or any other reference
to it.

LILIUM L.

Lilium martagon L. C.T.W. 127.9.1.   *Martagon Lily*
Woods
Rare: recorded for Herts., Hunts. and Northants.
OUSE: Pavenham*, where it was shown to me by G. H. Day (but a Hillhouse specimen in Herb. Birmingham Univ. is from Pavenham); *W.H. 1876, unlocalized*.

FRITILLARIA L.

Fritillaria meleagris L. C.T.W. 127.10.1.   *Snake's-head Fritillary*
Meadows   Native
Rare: this occurs as a native species in wet meadows, and as a garden escape in copses, etc., in which case it is rarely established. *E. Bot.* tab. 622, Aug. 1799, was drawn from a specimen sent by Abbot. Recorded for all neighbouring counties.
OUSE: Bromham; *F.B.*
OUZEL: Eaton Bray; *R. A. Chambers in W.F.B.*: meadow, Billington!*; *A. S. Johnston.*
IVEL: garden escape, copse, Ampthill.
COLNE: garden escape, Whipsnade.
First record: Abbot, 1798.

Tulipa sylvestris L., *Wild Tulip*, C.T.W. 127.10.1, was recorded from Whipsnade, 'on the border of Herts.', on the authority of D. Jenks in *Flora Bedfordiensis*. It was probably of garden origin. There is also a specimen in Herb. Watson collected 'near Leighton, Beds.' by E. and S. Warner, 1841. Recorded for all neighbouring counties except Cambs.

ORNITHOGALUM L.

Ornithogalum umbellatum L. C.T.W. 127.14.1.   *Star-of-Bethlehem*
Roadsides, rough pastures, etc.   ? Native
Rare, and growing on a wide range of soils, but usually an obvious garden escape. Recorded for all neighbouring counties except Hunts.
OUSE: Cotton End, as *O. pyrenaicum*; *J.S. in Herb. Beds. Nat. Hist. Soc.*: Queen's Park, Bedford; *R. Townsend.*
OUZEL: Totternhoe Mead, where Saunders thought it was probably native†; *W.F.B.*: Totternhoe Knolls; *H. B. Souster*: roadside, Heath and Reach.
IVEL: Everton Heath; *Plantae Bedford.*: Maulden, as *O. nutans!*; Flitwick, as *O. nutans*; *C.C.*: roadside, Clophill!; *L. W. Wilson*: Flitwick Moor; *V. H. Chambers*: rough chalk pasture, Pegsdon!; *H. and D. Meyer*: spinney, Millbrook; *H. B. Souster*: roadsides, Ampthill, Steppingley.
LEA: Park; *Davis*: near Luton; Limbury†; *S.B.P.L.*: Biscot Meadows; Luton Hoo†; *F.F.B.*: cart track, Wigmore Lane; *H. Cook.*
First record: Abbot, 1795.

O. nutans L. C.T.W. 127.14.2.   *Drooping Star-of-Bethlehem*
Roadsides, etc.   ? Garden escape
Rare: apparently well established in Cambs., and recorded, with some doubt, for all other neighbouring counties.
OUSE: 'a specimen was brought to me by a merchant at St. Neots, which he had found growing wild at Eatonford field'; *Sir T. G. Cullum in Bot. Guide.*
IVEL: near Ampthill, scarcely wild; *J. E. Leefe in New Bot. Guide II*: Maulden; *W.C. Herb.*: Potton; *Mr. Bond Smith in V.C.H.*
First record: T. G. Cullum, 1805.

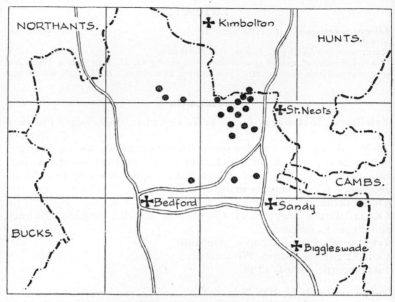

FIG. 22   DISTRIBUTION OF ORNITHOGALUM PYRENAICUM

O. pyrenaicum L. C.T.W. 127.14.3.   *Spiked Star-of-Bethlehem*
Roadsides, field borders, edges of woods, etc.   ? Native
One of our more interesting species. Abundant by the roadsides
between Keysoe and Eaton Socon it has apparently neither increased
nor diminished in its distribution since it was first observed by botanists.
Comital Flora lists it for Cambridgeshire, but I can find no justification
for its inclusion. In 1952 E. Milne-Redhead and I found a small
colony a few feet over the county boundary in Huntingdonshire, its
first record for that county. It almost reaches the county boundary
of Bucks. and Northants., but is apparently absent in each. The young
shoots were at one time collected and sold in the London markets as
asparagus. It is possible that this may account for its origin with us.
A similar colony of the species exists in the neighbourhood of Bath,
and it is said that here the young shoots were sold as Bath Asparagus.
The original drawing for *E. Bot.* tab. 499, Nov. 1798, is labelled
'sent by Rev. Thos. Orl[ebar] Marsh, F.L.S. brought from Keysoe
Park Woods, Beds., who also received it from Eaton Socon in the
same C[ounty] where they grow profusely'. Recorded also for Hunts.
OUSE: Thurleigh; *Plantae Bedford.*: Eaton Socon!*; *T. O. Marsh*:
between Eaton Socon and Thurleigh; *F.B.*: Basmead!; *C.R.O. List*:
Roxton!*; *G. H. Day*: Honeydon!; *A. W. Guppy*: Renhold; *L. W.*
Wilson: Tempsford.
KYM: Keysoe Park Wood!; *T. O. Marsh*.
CAM: Cockayne Hatley; *T.L.*
First record: Abbot, 1795.

ENDYMION Dum.

Endymion non-scriptus (L.) Garcke C.T.W. 127.16.1. *Bluebell*
  Hyacinthus non-scriptus L., *Scilla nutans* Sm., *S. non-scripta* (L.)
Hoffmanns. & Link
Woods   Native
Common in woods throughout the county, and seen occasionally on
heathy ground. White-flowered forms are frequent. Recorded for
all districts and neighbouring counties.
H.S. 2b, 3b, 18, 23, 32–34b, 51a, 52, 54, 55a, 78, 79b, 81, 83.
First record: Abbot, 1798.

COLCHICUM L.

Colchicum autumnale L. C.T.W. 127.19.1.   *Meadow Saffron*
Meadows   ? Native
Extinct as a truly wild plant. Recorded for all neighbouring counties, but some-
what doubtfully for Bucks.
OUSE: Barford; Thurleigh; *F.B.*
IVEL: Ampthill Park, planted; *C.C.*
First record: Abbot, 1798.

PARIS L.

Paris quadrifolia L. C.T.W. 128.1.1.   *Herb Paris*
Woods   Native
Frequent in the heavy boulder-clay woods, but absent or very rare
in woods on lighter soils. Forms with three or five parts to the
flowers, and a similar number of leaves, are not infrequent. Re-
corded for all neighbouring counties.
NENE: Great Hayes Wood; *R. R. B. Orlebar*: Wymington Scrub.
OUSE: Clapham and Hawnes Wood; *Plantae Bedford.*: Renhold;
F.B.: Thurleigh; *Abbot's annotated F.B.*: Exeter Wood; *M. L. Berrill
in J.H. Notes*: Twin Wood; Milton Wood; *J.H. Notes*: Hanger
Wood!; Astey Wood; *A. W. Guppy*: Wilstead Wood; Clapham
Park!*; Putnoe Wood; Bushmead; *L. W. Wilson*: Odell; Stagsden;
B. B. West: Harrold Park Wood!; Dungee Corner!; *G. H. Day*:
Windmill Wood, Souldrop; Holcot Wood.
KYM: Park Wood, Keysoe.
OUZEL: Salford Wood; *J. Burrell.*
IVEL: Maulden Wood!; *Plantae Bedford.*: Cainhoe Park Wood;
W.C. Herb.: Warden Wood; *J.McL., L.M.*: Sundon†; Barton!*;
S.B.P.L.: Snake's Wood; *M. L. Berrill in J.H. Notes*: Beadlow; *Miss
Harradine in J.H. Notes*: Pulloxhill; Silsoe; *C.C.*: Fancott; Potton
Wood.
CAM: Cockayne Hatley Wood.
LEA: Dumb Hills; *S.B.P.L.*
COLNE: Caddington; *F.F.B.*
First record: Abbot, 1795.

JUNCACEAE

JUNCUS L.

Juncus conglomeratus L. C.T.W. 129.1.10. *Clustered Rush*
Woods, heathy places, etc. Native
Frequent, but often confused with var. *compactus* of *J. effusus*. I have
seen the genuine species in all districts. Recorded for all neighbour-
ing counties.
H.S. 3a, 19a, 21, 30b, 34a, 35, 53a, 55b, 77a, 77b, 79a–80a.
First record: uncertain.

J. effusus L. C.T.W. 129.1.9. *Soft Rush*
   J. communis E. Mey.
Marshes, meadows, woods, etc. Native
Common, and growing generally in wetter places than the preced-
ing. Forms with compact cymes (var. *compactus* Lejeune & Court.)
are frequent. Recorded for all districts and neighbouring counties.
H.S. 2b, 3a, 15, 19a, 26, 34a, 35–39a, 40–43, 49, 53a, 55b, 74b, 77a,
77b, 80a, 83.
First record: Abbot, 1798.

J. inflexus L. C.T.W. 129.1.8. *Hard Rush*
   J. glaucus Sibth.
Marshes, meadows, wet places, etc. Native
Common; when growing in drier places usually a sign of water-
logging at some time of the year. Recorded for all districts except
Colne, and for all neighbouring counties.
H.S. 3a, 7–11, 13, 14b, 19a, 20a, 21, 56a–57a, 58a, 64, 84a, 85.
First record: Abbot, 1798.

J. inflexus × **effusus** is recorded, with no details, in *B.P.L.*, 1882. I have not
seen it in the county. Recorded for Herts.

J. pallidus R. Br. *Pale Rush*
Gravel pits Wool adventive
This rush, no doubt introduced with wool manure, has become well established in
a few of our gravel pits. It is equally well established in a gravel pit at East Bedfont
in Middlesex. In all cases it is accompanied by other rushes, but it is too early to
be certain whether these are of alien origin or the result of hybridization of *J.
pallidus* with the closely allied *J. effusus* and *J. inflexus*. The Australasian members of
the genus are badly in need of revision, and it seems probable that hybridization
is producing many forms in Australia and New Zealand. At present all that can
be stated is that we have in Britain plants of *J. pallidus*, and plants which appear
to be hybrids between *J. pallidus* and *J. effusus* and *J. pallidus* and *J. inflexus*.
Neither *J. effusus* nor *J. inflexus* appear in Australasia except perhaps as very recent
introductions. We have also plants which are superficially like plants to which
specific names have been given collected in Australasia.
Ouse: gravel pit, Eaton Socon, where it appeared with many other 'rushes' until
the pits were filled in*; Wyboston, not seen after 1946*; gravel pit, Chawston,
pure *J. pallidus* not seen here but other 'rushes' present*; gravel pit, Cople, with
probable hybrids with *J. inflexus**; gravel pit, Willington, a large and mixed colony.
First record: the author in *B.E.C. 1943–4 Rep.* 689.

J. squarrosus L. C.T.W. 129.1.1. *Heath Rush*
Damp heathy places Native
Rare: recorded for all neighbouring counties except Hunts.
H.S. 35.
OUZEL: Aspley Woods†; *S.B.P.L.*: New Wavendon Heath; v.c. 24
[Beds.]*; Aspley Heath.
IVEL: Ampthill Heath; *F.B.*: Flitwick; *J.McL., B.M.*
First record: Abbot, 1798.

J. compressus Jacq. C.T.W. 129.1.4. *Round-fruited Rush*
Wet meadows Native
Frequent in the Ouse water meadows, rare elsewhere. Considered
common by Abbot. Recorded for all neighbouring counties.
H.S. 7, 14b.
NENE: marshy meadows, Hinwick*; Wymington*; Podington.
OUSE: Cox's Pits†; J.S. in *Bot. Rec. Club 1883 Rep.*: water meadows,
Oakley Bottom*; *E.M.-R.!*: Tempsford Mill*; *P.T.!*: plentiful in
water meadows, Eaton Socon*.
OUZEL: Woburn; *W.H. 1875*: Totternhoe Mead*; *E.M.-R.!*
IVEL: Flitwick Moor*; *C.C.*: meadows, Fancott.
First record: Abbot, 1798.

J. bufonius L. C.T.W. 129.1.7. *Toad Rush*
Marshy places, gravel and sandpits, etc. Native
Common: a variable species. Recorded for all districts and neigh-
bouring counties.
H.S. 13, 16, 19a, 34a.
First record: Abbot, 1798.

J. subnodulosus Schrank C.T.W. 129.1.16. *Blunt-flowered Rush*
Marshy places Native
Frequent: recorded for all neighbouring counties.
H.S. 10a–10c, 39a.
NENE: Hinwick Lodge*.
OUSE: marshes, Stevington.
OUZEL: Hockliffe!*; *W.F.B.*: Cow Common.
IVEL: Harlington; *S.B.P.L.*: Flitwick, frequent on the Moor!†*;
J.McL., B.M.: Maulden; *M. L. Berrill in J.H. Notes*: Westoning;
W.F.B.: Southill; *J.E.L., J. Bot.*: below Radwell Mill; *J.E.L.,
F.F.B.*: Beckeringspark Moor.
LEA: Leagrave; *S.B.P.L.*
First record: Saunders, 1885.

J. acutiflorus Ehrh. ex Hoffm. C.T.W. 129.1.17.
J. sylvaticus auct. *Sharp-flowered Rush*
Marshes Native
More frequent than the preceding species. Recorded for all neigh-
bouring counties.
H.S. 7, 10c, 35, 38, 39a, 41, 43.
NENE: Hinwick*.

Ouse: Twin Woods, as *J. articulatus*; *J.McL.*, *L.M.*: Bromham Park; *J.H. Notes*: marsh, Stevington; water meadow, Eaton Socon; Oakley Bottom.
Kym: Knotting.
Ouzel: Woburn Sands; *R. A. Pryor in J. Bot. 1876*: Aspley Woods; *W.F.B.*: New Wavendon Heath, v.c. 24 [Beds.]*; Horsemoor Farm.
Ivel: Potton Marshes; *J.McL.*, *L.M.*: Flitwick†; *Herb. J.S.*: Westoning!; Harlington; Sutton Fen; *W.F.B.*: Washer's Wood*; marsh, foot of Cooper's Hill*.
Cam: Cockayne Hatley Wood*.
First evidence: J. McLaren, 1864.

J. articulatus L. C.T.W. 129.1.18.   *Jointed Rush*
Marshy places   Native
This is our common jointed rush. It is a variable species and proliferating forms are common. *E. Bot.* tab. 238, March 1795, was drawn from Bedfordshire material. Recorded for all districts and neighbouring counties.
H.S. 3a, 7, 9, 10a, 10b, 11, 13, 14b, 19a, 34a, 36, 38, 56a, 58b, 64.
First record: Abbot, 1798.

J. bulbosus L. C.T.W. 129.1.21.   *Bulbous Rush*
Marshy places   Native
Rare, and apparently limited to boggy places on acid soils. Proliferating forms are general. Recorded for all neighbouring counties.
H.S. 35.
Ouzel: Aspley Heath Wood†; Heath and Reach†; *Herb. J.S.*: New Wavendon Heath, v.c. 24 [Beds.]*; Aspley Heath.
Ivel: Ampthill Bogs, it still grows in the marsh at the foot of Cooper's Hill*; Potton Marshes, where it was still found by McLaren; *F.B.*: Flitwick!*; *S.B.P.L.*: Westoning; *C.C.*
First record: Abbot, 1798.

LUZULA DC.

Luzula pilosa (L.) Willd. C.T.W. 129.2.1. *Small Hairy Woodrush*
    Juncus pilosus L., *Juncoides pilosum* (L.) Kuntze
Woods   Native
Common in woods, especially those on light soils. Recorded for all districts except Nene, and for all neighbouring counties.
H.S. 2b, 19b, 30a, 51a, 79a, 79b, 80b.
First record: Abbot, 1798.

L. forsteri (Sm.) DC. C.T.W. 129.2.2.   *Forster's Woodrush*
    Juncoides forsteri (Sm.) Kuntze
Woods   Native
Rare: recorded for Herts. and Bucks.
H.S. 19b.
Ouse: Marston Thrift, a record doubted by Hamson, but plentiful in heathy parts of the wood (see author's note in *B.E.C. 1943–4 Rep.*, 812)*; *Ada Stimson in J.H.*

L. sylvatica (Huds.) Gaudin C.T.W. 129.2.3. *Great Hairy Woodrush*
Juncus sylvaticus Huds., *Juncoides sylvaticum* (Huds.) Kuntze
Woods   Native
Rare, and limited to woods on light soils. Considered common by
Abbot, but his herbarium specimen is represented by *L. pilosa.*
Recorded for all neighbouring counties except Hunts.
H.S. 30a.
KYM: West Wood, sparingly and probably introduced with young
trees.
OUZEL: Aspley Wood!†; *J.McL.*, *L.M.*: Woburn Woods, possibly
the same station; *S.B.P.L.*: King's Wood, in some quantity in
various parts of the wood!*; *B.P.L.*: Flitwick; *C.C.*
First record: Abbot, 1798: more certainly, J. McLaren, 1864.

L. campestris (L.) DC. C.T.W. 129.2.7.   *Field Woodrush,*
                                                        Good-Friday Grass
Juncus campestris L., *Juncoides campestre* (L.) Kuntze
Pastures, heaths, chalk downs, etc.   Native
Common: recorded for all districts and neighbouring counties.
H.S. 22b, 23, 26, 28, 37, 38, 47, 66b, 74a, 75, 76.
First record: Abbot, 1798.

L. multiflora (Retz.) Lejeune C.T.W. 129.2.8.   *Heath Woodrush*
Juncoides multiflorum (Retz.) Druce
Heaths, heathy woods, etc.   Native
Frequent on the Lower Greensand and Clay-with-Flints: rare
elsewhere. Our plants are mostly var. **congesta** Syme. Recorded
for all neighbouring counties.
H.S. 22a, 30a, 30b, 35, 40, 49, 77b, 79a, 79b.
OUSE: Cardington; *W.H. 1875*: Marston Thrift.
OUZEL: King's Wood!†; *S.B.P.L.*: Palmer's Shrubs; Aspley Wood*;
Heath and Reach; New Wavendon Heath, v.c. 24 [Beds.].
IVEL: Flitwick, plentiful on the Moor!*; *S.B.P.L.*: Ampthill;
J.McL., *B.M.*: Eversholt!*; *V. H. Chambers*: Daintry Wood; Weston-
ing Moor*; Sutton Fen.
LEA: Horsley's Wood.
COLNE: Deadmansey Wood*; Studham Common.
First record: W. Hillhouse, 1875.

AMARYLLIDACEAE

NARCISSUS L.

Narcissus pseudonarcissus L. C.T.W. 130.3.1.   *Wild Daffodil*
Woods   Native
Rare, and growing on a variety of soils. Possibly more plentiful at
one time, as it has frequently been uprooted. Recorded for all
neighbouring counties except Hunts.
OUSE: Clapham; *F.B.*: Lidlington; *W.H. 1875*: Ravensden; *C.C.*:
Odell Wood, possibly planted; *G. H. Day.*

KYM: Melchbourne Wood; *Lord St. John in Abbot's annotated F.B.*
IVEL: Warden; *F.B.*: Silsoe Park; *W.C. Herb.*: Barton; *S.B.P.L.*:
Warden Hill Plantation; *J.McL.*, *L.M.*: Maulden Green End;
M. L. Berrill in J.H. Notes: Steppingley Woods; *C.C.*: Tingrith;
V. H. Chambers.
LEA: Runley Wood, known locally as Daffodil Wood, where it is
now extinct!†; *Davis*: New Mill End; *S.B.P.L.*
COLNE: woods, Whipsnade Zoo, where it made a fine show when
the woods were first enclosed*; meadow, Studham.
First record: Abbot, 1798.

GALANTHUS L.

Galanthus nivalis L. C.T.W. 130.2.1.   *Snowdrop*
Woods   ? Native
Frequent in a number of woods, especially on the Lower Greensand, where it is
just possible that it may be native. Recorded for all neighbouring counties.
OUSE: Kempston Hill Plantation; *J.McL.*, *L.M.*
IVEL: Sandy Hills; *F.B.*: Southill; *J.McL.*, *B.M.*: Flitwick Wood; *W.C. Herb.*:
Flitwick Park!†; *Herb. J.S.*: Tingrith Park and neighbouring woods; Rowney
Warren*.
First record: Abbot, 1798.

ALLIUM L.

Allium vineale L. C.T.W. 127.18.5.   *Crow Garlic*
Arable fields, roadsides, etc.   Colonist
Common in the eastern parts of the county, where it was once a
serious weed. It invariably appears with us in the form of var. **com-
pactum** (Thuill.) Bor. Recorded for all neighbouring counties.
NENE: Wymington.
OUSE: common, especially in the eastern part.
KYM: side of West Wood; Tilbrook, v.c. 30 [Hunts.].
OUZEL: Potsgrove, one plant.
IVEL: common, especially in the eastern part.
CAM: common.
LEA: Luton, one plant.
COLNE: Caddington; *H. B. Souster*!
First record: Abbot, 1798.

A. ursinum L. C.T.W. 127.18.12.   *Ramsons*
Woods   Native
Frequent, well distributed in the county, and growing usually in
woods on light clay soils. Recorded for all neighbouring counties.
OUSE: Marston; *G. Squire in J.H. Notes*: Carlton Hall Wood;
A. W. Guppy: Roxton; *G. H. Day*: Holcot Wood.
KYM: Dean; *M. Dalton*.
IVEL: Cainhoe Park Wood; *W.C. Herb.*: Warden Great Wood!;
J.McL., *L.M.*: Southill; *J.McL. in J.H. Notes*: Silsoe; Buckle
Grove; *C.C.*: Chalton Springs†; *W.F.B.*: Maulden Wood; Fancott;
Flitwick Wood; Harlington.
LEA: Dumb Hill Wood†; *S.B.P.L.*: East Hyde!; *W.F.B.*: Arden Dell
Wood; *H. B. Sargent*.

COLNE: ditches beyond the Parsonage, Whipsnade; *Plantae Bedford.*: wood at foot of Blow's Downs!; *H. B. Sargent*: Greencroft Barn*. First record: Abbot, 1795.

A. oleraceum L., C.T.W. 127.18.6, is listed in *Topographical Botany* and *Comital Flora* as of doubtful occurrence for the county.

IRIDACEAE

IRIS L.

Iris pseudacorus L. C.T.W. 131.2.4.   *Yellow Flag*
Riversides, marshes, wet meadows, etc.   Native
Common in all suitable habitats: 'var. *bastardi*' was recorded from Ampthill by C. Crouch and from Limbury by D. M. Higgins in *W.B.E.C. 1892–3 Rep.* Recorded for all districts and neighbouring counties.
H.S. 8, 13, 38.
First record: Abbot, 1798.

I. foetidissima L. C.T.W. 131.2.3.   *Stinking Iris, Gladdon*
Woods   Native
Frequent, and widely distributed, but rarely flowering. Recorded for all neighbouring counties.
H.S. 5.
OUSE: Stevington Park; Pavenham facing the church in a hedge, H. B. Souster reports that it is still there; *Plantae Bedford.*: Bromham; *F.B.*: Basmead; *C.R.O. List*: Salem Thrift; *C. Clarke in J.H. Notes*: Twin Wood; *A. B. Sampson*: Colesdon; *A. W. Guppy*: Harrold Park Wood; *G. H. Day*: Cranfield; *H. B. Souster*: Woodcraft; Judge's Spinney.
OUZEL: Houghton Regis; *S.B.P.L.*
IVEL: Flitwick Wood!†; *S.B.P.L.*: Haynes; *Mr. Halahan in J.H. Notes*: Priestly Farm, Flitwick*; *M. Holdsworth*: College Wood, Northill; *I. J. Allison*: Sharpenhoe Hill; *H. B. Souster*.
CAM: Cockayne Hatley Wood; *T.L.*
LEA: Luton Hoo, plentiful in woods near the lake!†*; *H. Brown*: Eaton Green; *F.F.B.*: George's Wood.
First record: Abbot, 1795.

DIOSCOREACEAE

TAMUS L.

Tamus communis L. C.T.W. 132.1.1.   *Black Bryony*
Hedgerows, wood borders, etc.   Native
Common throughout the county: recorded for all districts and neighbouring counties.
H.S. 1b, 5, 32, 42, 51, 69, 70a, 70b, 83.
First record: Abbot, 1798.

ORCHIDACEAE

CEPHALANTHERA Rich.

Cephalanthera damasonium (Mill.) Druce C.T.W. 133.2.1.
Serapias longifolia Huds. p.p., *C. grandiflora* Gray   *White Helleborine*
Beech woods   Native
Frequent in beech woods on the Chalk, rare elsewhere. Recorded for all neighbouring counties.
H.S. 5, 69, 70a, 70b.
OUSE: Milton Hill, probably Judge's Spinney, where it still appears; *J.McL., B.M.*
OUZEL: Roman Road, west of Dunstable!; *W. G. Smith in J. Bot. 1885.*
IVEL: Markham Hills!; *S.B.P.L.*: Sundon; *W.F.B.*: Pegsdon!; Barton!; *J.E.L., J. Bot.*
LEA: Luton Park; *J. Anderson in New Bot. Guide II,* 1837: near Luton Park (presumably New Mill End Road)!†*; *H. Brown*: near Warden Hill, plentiful in Maulden Firs!; *W.F.B.*
COLNE: Kensworth Lane!; *E.M.-R.*: woods near Dunstable; *W. D. Coales.*
First record: unlocalized, Hudson, *Flora Anglica Ed. II,* 1778, 394. It was not recorded by Abbot.

EPIPACTIS Sw.

Epipactis palustris (L.) Crantz C.T.W. 133.3.1.
                                                       Marsh Helleborine
Serapias palustris (L.) Mill., *Helleborine palustris* (L.) Schrank
Marshes   Native
Extinct: listed as *Serapias longifolia* in *Flora Bedfordiensis.* Recorded for all neighbouring counties.
OUSE: Stevington Moor; *Plantae Bedford.*
IVEL: Gravenhurst Moor; *W.C. Herb.*
First record: Abbot, 1795.

E. helleborine (L.) Crantz C.T.W. 133.3.2.
                                                 Broad-leaved Helleborine
Serapias latifolia Huds. p.p., *E. latifolia* (L.) All., *Helleborine helleborine* (L.) Druce
Woods   Native
Frequent, especially in the boulder-clay woods. It varies in size, and in the colour of the flowers, both of which are affected by the intensity of the shade. Recorded for all districts except Cam, and for all neighbouring counties.
H.S. 3b, 34b, 52.
First record: Abbot, 1798.

E. purpurata Sm.   *Violet Helleborine*
  E. sessilifolia Peterm. C.T.W. 133.3.3, *E. violacea* (Dur. Duq.)
Bor., *Helleborine purpurata* (Sm.) Druce
Woods   Native
Rare, and usually growing with us in woods on lighter soils than
E. helleborine. *E. Bot. Supp.* tab. 2775, July 1833, credits the discovery
of this in Britain to Abbot, but it is interesting to note that the
original drawing, made apparently by James Forbes (of Woburn), is
not in the collection at the British Museum (Natural History).
There are many records for the species, but as it has been confused
with *E. helleborine* only the more certain ones are included here.
Recorded for all neighbouring counties except Hunts.
H.S. 19b, 79b.
OUSE: King's Wood!; *R. H. Goode*: Holcot Wood; Marston Thrift.
OUZEL: 'several patches . . . growing under the shade of lime and
hazel at Woburn', where it was still found by Druce (see *J. Bot.
1909*, 28); *E. Bot. Supp. 1833.*
IVEL: Tingley Wood Plantation; *J.E.L.*, *F.F.B.*: Herne Green,
Toddington; *F. Mander.*
LEA: Luton; *D. M. Higgins in W.B.E.C. 1913-4 Rep.*: Luton Hoo.
COLNE: High Wood, near Studham; copse near Zouche's Farm,
v.c. 20 [Beds.]; *Fl. Herts.*: Deadmansey Wood*; Oldhill Wood.
First record: *E. Bot. Supp.*, 1833.

A record, by the author, of *E. leptochila* (Godf.) Godf. in *B.E.C. 1945 Rep.* 68, was
in error.

E. phyllanthes G. E. Sm.
  E. pendula C. Thomas non A. A. Eaton. C.T.W. 133.3.8.
Beech woods   Native
Rare: one of our more interesting orchids. For a closer study of
the species and the Bedfordshire material see D. P. Young, in
Watsonia I (1949) 102, *II (1952)* 253. Recorded for Bucks. and
Cambs.
IVEL: Streatley, where it was shown to me by Sir Frederick Mander
who had known it there for many years!*: near Barton, in greater
quantity, var. *vectensis* (T. & T. A. Stephenson) D. P. Young*;
A. J. Wilmott!
First record: the author in *B.E.C. 1943-4 Rep.* 756.

SPIRANTHES Rich.

Spiranthes spiralis (L.) Chevall. C.T.W. 133.5.1.
  Ophrys spiralis L.   *Autumn Lady's Tresses*
Pastures   Native
Rare: recorded for all neighbouring counties except Hunts.
OUSE: Thurleigh; *F.B.*: Crow Hills near Clapham between Woods
(i.e. Twin Woods); *Abbot's annotated F.B.*: Basmead; *C.R.O. List*:
Pavenham; *J.McL.*, *L.M.*: Exeter Wood; *M. L. Berrill in J.H. Notes*.

IVEL: Haynes; *Mr. Halahan in Herb. Beds. Nat. Hist. Soc.*: Pegsdon Barns, where it is locally plentiful in a restricted area!†*; *J.S. in Bot. Rec. Club 1880 Rep.*
LEA: Stockwood Lawn, a doubtful record; *Davis.*
First record: Abbot, 1798.

LISTERA R.Br.

Listera ovata (L.) R.Br. C.T.W. 133.6.1.   *Twayblade*
  Ophrys ovata L.
Woods, chalk hills, etc.   Native
Common, equally in boulder-clay woods and on the chalk downs. Recorded for all districts and neighbouring counties.
H.S. 2b, 51a.
First record: Abbot, 1798.

NEOTTIA Ludw.

Neottia nidus-avis (L.) Rich. C.T.W. 133.7.1. *Bird's-nest Orchid*
  Ophrys nidus-avis L.
Woods   Native
Frequent in old woods throughout the county and usually growing in deep shade. Recorded for all neighbouring counties.
H.S. 51a.
NENE: Great Hayes Wood!; *R. R. B. Orlebar*: Dungee Corner.
OUSE: Clapham Park Wood; *F.B.*: Basmead; *C.R.O. List*: Oakley; *B.P.L.*: Twin Wood; *Mr. Robinson in J.H. Notes*: Kempston Wood; King's Wood; Reading's Wood; *Mr. Halahan in J.H. Notes*: Hanger Wood; Astey Wood; *R. Townsend*: Harrold Park Wood*; *G. H. Day*: Oakley Little Wood; *C. C. Foss*: Holcot Wood*.
OUZEL: Aspley Wood; *J.McL., L.M.*: Salford Wood; *V.C.H.*: King's Wood.
IVEL: Cainhoe Park Wood*; *W.C. Herb.*: Southill†; *J.McL., B.M.*: Sundon Wood; *S.B.P.L.*: Herne Green; *F. Mander*: Ridgmont; *C.C.*: Washer's Wood; Potton Wood.
LEA: Horsley's Wood; *Plantae Bedford.*: near Dunstable; *F.B.*: Dumb Hill Wood; Daffodil Wood; *J.S., J. Bot. 1883.*
COLNE: wood by Studham Common, on county boundary.
First record: Abbot, 1795.

HAMMARBYA Kuntze

Hammarbya paludosa (L.) Kuntze C.T.W. 133.9.1. *Bog Orchid*
  Ophrys paludosa L., *Malaxis paludosa* (L.) Sw.
Bogs   Native
Extinct: *E. Bot.* tab. 72, April 1792, was drawn from a specimen sent by Abbot from Potton. Recorded for Hunts. and Cambs.
IVEL: bogs near Potton; *Abbot in E. Bot 1792.*

HERMINIUM R.Br.

Herminium monorchis (L.) R.Br. C.T.W. 133.12.1. *Musk Orchid*
Chalk hills Native
Rare, and appearing only in places where the soil has at some time
been disturbed. Recorded for Herts., Bucks. and Cambs.
H.S. 62.
OUZEL: Sewell, on slopes of old chalk quarries!; *F. Mander*: Totternhoe Knolls*.
IVEL: on a hill between Sundon and Harlington!; *R. M. Welch in
F.F.B.*
First record: R. M. Welch, 1911.

COELOGLOSSUM Hartm.

Coeloglossum viride (L.) Hartm. C.T.W. 133.13.1. *Frog Orchid*
  Satyrium viride L., *Habenaria viridis* (L.) R.Br.
Pastures Native
Rare: *E. Bot.* tab. 94, Nov. 1792, was drawn from a specimen sent
from near Luton Hoo by Mr. Sibley, 'a gentleman who has paid
much attention to the orchis tribe'. Recorded for all neighbouring
counties.
OUSE: Stevington; Thurleigh; Bletsoe; *F.B.*: Basmead; *C.R.O. List*.
OUZEL: Dunstable Downs, it appears at intervals along the stretch
of the hills!*; *H. B. Sargent*: Totternhoe Knolls, one plant.
IVEL: Cainhoe; *W.C. Herb.*: Eversholt, as *Orchis ustulata*; *J. McL.,
B.M.*: Sundon; *S.B.P.L.*
LEA: near Luton Hoo (see above); *Sibley*: pastures near Someries,
from where there is a specimen collected by Saunders in Herb.
Beds. Nat. Hist. Soc.; *Plantae Bedford.*: Farley; *J.S. in J. Bot. 1883*:
Limbury†; *W.F.B.*
COLNE: Pepperstock; *S.B.P.L.*
First record: Sibley, 1792.

GYMNADENIA R.Br.

Gymnadenia conopsea (L.) R.Br. C.T.W. 133.14.1.
                                              Fragrant Orchid
  Orchis conopsea L., *Habenaria conopsea* (L.) Benth. non Reichb. f.,
H. gymnadenia Druce
Pastures Native
Common on the chalk downs, usually in well-established turf, rare
elsewhere. Recorded for all neighbouring counties.
H.S. 59, 60, 66a.
OUSE: Clapham; *F.B.*: Stevington; *Abbot's annotated F.B.*: Milton
Hill; *J.McL., B.M.*
OUZEL: Milton Bryan; *H. E. Pickering in J.H. Notes*: common on the
chalk hills.
IVEL: Gravenhurst Moor; *W.C. Herb.*: common on the chalk hills.
LEA: Luton Downs, probably Warden Hills, where it is common as
elsewhere on the Chalk!; *Plantae Bedford.*
COLNE: Zouche's Farm, v.c. 20 [Beds.]; *Fl. Herts.*
First record: Abbot, 1795.

PLATANTHERA Rich.

Platanthera chlorantha (Cust.) Reichb. C.T.W. 133.16.1.
Greater Butterfly Orchid
Habenaria chlorantha (Cust.) Bab. non Spreng., *H. virescens* Druce
non Spreng.
Woods   Native
Frequent in our boulder-clay woods, and growing more rarely in
open pasture. Most early records of *P. bifolia* referred to this species.
Recorded for all districts except Colne, and for all neighbouring
counties.
H.S. (33).
First record: Abbot, 1798, as *Orchis bifolia* L.

P. bifolia (L.) Rich. C.T.W. 133.16.2.   *Lesser Butterfly Orchid*
Habenaria bifolia (L.) R.Br.
Woods   Native
Much less common than the preceding, but no doubt awaiting
record in some stations. Recorded for all neighbouring counties.
OUSE: Hanger Wood; *C. C. Foss.*
IVEL: Holt Wood, Sundon!*; *F. Mander*: Potton Wood*; *F. L.
Chesham.*
First record: the author in *B.E.C. 1938 Rep.* 59.

OPHRYS L.

Ophrys apifera Huds. C.T.W. 133.18.1.   *Bee Orchid*
Pastures, etc.   Native
Common on the chalk hills, but appearing on calcareous soils
throughout the county. It is usually a plant of exposed habitats,
and grows both in established turf and in places where the soil has
been disturbed. Seen at its best in chalk scrub, it also occasionally
grows in open woodland where specimens with elongated lips and
narrow, pointed sepals are seen. Recorded for all districts and
neighbouring counties.
H.S. 21, 60, 62.
First record: Abbot, 1795.

O. insectifera L. C.T.W. 133.18.4.   *Fly Orchid*
O. muscifera Huds.
Woods, pastures, etc.   Native
Rare and limited to calcareous soils. Recorded for all neighbouring
counties except Hunts.
OUSE: Crapwell's Closes; Hollwell Bury Field; *Plantae Bedford.*:
Milton Ernest; *B.P.L.*: Kempston Wood; Wilstead Wood; *Mr.
Halahan in J.H. Notes*: Stagsden Wood, probably Hanger Wood,
where it was found by R. Townsend; *A. Beagley in J.H. Notes*:
Clapham; *F. G. R. Soper.*
OUZEL: *V.C.H.* with no details: Totternhoe; *H. B. Souster.*

IVEL: near Shefford; *F.B.*: Cain Hill, ? Cainhoe; *W.C. Herb.*: Streatley, where it grows both in a beech wood and on the chalk hill slope!†*; *S.B.P.L.*: Sundon Hills; *J.S. in J. Bot. 1853*: Barton *W.F.B.*: Barton Leete Wood; Pegsdon; *J.E.L., J. Bot.*
LEA: near Hoo Lodge, Hyde; *W.F.B.*: Daffodil Wood; *F. C. Minns*: Runley Wood; *H. B. Souster*.
First record: Abbot, 1795.

O. sphegodes Mill. (*O. aranifera* Huds.), C.T.W. 133.18.3, *Early Spider Orchid*, is of doubtful record for the county. Abbot recorded it in an appendix to *Flora Bedfordiensis* from old sand and gravel pits at Southill, on the authority of P. Walker. Both the habitat and station are unlikely. It was listed from Streatley by Davis, a more likely station, but to be doubted without more evidence. It was further recorded in the *Hillhouse List, 1876*, but with no details. Recorded also for Cambs. and Northants.

HIMANTOGLOSSUM Spreng.

Himantoglossum hircinum (L.) Spreng. C.T.W. 133.19.1.
   Orchis hircina (L.) Crantz   *Lizard Orchid*
Chalk hills   Native
Rare: recorded for Herts., Bucks. and Cambs.
OUZEL: Dunstable Downs; *S. Tearle in J. Bot. 1932*, 114.
IVEL: near Harlington, where it re-appeared over a number of years in scattered communities until the scrub was cleared in 1948!*; *Miss E. Stansfield (Mrs. Smith) in Herb. E. B. Bishop*, 1928.
First record: S. Tearle, 1932.

ORCHIS L.

Orchis ustulata L. C.T.W. 133.20.4.   *Burnt* or *Dwarf Orchid*
Chalk hills   Native
Rare: one of our more interesting orchids, and found usually only in well-established turf. Recorded for all neighbouring counties except Hunts.
H.S. 59.
OUZEL: Dunstable Downs; *V.C.H.*
IVEL: Barton Hills, where it grows sparingly in two stations!*; *Plantae Bedford.*: near Pegsdon!; *H. Brown*: Streatley; *C.C.*: Sundon†; *Herb. J.S.*: Sharpenhoe Hills; *Muriel Hamson in J.H. Notes*.
LEA: Luton Downs; *F.B.*: Warden Hills; *J.S. in J. Bot. 1883*: Galley Hill, north aspect; *H. B. Souster*.
First record: Abbot, 1795.

O. morio L. C.T.W. 133.20.5.   *Green-winged Orchid*
Pastures   Native
Frequent, and well distributed on calcareous soils throughout the county. It is becoming scarcer owing to increased ploughing of grassland and the reduction of mown and well-grazed pasture. Recorded for all districts and neighbouring counties.
First record: Abbot, 1798.

O. mascula (L.) L. C.T.W. 133.20.7.   *Early Purple Orchid, Cuckoos*
Woods   Native
Frequent, well distributed in the county, and growing almost
invariably in woods on heavy clay soils. My attention has been
drawn by C. M. Crisp to an attractive white-flowered form in Flit-
wick Wood. Recorded for all districts and neighbouring counties.
H.S. 3b, 18, 51a, 52.
First record: Abbot, 1798.

DACTYLORCHIS (Klinge) Vermeulen

Dactylorchis incarnata (L.) Vermeulen   *Marsh Orchid*
  Orchis incarnata L., *O. strictifolia* Opiz C.T.W. 133.20.10,
O. latifolia auct.
Wet meadows   Native
Rare, and apparently limited to water meadows by the Ouse.
Early records in many cases refer to *D. praetermissa*. Recorded for
all neighbouring counties except Hunts.
H.S. 14b.
OUSE: Basmead, if correctly named; *R. A. Pryor in Newbould MS.*:
Bromham, det., with reservation, by C. B. Tahourdin; *J.H. Notes*:
Chellington*; Turvey; *G. H. Day*: water meadows, Eaton Socon*;
W. Durant!
First record: uncertain.

D. praetermissa (Druce) Vermeulen   *Marsh Orchid*
  Orchis praetermissa Druce C.T.W. 133.20.11.
Marshy places   Native
More frequent than the preceding, and wider in its habitats.
Recorded for all neighbouring counties.
H.S. 38, 39a, 41.
OUSE: water meadows, Eaton Socon; *V. S. Summerhayes*!
IVEL: Gravenhurst, as *O. latifolia*; *W. C. Herb.*: Flitton Marshes, as
O. latifolia!*; *J.McL.*, *L.M.*: Flitwick Moor, as *O. latifolia*!†*;
J.McL., *B.M.*: Cadwell Bridge; *J.E.L.*, *J. Bot.*: Greenfield!;
E.M.-R.: in chalk scrub, Streatley!*; *K. G. Bull*: meadow, Southill;
Westoning Moor*.
IVEL: Limbury Marsh, as *O. latifolia*†; *Herb. J.S.*: East Hyde; *P.T.*!
First evidence: W. Crouch, 1844.

D. maculata (L.) Vermeulen   *Heath Orchid*
  Orchis maculata L., *O. ericetorum* (E. F. Linton) E. S. Marshall
C.T.W. 133.20.9.
Marshes   Native
Rare: recorded for Herts., Bucks. and Northants.
H.S. 38.
IVEL: Flitwick Moor, locally abundant in two heathy parts of the
Moor!*; *J.E.L. in B.E.C. 1911 Rep.* 55.
    D. maculata × praetermissa
IVEL: Flitwick Moor, with both parents, and making handsome plants as much as
2ft. tall.

D. fuchsii (Druce) Vermeulen   *Spotted Orchid*
  Orchis fuchsii Druce, C.T.W. 133.20.8, *O. maculata* auct.
Woods, calcareous pastures, etc.   Native
Our commonest orchid, but very variable: an obtuse-leaved form,
which is common on chalk downs, is very distinct from the pointed-
leaved form in our woods. Druce recorded '*O. O'Kellyi* Druce' from
Dunstable Downs in *B.E.C. 1923 Rep.*, and Little recorded it from
Warden in his annotated copy of *Field Flowers of Bedfordshire*; but this
is only a white-flowered form of *D. fuchsii*. Most early records of
'*O. maculata*' should be referred to *D. fuchsii*, which is recorded for
all districts and neighbouring counties.
H.S. 1b–3a, 51a, 52, 55b (34b).
First record: Abbot, 1798.
First certain evidence: Cainhoe Park Wood; *W.C. Herb.*, 1843.
    D. fuchsii × **Gymnadenia conopsea**
IVEL: Streatley, with both parents*.

ACERAS R.Br.

Aceras anthropophorum (L.) Sm. C.T.W. 133.21.1.
                                    Man Orchid
Chalk hills   Native
Rare: recorded for all neighbouring counties except Hunts.
OUZEL: Totternhoe Knolls, limited to a restricted area*; *M. Brown
in B.E.C. 1926 Rep.* 135!

ANACAMPTIS Rich.

Anacamptis pyramidalis (L.) Rich. C.T.W. 133.22.1.
  Orchis pyramidalis L.   *Pyramidal Orchid*
Pastures   Native
Frequent on the chalk downs, and occurring most frequently in
overgrazed parts, and in places where the soil has at some time
been disturbed. It occurs less frequently on calcareous soils else-
where. Recorded for all neighbouring counties.
H.S. 10b, 59, 60, 66a, 67, 68.
OUSE: Milton Hill; Oakley Plantations; *Plantae Bedford.*: Basmead;
C.R.O. List: Turvey; *J.H. Notes*: Wilstead; *Mr. Halahan in J.H.
Notes*: Salem Thrift; *A. W. Guppy*: Felmersham; Pavenham!;
G. H. Day: Stagsden; *B. B. West*: railway embankment, Souldrop;
Clapham Park.
KYM: Dean; *M. Dalton.*
OUZEL: frequent on the Chalk.
IVEL: frequent on the Chalk: Gravenhurst; *W.C. Herb.*: Warden
Wood; *J.McL., B.M.*: Warden Tunnel; *M. L. Berrill in J.H. Notes*:
pastures, Shefford.
CAM: pastures, Dunton, Wrestlingworth.
LEA and COLNE: frequent on the Chalk.
First record: Abbot, 1795.

ARACEAE
ACORUS L.

Acorus calamus L. C.T.W. 134.1.1. *Sweet Flag*
Riversides, pondsides, etc. Denizen
Frequent by the Ouse, where it is evidently long established; rare
elsewhere. It should have a special interest to Bedfordshire natura-
lists, as a controversy about its status was largely responsible for
bringing into being the Bedfordshire Natural History Society of
1875. Recorded for all neighbouring counties.
H.S. 8, 14a.
OUSE: Hassocks Ditches; *Plantae Bedford.*: near Kempston Church!;
J.McL., L.M.: Bedford!; *W.H. 1875*: Roxton; Willington!; *G. H.
Day*: Biddenham!*; *W. D. Coales*: Bromham*; Cardington;
Pavenham; Fenlake; Eaton Socon.
IVEL: Tingrith Park†; *Mrs. Twidell in J.S., J. Bot. 1889.*
First record: Abbot, 1795.

ARUM L.

Arum maculatum L. C.T.W. 134.3.1. *Cuckoo-pint, Lords-and-Ladies*
Woods, hedgerows, shady places, etc. Native
Common: variable in the size of plants, colour of spikes, and
blotchiness of the leaves. Recorded for all districts and neighbouring
counties.
H.S. 2b, 3b, 5, 18, 32–34b, 51a, 52, 55a, 78, 81, 83, 84b.
First record: Abbot, 1798.

LEMNACEAE
LEMNA L.

Lemna polyrhiza L. C.T.W. 135.1.1. *Great Duckweed*
Ponds, ditches, backwaters, etc. Native
Frequent, but variable in size. Recorded for all neighbouring
counties.
OUSE: Ford End; *F.B.*: Fenlake; *J.McL., B.M.*: Goldington; *J.H.
Notes*: Marston; *Ada Stimson in J.H. Notes*: backwaters, Eaton Socon*.
OUZEL: pond, Heath and Reach*; *J. E. Dandy*!
IVEL: Ridgmont; *W.C. Herb.*: Biggleswade.
CAM: Cockayne Hatley; *I. J. Allison.*
LEA: Luton Hoo Park†; *S.B.P.L.*: Leagrave; *M. Brown in W.F.S.
Diary.*
First record: Abbot, 1798.

L. trisulca L. C.T.W. 135.1.2. *Ivy-leaved Duckweed*
Ponds, ditches, etc. Native
Common, and well distributed. *E. Bot.* tab. 926, Sept. 1801, was
probably drawn from a specimen sent by Abbot. Recorded for all
districts except Colne, and for all neighbouring counties.
H.S. 7.
First record: Abbot, 1795.

L. minor L. C.T.W. 135.1.3. *Lesser Duckweed*
Ponds, etc. Native
Common almost everywhere where there is stagnant water.
Recorded for all districts and neighbouring counties.
H.S. 7, 21, 64, 84b.
First record: Abbot, 1795.

L. gibba L. C.T.W. 135.1.4. *Thick-leaved Duckweed*
Ponds, ditches, backwaters, etc. Native
Frequent in ditches and backwaters associated with the Ouse and
Ivel: absent or awaiting record in the Nene, Cam and Colne dis-
tricts. Recorded for all neighbouring counties.
H.S. 7.
First record: Abbot, 1798.

SPARGANIACEAE

SPARGANIUM L.

Sparganium erectum L. *Branched Bur-reed*
    S. ramosum Huds. C.T.W. 136.1.1.
    Subsp. **erectum**
Wet places Native
Common in all likely places: recorded for all districts and neigh-
bouring counties.
H.S. 6–8, 13, 42, 56a, 84b.
First record: Abbot, 1798.
    Subsp. **neglectum** (Beeby) Schinz & Thell.
OUZEL: Leighton Buzzard; *G. C. Druce in J. Bot. 1897*: Salford;
V.C.H.
IVEL: Flitwick; Biggleswade; *J.E.L., F.F.B.*: Fancott*; *A. J. Wilmott.*
CAM: Cockayne Hatley; *J.E.L., F.F.B.*

S. simplex Huds. ex With. C.T.W. 136.1.2. *Unbranched Bur-reed*
Pondsides, rivers, etc. Native
Frequent: recorded for all neighbouring counties.
H.S. 9.
NENE: Hinwick.
OUSE: Elstow; *Plantae Bedford.*: Medbury; *Abbot's annotated F.B.*:
Cardington Brook; *J. McL., L.M.*: Newnham Bridge; *A. B. Sampson*:
Harrold; *C. C. Foss*: Felmersham.
KYM: Melchbourne Park*.
IVEL: Flitwick!*; *M. L. Berrill in J.H. Notes*: Gravenhurst;
C.C.: Biggleswade Common; *J.E.L., J. Bot.*: Campton; *J.E.L.,
F.F.B.*
LEA: Biscot; New Mill End†; *S.B.P.L.*: Luton; *W.F.B.*: Leagrave
Marsh*.
First record: Abbot, 1795.

TYPHACEAE

TYPHA L.

Typha latifolia L. C.T.W. 137.1.1.   *Bulrush, Great Reedmace*
Wet places   Native
Common, especially at the sides of lakes in parks and ponds in pits
left after mineral working. Plants were recorded as var. *media*
Syme in *J.E.L., J. Bot.* and by *H. Phillips* in *B.E.C. 1934 Rep.*
Recorded for all districts and neighbouring counties.
H.S. 13, 16, 21, 58a, 64.
First record: Abbot, 1798.

T. angustifolia L. C.T.W. 137.1.2.   *Lesser Bulrush*
Wet places   Native
In similar situations to the preceding species, but much rarer.
Recorded for all neighbouring counties.
H.S. 21.
OUSE: Lidlington; *W.C. Herb.*: gravel-pits, Sharnbrook; Bletsoe;
E.M.-R.!: clay-pit, Brogborough; *H. B. Souster*!
KYM: Knotting; *F.B.*
IVEL: Clophill; *C.C. in J. Bot. 1889*: clay-pit, Arlesey!*; *J.E.L.,*
J. Bot.
First record: Abbot, 1798.

CYPERACEAE

ELEOCHARIS R.Br.

Eleocharis palustris (L.) Roem. & Schult. C.T.W. 138.3.5.
  Scirpus palustris L.   *Spike-rush*
    Subsp. **palustris**
Marshy places   Native
Common, but variable in the size of the individual plants. Recorded
for all districts and neighbouring counties.
H.S. 14b, 35, 56a, 57b, 84a.
First record: Abbot, 1798.
    Subsp. **microcarpa** S. M. Walters
Rare: recorded also for Cambs.
OUSE: water meadow, Eaton Socon (H.S. 14b)!*; *S. M. Walters in*
Watsonia II (1951) 50: Moor End, Felmersham.

E. uniglumis (Link) Schult. C.T.W. 138.3.6.
Marshy places   Native
Rare: recorded for Cambs. and Hunts.
H.S. 14b.
OUSE: water meadow, Eaton Socon, with hybrids with the above!*;
S. M. Walters in Watsonia II (1951) 50.

E. multicaulis (Sm.) Sm., C.T.W. 138.3.4, was reported by Abbot from Ampthill in a letter to J. E. Smith, 27 Aug. 1801. He did not include it in his annotated copy of *Flora Bedfordiensis* and it is probable that Smith did not agree with his determination. Recorded also by Saunders from Woodside in *B.P.L.*, and from Pepperstock, probably the same station, in *J. Bot. 1883*. As he did not include it in later lists and there is no specimen the records must be doubted.

E. acicularis (L.) Roem. & Schult., C.T.W. 138.3.2, was reported 'in great abundance above the town of Bedford in the meadows' by Abbot, in a letter to J. E. Smith, 3 Sept. 1802. He gave the station as Ford End in his annotated copy of *Flora Bedfordiensis*. Listed for v.c. 30 in *Comital Flora*, the manuscript of which has the entry '30 D', which could be interpreted as meaning that Druce had seen it in the county. Recorded for all neighbouring counties.

SCIRPUS L.

Scirpus sylvaticus L. C.T.W. 138.4.2. *Wood Club-rush*
Marshes Native
Locally plentiful in the marshes associated with the Lower Greensand. It is readily eaten by cattle. Recorded for Herts., Bucks. and Northants.
H.S. 39a, 42.
IVEL: Flitwick Park, in ditches by roadside!; Flitwick Moor, plentiful in wetter parts!*; *S.B.P.L.*: Westoning Moor, in some quantity!*; *C.C.*: meadow near Cainhoe Castle; *H. B. Souster*!: marshy places, Washer's Wood*; Moors Plantation.
First record: Saunders, 1885.

S. maritimus L. C.T.W. 138.4.1. *Sea Club-rush*
Gravel pits ? Native
Rare: this species appears some distance up the Thames and there is no reason why it should not be native in the county. Recorded for Cambs. and Hunts.
OUSE: gravel pit, Eaton Socon, in limited quantity and where I have failed to find it again*; *E.M.-R. and the author in B.E.C. 1945 Rep. 71.*

S. lacustris L. *Basket-rush*
Schoenoplectus lacustris (L.) Palla C.T.W. 138.7.3.
Riversides, etc. Native
Common by the Ouse, rare elsewhere. It was well utilized by the Pavenham basket-makers, whose craft had more than a local reputation. Recorded for all neighbouring counties.
H.S. 6, 7, 8, 13.
OUSE: common in and by the river.
KYM: Melchbourne Park.
OUZEL: Woburn Park; *S.B.P.L.*: river, Heath and Reach.
IVEL: Flitwick Park; *S.B.P.L.*: pond near Hawnes†; *Herb. J.S.*: Shefford!; Clifton; Campton; *J.E.L., F.F.B.*: Clophill; Biggleswade.
CAM: pond, Cockayne Hatley House, probably planted.
First record: Abbot, 1798.

S. tabernaemontani C. C. Gmel. *Glaucous Club-rush*
Schoenoplectus tabernaemontani (C. C. Gmel.) Palla C.T.W. 138.7.4.
Gravel pits, sides of ponds, etc. ? Native
Rare, and possibly introduced in all its known stations. Recorded for Bucks., Cambs. and Hunts.
OUSE: gravel pit, Eaton Socon*; gravel pit, Cople*.
OUZEL: Lower Drakelow Pond*.

S. cespitosus L.   *Deer-grass*
   Trichophorum cespitosum (L.) Hartm. C.T.W. 138.2.2.
Extinct: recorded for Bucks. and Cambs.
IVEL: Ampthill Moor; Flitton Moor; *F.B.*

S. setaceus L.   *Bristle Club-rush*
   Isolepis setacea (L.) R.Br. C.T.W. 138.8.1.   Native
Rare, and usually found in marshy places associated with sand or
gravel. Recorded for all neighbouring counties except Hunts.
OUSE: Bromham; *F.B.*: gravel pit, Sharnbrook*; *E.M.-R.*!
OUZEL: Totternhoe†; Aspley†; *S.B.P.L.*: Rushmere*; Baker's
Wood; sandpit near Fox and Hounds, Potsgrove; *E.M.-R.*!
IVEL: Ampthill Moor; *F.B.*: Flitwick; *J.McL., B.M.*: Chalton;
W.F.B.: Westoning; *C.C.*
LEA: Biscot†; *S.B.P.L.*
First record: Abbot, 1798.

BLYSMUS Panz.

Blysmus compressus (L.) Panz. ex Link C.T.W. 138.6.1.
   Scirpus compressus (L.) Pers. non Moench   *Compressed Club-rush*
Marshes   Native
Rare: recorded for all neighbouring counties except Hunts.
H.S. 56a.
OUSE: Fenlake Bogs; *Abbot's annotated F.B.*
OUZEL: Cow Common*; *A. J. Wilmott and V. H. Chambers in B.E.C.
1945 Rep.*!: Totternhoe Mead, scattered and locally abundant*;
E.M.-R.!
IVEL: Potton Marshes; *Abbot's annotated F.B.*
First record: Abbot in letter to J. E. Smith, 3 Sept. 1802.

Cyperus longus L., C.T.W. 138.10.1, was found by J. E. Dandy, E.M.-R. and
the author on an island in the river at Bedford, where it was possibly planted: see
B.E.C. 1943–4 Rep. 812.

SCHOENUS L.

Schoenus nigricans L. C.T.W. 138.11.1.   *Bog-rush*
Marshes   Native
Presumably extinct: recorded for all neighbouring counties.
IVEL: Ampthill Moor; Potton Marshes; *F.B.*

RHYNCHOSPORA Vahl

Rhynchospora alba (L.) Vahl C.T.W. 138.12.1.   *White Beak-sedge*
   Schoenus albus L.
Marshes   Native
Presumably extinct: recorded for Bucks., Cambs. and Hunts.
OUZEL: Aspley; *F.B.*
IVEL: Ampthill Moor; Potton; *F.B.*

ERIOPHORUM L.

Eriophorum angustifolium Honck. C.T.W. 138.1.1.   *Cotton-grass*
  E. polystachion p.p.
Marshes   Native
Rare, as there are few stations remaining in which it could grow.
Recorded for all neighbouring counties.
H.S. 36, 39a.
OUSE: Stevington Bogs; *F.B.*
OUZEL: Woburn Park; *Hortus Gramineus Woburnensis*: Aspley, where
it still grows by Mermaid's Pond!; *J.McL., L.M.*
IVEL: Ampthill Moor; *F.B.*: Flitwick Moor, plentiful in wetter
parts!†*; *S.B.P.L.*: marsh near Woodcock's, Fancott; *H. B. Souster*:
sparingly, Westoning Moor; Biggleswade Common.
LEA: Biscot; *S.B.P.L.*: Limbury; New Mill End; *W.F.B.*
First record: Abbot, 1798.

E. latifolium Hoppe (*E. paniculatum* Druce), C.T.W. 136.1.3, was recorded for v.c.
30 in *Comital Flora*, but I can find no evidence to support its inclusion. Recorded
for all neighbouring counties except Bucks.

CAREX L.

I have had considerable help with this genus from E. Nelmes, who
has checked all the available material and made visits to the county.

Carex pseudocyperus L. C.T.W. 138.15.14.   *Cyperus Sedge*
Marshy places   Native
Very rare: recorded for all neighbouring counties.
OUZEL: side of Battlesden Lake, in limited quantity!*; *E.M.-R. in
B.E.C. 1943-4 Rep.* 812.

C. riparia Curt. C.T.W. 138.15.19.   *Great Pond-sedge*
Riversides, etc.   Native
Frequent: recorded for all neighbouring counties.
H.S. 13.
NENE: Hinwick Lodge.
OUSE: frequent by the river.
KYM: Melchbourne Park.
OUZEL: Leighton Buzzard*; Salford Mill*; Cow Common.
IVEL: Southill Lake; *J.McL., L.M.*: Cadwell; Campton; *J.E.L.,
F.F.B.*: Shefford!*; *C.C.*: Arlesey; *J.E.L., Diary*: Flitwick Moor.
CAM: Dunton.
LEA: New Mill End!†*; *Herb. J.S.*
First record: Abbot, 1798, but more certainly J. McLaren, 1864,
as Abbot did not distinguish this from the following species.

C. acutiformis Ehrh. C.T.W. 138.15.20.   *Lesser Pond-sedge*
Marshy places   Native
More frequent than the preceding. Abbot's specimen of *C. acuta* is
represented by this. Recorded for all neighbouring counties.
H.S. 39a, 40, 42, 43.
NENE: Hinwick*.
OUSE: Bedford!*; *J.H. Notes*: frequent in marshy places by the river.

DD

OUZEL: Totternhoe, as *C. stricta**; *C.C.*: Milton Bryan*; *E.M.-R.*: King's Wood.
IVEL: Stotfold; *R. A. Pryor in J. Bot. 1876*, 24: Flitwick; *V.C.H.*: Barton!; Tingrith; Radwell; Southill; *J.E.L., F.F.B.*: Shefford!*; *C.C.*: Moors Plantation; marsh, foot of Cooper's Hill.
CAM: Wrestlingworth.
LEA: Marslets†; New Mill End!*; *S.B.P.L.*
First record: R. A. Pryor, 1876.

C. vesicaria L. was recorded by McLaren, and subsequently by Saunders, in error: see author's note in *B.E.C. 1943–4 Rep.* 812.

C. rostrata Stokes C.T.W. 138.15.15.   *Bottle Sedge*
    C. ampullacea Gooden., *C. inflata* auct.
Marshes   Native
Frequent in marshy places on acid soils. Recorded for all neighbouring counties except Hunts.
H.S. 36, 39a, 40.
OUZEL: Mermaid's Pond!*; Heath and Reach; *S.B.P.L.*: Totternhoe; *C.C.*
IVEL: Ampthill; Potton; *F.B.*: in a marshy meadow near Toddington; *E. Forster in Bot. Guide*: Flitwick, frequent on the Moor!†*; *J.McL., B.M.*: Shefford, as *C. vesicaria*; *J.McL., L.M.*: Westoning; *C.C.*: marshy ground, Ridgmont*; Tingrith Park*.
LEA: Marslets†; *S.B.P.L.*
First record: Abbot, 1798.

C. hirta L. C.T.W. 138.15.31.   *Hammer* or *Hairy Sedge*
Marshes, damp meadows, etc.   Native
One of the more common of our sedges, but variable: the form *hirtiformis* (Pers.) Kunth has been found at Tempsford* and Cow Common*. Recorded for all districts except Colne, and for all neighbouring counties.
H.S. 8, 9, 10a, 10c, 11, 14b, 19a, 21, 37, 56a, 57b, 85, [53b].
First record: Abbot, 1795.

C. pendula Huds. C.T.W. 138.15.21.   *Pendulous Sedge*
Woods   Native
Frequent in woods on the heavier clay soils in the north and middle of the county; rare elsewhere. Recorded for all neighbouring counties.
H.S. 19a, 34a, 34b, 42.
OUSE, KYM, OUZEL and IVEL: common in all likely woods.
LEA: Luton Hoo; *J.S. in Bot. Rec. Club 1884–6 Rep.*: Stockwood!; *J. Cowley.*
First record: Abbot, 1798.

C. sylvatica Huds. C.T.W. 138.15.11.   *Wood Sedge*
Woods   Native
Common: recorded for all districts and neighbouring counties.
H.S. 3a, 3b, 18, 19b, 32–34a, 51, 52, 53a, 55b, 77b, 78, 79b, 83.
First record: Abbot, 1798.

C. strigosa Huds. C.T.W. 138.15.22.
Woods   Native
Rare: *E. Bot.* tab. 944, March 1802, was drawn from a specimen
sent by Hemsted. Recorded for all neighbouring counties except
Hunts.
OUSE: Putnoe Wood; Renhold; *F.B.*
OUZEL: Nun Wood*.
First record: Abbot, 1798.

C. binervis Sm. C.T.W. 138.15.5.   *Green-ribbed Sedge*
Marshes   Native
Rare: a species of acid bogs. Records by Saunders from Markham
Hills in *Bot. Rec. Club 1883 Rep.*, and from Biscot in *Bot. Rec. Club
1884–6 Rep.*, and McLaren from Stevington, are all probably
errors for *C. distans*, which is recorded from all three stations.
Recorded for Herts., Bucks. and Northants.
OUZEL: New Wavendon Heath, v.c. 24 [Beds.]!*; *P.T. in Watsonia I
(1949)* 57.

C. distans L. C.T.W. 138.15.2.   *Distant Sedge*
Marshes   Native
Rare: a species of marshes on alkaline soils. Recorded for all neigh-
bouring counties.
H.S. 10b, 56a.
OUSE: Stevington!*; *F.B.*: water meadows, Eaton Socon*; *E.M.-R.*!
OUZEL: Totternhoe, as *C. binervis*, it is plentiful both on the Mead
and on Cow Common!†*; *Herb. J.S.*
IVEL: Markham Hills†; Chalton†, as *C. fulva*; *Herb. J.S.*: Pullox-
hill*; *C.C.*: Fancott*; *A. J. Wilmott*!: Tingrith; *M. Brown in W.F.S.
Diary.*
LEA: Biscot, as *C. fulva*†*; *Herb. J.S.*
First record: Abbot, 1798.
　　C. distans × lepidocarpa
OUZEL: Totternhoe Mead, det. *C. fulva* by A. Bennett†*; *Herb. J.S.*: Cow
Common*.

C. hostiana DC. (*C. fulva* auct.), C.T.W. 138.15.4, was recorded by Saunders in
S.B.P.L. from Totternhoe and Biscot in error. Recorded for all neighbouring
counties except Hunts.

C. lepidocarpa Tausch C.T.W. 138.15.7.   *Yellow Sedge*
Marshes   Native
Rare: a species of marshes on alkaline soils. Recorded for all
neighbouring counties except Hunts.
H.S. 56a.
OUZEL: Totternhoe Mead, as *C. flava*†*; *Herb. J.S.*: Cow Common*;
E.M.-R.!
First evidence: Saunders, 1882.

C. demissa Hornem. C.T.W. 138.15.8.
    C. tumidicarpa Anderss.
Marshes   Native
Rare: a species of acid bogs. It is possible that Abbot's records
of *C. flava* from Ampthill and Stevington should be referred to this.
Recorded for all neighbouring counties.
OUZEL: Aspley, as *C. flava*; *J.McL.*, *L.M.*: Heath and Reach, det.
C. flava by Arthur Bennett†; *Herb. J.S.*: Stockgrove*; *R. B. Drum-
mond*!
First evidence: J. McLaren, 1864.

C. serotina Mérat (*C. oederi* auct.), C.T.W. 138.15.9. I cannot find the basis of
the record of this for v.c. 30 in *Comital Flora*. A. J. Wilmott, however, named as
C. oederi a specimen in Herb. J. McLaren, British Museum, which was labelled
'*C. flava*, Flitwick, Potton'. It is recorded for all neighbouring counties except
Cambs.

C. caryophyllea Latourr. C.T.W. 138.15.35.   *Vernal* or *Spring Sedge*
    C. praecox auct.
Pastures   Native
Frequent and well distributed, a species of alkaline and neutral
soils it is common on the chalk hills. Recorded for all districts
except Kym, and for all neighbouring counties.
H.S. 59.
First record: Abbot, 1798.

C. montana L. was recorded in *Flora Bedfordiensis* from Clapham Park, but Abbot's
herbarium specimen represents *C. pilulifera*. It is unlikely that *C. montana* would
appear in the county, or that *C. pilulifera* would grow at Clapham Park.

C. pilulifera L. C.T.W. 138.15.33.   *Pill-headed Sedge*
Heaths, heathy woods, etc.   Native
Replacing *C. caryophyllea* on acid soils, where it is frequent. Recorded
for all neighbouring counties except Hunts.
H.S. 28.
OUZEL: Heath and Reach, common in many parts of the parish!†*;
S.B.P.L.: Woburn, on heathy pastures in the Park!; *V.C.H.*: rides,
Aspley Wood*.
IVEL: Eversholt; *J. Hemsted in F.B.*: Clophill; *J.McL.*, *L.M.*:
Sandy Warren; Rowney Warren; *J.E.L.*, *J.Bot.*
LEA: Chiltern Green; *F.F.B.*: Kidney Wood*.
COLNE: Deadmansey Wood*; *A. J. Wilmott*!
First record: Abbot, 1798.

C. flacca Schreb. C.T.W. 138.15.30.   *Glaucous Sedge*
    C. diversicolor Crantz p.p., *C. recurva* Huds.
Pastures, claypits, etc.   Native
Our most common sedge, and especially abundant on calcareous
soils. Recorded for all districts and neighbouring counties.
H.S. 3a, 10a, 19a, 21, 56b, 59, 60, 62, 65–69, 75.
First record: Abbot, 1798.

C. pallescens L. C.T.W. 138.15.23.   *Pale Sedge*
Woods   Native
Not infrequent in woods on light soils. It was considered rare by
Abbot. Recorded for all neighbouring counties.
OUSE: Putnoe Wood; *F.B.*
KYM: West Wood*.
OUZEL: King's Wood!†*; *J.S. in Bot. Rec. Club 1880 Rep.*
IVEL: Warden Little Wood; *J.McL., L.M.*: Cainhoe Park Wood*;
C.C.: Warden Great Wood; *J.E.L., B.E.C.*: Daintry Wood; *P.T.*:
Washer's Wood.
CAM: Cockayne Hatley Wood.
LEA: Kidney Wood.
COLNE: Deadmansey Wood*; *A. J. Wilmott*: Oldhill Wood.
First record: Abbot, 1798.

C. panicea L. C.T.W. 138.15.25.   *Carnation Sedge*
Marshes   Native
Not infrequent: considered common by Abbot, but his herbarium
specimen is *C. remota*. Recorded for all neighbouring counties.
H.S. 10a.
OUSE: Marston; *Ada Stimson in J.H. Notes*: Stevington; *W. Durant.*
OUZEL: Totternhoe, sparingly on the Mead and Cow Common!*;
C.C.
IVEL: Potton Marshes; *J.McL., L.M.*: Eversholt; *J.McL., B.M.*:
Toddington†; *S.B.P.L.*: Markham Hills†*; *Herb. J.S.*: Fancott*;
P.T.!: Flitwick Moor*.
LEA: Marslets; Limbury; *S.B.P.L.*: Leagrave Marsh; *J.S. in Herb.
Beds. Nat. Hist. Soc.*
COLNE: Pepperstock; *M. Brown in W.F.S. Diary.*
First record: Abbot, 1798.

C. limosa L. was recorded in *Hillhouse List, 1876*, on the authority of A. Poulton, and
by Ada Stimson, from Marston, in Hamson's Notes. Both records are to be
doubted.

C. elata All. C.T.W. 138.15.45.   *Tufted Sedge*
Marshes   Native
Rare: recorded for all neighbouring counties.
OUSE: Bromham; *Abbot in letter to J. E. Smith*, 3 Sept. 1802.
OUZEL: Totternhoe, det. F. G. Baker; *F.F.B.*
First record: Abbot, 1802.

C. acuta L. C.T.W. 138.15.46.   *Slender-spiked Sedge*
  C. gracilis Curt.
Riversides   Native
Probably not infrequent by the Ouse. Considered common by
Abbot, but he may have confused it with *C. acutiformis*. Recorded
for all neighbouring counties.
OUSE: Fenlake†; *J.McL. in Herb. J.S.*: Sharnbrook; *E.M.-R.*:
Radwell!*; *J. P. M. Brenan and N. Y. Sandwith*: Bedford*.
OUZEL: Totternhoe!*; *C.C.*
First record: Abbot, 1798; more certainly, *Newbould MS.* 1874.

C. nigra (L.) Reichard C.T.W. 138.15.49.   *Common Sedge*
   C. goodenowii Gay, *C. cespitosa* auct.
Marshes   Native
Not infrequent in wet places on acid soils: considered common by
Abbot. Recorded for all neighbouring counties.
H.S. 35, 37–39a, 40.
OUZEL: Totternhoe Mead*; *J.S.*: Cow Common*; Aspley Heath;
New Wavendon Heath, v.c. 24 [Beds.]*.
IVEL: Flitwick Moor!*; *J.McL.*, *B.M.*
LEA: Limbury†*; Biscot; New Mill End; *F.F.B.*: Marslets*; *J.S.*
First record: Abbot, 1798.

C. ovalis Gooden. C.T.W. 138.15.71.   *Oval Sedge*
   C. leporina auct.
Marshes   Native
Frequent in marshes on acid soils. Recorded for all neighbouring
counties.
H.S. 37–39a, 41.
OUSE: marsh, Ampthill Park; *H. B. Souster*!
OUZEL: Aspley Heath!; *S.B.P.L.*: Heath and Reach!†*; *Herb. J.S.*:
Horsemoor Farm; New Wavendon Heath, v.c. 24 [Beds.]*;
Baker's Wood*.
IVEL: Potton Marshes; *F.B.*: Clophill; *J.McL.*, *L.M.*: Flitwick
Moor!*; *S.B.P.L.*: Steppingley; *C.C.*: marsh, foot of Cooper's Hill*;
V. H. Chambers!: Westoning Moor.
LEA: Limbury; *F.F.B.*: Chiltern Green†*; *Herb. J.S.*
First record: Abbot, 1798.

C. echinata Murr. C.T.W. 138.15.67.   *Star Sedge*
   C. stellulata Gooden.
Marshes   Native
Rare, and limited to a few marshy places on the Lower Greensand.
Recorded for all neighbouring counties.
H.S. 35.
OUZEL: Heath and Reach; Aspley Heath Wood, probably New
Wavendon Heath, v.c. 24 [Beds.], where it is locally plentiful!*;
S.B.P.L.: near Lower Drakelow Pond*; Aspley Heath.
IVEL: Flitwick†; *S.B.P.L.*
First record: Saunders, 1885.

C. remota L. C.T.W. 138.15.68.   *Distant-spiked* or *Remote Sedge*
Woods, shady places, etc.   Native
Frequent throughout the county. Recorded for all districts except
Nene, and for all neighbouring counties.
H.S. 19a, 42, 53b.
First record: Abbot, 1798.

C. curta Gooden. C.T.W. 138.15.69. *White Sedge*
  C. canescens auct.
Marshes Native
Rare, and limited to a few boggy places in association with the
Lower Greensand. Recorded for Bucks. and Cambs.
H.S. 36.
OUZEL: Mermaid's Pond, where it is plentiful!†*; *Herb. J.S.*:
Heath and Reach; *J.S. in J. Bot. 1883.*
IVEL: Flitwick, plentiful in and near Folly Wood!†*; *J.S. in Bot.
Rec. Club 1880 Rep.*: Flitton; *J.McL.*, *B.M.*: Little Park, Ampthill;
near Lower Gravenhurst Church; *C.C.*
First record: Abbot, in letter to J. E. Smith, 3 Sept. 1802.

C. otrubae Podp. C.T.W. 138.15.55. *Fox Sedge*
  C. vulpina auct.
Marshes, ditches, etc. Native
Common: the true *C. vulpina* L. has not been recorded for the county.
Recorded for all districts except Colne, and for all neighbouring
counties.
H.S. 3a, 7, 9, 13a, 19a, 34a, 58a.
First record: Abbot, 1798.

  C × pseudoaxillaris Richt. (*C. otrubae × remota*)
  C. axillaris Gooden. non L.
  Recorded for all neighbouring counties except Hunts.
  OUSE: Fenlake†*; *J.McL.*, *L.M.*

C. spicata Huds. *Spiked Sedge*
  C. muricata L. p.p., *C. contigua* Hoppe C.T.W. 138.15.64.
Rough pastures, etc. Native
Frequent: a species of alkaline and neutral soils, which is well
distributed in the county. Abbot listed both *C. spicata* and *C.
muricata* as common, but it is difficult to know what he intended.
Recorded for all districts and neighbouring counties.
H.S. 57b.
First record: Abbot, 1798.

C. pairaei F. W. Schultz C.T.W. 138.15.65. *Prickly Sedge*
Heathy pastures Native
Less common than the preceding, but replacing it on acid soils.
Recorded for all neighbouring counties.
OUZEL: Baker's Wood!; *E.M.-R.*!: Rushmere*.
IVEL: Pulloxhill, as *C. vulpina**; *C.C.*: Tingrith*; Rowney Warren*;
Cainhoe*; *E.M.-R.*: Warren Wood, Clophill*; *P.T.*
LEA: Luton Hoo*.
First certain record: J. E. Little, 1919.

C. polyphylla Kar. & Kir. C.T.W. 138.15.63.
Woods, roadside verges, etc.
A species closely allied to *C. divulsa*, the distribution of which is not
yet fully known. Recorded for Bucks., Cambs. and Northants.

IVEL: Thrift Wood*.
COLNE: Whipsnade*.
First record: the author in *Watsonia I* (*1949*) 57.

C. divulsa Stokes C.T.W. 138.15.62.   *Grey Sedge*
Woods, roadside verges, etc.   Native
Frequent: recorded for all neighbouring counties.
OUSE: Clapham; *F.B.*: Putnoe, where it was still found by
McLaren†; *Abbot's annotated F.B.*: Oakley; *J.H. Notes*: Roxton;
Sheerhatch Wood*; King's Wood*.
OUZEL: Aspley, as *C. canescens*; *J. McL., L.M.*: King's Wood!*;
F.F.B.: Stanbridge*; Woburn Park*.
IVEL: Sundon Woods*; *F.F.B.*
LEA: Farley Hill; *J.S. in J. Bot. 1889*: Dunstable Road†; *Herb. J.S.*:
Luton Hoo*.
COLNE: between Zouche's Farm and Caddington; *Fl. Herts.*: Folly
Wood; Kensworth; Long Wood, v.c. 20 [Beds.]*.
First record: Abbot, 1798.

C. paniculata L. C.T.W. 138.15.52.   *Tussock or Panicled Sedge*
Marshes   Native
Frequent in the larger marshes of the county. Recorded for all
neighbouring counties.
H.S. 38, 39a, 41–43.
OUSE: Holywell, Millbrook; *H. B. Souster*!
OUZEL: Heath and Reach; *V. H. Chambers*: Husborne Crawley;
Milton Bryan.
IVEL: Southill Lake; *J.McL., L.M.*: Flitwick Moor, forming large
tussocks!†*; *J. McL. in Bot. Rec. Club 1884–6 Rep.*: Stotfold; *J.E.L.*,
F.F.B.: Steppingley*; *C.C.*: Westoning Moor, forming large tussocks;
Tingrith; Fancott; Harlington; Silsoe; Flitton Moor; Beckerings-
park Moor; Moors Plantation; marsh, foot of Cooper's Hill.
LEA: Leagrave Marsh; East Hyde.
First record: Abbot, 1798.

C. paniculata × *remota* (*C.* × *boenninghausiana* Weihe) is recorded for the county,
with no details, in *Flora of Buckinghamshire*. Recorded for Herts.

C. diandra Schrank was recorded with no details in *Hillhouse List, 1876*. Failing
further evidence the record is to be doubted. Recorded for Herts. and Cambs.

C. disticha Huds. C.T.W. 138.15.57.   *Brown Sedge*
Marshes   Native
One of our more common sedges, but considered rare by Abbot.
Recorded for all districts except Kym, Cam and Colne, and for all
neighbouring counties.
H.S. 9, 14b, 15, 37, 56a, 56b, 74b, 85.
First record: Abbot, 1798.

C. divisa Huds. C.T.W. 138.15.59.   *Divided Sedge*
Rough pastures   Native
Rare: a species usually limited to maritime pastures.
OUZEL: Woburn Park*; *A. B. Jackson in J. Bot. 1920*, 91.

C. pulicaris L. C.T.W. 138.15.76. *Flea Sedge*
Marshes Native
Probably extinct: a species of especial interest, as it grew on one of
our chalk hills with other marsh plants. Recorded for all neigh-
bouring counties.
IVEL: Eversholt; *F.B.*: Streatley, where it grew on a chalk hill with
Parnassia palustris and *Pinguicula vulgaris**; *C.C. in J. Bot. 1898*:
Flitwick Moor—it would be interesting to know more of this record;
V.C.H.
First record: Abbot, 1798.

C. dioica L. C.T.W. 138.15.77. *Dioecious Sedge*
Marshes Native
Extinct: recorded for Cambs. and Northants.
IVEL: Ampthill Bogs; *F.B.*: near Toddington; *E. Forster in Bot.
Guide.*

GRAMINEAE

I have received considerable assistance from C. E. Hubbard with
the grasses. He has named all my specimens and all the doubtfully
named ones in the Bedfordshire herbaria. In addition, he has visited
the county and made many records.

PHALARIS L.

Phalaris arundinacea L. C.T.W. 139.52.1. *Reed Canary-grass*
Riversides, pondsides, etc. Native
Common: var. **picta** L., the Ribbon Grass of gardens, has appeared,
as a garden escape, on a rubbish dump at Luton*. Recorded for all
districts and neighbouring counties.
H.S. 6, 8, 13, 21, 39a, 84b.
First record: Abbot, 1798.

ANTHOXANTHUM L.

Anthoxanthum odoratum L. C.T.W. 139.51.1. *Sweet Vernal-grass*
Pastures, woodland rides, heaths, etc. Native
Common especially on light soils. It is variable in its size and
pubescence. The name var. *villosum* Lois was applied by C. E.
Hubbard to a hirsute form found by him at Heath and Reach.
Recorded for all districts and neighbouring counties.
H.S. 19b, 23, 28–30b, 37, 38, 39b, 40, 66b, 74a, 75, 76, 79a, 80a.
First record: Abbot, 1798.

A. puelii Lecoq & Lamotte C.T.W. 139.51.2. *Early Vernal-grass*
   A. aristatum auct.
Arable fields, etc. Colonist
Rare: recorded for all neighbouring counties except Hunts.
OUZEL: Heath Wood†; *Mrs. J. Tindall in W.F.B.*
IVEL: Everton; *J.E.L., J. Bot.*
First record: Mrs. Tindall, 1897.

ALOPECURUS L.

Alopecurus pratensis L. C.T.W. 139.48.2. *Meadow Foxtail*
Rough pastures   Native
Common, but usually absent on well-established downland.
Recorded for all districts and neighbouring counties.
H.S. 7–9, 13, 31a, 37, 41, 55b, 83.
First record: Abbot, 1798.

A. myosuroides Huds. C.T.W. 139.48.1. *Field* or *Slender Foxtail*,
A. agrestis L.   *Black Twitch*
Arable land, waste places, etc.   Colonist
Common: recorded for all districts and neighbouring counties.
H.S. 4, 12, 71, 72, 86.
First record: Abbot, 1798.

A. geniculatus L. C.T.W. 139.48.3.   *Marsh Foxtail*
Marshes, pondsides, etc.   Native
Common: recorded for all districts and neighbouring counties.
H.S. 9, 14b, 56a, 84b.
First record: Abbot, 1798.

A. aequalis Sobol. C.T.W. 139.48.4.   *Orange Foxtail*
A. fulvus Sm.
Marshes, pondsides, etc.   Native
Rare: limited with us to marshes on acid soils. Recorded for all
neighbouring counties.
OUZEL: near Heath and Reach, it is locally plentiful at Rushmere!*;
V.C.H.: marsh to north of Leighton Buzzard; C. E. Hubbard!
IVEL: marsh, Long Lane, Tingrith.
First record: G. C. Druce, 1904.

MILIUM L.

Milium effusum L. C.T.W. 139.49.1.   *Wood Millet*
Woods   Native
Frequent: recorded for all neighbouring counties.
H.S. 32,78.
OUSE: Souldrop; Wilstead Wood; Sheerhatch Wood*; Roxton;
King's Wood; Woodcraft.
KYM: West Wood; Swineshead Wood.
OUZEL: Salford Wood; Palmer's Shrubs; King's Wood; Bush
Pastures; Apesfield Spring.
IVEL: Warden Wood; *J.McL., L.M.*: Washer's Wood; Flitwick
Wood; Potton Wood*.
CAM: Cockayne Hatley Wood.
LEA: Luton Hoo†; *Herb. J.S.*: East Hyde; Whitehill Wood.
COLNE: Deadmansey Wood; Oldhill Wood; Kensworth; Folly
Wood.
First record: Abbot, 1798.

PHLEUM L.

Phleum pratense L. C.T.W. 139.47.2. *Cat's-tail, Timothy*
Pastures, roadsides, etc. Native
Frequent: often sown in leys. Seen at its best as a cornfield weed, in pasture recently turned to arable. Recorded for all districts and neighbouring counties.
H.S. 6, 11, 13, 41, 56b, 85.
First record: Abbot, 1798.

P. nodosum L. C.T.W. 139.47.1. *Smaller Cat's-tail*
Pastures Native
More common than the preceding and more often a constituent of established pastures. Recorded for all districts and neighbouring counties.
H.S. 2a, 4, 7–9, 19a, 29, 61, 64, 72.
First record: Abbot, in letter to J. E. Smith, 3 Sept. 1802.

P. phleoides (L.) Karst. C.T.W. 139.47.4. *Purple-stalked Timothy*
Pastures Native
Rare: our Bedfordshire station is the western limit of the species as a native in Britain. Recorded for Herts. and as a casual for Bucks.
IVEL: Wilbury Hills, locally plentiful over a limited area!*; *T. B. Blow and H. Grove in J. Bot. 1876*, 244.

AGROSTIS L.

Agrostis stolonifera L. C.T.W. 139.40.5. *Creeping Bent-grass,*
  A. alba auct. *Fiorin*
Pastures, heaths, etc. Native
Common, but variable: var. **stolonifera**, to be found usually in drier situations, has been found in all districts except Colne, and var. **palustris** (Huds.) Farw., found generally in wetter situations, such as pond margins, has been seen in all districts except Lea. Recorded for all neighbouring counties.
H.S. 3a, 7, 10a, 13, 16, 19a, 21, 26, 31a, 37, 39a, 41, 58b, 61, 68, 69, 74a, 84a, 85a, [40].
First record: Abbot, 1798.

A. tenuis Sibth. C.T.W. 139.40.3. *Fine* or *Common Bent-grass*
  A. capillaris auct.
Pastures, heaths, etc. Native
Common, and growing more plentifully on light soils than *A. stolonifera*. A diseased condition formerly named *A. pumila* L. has been found by E.M.-R. near King's Wood, Heath and Reach. Recorded for all districts except Cam, and for all neighbouring counties.
H.S. 2a, 10c, 15, 19a, 21, 22b–31a, 37, 38, 41, 46–49, 66b, 74a, 75–77a, 79a, 80a.
First record: Abbot, 1798.

A. gigantea Roth C.T.W. 139.40.4.   *Black Bent-grass*
A. nigra With., *A. alba* auct.
Arable land, disturbed soils, etc.   Colonist
Frequent, and usually a weed of arable land. It is a variable species;
var. *ramosa* (Gray) Philipson has been found in a ride in Aspley
Wood* and an awned form at Pepperstock. For a study of the
behaviour of the species in the county see H. H. Mann in *Annals of
Applied Biology*, 1949, 273. Recorded for all districts except Cam,
and for all neighbouring counties.
H.S. 25.
First record: the author in *B.E.C. 1941–3 Rep.* 813.

A. canina L. C.T.W. 139.40.2.   *Bent-grass*
Heaths, heathy places, etc.   Native
Frequent on acid soils. Recorded for all neighbouring counties.
H.S. 23, 25, 35, 74b.
OUZEL: frequent on the Lower Greensand.
IVEL: Ampthill Warren; Sandy Heath; *F.B.*: Rowney Warren;
J.McL., *L.M.*: Sandy, f. *mutica* (Gaud.); *J.E.L.*, *J. Bot.*: Rowney
Warren; *J.E.L.*, *J. Bot.*: Clophill*.
LEA: Chiltern Green Common*.
COLNE: Studham Common; *C. E. Hubbard*.
First record: Abbot, 1798.
    Var. **canina** (var. *fascicularis* (Curt.) Sincl.)   *Velvet Bent-grass*
Found generally in moist places and woodland rides
OUZEL: Heath and Reach, with f. *mutica**; *C. E. Hubbard*!: New Wavendon
Heath, v.c. 24 [Beds.]*.
    Var. **arida** Schlecht.   *Brown Bent-grass*
Found generally on acid heaths
OUZEL: Aspley Wood*.

                    CALAMAGROSTIS Adans.

Calamagrostis epigejos (L.) Roth C.T.W. 139.39.1.
Arundo epigejos L.   *Wood Smallreed, Bushgrass*
Woodland verges, rough pastures, etc.   Native
Common on the heavier clay soils in the north and middle of the
county. The furthest south I have seen it in the county is at Tottern-
hoe Mead. Abbot considered it rare, but his herbarium specimen
is *C. canescens*, which could have grown at the station he cited
(Maulden Wood). Saunders also considered it rare. Recorded for
all districts except Lea and Colne, and for all neighbouring counties.
H.S. 19a, 35.
First record: Abbot, 1798; more certainly, Warden Little Wood,
J. McLaren, 1864.

C. canescens (Web.) Roth C.T.W. 139.39.2.   *Purple Smallreed*
    Arundo calamagrostis L., *C. lanceolata* Roth, *C. calamagrostis* (L.) Karst.
A species of wet woods and fens and of doubtful occurrence in the county. Abbot
recorded it from Sheerhatch Wood, an unlikely station, but see note to previous
species. Records by McLaren from Warden Wood in *J. Bot. 1884* and C. Crouch
from Chicksands are, failing specimens, to be doubted. Recorded for all neighbouring counties except Bucks.

APERA Adans.

Apera spica-venti (L.) Beauv. C.T.W. 139.41.1.   *Loose Silky Bent-grass*
  Agrostis spica-venti L.
Arable fields, waste places, etc.   Colonist or Casual
Rare: probably introduced recently as a wool adventive. Recorded for all neigh-
bouring counties except Northants.
OUSE: Ford End Farm; *F.B.*
IVEL: Streatley†*; *S.B.P.L.*: railway siding, Flitwick*.
LEA: Luton Downs; *F.B.*
First record: Abbot, 1798.

AIRA L.

Aira caryophyllea L. C.T.W. 139.35.2.   *Silvery Hair-grass*
Heaths, heathy places, etc.   Native
Frequent on the Lower Greensand and Clay-with-Flints, but
appearing also on gravelly soils. Recorded for all neighbouring
counties.
H.S. 25, 44, 48.
OUSE: gravel pit, Eaton Socon*; *P.T.*!
OUZEL: Husborne Crawley; *F.B.*: Aspley; *J.McL.*, *B.M.*: Heath
and Reach!†*; *Herb. J.S.*: King's Wood.
IVEL: Ampthill!; *F.B.*: Sandy!; *J. Pollard*: Rowney Warren;
Maulden!; *J.E.L.*, *F.F.B.*: Barton Hills, on glacial gravel!; *P.T.*:
Flitwick*.
COLNE: Pepperstock†; *S.B.P.L.*: Whipsnade Common!*; *V. H.
Chambers*: Deadmansey Wood.
First record: Abbot, 1798.

A. praecox L. C.T.W. 139.35.1.   *Early Hair-grass*
Heaths, heathy ground, etc.   Native
Frequent on the Lower Greensand, rare on the Clay-with-Flints.
E. Bot. tab. 1296, April 1804, was drawn from a specimen sent by
Abbot. Recorded for all neighbouring counties.
H.S. 22b–27, 46, 47.
OUZEL: Aspley Heath†; *Herb. J.S.*: Heath and Reach*; *C. E.
Hubbard*!: Leighton Buzzard; New Wavendon Heath, v.c. 24
[Beds.]*; King's Wood.
IVEL: Ampthill Heath; Rowney Warren!; *F.B.*: Sandy; *J. Pollard*:
Sandy Heath!; Sutton; Maulden!; *J.E.L.*, *F.F.B.*: Galley Hill!;
Portobello Farm; *J.E.L.*, *Diary*: Flitwick; Clophill; Cooper's Hill;
Bunker's Hill.
COLNE: Pepperstock*; *S.B.P.L.*
First record: Abbot, 1798.

DESCHAMPSIA Beauv.
Deschampsia cespitosa (L.) Beauv. C.T.W. 139.34.1.
  Aira cespitosa L.   *Tufted Hair-grass*
Rough pastures, roadside verges, woodland rides, etc.   Native
Common: a handsome species, var. **parviflora** (Thuill.) Coss. &
Germ. has been found by E.M.-R. in West Wood and by the author

in Cockayne Hatley Wood*. Recorded for all districts and neighbouring counties.

H.S. 2a–3a, 9, 10c, 13, 19a, 21, 32, 34a, 37, 38, 41, 51a, 52, 53a, 55a, 55b, 56b, 57b, 58b, 74b, 77a, 78, 79b, 80b, 81, 83, 85.

First record: Abbot, 1798.

D. flexuosa (L.) Trin. C.T.W. 139.34.3.   *Wavy Hair-grass*
 Aira flexuosa L.

Heaths, heathy pastures, etc.   Native

Frequent on the Lower Greensand, and often the dominant species in places where the soil has been previously disturbed. In these conditions it is very attractive. Recorded for all neighbouring counties except Hunts.

H.S. 22a, 24, 27, 31a, 39b, 47, 49.

OUZEL and IVEL: common in all likely situations.

LEA: on site of disused camp, Luton Hoo*.

First record: Abbot, 1795.

HOLCUS L.

Holcus lanatus L. C.T.W. 139.33.1.   *Yorkshire Fog*

Rough pastures, waste ground, etc.   Native

Common: recorded for all districts and neighbouring counties.

H.S. 1b, 2a, 3a, 9–11, 14b–16, 19a, 22b, 25, 26, 28, 29, 32, 37, 38–45a, 46, 48, 53a, 55b–57b, 58b, 61, 64, 68, 69, 74a–77a, 78—81, 83, 84a, 85.

First record: Abbot, 1798.

H. mollis L. C.T.W. 139.33.2.   *Creeping Soft-grass*

Heathy pastures, cleared woodland, etc.   Native

Not infrequent on the Lower Greensand and Clay-with-Flints. Recorded for all neighbouring counties.

H.S. 23, 26, 27, 30a, 46, 47, 49.

OUZEL: Aspley; *Plantae Bedford.*: King's Wood; *A. J. Wilmott!*

IVEL: Ampthill; Potton; *Plantae Bedford.*: Clophill; Rowney Warren!; *J.McL., L.M.*: Maulden!; Flitwick; Sandy; *J.E.L., F.F.B.*: Cooper's Hill; Galley Hill; Bunker's Hill; Sutton Fen.

LEA: Luton†; *Herb. J.S.*: Kidney Wood.

COLNE: Dunstable East Hill; *Fl. Herts.*: Studham; *C. E. Hubbard.*

First record: Abbot, 1795.

TRISETUM Pers.

Trisetum flavescens (L.) Beauv. C.T.W. 139.29.1.   *Yellow Oat*
 Avena flavescens L.

Pastures, roadsides, etc.   Native

Common: recorded for all districts and neighbouring counties.

H.S. 2a, 21, 29, 45a–46, 56b, 61, 64, 66b–69, 75, 76.

First record: Abbot, 1798.

AVENA L.

Avena fatua L. C.T.W. 139.30.1.   *Common Wild Oat*
Arable fields   Colonist
Common: a very variable species. Recorded for all districts except
Cam, and for all neighbouring counties.
First record: Abbot, 1798.
> Var. **fatua** (var. *pilosissima* Gray)
NENE: Wymington.
KYM: Thurleigh; *E.M.-R.*
IVEL: Harlington*.
> Var. **pilosa** Syme
OUSE: Eaton Socon; Tempsford*.
KYM: Keysoe.
OUZEL: Tilsworth*.
LEA: Luton*.
> Var. **glabrata** Peterm.
OUSE: Eaton Socon.
KYM: Keysoe.
IVEL: Harlington*; Sandy; Sundon*.
COLNE: Studham, v.c. 20 [Beds.]; *C. E. Hubbard.*

A. ludoviciana Durieu C.T.W. 139.30.2.   *Wild Oat*
Arable fields   Colonist
Possibly more common than the records indicate. Recorded for Northants.
NENE: Wymington; *J. P. M. Brenan!*
OUZEL: Tilsworth!*; *C. E. Hubbard in B.E.C. 1946–7 Rep. 319.*
CAM: Dunton*.
COLNE: near Long Wood, v.c. 20 [Beds.]*; *C. E. Hubbard.*
First record: C. E. Hubbard, 1947.

HELICTOTRICHON Bess.

Helictotrichon pratense (L.) Pilger C.T.W. 139.31.1.
    Avena pratensis L.   *Meadow Oat*
Pastures, railway banks, etc.   Native
Common on the Chalk and Limestone, and appearing less frequently
on calcareous soils elsewhere. Recorded for all districts except Cam,
and for all neighbouring counties except Hunts.
H.S. 2a, 41, 59, 62, 63b, 66a–67, 75.
First record: Abbot, 1798.

H. pubescens (Huds.) Pilger C.T.W. 139.31.2.   *Downy* or *Hairy Oat*
    Avena pubescens Huds.
Pastures, etc.   Native
Frequent on the Chalk and on calcareous soils. As widely distri-
buted as the preceding species, but less common. Recorded for all
districts except Cam, and for all neighbouring counties.
H.S. 65.
First record: Abbot, 1798.

ARRHENATHERUM Beauv.

Arrhenatherum elatius (L.) Beauv. ex J. & C. Presl C.T.W.
139.32.1.   *Oat-grass, False Oat*
    Avena elatior L., *Arrhenatherum tuberosum* F. W. Schultz
Roadside verges, sides of woods, etc.   Native

Common: the form *biaristatum* (Peterm.) Bertram has been observed
by C. E. Hubbard at Heath and Reach and by the roadside at
Studham, v.c. 20 [Beds.]*. Recorded for all districts and neighbouring counties.

H.S. 1b, 2a, 4, 10c, 13, 15, 21, 25, 27, 29, 31a, 32, 38, 39a, 41, 42,
44, 45a, 45b, 51b, 57a, 57b, 58b, 63a–64, 70b, 74a, 75, 85.
First record: Abbot, 1798.

SIEGLINGIA Bernh.

Sieglingia decumbens (L.) Bernh. C.T.W. 139.4.1. *Heath Grass*
    Festuca decumbens L.

Pastures   Native
Frequent on the Lower Greensand and rare on the Clay-with-
Flints. It appears also on the tops of chalk hills, where it is
probably an indication of leaching. Recorded for all neighbouring
counties.

H.S. 28, 37, 56b, 66b.
OUSE: Mouse's Pasture, near Bromham; *F.B.*
OUZEL: Woburn Park; *Hortus Gramineus Woburnensis*: in drying out
portions of Cow Common!*; *P.T.*: edge of marsh, Horsemoor
Farm*.
IVEL: Sandy; Everton; *Plantae Bedford.*: Warden Wilderness;
J.McL., *L.M.*: Eversholt; *J.McL.*, *B.M.*: Flitwick Moor†; *S.B.P.L.*:
Sharpenhoe Clappers, on chalk*; *E.M.-R.*: Knocking Hoe, on
chalk; *J. F. Hope-Simpson.*
LEA: Galley Hill, in clay pocket*; *J. F. Hope-Simpson.*
COLNE: Pepperstock*; *S.B.P.L.*
First record: Abbot, 1795.

PHRAGMITES Adans.

Phragmites communis Trin. C.T.W. 139.2.1.   *Reed*
    Arundo phragmites L., *P. phragmites* (L.) Karst.

Marshy places   Native
Frequent, especially in drying-out waterlogged places, as at
Flitwick Moor, where it is dominant in considerable areas.
Recorded for all districts except Colne, and for all neighbouring
counties.

H.S. 7, 13.
First record: Abbot, 1798.

CYNOSURUS L.

Cynosurus cristatus L. C.T.W. 139.15.1.   *Crested Dog's-tail*
Pastures   Native
Common: recorded for all districts and neighbouring counties.
H.S. 2a, 3a, 8, 21, 27, 28, 37, 56b, 57b, 62, 66b, 75, 76.
First record: Abbot, 1798.

C. echinatus L. C.T.W. 139.15.2.   *Rough Dog's-tail*
Waste places, etc.   Colonist
Rare, but often well established. Recorded for all neighbouring counties.
OUSE: Bedford; *J.H. Notes*: gravel pit, Eaton Socon, common over a relatively large area*.
IVEL: Ridgmont; Ampthill; *C.C. in J.E.L., B.E.C.*: Flitwick, regular in its appearance at sides of cart tracks†*; Deepdale.
LEA: roadside, Skimpot; *H. B. Souster*: roadsides, Leagrave, Luton.
First record: J. Hamson, 1916.

KOELERIA Pers.

Koeleria gracilis Pers. C.T.W. 139.28.1.   *Crested Hair-grass*
   K. britannica Domin, *Poa cristata* auct.
Pastures   Native
Common on the Chalk, frequent on the Oolitic Limestone and rare on sandy soils. Recorded for all neighbouring counties.
H.S. 56b, 59, 60, 62, 63b, 65, 66a, 67, 74a, 75.
NENE: railway embankment, Wymington*.
OUSE: Biddenham; *J.McL., L.M.*: Park Lane; *J.McL., B.M.*: pasture, Stevington.
OUZEL: common on the Chalk: roadside, Heath and Reach.
IVEL: common on the Chalk: Silsoe; *C.C.*: roadside, Clophill; wall, Potton Church.
LEA and COLNE: common on the Chalk.
First record: Abbot, 1798.

MOLINIA Schrank

Molinia caerulea (L.) Moench C.T.W. 139.3.1. *Purple Moor-grass*
   Melica caerulea L.
Heathy marshes   Native
Rare, and limited to marshy places on acid soils. Recorded for all neighbouring counties.
H.S. 35, 40.
OUZEL: Aspley; *S.B.P.L.*: New Wavendon Heath, v.c. 24 [Beds.]*.
IVEL: Ampthill, in limited quantity to north of Cooper's Hill!; *F.B.*: Potton Marshes; *J.McL., L.M.*: Flitwick, locally abundant near Folly Wood!†*; *S.B.P.L.*
First record: Abbot, 1798.

CATABROSA Beauv.

Catabrosa aquatica (L.) Beauv. C.T.W. 139.13.1.
   Aira aquatica L.   *Water Whorl-grass*
Ditches, wet meadows, etc.   Native
Rare: recorded for all neighbouring counties.
H.S. 14b.
NENE: marsh, Wymington.
OUSE: Cardington; *J.H. in Herb. Beds. Nat. Hist. Soc.*: water meadow, Willington; Eaton Socon.
OUZEL: Totternhoe; *F.F.B.*: marsh, Hockliffe; marsh, Tebworth.
IVEL: Potton; *Plantae Bedford.*: Biggleswade; *J.E.L., J. Bot.*:

Cadwell; *J.E.L.*, *F.F.B.*: Langford; *E.M.-R.*: Flitwick Moor*;
Henlow; meadow near Cainhoe Castle.
LEA: New Mill End; Biscot!; near Luton†*; *S.B.P.L.*: Leagrave
Marsh.
First record: Abbot, 1795.

MELICA L.

Melica uniflora Retz. C.T.W. 139.17.1.   *Wood Melick*
Woods   Native
Frequent in woods on the lighter soils, and often associated with
Asperula odorata. Recorded for all neighbouring counties.
OUSE: Hawnes and Renhold Woods; *Plantae Bedford.*: Twin Wood;
J.H.: Oakley; *J.H. in Herb. Beds. Nat. Hist. Soc.*: Hanger Wood;
A. W. Guppy: South Wood, Roxton; Sheerhatch Wood; King's
Wood; Marston Thrift.
OUZEL: Aspley Wood; *F.F.B.*: King's Wood*.
IVEL: Warden Wood; Flitwick†*; *Herb. J.S.*: Clophill; *C.C.*:
Washer's Wood.
LEA: Luton Hoo!; New Mill End Road!; *S.B.P.L.*: George's
Wood; Arden Dell Wood.
COLNE: Woodside!; *F.F.B.*: Whipsnade; Deadmansey Wood;
Oldhill Wood; Kidney Wood.
First record: Abbot, 1795.

DACTYLIS L.

Dactylis glomerata L. C.T.W. 139.14.1.   *Cock's-foot*
Roadsides, rough pastures, etc.   ? Native
One of our commonest grasses, but rarely found on well-established
downland. A slender form is found in woodland rides on light soils.
Recorded for all districts and neighbouring counties.
H.S. 1a–2a, 9, 11, 15, 19a, 21, 25–29, 31a, 32, 38, 42, 44, 45b, 46,
48, 55b, 56b–57b, 62, 64, 66b–70a, 74a, 75, 76, 78, 80a, 80b, 83, 85.
First record: Abbot, 1798.

BRIZA L.

Briza media L. C.T.W. 139.16.1.   *Quaking Grass*
Dry pastures   Native
Well distributed, but probably more common on the Chalk than
elsewhere. Recorded for all districts and neighbouring counties.
H.S. 1a, 1b, 3a, 56b, 59, 60, 62, 63a, 65, 66a–67, 75.
First record: Abbot, 1798.

POA L.

Poa annua L. C.T.W. 139.12.1.   *Annual Meadow-grass*
Arable land, roadsides, etc.
Common, but rarely a constituent of established pastures. It is a
variable species, of which perennial and biennial forms merit some
study. Recorded for all districts and neighbouring counties.
H.S. 6, 17, 18, 19a, 20a, 23, 26–28, 34a, 44, 48, 50, 53a, 53b, 54,
55b, 70a, 77a, 80a, 82, 83.
First record: Abbot, 1798.

P. nemoralis L. C.T.W. 139.12.7.   *Wood Meadow-grass*
Woods   Native
Frequent throughout the county, but considered rare by Abbot, who gave only one station for it. His herbarium specimen, named '*P. nemoralis*', is probably *P. compressa*. Recorded for all districts and neighbouring counties.
H.S. 19b, 31a, 70a.
First record: Abbot, 1798; more certainly, J. McLaren, 1864.

P. compressa L. C.T.W. 139.12.10.   *Flattened Meadow-grass*
Old walls, chalk quarries, arable land, etc.   Native
Frequent and well distributed. Recorded for all districts and neighbouring counties.
H.S. 64, 72, 73.
First record: Abbot, 1798.

P. pratensis L. C.T.W. 139.12.11.   *Meadow-grass*
   Subsp. **pratensis**
Pastures, roadsides, etc.   Native
Common, but very variable. It is frequently sown in leys. Recorded for all districts and neighbouring counties.
H.S. 2a, 3a, 9, 11, 12, 14b, 16, 21, 25, 26, 31a, 31b, 38, 41, 43, 46, 56b, 57b, 58b, 64, 69, 70a, 74b, 77a, 78, 82, 84a, 85, 86.
First record: Abbot, 1798.
   Subsp. **irrigata** (Lindm.) Lindberg f.   *Spreading Meadow-grass*
Marshes
Rare: C. E. Hubbard thinks that the records of *P. subcaerulea* Sm. from Flitwick Moor and the Ouzel district in *V.C.H.* may refer to this.
IVEL: Flitwick Moor!; *E.M.-R.*
   Subsp. **angustifolia** (L.) Lindberg f.
Roadsides, railway embankments, etc.
Frequent: Abbot recorded it from Clapham Lane but his herbarium specimen is *P. nemoralis*. I have found subsp. *angustifolia* in all districts.
H.S. 1a, 1b.

P. trivialis L. C.T.W. 139.12.12.   *Rough Meadow-grass*
Pastures, roadsides, arable fields, etc.   Native
Almost as common as *P. pratensis*. Extreme forms are found as weeds on dry arable land and in wet ditches. Recorded for all districts and neighbouring counties.
H.S. 4, 6, 7, 10a, 18, 19a, 20a, 32, 34a, 34b, 37, 39a, 52, 55b, 56a, 57b, 72, 79b, 80a, 83, 84a, 85.
First record: Abbot, 1798.

GLYCERIA R.Br.

Glyceria fluitans (L.) R.Br. C.T.W. 139.5.1. *Floating Sweet-grass,*
   Festuca fluitans L.   *Flote-grass*
Pondsides, marshy places, etc.   Native
Common: recorded for all districts and neighbouring counties.
H.S. 8, 9, 14b, 37, 56a, 74b.
First record: Abbot, 1798.

G. × pedicellata Towns. (*G. fluitans* × *plicata*) C.T.W. p. 1420.
Pondsides: recorded for Herts., Cambs. and Hunts.
H.S. 14b, 57a.
NENE: Podington; Hinwick*.
OUSE: water meadow, Eaton Socon.
KYM: Knotting; *E.M.-R.*!: pond, West Wood.
OUZEL: Totternhoe; *C. E. Hubbard*!: Houghton Regis; *E.M.-R.*: Well Head.
IVEL: Greenfield*.
LEA: Leagrave Common*.

G. plicata Fr. C.T.W. 139.5.2.   *Plicate Sweet-grass*
Pondsides, etc.   Native
Frequent: recorded for all neighbouring counties.
H.S. 14b, 34a, 57a, 84b.
NENE: Wymington.
OUSE: water meadow, Eaton Socon; Wilstead Wood.
OUZEL: Tilsworth; Heath and Reach*; *C. E. Hubbard*!: Milton
Bryan*; *E.M.-R.*: Totternhoe; Well Head.
IVEL: below Cadwell Bridge; *J.E.L.*, *F.F.B.*: Flitwick; Shefford;
E.M.-R.: Fancott; Greenfield*.
CAM: Cockayne Hatley.
LEA: Marslets, as *G. fluitans*†; *Herb. J.S.*: Leagrave Marsh*.
COLNE: Whipsnade!*; *E.M.-R.*
First record: *B.E.C. 1914 Rep.* 78, with no details.
First evidence: Herb. Saunders, 1883.

G. declinata Bréb. C.T.W. 139.5.3.   *Glaucous Sweet-grass*
Pondsides, etc.   Native
Rare: recorded also for Bucks. and Northants.
OUZEL: Heath and Reach*; *C. E. Hubbard*!
IVEL: Greenfield*; *E.M.-R.*
COLNE: Woodside.
First record: C. E. Hubbard and the author in *B.E.C. 1943–4 Rep.* 813.

G. maxima (Hartm.) Holmb. C.T.W. 139.5.4.   *Reed Sweet-grass*
    Poa aquatica L., *G. aquatica* (L.) Wahlb. non J. & C. Presl
Riversides, ditches, etc.   Native
Common by rivers and often the dominant plant in limited areas.
Recorded for all neighbouring counties.
H.S. 6–9, 13, 14a, 21, 38, 39a, 40, 41.
OUSE: common by the river; claypit, Brogborough.
KYM: Tilbrook, v.c. 30 [Hunts.].
OUZEL: river, Heath and Reach; *V. H. Chambers*: ditch, Tilsworth.
IVEL: Shefford!; Chicksands; Flitwick!; *J.E.L.*,*F.F.B.*: Biggleswade;
J.E.L., *Diary*: Shillington; *V. H. Chambers*: Silsoe; Gravenhurst;
Flitton Moor; Westoning Moor.
CAM: Cockayne Hatley; Dunton.
LEA: Marslets†; *Herb. J.S.*: East Hyde; Limbury.
First record: Abbot, 1798.

Puccinellia distans (L.) Parl. (*Poa distans* L.) was recorded by Abbot in *Flora
Bedfordiensis* from Clapham Lane. The record is to be doubted, although he had
a herbarium specimen correctly named as *Poa retroflexa* Curt. It is recorded for
Bucks. and Northants.

CATAPODIUM Link

Catapodium rigidum (L.) C. E. Hubbard[1]   *Fern Grass*
   Poa rigida L., *Festuca rigida* (L.) Rasp. non Roth, *Scleropoa rigida*
(L.) Griseb., *Desmazeria rigida* (L.) Tutin C.T.W. 139.11.1.
Dry pastures, railway ballast, old walls, etc.   Colonist
Common: recorded for all districts and neighbouring counties.
H.S. 25, 45a, 56b, 61, 64, 69.
First record: Abbot, 1798.

VULPIA C. C. Gmel.

Vulpia bromoides (L.) Gray C.T.W. 139.9.2. *Squirrel's-tail Grass*
   Festuca bromoides L.
Dry pastures, woodland rides, etc.   Native
Frequent on light soils. It has recently been introduced also as a
wool adventive. *E. Bot.* tab. 1411, March 1805, was apparently
drawn from material sent by John Hemsted. Recorded for all
neighbouring counties.
H.S. 22b, 23, 25, 26, 46, 48.
OUSE: Cardington; *J.McL. in J.H. Notes*: Stevington; *J.McL., B.M.*:
Millbrook*; railway siding, Willington; Sheerhatch Wood*.
KYM: West Wood; *E.M.-R.*
OUZEL: Aspley Heath†; *S.B.P.L.*: Heath and Reach*; Leighton
Buzzard; Sheep Lane*; King's Wood.
IVEL: Flitwick!; *F.F.B.*: Sandy!; *J.E.L., J. Bot.*: Galley Hill;
Portobello Farm; *J.E.L.,Diary*: Shefford; Clophill; Rowney Warren.
LEA: Luton; *P.T.*: railway bank, East Hyde.
COLNE: Pepperstock†; *J.S. in J. Bot. 1883*: Caddington.
First record: Abbot in letter to J. E. Smith, 3 Sept. 1802.

V. myuros (L.) C. C. Gmel. C.T.W. 139.9.3.   *Rat's-tail Grass*
   Festuca myuros L.
Sandy soils, waste places, etc.
Less common than the preceding, but also introduced recently as a
wool adventive. Found mainly in places where the soil has been
disturbed. *E. Bot.* tab. 1412, March 1805, was drawn from material
sent by Hemsted. Recorded for all neighbouring counties.
H.S. 16.
OUSE: Oakley Walls; *F.B.*: Stevington; *J.McL., B.M.*: Lidlington;
Herb. Beds. Nat. Hist. Soc.: gravel pit, Eaton Socon*; waste ground,
Ford End; gravel pit, Willington.
OUZEL: Aspley!†; *J.McL., B.M.*: old wall, Heath and Reach, as
Festuca ambigua Le Gall†; *J.S. in Bot. Rec. Club 1883 Rep.*: Heath and
Reach!; *S.B.P.L.*: Sheep Lane*.
IVEL: Potton, on garden walls; *Plantae Bedford.*: Everton Heath;
J.McL., L.M.: Silsoe; Clophill!*; Flitwick!; *F.F.B.*: Shefford*;
E.M.-R.!: Sundon Rubbish Dump*; railway siding, Southill;
arable field, Flitton*; Rowney Warren.
First record: Abbot, 1795.

[1] Comb. nov. Based on *Poa rigida* L., *Cent. I Pl.* 1755, 5.

FESTUCA L.

Festuca pratensis Huds. C.T.W. 139.6.1.   *Meadow Fescue*
   F. elatior L. p.p.   .
Pastures   Native
Common: recorded for all districts and neighbouring counties.
H.S. 8, 9, 14b, 37, 56b, 57b, 84a, 85.
First record: for the aggregate, Abbot, 1798.

F. arundinacea Schreb. C.T.W. 139.6.2.   *Tall Fescue*
Rough pastures   Native
Common: recorded for all districts except Lea, and for all neigh-
bouring counties.
H.S. 2a, 10a, 10c, 57b.
First record: by the river, Herb. J. McLaren, British Museum
c. 1880.

F. gigantea (L.) Vill. C.T.W. 139.6.3.   *Giant Fescue*
   Bromus giganteus L.
Woodland rides, hedgerows, etc.
Common: recorded for all districts and neighbouring counties.
H.S. 5, 42, 43, 51b, 52, 78, 83.
First record: Abbot, 1798.

F. rubra L. C.T.W. 139.6.6.   *Red* or *Creeping Fescue*
Pastures, old walls, etc.   Native
Common, but very variable: var. *barbata* Hack. was found by E.
Milne-Redhead and the author at Flitwick*. Recorded for all
districts and neighbouring counties.
H.S. 2a, 9, 14b, 23, 25, 28, 29, 38, 41, 45b, 48, 56b, 62, 63a, 63b,
64, 66a, 67, 68–70a, 74a, 75, 76, 85.
First record: Abbot's records of both *F. duriuscula*, common, and
F. rubra, rare, are probably to be referred to this.

F. fallax Thuill.   *Chewing's Fescue*
Pastures, etc.   ? Native
Frequently sown as a lawn grass.
OUSE: gravel pit, Eaton Socon*.
KYM: old wall, Knotting*.
OUZEL: Heath and Reach; *C. E. Hubbard*!: Sewell; *E.M.-R.*:
railway bank, Aspley Guise.
IVEL: Castle Hill, Sandy*.
LEA: Luton.

F. ovina L. C.T.W. 139.6.8.   *Sheep's Fescue*
Pastures   Native
Common on well-grazed pastures on the Chalk and Lower Green-
sand. Recorded for all neighbouring counties.
H.S. 22a, 28, 29, 56b, 59, 60–63b, 65–67, 75.

Nene: railway embankment, Wymington.
Ouse: Jackdaw Hill.
Ouzel: common on the chalk hills: Heath and Reach!†*; *Herb.*
J.S.: Horsemoor Farm.
Ivel: common on the chalk hills: Rowney Warren; *J.McL., L.M.*:
Flitwick†; *J. McL., B.M.*
Lea: common on the chalk hills: Luton Hoo.
Colne: Whipsnade Common*.
First record: Abbot, 1798.
 Var. **hispidula** (Hack.) Hack. ex Richt.
Ouse: railway embankment, Souldrop*.
Ouzel: Houghton Regis; *E.M.-R.*
Lea: Leagrave*; *E.M.-R.*

F. tenuifolia Sibth.   *Fine-leaved Sheep's Fescue*
 F. capillata Lam. p.p., *F. ovina* subsp. *tenuifolia* (Sibth.) Tutin
C.T.W. 139.6.8.
Pastures   Native
Not infrequent on the Lower Greensand. With us a species of poor
grassland on acid soils. Recorded for Bucks. and Northants.
H.S. 22a, 22b, 24, 27.
Ouse: Milton Ernest; *Abbot's annotated F.B.*
Ouzel: Heath and Reach*; *C. E. Hubbard*!
Ivel: Flitwick Moor!; *J.E.L., F.F.B.*: Cooper's Hill; Clophill*;
Rowney Warren.
Colne: roadside, Caddington*; Whipsnade Heath*.
First record: Abbot in letter to J. E. Smith, 3 Sept. 1802.

F. longifolia Thuill.   *Hard Fescue*
 F. trachyphylla (Hack.) Krajina non Hack. C.T.W. 139.6.10.
Pastures
Sown frequently, especially on railway embankments and roadsides.
H.S. 2a.
Nene: railway embankment, Wymington, with var. **trachyphylla** (Hack.)
Howarth*.
Ouse: Millbrook, var. *trachyphylla*; railway bank, Souldrop, with var. *trachyphylla*
and var. **villosa** (Schrad.) Howarth*.
Kym: West Wood.
Ouzel: railway bank, Sewell, var. *trachyphylla**; *E.M.-R.*
Lea: railway bank, East Hyde*.
Colne: Blow's Downs, v.c. 20 [Beds.], var. *villosa**.
First record: the author in *B.E.C. 1946–7 Rep.* 320.

   × **FESTULOLIUM** Aschers. & Graebn.

× **Festulolium loliaceum** (Huds.) P. Fourn. C.T.W. 139.7.1.
 Festuca pratensis × *Lolium perenne, Festuca loliacea* Huds., *F. adscendens*
Retz.
Pastures   Native
Probably more common than the records indicate. Recorded for all
neighbouring counties except Hunts.
H.S. 69.

OUSE: Friars Pasture Meadows near Oakley House; *Abbot's annotated*
F.B.: meadows by the Ouse; *J.McL. in J. Bot. 1884*: meadow, Eaton
Socon*; *J. P. M. Brenan.*
OUZEL: Cow Common; *E.M.-R.*
IVEL: Sharpenhoe*; *J.S.*: clearing in Leete Wood*.
LEA: Warden Hills.
First record: Abbot, *c.* 1800.

BROMUS L.

Bromus erectus Huds.   *Upright Brome*
 Zerna erecta (Huds.) Gray C.T.W. 139.19.1.
Pastures   Native
Common on rough and ungrazed pastures on calcareous soil. It
has increased greatly on the chalk downs in the past thirty years.
It is well distributed and I have seen it in all districts. Recorded
for all neighbouring counties.
H.S. 1a, 59, 60, 61, 65, 66a, 68, 70a.
First record: Sinclair, *Hortus Gramineus Woburnensis*, 1816, 96.

B. ramosus Huds.   *Hairy Brome*
 B. hirsutus Curt., *Zerna ramosa* (Huds.) Lindm. C.T.W. 139.19.2.
Woodland rides, hedgerows, etc.   Native
Common: recorded for all districts and neighbouring counties.
H.S. 2a, 2b, 5, 31a, 32, 51a, 52, 55b, 78, 80b, 83.
First record: Abbot, 1798.

B. sterilis L.   *Barren Brome*
 Anisantha sterilis (L.) Nevski C.T.W. 139.20.1.
Waste places, edges of arable fields, etc.   Colonist
Common: recorded for all districts and neighbouring counties.
H.S. 12, 17, 70a.
First record: Abbot, 1798.

B. mollis L. C.T.W. 139.21.1.   *Lop Grass, Soft Brome*
 B. hordeaceus auct.
Pastures, roadsides, etc.   Native
A common, but very variable species. Recorded for all districts
and neighbouring counties.
H.S. 9, 25, 44, 48.
First record: Abbot, 1798.
 Var. **leiostachys** Hartm.
OUSE: Houghton Conquest*.
OUZEL: Heath and Reach*; *C. E. Hubbard*!: Houghton Regis*; *E.M.-R.*: Leighton
Buzzard; *P.T.*: Totternhoe.
LEA: Leagrave.

B. thominii Hardouin C.T.W. 139.21.3.
 B. hordeaceus L., nom. ambig.
Roadsides, pastures, etc.   Colonist
Common, and frequently sown in short-term leys with *Lolium perenne*. Recorded
for all districts except Cam, and for all neighbouring counties.
First record: E. Milne-Redhead and the author in *B.E.C. 1943–4 Rep.* 776.

B. lepidus Holmb. C.T.W. 139.21.4. *Slender Brome*
    B. britannicus I. A. Williams
Roadsides, pastures, etc. Native
Frequent on the lighter soils and possibly sometimes sown in leys.
Recorded for Cambs. and Northants.
OUSE: Stevington*; *E.M.-R.*: side of King's Wood; arable field,
Wyboston.
KYM: Knotting.
OUZEL: roadside, Heath and Reach*; *A. J. Wilmott*!
IVEL: Clophill; Wilbury*; Tingrith.
LEA: Dunstable*; *E.M.-R.*: Warden Hills.
COLNE: Studham*; Kensworth, v.c. 20 [Beds.].
First record: A. J. Wilmott and the author in *B.E.C. 1943–4 Rep.* 813.
    Var. **micromollis** (Krösche) C. E. Hubbard
IVEL: Tingrith; *E.M.-R.*!
COLNE: Studham*.

B. racemosus L. C.T.W. 139.21.5. *Smooth Brome*
Meadows Native
Probably more common than the records indicate. Recorded for
all neighbouring counties.
H.S. 9, 14b.
OUSE: Tempsford*; *P.T.*!: Houghton Conquest*; water meadows,
Eaton Socon*, Felmersham*.
OUZEL: meadow, Tilsworth.
IVEL: meadows, Fancott, Westoning.
First record: *Newbould MS.*, 1874.

B. commutatus Schrad. C.T.W. 139.21.6. *Meadow Brome*
    B. pratensis Ehrh. ex Hoffm. non Lam.
Arable fields, pastures, etc.
Common and found mainly in cornfields. Recorded for all districts
and neighbouring counties.
H.S. 4, 17.
First record: Barton, *South Bedfordshire Plant List*, 1885.

B. secalinus L. C.T.W. 139.21.8. *Rye Brome*
Arable fields
Frequent: the Bedfordshire material appears to be always var.
hirtus (F. W. Schultz) Aschers. & Graebn. ex Hegi. It was considered rare by Abbot. Recorded for all districts except Cam, and
for all neighbouring counties.
First record: Abbot, 1798.

B. arvensis L. was recorded from Bromham Grange Farm by Abbot in *Flora Bed-*
fordiensis, but this was erased in his annotated copy and Clapham Hill and
Stevington Churchyard Wall, 1799, were added. It was recorded from Houghton
Conquest Park, in the name of E. Forster, in *Bot. Guide*, 1805. Saunders considered
it 'frequent near Luton', but his herbarium specimen from Heath and Reach is
B. secalinus. J. E. Little listed it from Wilbury Hill in his annotated copy of *F.F.B.*,
but I know of no specimen of Little's from here. Without further evidence all the
records must be suspect.

BRACHYPODIUM Beauv.

Brachypodium sylvaticum (Huds.) Beauv. C.T.W. 139.23.1.
Festuca sylvatica Huds.   *Slender False-brome*
Woodland rides, hedgerows, etc.   Native
Common: recorded for all districts and neighbouring counties.
H.S. 2b, 3a, 5, 31a, 31b, 32, 42, 51a, 52, 55b, 58b, 61, 68–70a, 78,
80b, 83.
First record: Abbot, 1798.

B. pinnatum (L.) Beauv. C.T.W. 139.23.2.   *Spiked False-brome,*
Festuca pinnata (L.) Huds.   *Tor Grass*
Pastures   Native
Common on and near Oolitic exposures in the north of the county,
rare elsewhere. Recorded for all neighbouring counties.
H.S. 1a, 1b, 10b.
NENE: common by roadsides and on railway banks.
OUSE: Clapham; *F.B.*: Hillfoot; *J.McL., B.M.*: Houghton Con-
quest; *Botanist's Guide*, 1805: Twin Wood; *A. Ransome in J.H. Notes*:
Sheerhatch Wood, var. *pubescens* Gray; *J.E.L., J. Bot.*: Souldrop;
Harrold; Odell; Pavenham; Stevington.
KYM: Knotting; Swineshead.
OUZEL: Heath and Reach; *C. E. Hubbard!*: roadside, Hockliffe.
IVEL: Warden Wood; *J.McL., L.M.*: Ampthill*; *C.C.*: Arlesey.
LEA: Winsdon Hills, Luton, one plant only†; *F.F.B.*
First record: Abbot, 1798.

LOLIUM L.

Lolium perenne L. C.T.W. 139.8.1.   *Perennial Rye Grass,*
                                                               Ray Grass
Pastures, roadsides, etc.   Native
Common, but the most frequently sown grass species. It is very
variable, and proliferating forms are frequent. Recorded for all
districts and neighbouring counties.
H.S. 2a, 3a, 8, 9, 11, 13, 19a, 21, 25, 26, 28, 56b, 63a, 64, 66b, 75,
80a.
First record: Abbot, 1798.

L. multiflorum Lam. C.T.W. 139.8.2.   *Italian Rye-grass*
     L. italicum A. Braun
Pastures, waste places, etc.   ? Colonist
Common, and frequently planted in short-term leys. Like *L. perenne* it is variable.
Recorded for all districts and neighbouring counties.
H.S. 4, 73.
First record: Saunders in *South Bedfordshire Plant List*, 1885.
     Var. **compositum** (Thuill.) Mutel
IVEL: Shefford, my specimen is in Herb. Kew.

L. temulentum L. C.T.W. 139.8.3.   *Darnel*
Arable fields, waste places, etc.
Rare: but probably more common in earlier times. It is possible that a record of
L. arvense by Abbot, in a letter to J. E. Smith, 3 Sept. 1802, refers to a form of this.
Recorded for all neighbouring counties.

Ouse: Fenlake, amongst corn; *Plantae Bedford.*: Crow Hill; *Abbot's annotated F.B.*
Ivel: Flitwick; *M. L. Berrill in J.H. Notes*: railway siding, Flitwick, var. **arvense**
Liljebl.; railway siding, Ampthill.
First record: Abbot, 1795.

AGROPYRON Gaertn.

Agropyron caninum (L.) Beauv. C.T.W. 139.24.1.
Triticum caninum L.   *Bearded Couch-grass*
Wood borders, hedgerows, etc.   Native
Common: recorded for all districts and neighbouring counties.
First record: Abbot, 1798.
 Var. **glaucescens** Lange
Ivel: Flitwick Moor; *N. Y. Sandwith.*

A. repens (L.) Beauv. C.T.W. 139.24.3.   *Twitch, Couch-grass*
Triticum repens L.
Arable land, waste places, etc.   Colonist
Common but very variable: recorded for all districts and neigh-
bouring counties.
H.S. 7, 13, 16, 21, 50, 64.
First record: Abbot, 1798.
 Var. **aristatum** Baumg.
Ouzel: Heath and Reach; *C. E. Hubbard!*
Ivel: Sundon Rubbish Dump*.
Lea: Round Green*.
 Var. **pubescens** Doell
Ivel: Sundon Rubbish Dump*.
 Var. **dumetorum** (Schweigg. & Koerte) Roem. & Schult.
Ivel: Flitwick Moor; *N. D. Simpson!*

NARDUS L.

Nardus stricta L. C.T.W. 139.54.1.   *Mat-grass, Matweed*
Heaths   Native
Rare: Abbot, no doubt mistakenly, considered it common. Re-
corded for all neighbouring counties except Hunts.
Ouzel: Rushmere*; *C. E. Hubbard!*
Ivel: Potton; *J.McL., L.M.*: Flitwick Moor; *J.McL., B.M.*
First record: Abbot, 1798.

HORDELYMUS (Jessen) Harz

Hordelymus europaeus (L.) Harz C.T.W. 139.27.1.   *Wood Barley*
Hordeum sylvaticum Huds., *H. europaeum* (L.) All.
Woods   Native
Rare: recorded for all neighbouring counties.
Ouse: Thurleigh; Putnoe Wood; *F.B.*: Bromham Park; *J.H. Notes*:
Twin Wood; *A. B. Sampson.*
Colne: Long Wood, v.c. 20 [Herts.]*.
First record: Abbot, 1798.

HORDEUM L.

Hordeum secalinum Schreb. C.T.W. 139.26.1.   *Meadow Barley*
   H. pratense Huds., *H. nodosum* auct.
Meadows   Native
Common in low-lying meadows in the north and middle of the
county, and usually evident in well-grazed pasture as it is avoided
by cattle. It is recorded for all districts except Colne, but I have not
seen it in the Lea district. Recorded for all neighbouring counties.
H.S. 8, 9, 13, 56b.
First record: Abbot, 1798.

H. murinum L. C.T.W. 139.26.2.   *Wall Barley*
Waste places, roadsides, etc.   Colonist
Common: recorded for all districts and neighbouring counties.
First record: Abbot, 1798.

CASUALS, GARDEN ESCAPES, ETC.

These are excluded from the general list as they are of a less permanent nature. Some may become permanently established, but is too early to consider them as such.

§ wool adventives.
‡ garden escapes.

RANUNCULACEAE (p. 200)

‡ERANTHIS HYEMALIS (L.) Salisb., C.T.W. 14.4.1 (*Cammarum hyemale* (L.) Greene), *Winter Aconite*, makes a fine show under trees in old gardens in New Bedford Road, Luton. Recorded for all neighbouring counties except Hunts.

‡ACONITUM NAPELLUS L., C.T.W. 14.6.1, *Monkshood*. LEA: Chiltern Green; *H. B. Souster*. Recorded for Herts. and Bucks.

‡DELPHINIUM ORIENTALE Gay, C.T.W. 14.7.2, *Eastern Larkspur*. OUSE: Lidlington*; *B. Verdcourt*.

‡ANEMONE RANUNCULOIDES L., C.T.W. 14.9.2, *Yellow Wood Anemone*. IVEL: Kitchen End; *C.C. in Herb. J. Pollard*. Recorded for Herts. and Bucks.

‡A. APENNINA L., C.T.W. 14.9.3, *Blue Anemone*. IVEL: Westoning*; *G. D. Nicholls*. LEA: 'in a wood at Lutton Hoe', as '*Ranunculus nemorosus*'; *T. Knowlton in Ray's Synopsis, Ed. III*, 1925, 259.

‡A. CORONARIA L. OUSE: Cardington, as *A. apennina*; *J.McL., B.M.*

BERBERIDACEAE (p. 207)

MAHONIA AQUIFOLIUM (Pursh) Nutt., C.T.W. 16.3.1 (*Berberis aquifolium* Pursh), *Oregon Grape*, is frequently planted in woods and spinneys. Recorded for all districts except Kym, Cam and Colne, and for all neighbouring counties.

PAPAVERACEAE (p. 209)

PAPAVER SOMNIFERUM L., C.T.W. 19.1.6, *Opium Poppy*. Casual. OUSE: Houghton Ruins; *W.C. Herb.*: Turvey; gravel pit, Chawston. OUZEL: Husborne Crawley; *W.H. 1875*. IVEL: Eversholt; north of Barton; *H. B. Souster*: Sundon Rubbish Dump. LEA: near Daffodil Wood†; *J.S. in J. Bot. 1889*. First record: W. Crouch, 1842.

§ARGEMONE MEXICANA L. IVEL: arable field, Clifton*; *J. Cowley*!

‡ESCHSCHOLZIA CALIFORNICA Cham., C.T.W. 19.6.1 (*E. douglasii* (Hook. & Arn.) Walp.). IVEL: Sundon Rubbish Dump*. Recorded for Herts. and Bucks.

‡GLAUCIUM CORNICULATUM (L.) Rudolph, C.T.W. 19.4.1. OUSE: Castle Mills; *E. M. Langley in J.H. Notes*. Recorded for Bucks.

FUMARIACEAE (p. 210)

‡CORYDALIS BULBOSA (L.) DC. (*C. solida* (L.) Sw., C.T.W. 20.1.1) was recorded in *Hillhouse List, 1876*, on the authority of A. P. Wise and A. Poulton, and in Hamson's Notes from Vicarage Garden, Milton Ernest, in the name of A. G. Nash. Recorded for Herts., Hunts. and Northants.

CRUCIFERAE (p. 212)

‡CHEIRANTHUS CHEIRI L., C.T.W. 21.47.1, *Wallflower*. OUSE: walls of Bedford Castle; *F.B.* IVEL: Ridgmont; *Herb. Miss Crouch*. Recorded for Cambs. and Northants. First record: Abbot, 1798.

BARBAREA INTERMEDIA Bor., C.T.W. 21.35.3. OUZEL: Heath and Reach*. IVEL: Flitwick; *H. Phillips in B.E.C. 1928 Rep.* 724: Pegsdon; *J.E.L., F.F.B.*: railway, Sundon*. LEA: railway, Luton*. Recorded for Herts., Bucks. and Hunts. First record: H. Phillips, 1928.

B. VERNA (Mill.) Aschers., C.T.W. 21.35.4 (*B. praecox* (Sm.) R.Br.). OUSE: Cardington; *W.H. 1875.* LEA: Round Green*; Stopsley*. Recorded for all neighbouring counties. First record: W. Hillhouse, 1875.

B. STRICTA Andrz., C.T.W. 21.35.2, is listed for Cambs., Beds. and Northants. in *Fl. of Bucks.* I cannot find the source of the record for Beds.

‡ARABIS CAUCASICA Willd., C.T.W. 21.37.3, *Garden Arabis*, appears frequently on the railway as a relic from platelayers' gardens. Recorded for Herts. and Bucks. First record: sandpit, Heath and Reach; G. C. Druce in *B.E.C. 1929 Rep.* 102.

‡LUNARIA ANNUA L., C.T.W. 21.26.1, *Honesty*. KYM: derelict garden, Dean*.

‡LOBULARIA MARITIMA (L.) Desv., C.T.W. 21.28.1 (*Alyssum maritimum* (L.) Lam.), *Sweet Alison*. IVEL: Sundon Rubbish Dump*. Recorded for Herts. and Bucks.

BERTEROA INCANA (L.) DC., C.T.W. 21.29.1 (*Alyssum incanum* L.). OUSE: Cardington Mill; *J.McL., B.M.* IVEL: Greenfield; *J.McL., B.M.*: in clover, Flitwick†; *J.McL. in Herb. J.S.* COLNE: Pepperstock; *F.F.B.* Recorded for Herts. and Bucks. First record: J. McLaren, *c.* 1884.

‡MALCOLMIA AFRICANA (L.) R.Br., C.T.W. 21.44 (*Wilckia africana* (L.) F. Muell.), *Stock*. IVEL: brickworks, Arlesey; *J.E.L., B.E.C.* Recorded for Bucks.

SISYMBRIUM LOESELII L., C.T.W. 21.49.2a. IVEL: Sundon Rubbish Dump*. Recorded for Herts.

§S. IRIO L., C.T.W. 21.49.2, *London Rocket*. IVEL: railway siding, Biggleswade*; arable field, Maulden*. Recorded for Hunts., records for Herts. and Bucks. need confirmation.

ERYSIMUM HIERACIFOLIUM L. subsp. DURUM (J. & C. Presl) Hegi. A specimen collected on Sundon Rubbish Dump* agrees with one in Herb. Kew collected at Mildenhall, W. Suffolk, 1913, by W. C. Barton and so named by O. E. Schulz.

CONRINGIA ORIENTALIS (L.) Dum., C.T.W. 21.13.1 (*Erysimum orientale* (L.) Crantz non Mill.), *Hare's-ear Cabbage*. Records by C. Crouch are in error. OUSE: Twin Wood; *J.McL., L.M.* Recorded for Herts., Cambs. and Northants.

CAMELINA MICROCARPA Andrz. ex DC., C.T.W. 21.51.2 (*C. sylvestris* Wallr.). IVEL: railway bank, Flitwick, established for some years!*; *E. T. Blundell*. Recorded for Herts.

BRASSICA ELONGATA Ehrh., C.T.W. 21.1.4. OUZEL: mill, Leighton Buzzard; *G. C. Druce in J. Bot. 1897.* Recorded for Bucks. and Northants.

B. JUNCEA (L.) Coss., C.T.W. 21.1.6. A very variable species. OUSE: Eaton Socon*; *H. B. Souster!*: flax field, Stagsden; *D. W. Elliott.* IVEL: between Edworth and Langford; *J.E.L., J. Bot.*: Little Park, Ampthill*; *C.C.*: railway sidings, Flitwick, Biggleswade. Recorded for Herts., Bucks. and Hunts. First record: J. E. Little, 1919.

B. NIGRA (L.) Koch, C.T.W. 21.1.8 (*Sinapis nigra* L.), *Black Mustard*. A frequent relic of cultivation, but never established. Recorded for all districts except Kym and Cam, and for all neighbouring counties. First record: Abbot, 1919.

§B. TOURNEFORTII Gouan, C.T.W. 21.1.7. IVEL: Maulden!; *N. Y. Sandwith*: Flitton*.

§B. GRIQUANA N. E. Br. IVEL: railway siding, Biggleswade*; *J. Codrington*!

HIRSCHFELDIA INCANA (L.) Lagrèze-Fossat, C.T.W. 21.5.1 (*Brassica adpressa* Boiss., *B. incana* (L.) Meigen non Ten.), *Hoary Mustard*. OUZEL: railway siding, Leighton Buzzard*; *E. T. Blundell*! Recorded for Herts.

ERUCASTRUM GALLICUM (Willd.) O. E. Schulz, C.T.W. 21.2.2 (*Brassica gallica* (Willd.) Druce). IVEL: brickyard, Henlow*; *E.M.-R.*!: brickyard, Arlesey*.

§VELLA ANNUA L. (*Carrichtera annua* (L.) Aschers. C.T.W. 21.12a.1). OUSE: railway siding, Tempsford*. Recorded for Herts.

LEPIDIUM RUDERALE L., C.T.W. 21.14.6, *Narrow-leaved Pepperwort*. Frequent in waste places in the Ouse, Ouzel, Ivel and Lea districts and usually established for a number of years. Recorded for all neighbouring counties. First record: Saunders in *S.B.P.L.*, 1885.

‡L. SATIVUM L., C.T.W. 21.14.1, *Garden Cress.* OUSE: Goldington Road; *J.McL.*, *B.M.*: gravel pit, Chawston. IVEL: Sundon Rubbish Dump, in two forms, one with curled leaves*; *E.M.-R.*!: railway siding, Flitwick. LEA: railway sidings, Luton, Chiltern Green. Recorded for all neighbouring counties except Cambs. First record: J. McLaren, *c.* 1880.

L. PERFOLIATUM L., C.T.W. 21.14.11. IVEL: Sandy; *Hon. S. Peel in B.E.C. 1923 Rep.* 170.

§L. HYSSOPIFOLIUM Desv. (there appears to be nothing to distinguish this from *L. linoides* Thunb.). OUSE: railway siding, Blunham. IVEL: arable field, Beeston*; *R. A. Graham*!: arable field, Clifton; railway siding, Langford.

§L. VIRGINICUM L., C.T.W. 21.14.7. IVEL: arable field, Flitwick*. Recorded for Herts., Bucks. and Hunts.

L. LATIFOLIUM L., C.T.W. 21.14.10. McLaren's specimens from Kempston Mill are *Cardaria draba* and a record by C. Crouch from Marston lacks confirmation. Recorded for Herts., Bucks. and Cambs.

‡IBERIS UMBELLATA L., C.T.W. p. 182, *Garden Candytuft,* is established near a platelayer's garden, north of Ampthill. Recorded for Bucks.

ISATIS TINCTORIA L., C.T.W. 21.17.1, *Woad,* was recorded in Ray's List in *Magna Britannia,* 1695. There is also a very doubtful record from Sundon in Davis: *History of Luton.* Odell was spelt Wadella in *Domesday Book (P.N.B.).* Recorded for, but extinct in, Bucks., Cambs. and Northants.

‡ARMORACIA RUSTICANA Gaertn., Mey. & Scherb., C.T.W. 21.32.1 (*Cochlearia armoracia* L., *A. lapathifolia* Gilib.), *Horse Radish.* Frequent in waste places as a relic of cultivation. Recorded for all districts and neighbouring counties. First record: *Hillhouse List, 1876.*

A record of *Cochlearia officinalis* L. in *Hamson's Notes* from Thurleigh, 1912, by J. Woods, is to be doubted.

NESLIA PANICULATA (L.) Desv., C.T.W. 21.25C.1 (*Vogelia paniculata* (L.) Hornem.). LEA: Luton; *R. A. Pryor in J. Bot. 1876,* 22. Recorded for Herts., Bucks. and Northants.

BUNIAS ERUCAGO L., C.T.W. 21.25d.1. OUSE: railway siding, Bedford*. OUZEL: Leighton Mill; *V.C.H.* LEA: waste ground, Luton*; *H. B. Souster.* Recorded for Bucks. and Northants. First record: Druce, 1904.

B. ORIENTALIS L., C.T.W. 21.25d.2. OUSE: railway yard, Bedford; *J. Lamb in J.E.L.,* *F.F.B.* IVEL: Arlesey; *J.E.L., J. Bot.* LEA: Chalton Cross*; *V. H. Chambers.* Recorded for Herts., Bucks. and Northants. First record: J. E. Little, 1919.

RAPISTRUM PERENNE (L.) All., C.T.W. 21.10.1. OUSE: waste ground, Cople*; *E.M.-R. and the author in B.E.C. 1945 Rep.* 22. OUZEL: railway siding, Leighton Buzzard*. Recorded for Bucks.

R. ORIENTALE (L.) Crantz (*R. rugosum* (L.) All., C.T.W. 21.10.2). A variable species. OUSE: railway siding, Blunham*; *H. B. Souster*! OUZEL: waste ground, Hockliffe; *H. B. Souster.* IVEL: Upper Caldecote*; *H. B. Souster*: plentiful at Beeston. Recorded for all neighbouring counties except Cambs.

‡RAPHANUS SATIVUS L., C.T.W. 21.8.3, *Garden Radish,* occurs frequently as a garden throw-out. Recorded for Bucks.

RESEDACEAE (p. 220)

‡RESEDA ALBA L., C.T.W. 22.1.3, *Upright Mignonette.* OUSE: roadside, near Oakley Station, 1914; *S. Crowsley in J.H. Notes*: roadside, Staploe*. Recorded for Herts. and Bucks.

HYPERICACEAE (p. 224)

HYPERICUM CALYCINUM L., C.T.W. 25.1.4, *Rose of Sharon.* Probably originally planted. OUZEL: Woburn; *W.H. 1875.* IVEL: Haynes; *J.McL. in J.H. Notes*: Sundon; *Davis*: Washer's Wood. LEA: Luton Hoo; *V.C.H.* Recorded for Bucks. and Northants. First record: W. Hillhouse, 1875.

H. ANDROSAEMUM L., C.T.W. 25.1.1, *Tutsan,* appears to be naturalized in Woburn Park. Recorded for Herts., Bucks. and Northants.

H. MONTANUM L., C.T.W. 25.1.13. There is an unlocalized record in *W.H. 1875* and a record from Maulden by M. L. Berrill in *J.H. Notes.* Recorded for Herts., Bucks. and Northants.

H. HIRCINUM L. C.T.W. 25.1.3. IVEL: Leete Wood, Barton†; *Herb. J.S.*

CARYOPHYLLACEAE (p. 226)

‡DIANTHUS BARBATUS L., C.T.W. 30.7.2, *Sweet William.* IVEL: Sundon Rubbish Dump*.

VACCARIA PYRAMIDATA Medic., C.T.W. 30.8.1 (*Saponaria vaccaria* L.), *Cow Basil.* OUSE: sandpits, Bromham; *J. Burgoyne in J.H. Notes:* Stoke Mills; *E. M. Langley in J.H. Notes:* Honey Hills; *C. Clarke in J.H. Notes:* Bedford Park; *L. Perry in J.H. Notes:* cornfield, Kempston*. IVEL: Flitwick†; *J. S. Thorne in W.F.B.*: Maulden *L. J. Tremayne.* LEA: garden weed, Luton, *F. Seymour Lloyd.* Recorded for all neighbouring counties except Cambs. First record: Saunders, 1897.

SILENE ANGLICA L., C.T.W. 30.1.6, *Small-flowered Catchfly.* OUSE: Biddenham; *J.McL., B.M.*: Harrowden; *F. W. Crick in J.H. Notes.* OUZEL: Woburn; *F.B.* IVEL: Barton; *F.B.*: Biggleswade; *M. C. Williams in J.E.L., B.E.C.* Recorded for all neighbouring counties. First record: Abbot, 1798.

S. DICHOTOMA Ehrh., C.T.W. 30.1.5. OUSE: Cardington Mill; *J.McL. in Bot. Rec. Club 1884–6 Rep.* Recorded for all neighbouring counties except Cambs.

‡S. PENDULA L., C.T.W. 30.1.7. OUSE: gravel pit, Clapham*.

‡S. COELI-ROSA (L.) Rohrb., C.T.W. p. 282. IVEL: Sundon Rubbish Dump*.

‡CERASTIUM TOMENTOSUM L., C.T.W. 30.12.3, *Snow in Summer.* Frequent near gardens: it makes a fine show on the railway banks north of Leagrave.

PORTULACACEAE (p. 235)

CLAYTONIA ALSINOIDES Sims, C.T.W. 31.2.2 (*C. sibirica* auct.). OUZEL: garden weed, Aspley Guise*; *R. A. Palmer.* Recorded for Herts.

AMARANTHACEAE

§AMARANTHUS RETROFLEXUS L., C.T.W. 33.1.1. OUSE: Cardington Mill; *J.McL. in Herb. Brit. Mus.*: railway siding, Blunham; *P.T.*! IVEL: arable field, Clophill*; *J. E. Lousley*!: Beeston; *R. A. Graham*!: arable fields, Maulden, Flitwick, Flitton. LEA: Luton; *D. M. Higgins in Herb. Brit. Mus.* Recorded for all neighbouring counties except Cambs. First record: J. McLaren, 1880.

§A. THUNBERGII Moq. IVEL: railway siding, Flitwick*; arable fields, Flitwick*, Flitton*. Recorded for Bucks.

§A. DINTERI Schinz var. UNCINATUS Thell. IVEL: arable field, Beeston*; *R. A. Graham*!: arable fields, Maulden!*; *N. Y. Sandwith.*

§A. HYBRIDUS L. var. CHLOROSTACHYS (Willd.) Thell. (*A. chlorostachys* Willd.). IVEL: railway siding, Flitwick*; arable fields, Maulden, Beeston, Flitwick.

§A. VIRIDIS L. IVEL: arable field, Maulden; *R. A. Graham and J. E. Lousley.*

‡A. CAUDATUS L., *Love Lies Bleeding.* IVEL: Sundon Rubbish Dump*. Recorded for Bucks. and Hunts.

CHENOPODIACEAE (p. 236)

CHENOPODIUM VULVARIA L., C.T.W. 34.1.3, is recorded for v.c. 30 in *Comital Flora*, but I do not know any basis for its inclusion. Recorded for all neighbouring counties.

§C. BERNBURGENSE J. Murr. IVEL: arable field, Maulden; *J. P. M. Brenan.*

C. CAPITATUM (L.) Aschers., C.T.W. 34.1.8. LEA: garden weed, Stockwood!*; *J. Cowley.* Recorded for Northants.

§C. CARINATUM R.Br. IVEL: railway siding, Flitwick*.

§C. PROBSTII Aellen. IVEL: railway sidings, Flitwick*, Biggleswade; arable field, Maulden. Recorded for Hunts.

§C. GIGANTEUM D Don is one of the more attractive wool adventives. Like *C. probstii* it does not flower in Britain. IVEL: arable fields, Flitton*, Clifton, Maulden.

§C. SCHRADERIANUM Schult. (*C. foetidum* Schrad. non Lam.). IVEL: arable field, Maulden*; *C. M. Goodman*!

§C. PUMILIO R.Br., a common wool adventive in the Ouse and Ivel districts.

§C. CRISTATUM (F. Muell.) F. Muell. IVEL: railway siding, Flitwick*; arable fields, Flitwick*, Maulden*, Flitton*.

§C. AURICOMIFORME Murr & Thell. IVEL: arable field, Maulden*; *J. P. M. Brenan*: arable field, Flitton.

§SCLEROBLITUM ATRIPLICINUM (F. Muell.) Ulbr. (this has been found previously in South Hants. by R. C. L. Burges, but not recorded). IVEL: arable field, Flitwick, my specimen is in Herb. Kew.

§MONOLEPIS NUTALLIANA (Schult.) Greene. IVEL: railway siding, Biggleswade*.

‡BETA VULGARIS L., C.T.W. p. 355, *Beet*, occurs frequently as a relic of cultivation.

‡SPINACIA OLERACEA L., *Spinach*. OUSE: river bank, Bedford; *J.H. Notes*. IVEL: Sundon Rubbish Dump*. Recorded for Bucks. and Northants.

‡ATRIPLEX NITENS Schkuhr. OUSE: roadside, Clapham. IVEL: roadside, Silsoe*.

‡A. HORTENSIS L., C.T.W. p. 358. OUSE: gravel pit, Chawston*. IVEL: waste ground, Maulden. Recorded for Herts., Bucks. and Northants.

§A. EARDLEYAE Aellen. IVEL: railway siding, Flitwick*.

§A. SEMIBACCATA R.Br. IVEL: railway siding, Flitwick*.

SALSOLA PESTIFER A. Nels. OUSE: gravel pit, Cople. IVEL: railway siding, Biggleswade. LEA: Dallow Road; *E.M.-R.! in B.E.C. 1946–7 Rep.* 307: North Station Dunstable. Recorded for Cambs., Hunts. and Northants.

TILIACEAE (p. 239)

TILIA TOMENTOSA Moench is planted in some quantity at Heath and Reach. Recorded for Herts and Bucks.

§TRIUMFETTA ANNUA L. IVEL: arable field, Maulden; *N. Y. Sandwith*.

MALVACEAE (p. 240)

§MALVA PUSILLA Sm., C.T.W. 36.1.4 (*M. rotundifolia* L., nom. ambig.). OUSE: gravel pit, Eaton Socon*; *B. Verdcourt in B.E.C. 1943–4 Rep.* 807. Recorded for Herts. and Bucks.

§M. PARVIFLORA L., C.T.W. 36.1.5. IVEL: arable fields, Maulden*, Flitton*. LEA: waste ground, Dunstable*; *H. B. Souster*. Recorded for Herts. and Bucks.

§M. NICAEENSIS All., C.T.W. p. 370. IVEL: arable field, Flitton; *R. A. Graham and J. E. Lousley*!: arable fields, Maulden*, Clifton*.

‡M. ALCEA L., C.T.W. p. 369, is established on Sundon Rubbish Dump!*; *H. B. Souster*.

‡ALTHAEA ROSEA (L.) Cav., C.T.W. 36.3.3, *Hollyhock*, appears on Sundon Rubbish Dump, etc.

A. OFFICINALIS L. was recorded in error in *Plantae Bedford*. from Wilshamstead. Recorded as a native for Cambs.

‡LAVATERA THURINGIACA L. CAM: Cockayne Hatley Wood, escape from churchyard*.

L. TRIMESTRIS L. IVEL: arable field, Arlesey!*; *P.T. in B.E.C. 1945 Rep.* 54.

§L. PLEBEIA Sims. IVEL: arable field, Flitton*.

§MALVASTRUM PERUVIANUM (L.) A. Gray. IVEL: arable field, Beeston*.

§ABUTILON THEOPHRASTI Medic. IVEL: arable field, Maulden*.

§SIDA GLOMERATA Cav. IVEL: shoddy heap, Deepdale*.

LINACEAE (p. 240)

LINUM USITATISSIMUM L., C.T.W. p. 375, *Flax*. Frequent as a relic of cultivation. Recorded for all districts and neighbouring counties. First record: Abbot, 1795.

L. ANGLICUM Mill., C.T.W. 37.1.2 (*L. perenne* L. p.p.). What appears to be this is found on Sundon Rubbish Dump*. Recorded as a native for Cambs., Hunts. and Northants.

Records of *L. bienne* Mill., C.T.W. 37.1.1, from railway yard, Bedford, *J.S. in J.H. Notes* and Wilden by G. H. Day lack confirmation.

GERANIACEAE (p. 241)

‡GERANIUM SANGUINEUM L., C.T.W. 38.1.7, *Bloody Cranesbill*. OUSE: Cardington Cross†; *J.McL*. OUZEL: Aspley Guise; *W.F.S. Diary*. IVEL: Potton; *Mr. Rugely in F.B.* Recorded for Bucks., Cambs. and Northants. First record: Abbot, 1798.

‡G. PHAEUM L., C.T.W. 38.1.6, *Dusky Cranesbill*. *E. Bot.* tab. 322, May 1795, was drawn from a specimen sent by Abbot. IVEL: between Eversholt and Steppingley; *Plantae Bedford*.: near kiln garden, Clophill*; Ridgmont†; *C.C.* LEA: Luton Park; *Davis*: Farr's Lane, East Hyde!*; *H. Cole*. Recorded for all neighbouring counties except Hunts. First record: Abbot, 1795.

‡G. ENDRESSI Gay, C.T.W. 38.1.3. IVEL: garden, Flitwick Manor, shown to me by Mrs. Rayner*. Recorded for Cambs.

§ERODIUM BOTRYS (Cav.) Bertol., C.T.W. p. 394. A common wool adventive with too many stations to merit listing. Recorded for Hunts. First record: the author in *B.E.C. 1945 Rep.* 54.

§E. CYGNORUM Nees—almost as common as *E. botrys*. Recorded for Herts. and Hunts. First record: the author in *Watsonia I* (1950) 247.

§E. OBTUSIPLICATUM (Maire, Weiller & Wilczek) J. T. Howell. OUSE: fields, north of Sandy*. IVEL: Beeston*; *J. E. Lousley*: Maulden; *N. Y. Sandwith*: Flitton*; Clifton.

§MONSONIA BREVIROSTRATA R. Knuth. IVEL: arable field, Flitton; *J. E. Lousley*!

§M. BIFLORA DC. IVEL: arable field, Flitton*; *J. E. Lousley*!: arable field, Maulden. Recorded for Hunts.

TROPAEOLACEAE

‡TROPAEOLUM PEREGRINUM L. IVEL: roadside, Tingrith; *B. Verdcourt*.

OXALIDACEAE (p. 243)

‡OXALIS CORNICULATA L., C.T.W. 39.1.2. *E. Bot.* tab. 1726, April 1807, was drawn from a specimen sent by Abbot from Oakley. It had been introduced by the Duchess of Bedford. Recorded for all neighbouring counties except Cambs.

ACERACEAE (p. 244)

ACER PLATANOIDES L., C.T.W. 41.1.2, *Norway Maple*, occurs rarely as a planted tree. Recorded for Herts. and Cambs.

STAPHYLEACEAE

STAPHYLEA PINNATA L., C.T.W. 42.1.1, *Bladdernut*. Hedgerows, usually near parks. IVEL: Silsoe; *C.C.*: Tingrith; Eversholt. LEA: Leagrave†; *Herb. J.S.*

HIPPOCASTANACEAE

AESCULUS HIPPOCASTANUM L., C.T.W. 43.1.1, *Horse Chestnut*, is common, but apparently always planted.

LEGUMINOSAE (p. 246)

TRIGONELLA PROCUMBENS (Bess.) Reichb. LEA: Kimpton Road, Luton; *D. M. Higgins in W.B.E.C. 1912–3 Rep.*

§MEDICAGO MINIMA (L.) Bartal., C.T.W. 49.8.4, *Small Medick*. This is variable and agrees little with the native form. IVEL: railway sidings, Henlow, Flitwick*, Biggleswade*, Arlesey; arable fields, Maulden, Flitton. Var. RECTA (Desf.) Burnat. The hardiest and most common of the wool adventives and flowering from early spring to late autumn. I have seen it in all neighbouring counties except Bucks. Var. VISCIDA Koch. IVEL: railway sidings, Langford*, det. N. Y. Sandwith.

§M. HISPIDA Gaertn., C.T.W. 49.8.5 (*M. lappacea* Desr.). IVEL: shoddy heap, Deepdale. COLNE: near Caddington and supposed to have been introduced with straw plait, if so it was unique; *R. A. Pryor in J. Bot. 1876*, 22.

§M. DENTICULATA Willd., one of the most common wool adventives and introduced also by other means. Recorded for the Ouse, Cam and Ivel districts, and for all neighbouring counties. First record: A. Poulton in *W.H. 1876*. Var. APICULATA (Willd.) Boiss. OUSE: gravel pit, Eaton Socon*; *J. P. M. Brenan*! IVEL: railway siding, Flitwick; *R. Melville in B.E.C. 1946 Rep.* 220.

§M. LACINIATA (L.) Mill. A common wool adventive in the Ouse and Ivel districts. Recorded for Herts. and Hunts. First record: J. P. M. Brenan in *B.E.C. 1946 Rep.* 222.

§M. ASCHERSONIANA Urb. Closely allied to *M. laciniata*, but much less common. I wish to thank N. Y. Sandwith for his assistance in naming my material. IVEL: railway sidings, Langford*, Shefford*; arable fields, Maulden*, Ampthill.

§M. PROECOX DC. A common wool adventive. Recorded for Herts. and Hunts. First record: E.M.-R. and the author in *B.E.C. 1946 Rep.* 222.

§M. ciliaris Krocker. ivel: arable field, Flitwick*; *R. A. Graham*.

§M. tribuloides Desr. (*M. truncatula* var. *longeaculeata* Urb.). ouse: railway siding, Blunham*. ivel: railway siding, Flitwick; *N. Y. Sandwith in Watsonia I (1950)* 248; railway siding, Biggleswade; arable fields, Maulden*, Flitton*.

Melilotus indica (L.) All., C.T.W. 49.9.4, *Small-flowered Melilot*. A recent introduction which has increased. ouse: gravel pit, Eaton Socon*; railway siding, Bedford. ouzel: allotments, Heath and Reach (H.S. 44); *W. D. Coales and S. P. Rowlands*! ivel: between Langford and Edworth; *J.E.L., J. Bot.*: railway sidings, Flitwick, Ampthill; arable fields, Old Warden, Sutton, Flitwick, Ampthill; Sundon Rubbish Dump. lea: waste ground, Biscot; *H. B. Souster*: railway siding, Luton. Recorded for all neighbouring counties. First record: J. E. Little, 1919.

§M. sulcata Desf. ivel: arable field, Broom*. Recorded for Bucks.

Trifolium incarnatum L., C.T.W. 49.10.5, *Crimson Clover*. A relic of cultivation. A white-flowered form has been seen at Aspley Guise. Recorded for all districts except Nene and Cam, and for all neighbouring counties. First record: railway banks, Herb. J. McLaren, British Museum, 1864.

§T. glomeratum L., C.T.W. 49.10.13, *Clustered Clover*. ouse: gravel pit, Eaton Socon!*; *E.M.-R. in B.E.C. 1946–7 Rep.* 289: railway sidings, Tempsford. ivel: railways sidings, Flitwick*, Arlesey, Sandy, Southill; arable field, Flitton. Recorded for Herts.

T. aureum Poll., C.T.W. 49.10.19 (*T. agrarium* L. p.p.), *Large Hop-trefoil*. ouse: gravel pit, Eaton Socon, established for some years*. Recorded for all neighbouring counties except Hunts. First record: pastures, Herb. J. McLaren, Luton Museum, 1864.

§T. angustifolium L. ouse: gravel pit, Eaton Socon*. ivel: railway sidings, Flitwick*, Sandy, Arlesey, Langford; arable fields, Maulden, Flitton; shoddy heap, Deepdale. First record: the author in *B.E.C. 1946–7 Rep.* 220.

§T. resupinatum L., C.T.W. p. 431, *Reversed Clover*. ivel: arable field, Flitton*; *W. Durant*! Recorded for all neighbouring counties except Hunts.

§T. tomentosum L. ivel: railway siding, Flitwick*; arable fields, Flitton, Maulden*, Flitwick; shoddy heap, Deepdale. Recorded for Herts. and Hunts.

‡Galega officinalis L., C.T.W. 49.13.1, *French Lilac*. Woods and waste places. nene: Wymington*. ouse: Marston Thrift*; gravel pit, Cople*; waste ground, Bedford. lea: Biscot*; *H. B. Souster*. Recorded for Herts., Bucks. and Cambs. First record: the author in *B.E.C. 1943–4 Rep.* 807.

Robinia pseudoacacia L., C.T.W. 49.14.1, *Acacia*, is planted in some woods on the Lower Greensand. Recorded for Herts. and Bucks.

Colutea arborescens L., C.T.W. 49.15.1, *Bladder Senna*, occurs frequently on waste ground by the L.M.R. Railway. Recorded for Herts., Hunts. and Northants.

‡Coronilla varia L., C.T.W. 49.19.1, *Crown Vetch*. ouse: Roxton; *A. West in J.H. Notes*: Cardington Mill!*; *L. W. Wilson*: Bletsoe; *E.M.-R.*!: gravel pit, Clapham; *H. B. Souster*!: railway bank, Lidlington. ouzel: Houghton Regis; *W. D. Coales*: old cottage, Sharpenhoe; *D. M. Higgins in W.B.E.C. 1894–5 Rep.* Recorded for all neighbouring counties except Hunts. First record: D. M. Higgins, 1895.

Vicia dasycarpa Ten. ouse: gravel pit, Eaton Socon!*; *H. K. Airy Shaw in B.E.C. 1946–7 Rep.* 220. ivel: railway siding, Harlington*; arable field, Beeston. Recorded for Bucks.

V. pannonica Crantz. ivel: Sundon Rubbish Dump*.

V. lutea L., C.T.W. 49.22.8, *Yellow Vetch*. ouse: railway bank, Cardington†*; *J.McL. in Bot. Rec. Club. 1884–6 Rep.* ivel: Flitwick; *C. H. Hemsley in J.H. Notes*. Recorded for Herts., Bucks. and Cambs. First record: *Hillhouse List, 1876*.

V. benghalensis L. ouse: arable field between Sandy and Tempsford*; *R. A. Graham and J. E. Lousley*! Recorded for Herts. and Bucks.

Lens culinaris Medic. (*L. lens* (L.) Huth), *Lentil*. ivel: Sundon Rubbish Dump*. Recorded for Bucks.

Lathyrus hirsutus L., C.T.W. 49.23.3. ivel: Sundon Rubbish Dump*. colne: in orchard, Whipsnade; *G. M. Vevers*. Recorded for Herts., Bucks. and Northants.

ROSACEAE (p. 259)

PRUNUS CERASIFERA Ehrh., C.T.W. 50.17.3, *Cherry-plum*. Frequently planted in hedgerows and observed in the Ouse, Ouzel, Ivel and Lea districts. Recorded for Herts., Cambs. and Northants. First record: the author in *B.E.C. 1945 Rep.* 55.

P. PERSICA (L.) Batsch, C.T.W. p. 531, *Peach*. IVEL: Sundon Rubbish Dump*.

P. PADUS L., C.T.W. 50.17.6, *Bird Cherry*. Planted. IVEL: Flitwick Wood; *J. McL., L.M.*: Tingrith; *J.E.L., F.F.B.* LEA: Luton Hoo!; *S.B.P.L.* Recorded for all neighbouring counties except Hunts. First record: J. McLaren, 1864.

P. LAUROCERASUS L., C.T.W. 50.17.7, *Laurel*. Planted occasionally in woods. Recorded for Bucks. and Northants.

§SPIRAEA SALICIFOLIA L., C.T.W. 50.1.1. OUSE: gravel pit, Clapham*. Recorded for Herts. and Northants.

MESPILUS GERMANICA L., C.T.W. 50.20.1, *Medlar*. Planted. IVEL: Tingrith Park; *W.F.B.* LEA: East Hyde Park; *W.F.B.* Recorded for Herts., Bucks. and Northants.

‡FRAGARIA × ANANASSA Duchesne, C.T.W. 50.6.3 (*F. chiloensis* auct.), *Garden Strawberry*. Frequent on railway banks on the L.M.R. line. COLNE: site of old garden, Deadmansey Wood. Recorded for all neighbouring counties.

‡POTENTILLA RECTA L., C.T.W. 50.4.7. OUSE: Oakley*; *G. H. Day*. LEA: waste ground between Luton and Dunstable!*; *H. B. Souster*. Recorded for Bucks. and Northants.

Var. SULFUREA (Lam.) DC. Found so far only in chalk scrub. IVEL: Noon Hill!*; *S. Bowden in B.E.C. 1945 Rep.* 56: foot of Barton Cutting; Ravensburgh Castle.

‡P. NORVEGICA L., C.T.W. 50.4.8. OUSE: Bedford; *J.H. Notes*. LEA: East Hyde†; *W.F.B.* Recorded for all neighbouring counties except Cambs.

‡P. BIFURCA L. OUSE: Cardington Mill†; *J.McL. in Herb. J.S.*

ROSA RUGOSA Thunb., C.T.W. 50.16.3. OUSE: plentiful on roadside north of Mill-brook Station!*; *H. B. Souster*.

ONAGRACEAE (p. 283)

OENOTHERA ERYTHROSEPALA Borbás, C.T.W. 62.4.2 (*O. lamarkiana* auct.). OUSE: railway sidings, Bedford*. Recorded for Bucks., Cambs. and Northants.

O. STRIATA Ledeb. ex Link, C.T.W. 62.4.3 (*O. odorata* auct.). IVEL: Flitwick Station, as *O. biennis*; *J.McL., B.M.*: railway bank, Potton*; *W. Durant*!: arable field, Maulden*; *H. B. Souster*! Recorded for Hunts. and Northants.

HALORAGACEAE (p. 286)

§MYRIOPHYLLUM VERRUCOSUM Lindl., a wool alien, appeared from 1944 to 1946 in gravel pits at Eaton Socon*, but failed to survive the severe winter of 1946-7: see J. P. M. Brenan and J. F. G. Chapple: The Australian *Myriophyllum verrucosum* in Britain; *Watsonia I (1949)* 63.

CORNACEAE (p. 289)

CORNUS MAS L., *Cornelian Cherry*. OUSE: Houghton Ruins!*; *B. F. Haylock*: Rectory Farm, Holcot; *W. J. Brown*.

UMBELLIFERAE (p. 290)

§AMMI MAJUS L., C.T.W. 71.20.1. OUSE: gravel pit, Eaton Socon*; *J. P. M. Brenan and J. F. G. Chapple in B.E.C. 1946-7 Rep.* 220. IVEL: arable field, Beeston; *J. E. Lousley*: arable field, Clifton; railway siding, Flitwick. Recorded for Herts. and Northants.

CORIANDRUM SATIVUM L., C.T.W. 71.10.1, *Coriander*. OUSE: Newnham; *W. Durant*: IVEL: Sundon Rubbish Dump*; arable field and railway siding, Harlington. CAM: arable field, Cockayne Hatley; *T.L.* Recorded for all neighbouring counties. First record: the author in *B.E.C. 1945 Rep.* 59.

PETROSELINUM CRISPUM (Mill.) Nym., C.T.W. 71.17.1 (*Carum petroselinum* (L.) Benth.), *Garden Parsley*. Frequent as a relic of cultivation in the market-gardening area. Recorded for all neighbouring counties. First record: J. McLaren.

ANTHRISCUS CEREFOLIUM (L.) Hoffm., C.T.W. 71.5.3 (*Cerefolium cerefolium* (L.) Schinz & Thell.), *Chervil.* OUSE: roadside, Goldington; *F.B.* Recorded for Herts. and Bucks.

§DAUCUS GLOCHIDIATUS (Labill.) Fisch., Mey. & Avé-Lall. IVEL: Beeston*; *R. A. Graham!*

CAUCALIS LATIFOLIA (L.) L., C.T.W. 71.9.2, *Great Bur Chervil.* OUSE: Oakley; *Plantae Bedford.*: Thurleigh; *F.B.*: Moggerhanger; *A. Ransome in J.H. Notes*: Bedford; *W. Odell in J.H. Notes.* Recorded for all neighbouring counties except Hunts., usually on the evidence of old records. First record: Abbot, 1795.

C. ROYENI (L.) Crantz (*C. lappula* Grande p.p., C.T.W. 71.9.1, *C. daucoides* auct.), *Small Bur Parsley.* OUSE: Oakley North Field; *Plantae Bedford.*: Oakley West Field; *F.B.* IVEL: Barton Hill; *Abbot's annotated F.B.*: Pegsdon; *I. Brown in Herb. Watson*: above Maulden Wood; *M. L. Berrill in J.H. Notes.* Recorded for all neighbouring counties except Hunts. First record: Abbot, 1795.

CUCURBITACEAE (p. 300)

‡CUCUMIS MELO L., *Melon.* IVEL: Sundon Rubbish Dump*.

POLYGONACEAE (p. 302)

POLYGONUM PATULUM Bieb. OUSE: arable field, north of Sandy*.

P. CUSPIDATUM Sieb. & Zucc., C.T.W. 71.1.16, '*Bamboo*'. A species increasing in waste places and recorded for all districts except Nene, Kym and Cam.

P. BALDSCHUANICUM Regel is frequent in hedgerows near Whipsnade Zoo, where it was originally planted.

FAGOPYRUM ESCULENTUM Moench, C.T.W. 75.2.1 (*Polygonum fagopyrum* L., *F. sagittatum* Gilib., *F. fagopyrym* (L.) Karst.), *Buckwheat.* Frequent as a relic of cultivation and recorded for the Ouse, Ouzel, Ivel and Lea districts. First record: Abbot, 1798.

§RUMEX BROWNII Campd., C.T.W. 75.4.21. IVEL: Beeston*; *J. E. Lousley*: railway sidings, Shefford*, Biggleswade*.

ARISTOLOCHIACEAE

‡ARISTOLOCHIA CLEMATITIS L., C.T.W. 73.2.1, *Birthwort.* IVEL: Sundon*; *J.S.* LEA: established in wood, Luton Hoo!†*; *S.B.P.L.*

URTICACEAE (p. 307)

URTICA PILULIFERA L., C.T.W. p. 714, *Roman Nettle.* Apparently established in a garden at Shefford for more than 50 years—specimens Herb. J.S. and Herb. J.McL., B.M.; *J.McL. in W.H., 1876.*

CANNABACEAE (p. 307)

CANNABIS SATIVA L., C.T.W. p. 715, *Hemp.* Usually introduced with bird-seed. IVEL: Sundon Rubbish Dump*; Flitwick. LEA: Sewage Farm, Luton, *c.* 1880†; *Herb. J.S.* Recorded for all neighbouring counties except Hunts.

FAGACEAE (p. 311)

QUERCUS ILEX L., C.T.W. 83.3.2, *Holm Oak.* Planted, Clapham Park!*; *E. Proctor.*
Q. ALBA L., *White Oak.* IVEL: Rowney Warren*.
Q. GEORGIANA M. A. Curt. OUZEL: New Wavendon Heath*.

SALICACEAE (p. 312)

SALIX PENTANDRA L., C.T.W. 84.2.1, *Bay Willow.* LEA: Leagrave Marsh†; *J.S. in J. Bot. 1889.* Recorded for all neighbouring counties except Hunts.
S. × SEPULCRALIS Simonkai (*S. alba* × *babylonica*), C.T.W. p. 754. OUZEL: planted, Woburn Park; *A. B. Jackson in B.E.C. 1931 Rep.* 721.

PLUMBAGINACEAE

‡LIMONIUM SUWOROWI (Regel) Kuntze. IVEL: Sundon Rubbish Dump*.
‡ARMERIA MARITIMA (Mill.) Willd., C.T.W. 89.21.1 (*Statice maritima* L.), *Thrift.* Relic from platelayer's garden, Sundon*. Native in Cambs.

LOGANIACEAE

BUDDLEJA DAVIDI Franch., C.T.W. 91.1.1. OUSE: gravel pits, Eaton Socon, Bromham.

OLEACEAE (p. 323)

SYRINGA VULGARIS L., C.T.W. 92.2.1, *Lilac*, occurs only in hedges near gardens.

POLEMONIACEAE

‡POLEMONIUM CAERULEUM L., C.T.W. 96.1.1, *Jacob's Ladder*. IVEL: edge of Flitwick Moor*; *E.M.-R. and the author in B.E.C. 1943–4 Rep.* 809.

HYDROPHYLLACEAE

NEMOPHILA MENZIESII Hook. & Arn. (*N. insignis* Dougl. ex Benth.). IVEL: Sandy; *Hon. S. Peel in B.E.C. 1928 Rep.* 748.

BORAGINACEAE (p. 327)

LAPPULA MYOSOTIS Moench (*L. lappula* (L.) Karst.). OUSE: Cardington Mill, det. *Echinospermum lappula* (L.) Lehm.†; *Herb. J.S.* Recorded for all neighbouring counties.

AMSINCKIA INTERMEDIA Fisch. & Mey. (*Benthamia intermedia* (Fisch. & Mey.) Druce). Near Knocking, Beds. (*sic*), so abundant in arable fields as to be a pest; *Mrs. Maldon in B.E.C. 1923 Rep.* 200. Recorded for Herts.

A. MENZIESII (Lehm.) Nels. & Macbr. (*Benthamia menziesii* (Lehm.) Druce). IVEL: Biggleswade; *C. Course and M. C. Williams in B.E.C. 1924 Rep.* 584. Recorded for Bucks. and Northants.

‡OMPHALODES VERNA Moench, C.T.W. 97.2.1 (*O. omphaloides* (L.) Voss), *Blue-eyed Mary*. Naturalized in a spinney near Houghton Conquest Rectory*. Recorded for Herts. and Bucks.

SYMPHYTUM ASPERUM Lepech., C.T.W. 94.4.2 (*S. asperrimum* Donn ex Sims), *Rough Comfrey*. IVEL: on land reclaimed from bog, Flitwick Moor; *V.C.H.* Recorded for Herts. and Northants.

‡S. ORIENTALE L., C.T.W. 97.4.4. KYM: locally abundant in and around Pertenhall*. Recorded for Herts. and Cambs.

BORAGO OFFICINALIS L., C.T.W. 97.5.1, *Borage*. OUSE: Duck Mill, Bedford; *F.B.*: Renhold; *W.H. 1875.* IVEL: Ampthill Warren; *F.B.*: near Barton; *Davis*: Flitwick Marsh; *S.B.P.L.*: Sandy; *A. B. Sampson.* LEA: garden weed, Dunstable*; *W. D. Coales.* First record: Abbot, 1798.

‡TRACHYSTEMON ORIENTALIS (L.) Don, C.T.W. 97.6.1. LEA: Luton Hoo†; *Herb. J.S.*

‡ANCHUSA OFFICINALIS L. Most early records, *e.g.*, by Hillhouse are in error for *Lycopsis arvensis*. OUSE: gravel pit, Bromham*; *C. C. Foss*! LEA: rubbish dump, Sundon. Recorded for Bucks. and Northants.

‡PULMONARIA OFFICINALIS L., C.T.W. 97.9.2, *Lungwort*. OUSE: Thurleigh; *Plantae Bedford.*: Basmead; *C.R.O. List*: Castle Close, Bedford; *R. Townsend.* Recorded for Herts., Hunts. and Northants. First record: Abbot, 1795.

‡LITHOSPERMUM PURPUROCAERULEUM L., C.T.W. 97.11.1. OUSE: garden, Colworth House; *W.H. 1875.* Recorded for Cambs.

SOLANACEAE (p. 333)

SOLANUM SARRACHOIDES Sendtn. Becoming more frequent. OUZEL: allotments, Heath and Reach. IVEL: arable fields, Sandy*, Broom, Potton, Maulden, Old Warden.

S. ROSTRATUM Dunal. IVEL: Beckeringspark*; *C.C.* Recorded for Cambs. and Northants.

§PHYSALIS IXOCARPA Brot. ex Hornem. IVEL: arable fields, Flitwick*, Beeston, Maulden.

‡P. ALKEKENGI L. IVEL: Sundon Rubbish Dump.

§NICANDRA PHYSALODES (L.) Gaertn. One of the more attractive wool-adventive species. IVEL: arable fields, Flitwick, Maulden*, Flitton. Recorded for Herts. and Northants.

SCROPHULARIACEAE **455**

‡LYCOPERSICON ESCULENTUM Mill. (*L. lycopersicum* (L.) Karst.), *Tomato*, appears frequently on sewage farms and on Sundon Rubbish Dump.
LYCIUM CHINENSE Mill., C.T.W. 99.1.2, *The Duke of Argyll's Tea-plant.* Frequent, but usually near houses. Recorded for all districts and neighbouring counties. First record: *Hillhouse List, 1876.*
§DATURA FEROX L. IVEL: arable fields, Flitwick, Beeston*.
§NICOTIANA SUAVEOLENS Lehm. IVEL: arable field, Beeston*; *R. A. Graham*!
N. RUSTICA L. OUZEL: allotment, Heath and Reach*; *W. D. Coales.*

SCROPHULARIACEAE (p. 335)

‡VERBASCUM BLATTARIA L., C.T.W. 100.1.5, *Moth Mullein.* OUSE: Harrowden; *J.McL., B.M.*: railway bank, Cardington; *J.McL. in J.H. Notes*: Ravensden; *J.H. Notes.* IVEL: *M. Brown in W.F.S. Diary.* LEA: Hart Lane, 1878†, an inadequate specimen; *F. Wiseman in Herb. J.S.* Recorded for all neighbouring counties except Hunts. First record: uncertain.
‡V. PULVERULENTUM Vill., C.T.W. 100.1.3, *Hoary Mullein.* OUSE: railway bank, Wootton; *W.H. 1875.* Recorded for Herts.
‡LINARIA PURPUREA (L.) Mill., C.T.W. 100.3.2, *Blue Toadflax.* IVEL: old chalk workings, Barton Cutting*. LEA: garden wall, Luton Hoo†; *Herb. J.S.*: chalk pit, foot of Dunstable Downs!; *P.T.*: waste ground on chalk, Round Green. Recorded for all neighbouring counties. First record: Saunders, 1880.
L. PURPUREA × REPENS. LEA: chalk pit, foot of Dunstable Downs, with both parents!*; *P.T.*
‡L. MAROCCANA Hook. f. IVEL: Sandy Warren*.
‡ANTIRRHINUM MAJUS L., C.T.W. 100.2.2, *Snapdragon.* OUSE: Elstow; Bedford!; *F.B.*: churchyard, Millbrook; *W.C. Herb.* IVEL: railway, between Flitwick and Harlington!†; *S.B.P.L.*: railway, Sundon. LEA: Luton Hoo; *S.B.P.L.*: St. Ann's Hill; *D. M. Higgins in W.B.E.C. 1912–3 Rep.*: railway, south of Luton*. Recorded for all neighbouring counties except Northants. First record: Abbot, 1798.
‡MIMULUS MOSCHATUS Dougl. ex Lindl., C.T.W. 100.8.3, *Musk.* OUZEL: muddy shore of Lower Drakelow Pond*; *E.M.-R. and the author in B.E.C. 1945 Rep.* IVEL: Biggleswade; *M. C. Williams in J.E.L., B.E.C.*
‡ERINUS ALPINUS L., C.T.W. 100.11.1. OUSE: old wall, Tempsford*; *P.T.*! Recorded for Bucks.
‡VERONICA FILIFORMIS Sm., C.T.W. 100.13.23. KYM: hedgebank, Tilbrook, v.c. 30 [Hunts.]*. LEA: garden weed, Luton Hoo.

LABIATAE (p. 346)

‡MENTHA × GENTILIS L. (*M. arvensis × spicata*), C.T.W. p. 935. OUSE: Elstow; *Abbot's annotated F.B.* OUZEL: Aspley Guise; *J.E.L., F.F.B.* IVEL: Sundon Rubbish Dump. COLNE: site of old garden, Deadmansey Wood. Recorded for all neighbouring counties. First record: Abbot, *c.* 1800.
‡M. × NILIACA Juss. ex Jacq. (*M. longifolia × rotundifolia*), C.T.W. p. 938. IVEL: Ridgmont, as *M. rotundifolia**; *C.C.*: by stream, Clophill; *J. E. Lousley.* Recorded for Herts., Cambs. and Northants.
M. PULEGIUM L., C.T.W. 105.1.2, *Pennyroyal*, was recorded from Salford by G. C. Druce in *B.E.C. 1907 Rep.* 306 and without locality by Abbot in a letter to J. E. Smith, 3 Feb. 1802. Recorded for all neighbouring counties.
M. CITRATA Ehrh. *E. Bot.* tab. 1025, May 1802, *M. odorata* Sole, was drawn from material sent by Abbot from Bedford. R. A. Graham thinks that it should be referred to this species.
‡MELISSA OFFICINALIS L., C.T.W. 105.10.1, *Balm.* OUSE: Wootton; *W.H. 1875.* OUZEL: Woburn; *W.H. 1875*: Battlesden!*; *H. E. Pickering.* Recorded for all neighbouring counties except Hunts. First record: *Hillhouse List, 1875.*
SALVIA VERTICILLATA L., C.T.W. 105.11.1. OUSE: Harrold!*; *G. H. Day.* IVEL: Flitwick Mill†; *C.C. in W.C. Herb.*: Arlesey; *J.E.L. in B.E.C. 1911 Rep.*: arable field, Barton*; *W. D. Coales*! Recorded for all neighbouring counties except Cambs. First record: J. E. Little, 1911.
‡S. SYLVESTRIS L. IVEL: Flitwick Mill, 1889†; *C.C. in W.C. Herb. and Herb. J.S.*

‡LAMIUM MACULATUM L., C.T.W. 105.17.6, *Spotted Dead-nettle*. IVEL: roadside, between Tempsford and Everton; *H. B. Souster*. LEA: waste ground, near Luton!*; *H. B. Souster*. Recorded for Herts., Bucks. and Northants. First record: Herb. J. McLaren, British Museum.

‡LEONURUS CARDIACA L., C.T.W. 105.18.1, *Motherwort*. OUSE: Ford End; *F.B.* Recorded for Cambs.

‡NEPETA MUSSINI Spreng. ex Henckel, *Garden Catmint*. IVEL: Sundon Rubbish Dump*.

‡TEUCRIUM CHAMAEDRYS L., C.T.W. 105.24.1. IVEL: Warden; *F.B.* Recorded for Bucks.

PLANTAGINACEAE (p. 354)

PLANTAGO OVATA Forsk. IVEL: Sundon Rubbish Dump*; *the author in B.E.C. 1945 Rep.* 67.

P. INDICA L. (*P. psyllium* L., nom. ambig., C.T.W. 106.1.6, *P. ramosa* Aschers.). OUSE: Cardington Mill†*; *J.McL. in Herb. J.S.* IVEL: Sundon Rubbish Dump*. Recorded for all neighbouring counties except Cambs.

CAMPANULACEAE (p. 355)

‡CAMPANULA PATULA L., C.T.W. 107.2.7. LEA: Luton Hoo†; *Mr. Hedge in S.B.P.L.* Recorded for Herts and Northants.

‡C. MEDIUM L., C.T.W. 107.2.9, *Canterbury Bell*, occurs frequently with colour forms on railway embankments (H.S. 45a). Recorded for Herts. and Bucks.

‡C. BONONIENSIS L. OUSE: spinney near Rectory, Houghton Conquest*.

CAPRIFOLIACEAE (p. 361)

‡SYMPHORICARPOS RIVULARIS Suksd., C.T.W. 110.3.1 (*S. albus* auct., *S. symphoricarpos* auct.), *Snowberry*. Common but always near gardens. Recorded for all districts and neighbouring counties. First record: G. C. Druce, 1904.

‡LONICERA XYLOSTEUM L., C.T.W. 110.5.1, *Fly Honeysuckle*. IVEL: Ridgmont; *C.C.*: Luton Hoo†; *S.B.P.L.* Recorded for all neighbouring counties except Hunts. and Cambs. First record: Saunders, 1885.

L. CAPRIFOLIUM L., C.T.W. 110.5.3, *Perfoliate Honeysuckle*. IVEL: Pulloxhill; Ridgmont (1889)†*; *C.C.* Recorded for all neighbouring counties except Hunts.

VALERIANACEAE (p. 362)

‡CENTRANTHUS RUBER (L.) DC., C.T.W. 112.3.1 (*Valeriana rubra* L.), *Red Valerian*. IVEL: Potton; *Abbot's annotated F.B.*: railway sidings, Sundon. LEA: railway banks between Luton and Chiltern Green. Recorded for all neighbouring counties. First record: Abbot in letter to J. E. Smith, 3 Sept. 1802.

VALERIANELLA ERIOCARPA Desv., C.T.W. 112.1.4. OUZEL: on old arable returning to heath, Heath and Reach (H.S. 22b)*; *the author in Watsonia II (1951)* 44. Recorded for Herts.

COMPOSITAE (p. 365)

‡SOLIDAGO CANADENSIS L., C.T.W. p. 1064, *Garden Golden-rod*, occurs frequently as a garden outcast.

§CALOTIS CUNEIFOLIA R.Br. OUSE: railway siding, Willington; *H. B. Souster*! IVEL: railway sidings, Ampthill, Arlesey*, Biggleswade*.

§C. DENTEX R.Br. IVEL: arable land, Flitwick!, my specimen is in Herb. Kew; *W. D. Coales*.

§C. LAPPULACEA Benth. OUSE: railway sidings, Willington*; *H. B. Souster*!

§C. HISPIDULA F. MUELL. IVEL: railway sidings, Southill*; *J. E. Lousley*!

‡FELICIA TENELLA (L.) Nees. OUSE: garden weed, Bedford*; *K. E. West*.

§ERIGERON BONARIENSIS L. IVEL: railway siding, Flitwick*.

AMBROSIA TRIFIDA L. OUSE: near Tempsford Mill*; *Mrs. B. Reynolds in Watsonia I (1949)* 45. Recorded for Herts.

‡HELIANTHUS ANNUUS L., C.T.W. 114.1.1, *Sunflower*. Frequent in waste places.

§XANTHIUM STRUMARIUM L., C.T.W. 114.6.1. IVEL: arable field, Beeston*; *R. A. Graham*!: Maulden; *H. B. Souster*! Recorded for Bucks.

§X. SPINOSUM L., C.T.W. 114.6.3, *Barbed-wire Weed, Thistle Weed*. The most well-known of the wool adventives and common in likely places in the Ouse and Ivel districts. Recorded for all neighbouring counties.

BIDENS FRONDOSA L. IVEL: railway sidings, Flitwick*. LEA: railway sidings, Luton*.

§B. BIPINNATA L. IVEL: arable fields, Flitwick, Maulden, Old Warden; railway siding, Flitwick*.

§B. PILOSA L. IVEL: arable fields, Sandy*, Maulden, Flitton, Flitwick.

GALINSOGA CILIATA (Raf.) Blake, C.T.W. 114.4.2 (*G. quadriradiata* auct.). IVEL: railway siding, Ampthill*; Sundon Rubbish Dump*. Recorded for Herts.

§MADIA SATIVA Molina, C.T.W. p. 1032. IVEL: railway siding, Ampthill*; *J. Russell*!

‡CHRYSANTHEMUM PARTHENIUM (L.) Bernh., C.T.W. 114.28.3 (*Matricaria parthenium* L.), *Feverfew*. Frequent in waste places and sometimes established there for a few years. Recorded for all districts except Nene, Kym and Cam, and for all neighbouring counties. First record: Abbot, 1798.

§MATRICARIA OCCIDENTALIS Greene, C.T.W. p. 1080. IVEL: arable field, Beeston*; *R. A. Graham*!

§COTULA AUSTRALIS (Spreng.) Hook. f. IVEL: railway siding, Flitwick*; *J. E. Lousley*!

‡PETASITES ALBUS (L.) Gaertn., C.T.W. 114.10.2, *White Butterbur*. IVEL: plantation, Ampthill*; *J. K. Horne*. Recorded for Northants.

‡P. FRAGRANS (Vill.) C. Presl, C.T.W. 114.10.3, *Winter Heliotrope*. OUSE: Bedford; *A. W. Guppy*. LEA: Luton Hoo†; *Herb. J.S.* COLNE: waste ground, Farley Estate; *H. B. Souster*. Recorded for all neighbouring counties. First record: Saunders, 1885.

‡DORONICUM PARDALIANCHES L., C.T.W. 114.8.1. OUSE: Cardington†; *J.McL. in Herb. J.S.* Recorded for Herts., Cambs. and Northants.

§SENECIO ARENARIUS Thunb. IVEL: arable field, Maulden*.

§S. ? INAEQUIDENS DC. IVEL: railway sidings, Biggleswade, Flitwick, Ampthill; arable field, Old Warden.

‡CALENDULA OFFICINALIS L., C.T.W. p. 1050, *Garden Marigold*. IVEL: Sundon Rubbish Dump.

‡C. ARVENSIS L., C.T.W. p. 1050. IVEL: Sundon Rubbish Dump.

§SCHKUHRIA PINNATA (Lam.) Thell. IVEL: railway siding, Flitwick*; arable fields, Flitwick, Old Warden, Maulden. First record: the author in *Watsonia I (1949)* 46.

§TAGETES MINUTA L. One of the more frequent wool adventives. IVEL: railway sidings, Flitwick, Biggleswade; arable fields, Flitwick*, Maulden, Beeston, Flitton, Old Warden. LEA: railway siding, Luton*. Recorded for Herts. and Hunts. First record: the author in *Watsonia II (1951)* 45.

ACHILLEA NOBILIS L., C.T.W. p. 1076. OUSE: Cardington Mill†; *J.McL. in Herb. J.S.*

A. MILLEFOLIUM subsp. TANACETIFOLIA Fiori & Paoletti (*A. tanacetifolia* All. non Mill., C.T.W. p. 1075). OUSE: Cardington Mill†; *J.McL. in Herb. J.S.*

§LASIOSPERMUM PEDUNCULARE Lag. IVEL: railway siding, Biggleswade*.

ANTHEMIS TINCTORIA L., C.T.W. 114.23.1, was recorded from Leagrave by Saunders in *J. Bot. 1884*, but I cannot find a specimen to support the record. Recorded for all neighbouring counties except Hunts.

§CENTAUREA MELITENSIS L., C.T.W. 114.39.9, *Maltese Star-thistle*. OUSE: gravel pit, Eaton Socon*; *P.T. and the author in B.E.C. 1946–7 Rep.* 220. Recorded for all neighbouring counties except Hunts.

§CARTHAMUS LANATUS L., C.T.W. p. 111. IVEL: railway sidings, Flitwick*; arable fields, Maulden*, Flitton, Beeston. Recorded for Herts.

C. TINCTORIUS L. IVEL: Sundon Rubbish Dump*. Recorded for Bucks.

SCOLYMUS HISPANICUS L., C.T.W. p. 1174. LEA: Vauxhall Works, Luton!*; *A. J. Newson in Watsonia II (1951)* 46. Recorded for Bucks.

‡CICHORIUM INTYBUS L., C.T.W. 114.41.1, *Chicory*. Frequent and sometimes established for a few years in waste places near towns. Otherwise a relic of cultivation. Recorded for all districts and neighbouring counties. First record: Abbot, 1798.

LAPSANA INTERMEDIA Bieb. OUZEL: well established on railway embankment near Totternhoe!*; *E.M.-R.*—see B. L. Burtt: *Lapsana intermedia* in Britain, *Watsonia I (1950)* 234–7.

CREPIS SETOSA Haller f., C.T.W. 114.54.3, *Bristly Hawk's-beard*. IVEL: Pulloxhill; *C.C. in W.F.B.*: Flitton*; *C.C.*: Wilbury Hill; *J.E.L.*,*F.F.B.* Recorded for Herts. and Northants. First record: C. Crouch, 1897.

C. SANCTA (L.) Babc. (*Lagoseris nemausensis* Bieb.). IVEL: in sainfoin, Wilbury Hill; *B.E.C. 1920–2 Rep.* 386.

C. NICAEENSIS Balb. IVEL: Flitton*; *C.C.*

SONCHUS TENERRIMUS L. IVEL: Biggleswade; *H.P. in B.E.C. 1933 Rep.* 532.

‡TRAGOPOGON PORRIFOLIUS L., C.T.W. 114.47.2, *Salsify*. OUSE: Bedford, 1908, where it has since been found again*; *J.H. Notes*. OUZEL: near A.C. Delco Works, Dunstable; *E.M.-R.* IVEL: Pennyfather's Moor; *H. B. Souster*. Recorded for all neighbouring counties except Hunts.

‡HIERACIUM PULMONARIOIDES Vill. LEA: walls, Luton Hoo*; *H. Catt*.

LILIACEAE (p. 394)

‡ASPARAGUS OFFICINALIS L., C.T.W. 127.7.1, *Garden Asparagus*, occurs in waste places in the market-gardening area of the county.

‡RUSCUS ACULEATUS L., C.T.W. 127.8.1, *Butcher's Broom*. NENE: Hinwick Lodge*. OUSE: Oakley Reynes; *Abbot in letter to J. E. Smith, 3 Feb. 1799*: Kempston Hill Plantation; *J.McL.*, *L.M.* OUZEL: Tebworth; *H. E. Pickering*: Houghton Regis; *W. D. Coales.* IVEL: Warren Wood; *H. B. Souster*. Recorded for all neighbouring counties except Hunts. First record: Abbot, 1799.

‡MUSCARI ATLANTICUM Boiss. & Reut. (*M. racemosum* auct., C.T.W. 127.17.1), *Grape Hyacinth*. IVEL: roadside, north of Potton*; *H. B. Souster*. LEA: Leagrave; *Miss Lye in F.F.B.* Recorded for Herts. and Northants.; native in Cambs. First record: Saunders, 1911.

‡M. COMOSUM (L.) Mill., C.T.W. 127.17.2. LEA: garden, Luton; *H. B. Souster*.

JUNCACEAE (p. 398)

JUNCUS TENUIS Willd., C.T.W. 129.1.2 (*J. macer* Gray). OUSE: gravel pit, Eaton Socon*; *A. J. Wilmott and the author in B.E.C. 1945 Rep.* 70. Recorded for Cambs.

J. GERARDI Lois., C.T.W. 129.1.5, *Mud Rush*. OUSE: gravel pit, Eaton Socon*; *E.M.-R. and the author in B.E.C. 1945 Rep.* 70. Recorded for Cambs.

PALMAE

PHOENIX DACTYLIFERA L., *Date*, occurs frequently in the seedling stage on rubbish dumps.

AMARYLLIDACEAE (p. 401)

‡LEUCOJUM AESTIVUM L., C.T.W. 130.1.2, *Summer Snowflake*. OUSE: Bolnhurst; *R. Townsend*. IVEL: Tingrith Park, where I have known it for a number of years*. Recorded for Herts. and Bucks.

‡NARCISSUS × BIFLORUS Curt., C.T.W. 130.3.5, *Primrose Peerless*. OUZEL: Heath; *G. C. Druce in B.E.C. 1929 Rep.* IVEL: Steppingley, 1845; *W. C. Herb.* COLNE: Whipsnade, established for some years!*; *H. B. Sargent*. Recorded for Herts., Bucks. and Northants.

‡ALLIUM MOLY L. LEA: Sundon Rubbish Dump*; *H. B. Souster*.

IRIDACEAE (p. 403)

‡SISYRINCHIUM BERMUDIANA L. (*S. angustifolium* Mill., C.T.W. 131.1.1), *Blue-eyed Grass*. OUSE: gravel pit, Eaton Socon*; *W. Durant*!

GRAMINEAE (p. 425)

PANICUM MILIACEUM L., *Common* or *Broomcorn Millet*. IVEL: appearing regularly on Sundon Rubbish Dump. Recorded for Herts., Bucks. and Northants.

P. CAPILLARE L. var. OCCIDENTALE Rydb. IVEL: arable field, Sandy*. Recorded for Herts.

§P. LAEVIFOLIUM Hack. IVEL: arable fields, Maulden*, Flitwick*.

DIGITARIA SANGUINALIS (L.) Scop., C.T.W. 139.58.2 (*Panicum sanguinale* L.), *Hairy Fingered-grass*. IVEL: Pulloxhill*; *C.C. in W.F.B.*: railway siding, Flitwick*; *P.T.*! LEA: Luton†; *W. Green in Herb. J.S.*: garden weed, Stockwood!*; *J. Cowley*. Recorded for Herts. and Bucks. First record: C. Crouch, 1897. First evidence: without station, as *P. humifusum*; Herb. J. McLaren, British Museum.

§Echinochloa crusgalli (L.) Beauv., C.T.W. 139.57.1 (*Panicum crusgalli* L.), *Cockspur*. Has increased recently as a wool adventive. ouze: Cardington; *J.McL. in Bot. Rec. Club 1884–6 Rep.*: Castle Mills; *J.McL., B.M.* ousel: allotments, Heath and Reach, var. *aristata* Gray*. ivel: arable fields, Flitton*, Flitwick*, Clophill, Beeston, Maulden, Potton. lea: garden weed, Luton!*; *H. B. Souster*. Recorded for all neighbouring counties. First record: J. McLaren, 1886.

§Setaria viridis (L.) Beauv., C.T.W. 139.59.1, *Green Bristle-grass*. Increasing as a wool adventive. ouse: Eaton Socon*. ouzel: Leighton Buzzard; *V.C.H.*: allotments, Heath and Reach!; *W. D. Coales.* ivel: Flitwick†; *J.S. in Bot. Rec. Club 1884–6 Rep.*: Sandy, var. *major* Gaudin; Biggleswade, Sandy, Maulden. Recorded for all neighbouring counties. First record: Saunders, 1886.

§S. verticillata (L.) Beauv., C.T.W. 139.59.2, *Rough Bristle-grass*. ivel: railway siding, Flitwick; arable fields, Maulden*, Old Warden. Recorded for Herts., Bucks. and Hunts.

S. italica (L.) Beauv., *Foxtail Millet*. ouzel: Leighton Buzzard; *W. D. Coales.* ivel: appearing regularly on Sundon Rubbish Dump. lea: railway, Luton. Recorded for Bucks.

§Rhynchelytrum villosum (Parl.) Chiov. ivel: railway siding, Flitwick, my specimen is in Herb. Kew.

Zea mays L., *Maize*, occurs near chicken runs.

§Tragus racemosus (L.) All. ivel: railway sidings, Flitwick*, Southill; arable field, Flitwick. Recorded for Hunts.

§T. australiensis S. T. Blake. ivel: railway sidings, Flitwick, my specimen is in Herb. Kew.

§T. koelerioides Aschers. ivel: railway sidings, Sandy; *R. A. Graham*!

§T. berteronianus Schult. ivel: railway sidings, Flitwick; *J. P. M. Brenan*!: arable field, Flitwick; *D. McClintock*!

‡Dactyloctenium radulans (R.Br.) Beauv. ivel: railway sidings, Flitwick*; *P.T.*! Recorded for Hunts.

§Eleusine indica (L.) Gaertn. ivel: railway sidings, Flitwick; arable field, Old Warden*.

§E. multiflora Hochst. ex A. Rich. ivel: arable field, Flitwick*; *J. E. Lousley*!

§Eragrostis cilianensis (All.) Link ex Vign. Lut. ivel: railway siding, Flitwick*; *N. Y. Sandwith and B. Verdcourt*: arable field, Beeston; *R. A. Graham and J. E. Lousley*: railway sidings, Arlesey; arable field, Maulden.

§E. dielsii Pilger. Ivel: railway sidings, Flitwick; *J. P. M. Brenan*!

E. parviflora (R.Br.) Trin. ivel: arable fields, Flitwick, Beeston*, Maulden; railway sidings, Sandy.

Phalaris canariensis L., C.T.W. 139.52.2, *Canary Grass*. Frequent in waste places and occasionally sown in small plots. Recorded for all districts except Cam, and for all neighbouring counties. First record: Abbot in letter to J. E. Smith, 3 Sept. 1802.

§P. minor Retz., C.T.W. 139.52.3. One of the more frequent wool adventives in the Ouse and Ivel districts. Recorded for Herts. and Bucks. First record: the author in *B.E.C. 1946–7 Rep*. 318.

§‡P. paradoxa L., C.T.W. 139.52.4. ivel: arable field, Flitton*. Recorded for Herts.

Var. appendiculata (Schult.) Chiov. (var. *praemorsa* auct.). ivel: arable field, Beeston; *R. A. Graham and J. E. Lousley*: railway sidings, Ampthill.

P. tuberosa L. (*P. bulbosa* auct.). ivel: in meadow land, Old Warden, originally planted; *E.M.-R. and the author in B.E.C. 1945 Rep*. 72. Recorded for Bucks.

§Sorghum halepense (L.) Pers. ouse: arable field, between Sandy and Tempsford, my specimen is in Herb. Kew; *J. C. Culshaw*!

Mibora minima (L.) Desv., C.T.W. 139.44.1. lea: established in gardens, Stockwood!*; *J. Cowley*.

§Agrostis lachnantha Nees. ivel: railway sidings, Biggleswade*.

§A. avenacea J. F. Gmel. (*Calamagrostis filiformis* (Forst. f.) Pilger non Griseb.). ivel: railway sidings, Flitwick*, Sandy.

§Polypogon monspeliensis (L.) Desf., C.T.W. 139.42.1, *Annual Beardgrass*. One of the more frequent wool adventives in the Ouse and Ivel districts. Recorded for Herts., Bucks. and Northants. First record: B. Verdcourt in *B.E.C. 1943–4 Rep.* 813.

P. maritimus Willd. ouzel: shore of Lower Drakelow Pond; *W.W.*!

§Cynodon hirsutus Stent. ivel: Maulden; Flitton; but in vegetative condition only.

§Chloris truncata R.Br., *Sunshade Grass*. ivel: arable field, Beeston!*; *G. Cope*: railway sidings, Flitwick*; *J. P. M. Brenan*!: railway sidings, Southill.

§C. virgata Sw. ivel: railway sidings, Flitwick*; *J. P. M. Brenan*!: arable fields, Beeston, Maulden; railway sidings, Langford.

§Danthonia sp. ivel: arable field, Beeston; *R. A. Graham and J. E. Lousley.*

§Diplachne fusca (L.) Beauv. ex Roem. & Schult. (*D. reptatrix* (L.) Druce). ivel: arable field, Beeston; *R. A. Graham and J. E. Lousley*: arable field, Flitton*.

§Koeleria phleoides (Vill.) Pers. ivel: railway sidings, Southill*.

Vulpia megalura (Nutt.) Rydb. ivel: railway sidings, Flitwick. lea: railway ballast, East Hyde*. First record: the author in *B.E.C. 1946–7 Rep.* 279.

§Bromus madritensis L. (*Anisantha madritensis* (L.) Nevski, C.T.W. 139.20.2), *Compact Brome*. It is possible that a record of Abbot's of *B. diandrus* Curt. non Roth in a letter to J. E. Smith, 3 Sept. 1802, may refer to this. ivel: railway siding, Flitwick, var. *ciliatus* Guss.

§B. diandrus Roth (*B. gussonii* Parl., *Anisantha gussonii* (Parl.) Nevski, C.T.W. 139.20.3, *B. rigens* auct., *B. maximus* auct.), *Great Brome*. ivel: Sundon Rubbish Dump*; *J. P. M. Brenan and N. Y. Sandwith in Watsonia I (1951)* 58: arable field, Maulden.

B. tectorum L. (*Anisantha tectorum* (L.) Nevski, C.T.W. 139.20.5), *Drooping Brome*. ouse: gravel pit, Eaton Socon, where it dominated large patches!*; *J. P. M. Brenan in B.E.C. 1946–7 Rep.* 220. Recorded for Herts. and Bucks.

§B. molliformis Lloyd. ivel: railway sidings, Flitwick*; *the author in B.E.C. 1946–7 Rep.* 321.

B. squarrosus L. ouse: Cardington; *J.McL., B.M.* Recorded for Bucks.

§B. unioloides Kunth (*Ceratochloa unioloides* (Willd.) Beauv., C.T.W. p. 1458). ivel: railway sidings, Flitwick*; arable fields, Flitton*, Old Warden. Recorded for Bucks. and Northants.

§B. rubens L. ivel: railway sidings, Flitwick; *J. P. M. Brenan in B.E.C. 1946–7 Rep.* 220.

§Trachynia distachya (L.) Link (*Brachypodium distachyon* (L.) Beauv.). ivel: railway sidings, Sandy, my specimen is in Herb. Kew; *R. A. Graham*! Recorded for Hunts.

§Hordeum leporinum Link. ouse: railway sidings, Willington*. ivel: railway sidings, Flitwick, Southill; shoddy heap, Deepdale.

§H. marinum Huds., C.T.W. 139.26.3, *Sea Barley*. ivel: railway sidings, Southill*; arable field, Maulden. Recorded for Bucks. and Herts.

§H. hystrix Roth (*H. gussonianum* Parl.). ouse: railway sidings, Willington. ivel: railway sidings, Flitwick*, Ampthill; arable fields, Flitwick, Maulden; shoddy heap, Deepdale.

§H. jubatum L., C.T.W. p. 1465. ivel: arable field, Maulden*. Recorded for Herts., Bucks. and Northants.

Elymus arenarius L., C.T.W. 129.25.1. colne: derelict garden, Whipsnade!*; *W. D. Coales.*

In addition to those listed above the following, included in the main body of the Flora, appear also as wool adventives: *Sisymbrium orientale, Coronopus didymus, Chenopodium murale, C. hybridum, Malva neglecta, Erodium moschatum, E. cicutarium, Medicago arabica, Trifolium subterraneum, Acaena anserinifolia, Lythrum hyssopifolia, Solanum nigrum, Datura stramonium, Marrubium vulgare, Carduus tenuiflorus, Silybum marianum, Centaurea solstitialis, Juncus pallidus, Apera spicaventi, Vulpia bromoides, Lolium temulentum.*

FUNGI

by D. A. Reid

THE FIRST list of fungi for the county is in Abbot's (*q.v.*) *Plantae Bedfordiensis* (1795). This was followed three years later by an extensive list of species from a wide area of the county in *Flora Bedfordiensis* (1798). Almost a century passed with no evidence of any study of the group. In the later years of the nineteenth century James Saunders (*q.v.*) made a thorough study of the *Myxomycetes*, particularly in the neighbourhoods of Luton and Flitwick. He worked very closely with Charles Crouch (*q.v.*), who also became interested, and they were indeed fortunate to be accompanied on some expeditions by Guilielma Lister (1860–1949), daughter of Arthur Lister, then engaged in the revision of this fascinating group of plants. D. Martha Higgins (*q.v.*) also made a collection of *Myxomycetes*, now at Luton Museum, but there is no evidence that she added anything to the information already collected by Saunders. Simultaneously John Hamson (*q.v.*) was giving some attention to fungi, more particularly the *Basidiomycetes*. He wisely sent most of his gatherings to either W. B. Grove or W. G. Smith (*q.v.*)[1] and it is certain that most of his records are correct. Both Grove and Smith made also a few independent records. He received a number of specimens from E. M. Langley (*q.v.*) and Mr. Ferraby, of whom I know nothing. The *Victoria County History* (Bedfordshire) contained, as a result of the work done by Saunders and Hamson, a comparatively full and accurate list of the fungi of the county.

More than forty years passed before there was evidence of further work. Hamson may have continued his work, but neither lists nor specimens can be found. In 1947 the Bedfordshire Natural History Society held its first fungi foray, and this has since been an annual event. The leader, except in one year, has been R. W. G. Dennis, of the Royal Botanic Gardens, Kew, an excellent mycologist, and under his leadership the forays have added much to our knowledge of the fungi of the county. Dr. Dennis has also made independent visits and E. Milne-Redhead (*q.v.*) on many visits has collected further specimens. In recent years further records have come from F. H. Warcup (see Bibliography) and Miss Holden of Harpenden. I am indebted to T. Laflin for records mainly from the Cockayne Hatley neighbourhood.

My own interest in the fungi began in 1947, since when I have collected extensively, but mainly in the neighbourhood of Leighton Buzzard.

The list which follows is the result of the work of very few mycologists and covers only small areas of the county. No attempt has been made to estimate the distribution of species, which, in many cases, must be much wider than the records given would indicate. There

---

[1] These are shown within brackets, after the record, in the text.

is some difficulty in interpreting Abbot's records in terms of modern nomenclature, but fortunately he cited one or more plates with each of the species he listed and it is usually possible to identify the species thus represented. This, combined with Abbot's notes on habitat, host, etc., makes it possible to reconcile the records with the present-day definitions of the species. This was done by W. B. Grove for the species listed in *Flora Bedfordiensis*, and I have found myself in almost complete agreement with his conclusions, and his paper has been of great assistance. With the many additions in Abbot's annotated copy of *Flora Bedfordiensis* I have had to use my own judgment. Many of Abbot's records cannot, however, be placed beyond the genus to which they belong, if at all, and they have been excluded from the present list.

The list is incomplete and many groups are poorly represented. It is hoped that its publication will stimulate a new interest in the fungi and that future records will be reported.

SEQUENCE. This follows that adopted by C. C. Ainsworth and G. R. Bisby: *A Dictionary of the Fungi*, Third Edition (1950), which is derived from G. W. Martin: *Outline of the Fungi* (1941). The species have been listed alphabetically within the genera, and the genera similarly listed within the families.

NOMENCLATURE. This is, in the main, based on the following standard works for the various groups concerned:

Myxomycetes: A. Lister, *A Monograph of the Mycetozoa*, Third Edition, revised by G. Lister (1925).

Pyrenomycetes: G. R. Bisby and E. W. Mason, List of Pyrenomycetes recorded for the British Isles, *Trans. Brit. Mycol. Soc.* 1951.

Discomycetes: J. Ramsbottom and F. L. Balfour-Browne, A List of Discomycetes recorded for the British Isles, *Trans. Brit. Mycol. Soc.* 1951.

Basidiomycetes: Carleton Rea, *British Basidiomycetae*, 1922. A. A. Pearson and R. W. G. Dennis, Revised list of British Agarics and Boleti, *Trans. Brit. Mycol. Soc.* 1948.
  Ustilaginales: G. C. Ainsworth and K. Sampson, *The British Smut Fungi*, 1951.
  Uredinales: W. B. Grove, *The British Rust Fungi*, 1913.

Fungi Imperfecti
  Hyphomycetes: E. M. Wakefield and G. R. Bisby, List of Hyphomycetes recorded for Britain, *Trans. Brit. Mycol. Soc.* 1941.
  Coelomycetes: W. B. Grove, *British Stem and Leaf Fungi*, Vol. I, 1935, Vol. II, 1937.

ABBREVIATIONS. In addition to those listed on pp. 144–6 the following are used in this section only:

D.A.R.   D. A. Reid.
Eng. Fungi. J. Sowerby, *Illustrations of English Fungi*, 1797–1815.
J.H., V.C.H. Fungi, J. Hamson in *Victoria County History* (Bedfordshire), 1904.

J.S., V.C.H.   Myxomycetes, J. Saunders in *Victoria County History* (Bedfordshire), 1904.

Trans. Brit. Mycol. Soc.   *Transactions of the British Mycological Society.*

BIBLIOGRAPHY (see also pp. 146–9).

Grove, W. B.   The Fungi of Abbot's Flora Bedfordiensis; *Midland Naturalist*, 1893.

Hamson, J.   Fungi; *Victoria County History* (Bedfordshire), 1904.

Saunders, J.   The Mycetozoa of South Beds. and North Herts.; *J. Bot. XXXI*, 1893.

Mycetozoa of the South Midlands; *Ibid.*, *XXXVIII*, 1900.

Mycetozoa or Myxomycetes; *Victoria County History* (Bedfordshire), 1904.

Warcup, F. C.   Effects of partial sterilization by steam or formalin on damping off of Sikta Spruce; *Trans. Brit. Mycol. Soc.* 1952.

MYXOMYCETES (Mycetozoa) *Slime Moulds*

CERATIOMYXACEAE

Ceratiomyxa fruticulosa Macbr. (*C. mucida Schroet.*)—Luton Hoo; Flitwick; *J.S. in J. Bot. 1900.*

PHYSARACEAE

Badhamia capsulifera Berk. (*B. hyalina* Berk.)—Heath; Caddington; *J.S. in J. Bot. 1893.*

B. foliicola Lister.—unlocalized; *J.S., V.C.H.*

B. lilacina Rost.—unlocalized; *J.S., V.C.H.*

B. macrocarpa Rost.—Flitwick; *C.C. in J. Bot. 1900.*

B. nitens Berk. (*B. inaurata* Currey)—Caddington; *J.S. in J. Bot. 1893*: Ridgmont; *C.C. in J. Bot. 1900.*

B. ovispora Racib.—Barton; Stopsley Common; Nether Crawley; *J.S. in J. Bot. 1900.* The first British record for the species.

B. panicea Rost.—Luton Hoo; *J.S. in J. Bot. 1893.*

B. papaveracea Berk. & Rav. (*B. hyalina* Berk. var. *papaveracea* Lister)—unlocalized; *J.S., V.C.H.*

B. rubiginosa Rost.—Heath; *J.S. in J. Bot. 1893.*

B. utricularis Berk.—Heath; Caddington; *J.S. in J. Bot. 1893.*

Craterium aureum Rost. (*C. mutabile* Fr.)—Flitwick; Chaul End; *J.S. in J. Bot. 1900.*

C. leucocephalum Ditm.—Pepperstock; Totternhoe; *J.S., V.C.H.*

C. minutum Fr. (*C. pedunculatum* Trentep., *C. vulgare* Ditm.)—Clapham Park Wood on *Hypnum triquetrum* as *Trichia minuta*; *F.B.* [Neither Berkeley nor Lister mention *Trichia minuta*, but it is probable that Grove was correct in assuming it to be this species]: Heath; Stopsley; Pepperstock; *J.S. in J. Bot. 1893.*

Diachaea leucopoda Rost. (*D. elegans* Fr.)—Flitwick; *J.S. in J. Bot. 1900.*

D. subsessilis Peck—Flitwick; *Guilielma Lister in J. Bot. 1900.* The first record for Europe.

Diderma effusum Morgan (*Chondrioderma reticulatum* Rost.)—Flitwick; *J.S. in J. Bot. 1900.*

D. hemisphericum Hornem. (*Chondrioderma michelii* Rost.)—Silsoe; *C.C. in J. Bot. 1900*: Flitwick; *J.S. in J. Bot. 1900.*

D. niveum Macbr. (*Chondrioderma niveum* Rost.)—Flitwick; *J.S. in J. Bot. 1900.*

D. radiatum Morgan (*Chondrioderma radiatum* Rost.)—Heath; Pepperstock; *J.S. in J. Bot. 1893.*

D. spumarioides Fr. (*Chondrioderma spumarioides* Rost.)—Ridgmont; *E. Crouch in J. Bot. 1900*: Flitwick; *J.S. in J. Bot. 1900.*

D. testaceum Pers. (*Chondrioderma testaceum* Rost.)—Stopsley (the first British record); *J.S. in J. Bot. 1893*: Flitwick, abundant in some seasons; *J.S. in J. Bot. 1900.*

Fuligo cinerea Morgan (*F. ellipsospora* Lister)—Stopsley Common—the first British record; *J.S. in J. Bot. 1900.*

F. septica Gmelin (*F. varians* Somm.)—common, as *Reticularia septica*; Warden, as *R. ovata*; *F.B.*: Kitchen End; *C.C. in J. Bot. 1893*: Luton Hoo; *J.S. in J. Bot. 1893*: Clophill; *Beds. N.H.S. Foray* 1952.

Leocarpus fragilis Rost. (*L. vernicosus* Link)—Silsoe, as *Lycoperdon fragile*; *F.B.*: Heath; Ampthill; Pepperstock; *J.S. in J. Bot. 1893.*

Physarum bitectum Lister (*P. diderma* Rost.)—Flitwick; *J.S. in J. Bot. 1900.*

P. cinereum Pers.—Clapham between Woods, as *Lycoperdon cinereum*; *F.B.*: Flitwick; Stopsley Common; *J.S. in J. Bot. 1900.*

P. citrinum Schum.—unlocalized; *J.S., V.C.H.*

P. conglomeratum Rost.—Flitwick; *J.S. in J. Bot. 1900.*

P. contextum Pers.—Flitwick; *J.S. in J. Bot. 1900.*

P. didermoides Rost.—Mead Hook; *C.C. in J. Bot. 1900*: Nether Crawley; *J.S. in J. Bot. 1900.*
 Var. **lividum** Lister—Chaul End; Nether Crawley; Stopsley Common; *J.S. in J. Bot. 1900.*

P. globuliferum Pers.—Clapham Park Wood, as *Trichia globulifera*; *F.B.*

P. leucopus Link—unlocalized, *J.S., V.C.H.*

P. nutans Pers.—Luton Hoo, as var. *violascens*; *J.S. in J. Bot. 1893.*
 Var. **leucophaeum** (Fr.) Lister (*P. leucophaeum* Fr.)—common; *J.S. in J. Bot. 1893.*

P. pusillum Lister (*P. calidris* Lister)—Pulloxhill; *J.S. in J. Bot. 1893.*

P. reniforme Lister (*P. compressum* Alb. & Schw.)—Luton Hoo; *J.S. in J.Bot. 1893*: Flitwick; *J.S. in J. Bot. 1900.*

P. sinuosum Weinm. ex Fr. (*P. bivalve* Pers.)—unlocalized; *J.S., V.C.H.*

P. straminipes Lister—Chaul End; Nether Crawley; Maiden Common; *J.S. in J. Bot. 1900.*

P. vernum Somm.—Kitchen End; *C.C. in J. Bot. 1900*: Bedford; Warden Hills; *J.S. in J. Bot. 1900.*

P. viride Pers.—Heath; Stopsley; Luton Hoo; *J.S. in J. Bot. 1893.*

DIDYMIACEAE

Didymium clavus Rabenh.—Flitwick; *C.C. in J. Bot. 1900.*

D. complanatum Rost. (*D. serpula* Fr.)—Flitwick; *C.C. in J. Bot. 1900.*

D. difforme Duby (*Chondrioderma difforme* Rost.)—Heath; Luton; *J.S. in J. Bot. 1893.*

D. melanospermum Macbr. (*D. farinaceum* Schrad.)—common, as *Trichia sphaerocephala*; *F.B.*: Heath; *J.S. in J. Bot. 1893.*

D. nigripes Fr. (*D. microcarpon* Rost.)—Kitchen End; *C.C. in J. Bot. 1893.*
 Var. **xanthopus** Lister (*D. pertusum* Berk.)—Clophill; *J.S. in J. Bot. 1893*: Flitwick; Stopsley Common; *J.S. in J. Bot. 1900.*

D. squamulosum Fr. (*D. effusum* Link)— Sundon; Luton Hoo; *J.S. in J.Bot. 1893*: Kitchen End; *C.C. in J. Bot. 1893*: on *Polyporus fumosus*, Ampthill (*W. B. Grove*); *J.H., V.C.H.*

D. vaccinum Buchet (*D. trochus* Lister)—Kitchen End; *C.C. in J. Bot. 1900*: Chaul End; Stopsley Common; *J.S. in J. Bot. 1900.*

Lepidoderma tigrinum Rost.—Heath; *Guilielma Lister in J.S., V.C.H.*

Mucilago spongiosa Morgan (*Spumaria alba* DC.)—Flitwick; Chalton; *J.S. in J. Bot. 1900.*

STEMONITACEAE

Comatricha laxa Rost.—Flitwick; *J.S. in J. Bot. 1900.*

C. nigra Schroet. (*C. friesiana* Rost., *C. obtusata* Preuss)—common, as *Mucor embolus*; *F.B.*: Heath; Leagrave; Pepperstock; *J.S. in J. Bot. 1893.*

C. pulchella Rost. (*C. persoonii* Rost.)—Flitwick; *J.S. in J. Bot. 1900.*

C. rubens Lister—Flitwick; Chaul End; *J.S. in J. Bot. 1900.*

C. typhoides Rost. (*C. typhina* Rost.)—Luton Hoo; Stopsley; *J.S. in J. Bot. 1893.*

Enerthenema papillatum Rost. (*E. elegans* Bowman)—Caddington; Luton Hoo; *J.S. in J. Bot. 1893.*

Lamproderma columbinum Rost. (*L. physaroides* Rost.)—Heath; *J.S. in J. Bot. 1893.*
L. scintillans Morgan (*L. irideum* Massee)—unlocalized; *J.S., V.C.H.*
L. violaceum Rost.—on decayed poplar, Luton Hoo; *J.S. in J. Bot. 1900.*
Stemonitis ferruginea Ehrenb.—common, as *Trichia nuda*; *F.B.* [Lister gives *T. nuda* as a synonym of *S. fusca*, but the plate cited by Abbot (Bulliard tab. 477 fig. 1) is that of *T. axifera* which Lister considered to be synonymous with *S. ferruginea*]: Chalton; Pepperstock; Kitchen End; *J.S. in J. Bot. 1893.*
    Var. **smithii** Lister (*S. smithii* Macbr.)—unlocalized; *J.S., V.C.H.*
S. fusca Roth—Silsoe, Southill, as *T. typhaeformis*, considered by W. B. Grove to be synonymous with *S. fusca*, *F.B.*: Heath; Luton Hoo; Sundon; *J.S. in J. Bot. 1893.*
    Var. **confluens** Lister—unlocalized; *J.S., V.C.H.*
S. splendens Rost.—unlocalized; *J.S., V.C.H.*

AMAUROCHAETACEAE

Brefeldia maxima Rost.—Sewell; Luton Hoo; *J.S. in J. Bot. 1900.*

HETERODERMACEAE

Cribraria argillacea Pers.—Heath; Luton Hoo; *J.S. in J. Bot. 1893.*
C. vulgaris Schrad. var. **aurantiaca** Pers. (*C. aurantiaca* Schrad.)—Heath; Luton Hoo; *J.S. in J. Bot. 1893.*
Dictydium cancellatum Macbr. (*D. umbilicatum* Schrad., *D. cernuum* Nees)—common, as *Trichia recutita*; *F.B.*: Luton Hoo; Chalton; *J.S. in J. Bot. 1893.*
Lindbladia effusa Rost. (*L. tubulina* Fr.)—unlocalized; *J.S., V.C.H.*

LICEACEAE

Licea flexuosa Pers.—Ruxox Farm, Flitwick; *J.S. in J. Bot. 1900.*

TUBULINACEAE

Tubifera ferruginosa Gmel. (*Tubulina cylindrica* DC., *T. fragiformis* Pers.)—Kitchen End; *C.C. in J. Bot. 1893*: Luton Hoo; *J.S. in J. Bot. 1900*: Sandy, 1887 (W. B. Grove); *J.H., V.C.H.*

RETICULARIACEAE

Dictydiaethalium plumbeum Rost.—Kitchen End; *C.C. in J. Bot. 1900.*
Enteridium olivaceum Ehrenb.—Heath; *J.S. in J. Bot. 1893.*
Liceopsis lobata Torrend (*Reticularia lobata* Lister, *R. rozeana* Lister)—Heath; *J.S. in J. Bot. 1893.*
Reticularia lycoperdon Bull. (*R. umbrina* Fr.)—Luton Hoo; *J.S. in J. Bot. 1893*: Ampthill (W. B. Grove); *J.H., V.C.H.*

LYCOGALACEAE

Lycogala epidendrum Fr. (*L. miniatum* Pers.)—common, as *Lycoperdon epidendrum*; *F.B.*: Luton Hoo; Kitchen End; Sharpenhoe; *J.S. in J. Bot. 1893.*
L. flavo-fuscum Rost.—Kitchen End, the first British record; *C.C. in J. Bot. 1893.*

TRICHIACEAE

Hemitrichia clavata Rost.—unlocalized; *J.S., V.C.H.*
H. intorta Lister—unlocalized; *J.S., V.C.H.*
H. vesparium Macbr. (*Hemiarcyria rubiformis* Rost., *Hemitrichia rubiformis* Lister)—Kitchen End; *C.C. in J. Bot. 1893*: Barton Springs, as var. *neesiana*; *J.S. in J. Bot. 1893.*
Trichia affinis de Bary—Heath; Sundon; near Luton; *J.S. in J. Bot. 1893.*
T. botrytis Pers. (*T. fragilis* Rost.)—Heath; Pepperstock; *J.S. in J. Bot. 1893.*
    Var. **munda** Lister—Pepperstock; *J.S. in J. Bot. 1900.*
T. contorta Rost.—Caddington; *J.S. in J. Bot. 1893.*
    Var. **inconspicua** Lister—unlocalized; *J.S., V.C.H.*
T. decipiens Macbr. (*T. fallax* Pers.)—common, as *T. pyriformis*; *F.B.*: Heath; Sundon; near Luton; Luton Hoo; *J.S. in J. Bot. 1893.*

T. persimilis Karst. (*T. jackii* Rost.)—Heath; Pepperstock; near Luton; *J.S. in J. Bot. 1893.*
T. scabra Rost.—Sewell; *J.S. in J. Bot. 1893.*
T. varia Pers. (*T. nigripes* Pers.)—Renhold Wood, as *T. olivacea*; common, as *T. turbinata*, probably this species (Grove); *F.B.*: Heath; near Luton; Leagrave; *J.S. in J. Bot. 1893*: Ampthill, as var. *genuina* (not listed by Lister) (W. B. Grove); *J.H., V.C.H.*: Clophill; *Beds. N.H.S. Foray*, 1952.

ARCYRIACEAE

Arcyria cinerea Pers.—common, as *Trichia cinerea*, *F.B.*: Luton Hoo; Stopsley; *J.S. in J. Bot. 1893.*
A. denudata Wettstein (*A. punicea* Pers.)—Southill, as *Trichia denudata*; *F.B.*: common; *J.S. in J. Bot. 1893.*
A. ferruginea Sauter—Heath; *J.S. in J. Bot. 1893.*
A. incarnata Pers.—Heath; Barton Springs; Caddington; *J.S. in J. Bot. 1893.*
A. nutans Grev. (*A. flava* Pers.)—Caddington; Luton Hoo; *J.S. in J. Bot. 1893.*
A. pomiformis Rost. (*A. albida* Pers. var. *pomiformis* Lister)—unlocalized; *J.S., V.C.H.*
Perichaena corticalis Rost. (*P. populina* Fr.)—Friar's Walls, as *Trichia fusco-ater*; *F.B.*: Luton Hoo; *C.C. in J. Bot. 1900*: East Hyde; *J.S. in J. Bot. 1900*: Clophill; *Beds. N.H.S. Foray*, 1952.
P. depressa Libert—Upbury; *C.C. in J. Bot. 1900*: Streatley; *J.S. in J. Bot. 1900.*
P. vermicularis Rost. (*P. variabilis* Rost.)—Leagrave; Chiltern Green; Maiden Common; *J.S. in J. Bot. 1900.*

MARGARITACEAE

Margarita metallica Lister—Ridgmont; *E. Crouch in J. Bot. 1900.*
Prototrichia metallica Massee (*P. flagellifera* Rost.)—Heath; *J.S. in J. Bot. 1893.*

PHYCOMYCETES

PLASMODIOPHORALES: PLASMODIOPHORACEAE

Plasmodiophora brassicae Woron.—Potton; *T.L.*

PERONOSPORALES: PYTHIACEAE

Pythium dissotocum Drechsler—Ampthill; *F. H. Warcup* 1952.
P. intermedium de Bary—Ampthill; *F. H. Warcup* 1952.
P. rostratum Butler—Ampthill; *F. H. Warcup* 1952.
P. ultimum Trow.—Ampthill; *F. H. Warcup* 1952.

PERONOSPORALES: ALBUGINACEAE

Cystopus candidus (Pers. ex Chev.) Lév.—on *Capsella bursa-pastoris*, Leighton Buzzard; *D.A.R.* 1953: widespread, *T.L.*

PERONOSPORALES: PERONOSPORACEAE

Peronospora alta Fuckel—on *Plantago*, Leighton Buzzard; *D.A.R.* 1953.
P. destructor (Berk.) Casp.—on onions, unlocalized; *Min. of Ag. and Fish Bulletin*, 126, 48: Cockayne Hatley; *T.L.*
P. effusa (Grev. ex Desm.) Rabenh.—on spinach, 1934, unlocalized; *Min. of Ag. and Fish. Bulletin*, 126: Cockayne Hatley; *T.L.*
P. galligena Blumer—common on *Alyssum saxatile* L., Leighton Buzzard; *D.A.R.* 1953.
P. niessleana Berlese—on *Alliaria petiolata*, Leighton Buzzard; *D.A.R.* 1953.
P. schachtii Fuckel—on red beetroot, unlocalized; *Min. of Ag. and Fish. Bulletin*, 139, 31.
P. trivialis Gaüm.—on *Cerastium* sp., Heath and Reach; *D.A.R.* 1953.
P. viciae (Berk.) de Bary—on *Vicia hirsuta*, and probably not uncommon, Leighton Buzzard; *D.A.R.* 1953.

P. viciae-sativae Gaüm.—on *Vicia sativa*, common, Leighton Buzzard; *D.A.R.* 1953.

Phytophthora cryptogea Pethybr. and Laff.—on *Petunia* sp., unlocalized; *Min. of Ag. and Fish. Bulletin*, 126, 87: Cockayne Hatley; *T.L.*

P. erythroseptica Pethybr.—on potato, unlocalized; *Min. of Ag. and Fish. Bulletin*, 139, 22.

P. infestans (Mont.) de Bary—widespread; *T.L.*

P. parasitica Dast.—Cockayne Hatley; *T.L.*

Plasmopara aegopodii (Casp.) Trotter—on *Aegopodium podagraria*, Leighton Buzzard; *D.A.R.* 1953.

Pseudoperonospora humuli (Miyabe & Tak.) G. W. Wilson—on wild *Humulus lupulus*, Leighton Buzzard; *D.A.R.* 1953.

MUCORALES: MUCORACEAE

Mucor mucedo L.—common; *F.B.*

Rhizopus nigricans Ehrenb.—on rotting flower-head of *Dracunculus vulgaris* Schott, Leighton Buzzard; *D.A.R.* 1952.

Spinellus fusiger (Fr.) van Tiegh. (*Mucor fusiger* Link)—on *Agaricus fusipes*; *W. B. Grove in J.H.*, *V.C.H.*

Syzygites megalocarpus Ehrenb. (*Sporodinia grandis* Link)—growing on *Lactarii* from Sandy (W. B. Grove) *J.H.*, *V.C.H.*

MUCORALES: PILOLOLACEAE

Pilobolus crystalinus (Tode) van Tiegh.—Paradise, near Bedford as *Mucor roridus* and as *M. ureolatus*; *F.B.* (Abbot cited *Dicks. Crypt. 1* tab. 3, fig. 6 for the latter. This was regarded as *P. crystalinus* by Berkeley but Grove thought that it might be *P. kleinii* van Teigh.).

ASCOMYCETES

TAPHRINALES: PROTOMYCETACEAE

Protomyces macrosporus Unger—on *Aegopodium podagraria*, Leighton Buzzard; Heath and Reach; *D.A.R.* 1952.

TAPHRINALES: TAPHRINACEAE

Taphrina aurea (Pers.) Fr. (*T. populina* Fr.)—on poplars, Whipsnade and Leighton Buzzard; *D.A.R.* 1952.

T. betulina Rost. (*T. turgida* (Sadeb.) Giesenhag)—on silver birch causing witches' broom, common, Heath and Reach, Woburn, Leighton Buzzard; *D.A.R.* 1948.

T. deformans (Berk.) Tul.—very common on peach and almond trees causing leaf curl, Leighton Buzzard, Heath and Reach, etc.; *D.A.R.* 1948.

PYRENOMYCETES *Flask Fungi*

EUROTIALES: ONYGENACEAE

Onygena equina Willd. ex Fr.—old horses' hoofs, Clapham, Sandy, as *Lycoperdon gossypinum*; *Plantae Bedford.*: Potton Wood, as *L. gossypinum*; *F.B.*: Everton, Clapham, as *L. equinum*; *F.B.*

EUROTIALES: ELAPHOMYCETACEAE

Elaphomyces cervinus (L. ex Fic. & Schub.) Schlect.—rare, as *Tuber radicatum*; *F.B.*

ERYSIPHALES: ERYSIPHACEAE

Erysiphe cichoracearum DC.—on marrow, 1941 and 1942, unlocalized; *Min. of Ag. and Fish. Bulletin*, 126: common on wide variety of hosts, Leighton Buzzard, Heath and Reach, etc.; *D.A.R.* 1948: Cockayne Hatley; *T.L.*

E. graminis DC.—common on wide variety of grasses, Leighton Buzzard, etc.; *D.A.R.* 1953: widespread; *T.L.*

E. polygoni DC.—common on wide variety of hosts, Leighton Buzzard, Heath and Reach, etc.; *D.A.R.* 1948: widespread; *T.L.*
E. tortilis (Wallr.) Fr.—on *Thelycrania sanguinea*, Heath and Reach; *D.A.R.* 1952.
Microsphaera alphitoides Griff. and Maubl.—on oak, perithecia found 1948, the oidium stage is exceedingly common, Leighton Buzzard, Heath and Reach, etc.; *D.A.R.* 1948.
Podosphaera leucotricha (Ell. and Everh.) Salm.—on apple trees, Leighton Buzzard; *D.A.R.* 1953; Cockayne Hatley; *T.L.*
Sphaerotheca humuli (DC.) Burrill—very common on *Filipendula ulmaria*, Leighton Buzzard; *D.A.R.* 1948: Cockayne Hatley; *T.L.*
S. mors-uvae (Schw.) Berk. & Curtis—Cockayne Hatley; *T.L.*
S. pannosa (Wallr.) Lév.—very common on wild and cultivated roses, Leighton Buzzard, etc.; *D.A.R.* 1948: Cockayne Hatley; *T.L.*
Uncinula necator (Schw.) Burrill—Cockayne Hatley; *T.L.*

HYPOCREALES: NECTRIACEAE

Dialonectria sanguinea (Bolt. ex Fr.) Cooke (*Nectria sanguinea* Fr.)—Clapham Park Wood, as *Sphaeria sanguinea*; *F.B.*
Nectria cinnabarina (Tode ex Fr.) Fr.—exceedingly common on dead twigs, Leighton Buzzard, etc.; *D.A.R.* 1948: widespread; *T.L.*
N. coccinea (Pers. ex Fr.) Fr.—Renhold Wood, as *Sphaeria mori*; *F.B.*

HYPOCREALES: HYPOCREACEAE

Claviceps purpurea (Fr.) Tul.—on barley, 1937, unlocalized; *Min. of Ag. and Fish. Bulletin*, 126, 9: Cockayne Hatley; *T.L.*
Cordyceps militaris (L. ex Fr.) Link (*Torrubia militaris* Fr.)—Ampthill Bogs, as *Clavaria militaris*; *F.B.*
Epichloe typhina (Pers. ex Fr.) Tul.—on *Agrostis*, *Dactylis* and *Holcus*, Leighton Buzzard, etc.; *D.A.R.* 1948.
Hypocrea pulvinata Fuckel—on *Polyporus betulinus*, Heath and Reach; *D.A.R.* 1953.

SPHAERIALES: SPHAERIACEAE: ALLANTOSPORAE

Diatrype bullata (Hoffm. ex Fr.) Fr.—common, as *Sphaeria depressa*; *F.B.*
D. disciformis (Hoffm. ex Fr.) Fr.—Clapham between Woods, as *Sphaeria echinata*; *F.B.*
D. stigma (Hoffm. ex Fr.) Fr.—common as *Sphaeria stigma*; *F.B.*: common, as *S. decorticans*; *Abbot's annotated F.B.*
Nitschkia cupularis (Pers. ex Fr.) Karst.—common, as *Sphaeria corticalis*; *F.B.*

SPHAERIALES: SPHAERIACEAE: PHAEOSPORAE

Bombardia fasciculata Fr. (*Sphaeria bombarda* Batsch)—Clapham between Woods, as *Sphaeria clavata*; *F.B.*
Daldinia concentrica (Bolt. ex Fr.) Ces. & de Not.—Thurleigh Wood, as *Sphaeria fraxinea*; *Plantae Bedford*: Clapham Lane, as *S. fraxinea*; *F.B.*: Barton Hills; *Beds. N.H.S. Foray*, 1952.
Hypoxylon coccineum Bull.—common, as *Sphaeria lycoperdoides*; *F.B.*
H. multiforme (Fr.) Fr.—on birch, Heath and Reach, Woburn, etc.; *D.A.R.* 1952.
Poronia punctata (L. ex Fr.) Fr.—on horse-dung, Clapham Closes, as *Peziza punctata*; *Plantae Bedford*.
Rosellinia aquila (Fr.) de Not.—common, as *Sphaeria mammosa* and *S. globularis*; *F.B.* (W. B. Grove considered the former record to refer to *S. mammaeformis* Pers. (*Rosellinia mammiformis* (Pers. ex Fr.) Ces. and de Not.) but I consider it to be this species): Renhold, as *S. papillosa*; *Abbot's annotated F.B.*
Ustulina vulgaris Tul.—on fallen and live trunks of trees, especially beech, Heath and Reach; *D.A.R.* 1952.
Xylaria hypoxylon (L. ex Fr.) Grev.—common, as *Clavaria hypoxylon*; *F.B.*: common (W. B. Grove); *J.H.*, *V.C.H.*: Clophill; *Beds. N.H.S. Foray*, 1949; Rowney Warren; *Beds. N.H.S. Foray*, 1950: Barton Hills; *Beds. N.H.S. Foray*, 1951; common, Heath and Reach, etc.; *D.A.R.*

X. pedunculata (Dicks. ex Berk.) Fr.—*Eng. Fungi* tab. 437, 1814 was drawn from a specimen sent by Abbot.
X. polymorpha (Pers. ex Fr.) Grev.—common, as *Clavaria digitata*; *F.B.*: Ampthill, 1887 (W. B. Grove); *J.H.*, *V.C.H.*: Barton Hills; *Beds. N.H.S. Foray*, 1951: common, Heath and Reach; *D.A.R.*

SPHAERIALES: SPHAERIACEAE: HYALODIDYMAE

Didymella applanata (Niessl) Sacc.—Cockayne Hatley; *T.L.*
D. lycopersici Kleb.—on tomato, unlocalized; *Min. of Ag. and Fish. Bulletin*, 139, 56.
Mycosphaerella fragariae (Tul.) Lindau—Cockayne Hatley; *T.L.*

SPHAERIALES: SPHAERIACEAE: HYALOPHRAGMIAE

Lasiosphaeria spermoides (Hoffm. ex Fr.) Ces. and de Not. (*Sphaeria spermoides* Hoffm. ex Fr.)—common, as *Sphaeria aggregata*; *F.B.*

SPHAERIALES: SPHAERIACEAE: PHAEOPHRAGMIAE

Leptosphaeria acuta (Hoffm. ex Fr.) Karst. (*Sphaeria acuta* Hoffm. ex Fr.)— Bedford; Oakley; *Plantae Bedford.*: common; *F.B.*

SPHAERIALES: SPHAERIACEAE: SCOLICOSPORAE

Ophiobolus graminis Sacc.—widespread; *T.L.*

HYSTERIALES

Hysterographium fraxini (Pers. ex Fr.) de Not.—common, as *Sphaeria sulcata*; *F.B.*
Lophium mytilinum (Pers. ex Fr.) Fr.—common, as *Sphaeria ostreacea*; *F.B.*

PHACIDIALES: PHACIDIACEAE

Coccomyces coronatus (Schum. ex Fr.) de Not.—rare, as *Peziza viridis*; *F.B.*
Lophodermium pinastri (Schrad. ex Fr.) Chev.—Woburn Sands; *R. W. G. Dennis*, 1948: Heath and Reach; *D.A.R.* 1953.
Rhytisma acerinum (Pers.) Fr. (*Melasmia acerina* Lév.)—common, as *Mucor granulosus*; *F.B.* (Abbot cites Bulliard tab. 504, 5, but this plate does not show a fungus on sycamore, a host to which the above is specific. He no doubt saw *R. acerinum*, as he records it as common): common, Ampthill; *J.H.*, *V.C.H.*: common at Leighton Buzzard, Heath and Reach, Woburn, etc.; *D.A.R.*: Cockayne Hatley; *T.L.*

PHACIDIALES: STICTIDIACEAE

Stictis radiata (L.) Pers. ex Fr.—Clapham Wood, as *Peziza marginata*; *Plantae Bedford.*: common; *F.B.*

DISCOMYCETES *Disc Fungi*

HELOTIALES

Calycella citrina (Hedw. ex Fr.) Quél. (*Helotium citrinum* Fr.)—common, as *Peziza cyathoides*; *F.B.*
C. lenticularis (Bull. ex Fr.) Boud. (*Helotium lenticularis* Bull. ex Fr.)—common, as *Peziza lenticularis*; *F.B.*
Coryne sarcoides (Jacq. ex Fr.) Tul. (*Bulgaria sarcoides* Fr.)—Southill Wood, Clapham Wood, as *Helvella sarcoides*; common, as *Peziza tremelloidea*; *F.B.*
Cudoniella acicularis (Bull. ex Fr.) Schroet. (*Helotium aciculare* Fr.)—Clapham Wood, Twin Wood, as *Helvella agariciformis*; *Plantae Bedford.*: common; *F.B.*: Aspley Wood; *Beds. N.H.S. Foray*, 1948: Clophill; *Beds. N.H.S. Foray*, 1952: Heath and Reach; *D.A.R.* 1952.

Dasyscypha bicolor (Bull. ex Fr.) Fuckel (*Peziza bicolor* Bull. ex Fr.)—(Abbot's record of *Peziza bicolor*, common, could be referred, according to Grove, to either *Dasyscypha bicolor* (Bull. ex Fr.) Fuckel or *Peziza calycina* Schum.).

D. brevipila Le Gal—on holly twig, Leighton Buzzard; *D.A.R.* 1953.

D. nivea (Hedw. fil ex Fr.) Sacc.—on oak, Heath and Reach; *D.A.R.* 1953.

D. virginea (Batsch ex Fr.) Fuckel (*Peziza virginea* Batsch ex Fr.)—common, as *Peziza nivea*; *F.B.*

Geoglossum elongatum Starb.—Rowney Warren; *Beds. N.H.S. Foray*, 1950.

Haglundia perelegans Nannf.—Aspley Wood; *Beds. N.H.S. Foray*, 1948 (the first British record).

Helotium amenti (Batsch ex Fr.) Fuckel—very common on old male catkins of *Salix* spp., Leighton Buzzard; *D.A.R.* 1952.

H. calyculus (Sow. ex Fr.) Berk.—Silsoe New Woods; Clapham Park, as *Peziza calyculus*; *Plantae Bedford.*: common; *F.B.* (W. B. Grove thought these records might refer to *H. virgultorum* (Wahl. ex Fr.) Karst.)

H. cyathoideum (Bull. ex Fr.) Karst.—on herbaceous stems, Leighton Buzzard; *D.A.R* 1953.

H. fructigenum (Bull. ex Fr.) Fuckel—common, as *Peziza fructigena*; *F.B.*

Hyaloscypha hyalina (Pers. ex Fr.) Boud.—Clophill; *Beds. N.H.S. Foray*, 1952: on dead wood and twigs, etc., Leighton Buzzard; *D.A.R.*

Lachnella spadicea (Pers. ex Fr.) Phill.—Potton Wood, as *Peziza spadicea*; *F.B.*

L. sulphurea (Pers. ex Fr.) Quél.—Renhold, Clapham Park, between Woods, as *Peziza hydnoides*; *Abbot's annotated F.B.*

Lecanidion atratum (Hedw. ex Fr.) Rabenh. (*Patellaria atrata* Fr.)—Ampthill, as *Peziza atra*; *F.B.* (it is probable that this species was intended).

Leotia lubrica (Scop.) Pers. ex Fr.—Southill, Silsoe, as *Helvella gelatinosa*; *F.B.*: Clophill; *Beds. N.H.S. Foray*, 1952.

Mitrula abietis Fr.—Clophill; *Beds. N.H.S. Foray*, 1952.

M. phalloides (Bull.) Chev.—Silsoe New Wood, Ampthill Heath, as *Clavaria epiphylla*; *Abbot's annotated F.B.*

Mollisia cinerea (Batsch ex Fr.) Karst. (*Peziza cinerea* Batsch)—Sheerhatch Wood; Clapham Wood, as *Peziza cinerea*; *Plantae Bedford.*: common; *F.B.*

Orbilia xanthostigma Fr.—common on rotting wood, Heath and Reach, etc.; *D.A.R.* 1952.

Patellaria abbotiana (Sow. ex Fr.) Sacc. (*Peziza abbotiana* Sow. ex Fr.)—*Eng. Fungi* tab. 389, Jan. 1803, was drawn from a specimen sent by Abbot in whose honour the species was named.

Phialea firma (Pers. ex Fr.) Gillet (*Peziza firma* Pers. ex Fr.)—Clapham Park Wood, as *Peziza ochroleuca*; *F.B.*

Pseudohelotium pineti (Batsch ex Fr.) Fuckel (*Peziza pineti* Batsch ex Fr.)—Warden; Ampthill; *F.B.*

Pseudopeziza ribis Kleb.—Cockayne Hatley; *T.L.*

Sclerotinia tuberosa (Hedw. ex Fr.) Fuckel (*Peziza tuberosa* Hedw. ex Fr.)—Renhold Wood; *F.B.*

S. fructigena Aderh. & Ruhl.—widespread; *T.L.*

S. laxa Aderh. & Ruhl.—Cockayne Hatley; *T.L.*

S. serica Keay—on *Gyposohila elegans* Bieb., Biggleswade; *J. Bot.* 1937, the first record for Britain.

S. trifoliorum Erikss.—Cockayne Hatley; *T.L.*

Spathularia clavata (Schaeff.) Sacc. (*S. flavida* Pers. ex Fr.)—Southill; Silsoe New Wood, as *Helvella clavata*; *Plantae Bedford.*: Ampthill; as *Helvella spathulata*; *F.B.*

Trichoscypha calycina (Schum ex Fr.) Boud. (*Peziza calycina* Schum ex Fr.)—on larch twigs (W. B. Grove); *J.H.*, *V.C.H.*: common on larch twigs, Heath and Reach; *D.A.R.*

Urceolella leuconica (Cooke) Boud.—fir cone, Woburn; *D.A.R.* 1951.

U. papillaris (Bull. ex Fr.) Boud. (*Peziza papillaris* Bull. ex Fr.)—common; *F.B.*

PEZIZALES

Acetabula vulgaris Fuckel (*Peziza acetabulum* L. ex Fr.)—Ampthill; Aspley; *F.B.*

Aleuria amplissima Boud.—on experimental plots treated with sawdust and in one case, formalin, Ampthill; *Miss M. Holden*, 1953.

A. repanda (Karst.) Boud.—on beech log, Deadmansey Wood; *E.M.-R.*

A. vesiculosa (Bull. ex Fr.) Boud. (*Peziza vesiculosa* Bull. ex Fr.)—Kempston (W. G. Smith); *J.H., V.C.H.*

Ascobolus stercorarius (Bull.) Schroet. (*A. furfuraceus* Pers. ex Fr.)—common, as *Peziza stercoraria*; *F.B.*

Bulgaria inquinans (Pers.) Fr.—Silsoe New Wood, Clapham Park, as *Peziza turbinata*; common, as *P. polymorpha*; *F.B.*

Ciliaria scutellata (L. ex Fr.) Quél. (*Peziza scutellata* L. ex Fr., *Lachnea scutellata* (L. ex Fr.) Gill.)—Maulden; Clapham; *Plantae Bedford.*: common; *F.B.*: Barton Hills; *Beds. N.H.S. Foray*, 1951.

Coprobia granulata (Bull. ex Fr.) Boud. (*Peziza granulata* Bull. ex Fr., *Humaria granulata* (Bull ex Fr.) Sacc.)—Bedford Meadows, Clapham Closes, as *P. fulva*; *Plantae Bedford.*: common, *F.B.*

Galactinia badia (Pers. ex Fr.) Boud. (*Peziza badia* Pers. ex Fr.)—Pavenham (W. G. Smith); *J.H., V.C.H.*

Helvella crispa Scop. ex Fr.—Clapham; Houghton Conquest, as *Helvella mitra*; *Plantae Bedford.*: Barton Hills; *Beds. N.H.S. Foray*, 1951: Woburn; *D.A.R.*

Humaria humosa (Fr.) Sacc.—Rowney Warren; *Beds. N.H.S. Foray*, 1950: with *Polytrichum*, sandpits, Grovebury Road, Leighton Buzzard; *D.A.R.*

Lachnea hemispherica (Wigg. ex Fr.) Gillet (*Peziza hemispherica* Wigg. ex Fr.)—Silsoe New Wood, as *Peziza hispida*; *Plantae Bedford.*: common; *F.B.*

Macropodium macropus (Pers. ex Fr.) Fuckel—Heath and Reach; *D.A.R.* 1951.

Morchella crassipes Krombh.—Ampthill, 1877 (W. B. Groves); *J.H., V.C.H.*

M. rotunda (Pers.) Boud.—Clapham Closes, Medbury Closes, as *Phallus esculentus*; *Plantae Bedford.*: common, as *Helvella esculenta*; *F.B.* (Abbot cited Sowerby tab. 51, but this depicts two species, *Morchella rotunda* and *M. patula* (Pers. ex Fr.) Boud. It is more likely that he found the former): Ampthill; Sandy; *J.H., V.C.H.*

Otidea cochleata (L. ex Fr.) Fuckel (*Peziza cochleata* L. ex Fr.)—Maulden, Clapham, as *Peziza cochleata*; *Plantae Bedford.*: common; *F.B.*

O. onotica (Pers. ex Fr.) Fuckel (*Peziza onotica* Pers. ex Fr.)—Silsoe, as *P. leporina*; *F.B.*

Peziza aurantia Pers. ex Fr., *Orange-peel fungus*—Silsoe New Wood, Clapham, as *P. coccinea*; *Plantae Bedford.*: Clapham Springs; *F.B.*: Clophill; *Beds. N.H.S. Foray*, 1952.

P. rutilans Fr.—with *Polytrichum*, sandpits, Grovebury Road, Leighton Buzzard; *D.A.R.* 1951.

Sarcoscypha coccinea (Jacq. ex Fr.) Cooke (*Peziza coccinea* Jacq. ex Fr.)—Clapham, Southill, as *P. epidendra*; *F.B.*

Verpa conica Mull. ex Fr. var. **relhani** (Sow. ex Fr.) Boud.—rare, as *Helvella relhani*; *F.B.*

TUBERALES (*Truffles*)

Genea hispidula Berk.—listed for the county in Cooke, *Handbook of British Fungi*, 1871.

Tuber aestivum Vitt.—Renhold Wood, Willington Pastures, as *T. cibarium* *F.B.* (Abbot cites Bulliard tab. 356 and in his annotated copy, Sowerby tab. 309. The latter is referred to *T. aestivum* in *Trans. Brit. Mycol. Soc.*, XXXIV, 65, and Grove refers Abbot's record to *T. brumale* Vitt.): Ampthill, Flitwick, common; *Mr. Ferraby in J.H., V.C.H.*

BASIDIOMYCETES

USTILAGINALES: USTILAGINACEAE *Smut Fungi*

Sphacelotheca hydropiperis (Schum.) de Bary—on *Polygonum hydropiper*, Heath and Reach; *D.A.R.* 1952.

Ustilago avenae (Pers.) Rostr.—on *Arrhenatherum elatius*, common everywhere; *D.A.R.* 1948.

U. hypodytes (Schlecht.) Fr.—on *Bromus erectus*, Totternhoe; *D.A.R.* 1952.

U. longissima (Sow. ex Schlecht.) Meyen.—on *Glyceria aquatica*, Leighton Buzzard; *D.A.R.* 1953.

U. nuda (Jens.) Rostr. (*U. carbo* (DC.) Tul.)—common, as *Reticularia segetum*; *F.B.*

U. tritici (Pers.) Rostr.—widespread; *T.L.*

U. violacea (Pers.) Fuckel—Rowney Warren; *Beds. N.H.S. Foray*, 1950; common on *Melandrium album* and *M. dioicum*, Woburn, Leighton Buzzard, Heath and Reach; *D.A.R.* 1948.

USTILAGINALES: TILLETIACEAE

Urocystis cepulae Frost—on onion, Sandy; Biggleswade; *Min. of Ag. and Fish. Bulletin*, 139, 52: Cockayne Hatley; *T.L.*

UREDINALES: MELAMPSORACEAE *Rust Fungi*

Coleosporium rhinanthacearum Lév.—on *Odonites verna*, Totternhoe; *D.A.R.* 1952.

C. senecionis Fr.—on *Senecio vulgaris*, very common, Leighton Buzzard, Heath and Reach, etc.; *D.A.R.* 1948.

C. sonchi Lév.—on *Sonchus arvensis* and *S. oleraceus*, Leighton Buzzard, Stanbridge, etc.; *D.A.R.* 1948.

C. tussilaginis Tul.—on *Tussilago farfara*, Leighton Buzzard; *D.A.R.* 1948.

Cronartium ribicola F. de Waldh.—rare on *Pinus* and *Ribes*, Woburn; *W. B. Grove in Brit. Rust. Fungi*, 1913.

Melampsora larici-caprearum Kleb.—on *Salix caprea*, Leighton Buzzard; *D.A.R.* 1952.

Melampsoridium betulinum Kleb.—on silver birch, Heath and Reach, Woburn, Leighton Buzzard; *D.A.R.*

UREDINALES: PUCCINIACEAE

Gymnosporangium clavariaeforme DC.—St. Peter's, Bedford, as *Aecidium laceratum*; *Abbot's annotated F.B.*

G. sabinae Wint.—on sabine, Oakley Plantations, as *Tremella sabinae*; *Abbot's annotated F.B.*

Ochropsora sorbi Diet. (*Aecidium leucospermum* DC.)—Aspley Wood; Clapham Wood; as *Lycoperdon anemones*; *Plantae Bedford.*: unlocalized, as *Lycoperdon innatum*; *F.B.* (W. B. Grove lists this as *Aecidium leucospermum*).

Phragmidium mucronatum (Pers.) Schlecht.—on *Rosa* sp., Barton Hill; *T.L.*

P. violaceum Wint.—very common on *Rubus fruticosus* (agg.), Leighton Buzzard, Heath and Reach, Woburn, etc.; *D.A.R.* 1948: Cockayne Hatley; *T.L.*

P. sanguisorbae (DC.) Schrot.—on *Poterium sanguisorba*, Deacon Hill, *T.L.*

Puccinia adoxae Hedw.—on *Adoxa moschatellina*, Heath and Reach; *D.A.R.* 1953.

P. aegopodi Mart.—on *Aegopodium podagraria*, Heath and Reach; *D.A.R.* 1953.

P. antirrhini Diet. and Holw.—very common on cultivated antirrhinums, Leighton Buzzard, etc.; *D.A.R.* 1948: Cockayne Hatley; *T.L.*

P. asparagi DC.—on *Asparagus*, 1933; *Min. of Ag. and Fish. Bulletin*, 126, 41: Cockayne Hatley; *T.L.*

P. bulbocastani Fuckel—very rare, Dunstable; *W. G. Smith in W. B. Grove, British Rust. Fungi*, 1913.

P. caricis Reb.—on *Carex riparia*, with aecidial stage on *Urtica*, Leighton Buzzard; *D.A.R.* 1953.

P. centaureae DC.—on *Centaurea nigra*, Leighton Buzzard; *D.A.R.* 1952.

P. chrysanthemi Roze—on cultivated chrysanthemums, Leighton Buzzard; *D.A.R.* 1948: Cockayne Hatley; *T.L.*
P. cirsii-palustris (Desm.) M. Wilson—on *Cirsium palustris*, Leighton Buzzard; *D.A.R.* 1953.
P. graminis Pers.—on various grasses, Leighton Buzzard; *D.A.R.* 1948.
P. holcina Erikss.—on *Holcus lanatus*, Leighton Buzzard; *D.A.R.* 1952.
P. lapsanae Fuckel—on *Lapsana communis*, Totternhoe; *D.A.R.* 1952.
P. lychnidearum Link—on *Melandrium album*, Leighton Buzzard; *D.A.R.* 1948.
P. malvacearum Mont.—on *Malva sylvestris* and *Althea rosea*, Leighton Buzzard, etc.; *D.A.R.* 1948: Cockayne Hatley; *T.L.*
P. menthae Pers.—Cockayne Hatley; *T.L.*
P. obtegens Tul.—on *Cirsium arvense*, Leighton Buzzard, *D.A.R.* 1948: Cockayne Hatley; *T.L.*
P. poarum Niels.—aecidia on *Tussilago farfara*, Leighton Buzzard; *D.A.R.* 1948.
P. pruni-spinosae Pers.—Cockayne Hatley; *T.L.*
P. sonchi Rob.—on *Sonchus oleraceus*, Leighton Buzzard; *D.A.R.* 1948.
P. variabilis Grev.—on *Taraxacum officinale*, Leighton Buzzard; *D.A.R.* 1952.
P. violae DC.—on *Viola riviniana* and *V. hirta*, Heath and Reach; *D.A.R.* 1952.
Uromyces dactylidis Otth.—aecidia on *Ranunculus acris*, Leighton Buzzard; *D.A.R.* 1953.
U. dianthi (Pers.) Niessl.—Cockayne Hatley; *T.L.*
U. ficariae Lév.—on *Ranunculus ficaria*, Leighton Buzzard; *D.A.R. 1952.*
U. phaseolorum de Bary—on dwarf bean, 1941; *Min. of Ag. and Fish. Bulletin*, 126, 33: Cockayne Hatley; *T.L.*
U. rumicis Wint.—on *Rumex crispus*, Leighton Buzzard, *D.A.R.* 1952.

TREMELLALES: DACRYOMYCETACEAE

Calocera cornea (Batsch) Fr.—timber yards, Bedford, as *Clavaria aculeiformis*; *F.B.*: common on trunks of broad-leaved trees, especially beech, Heath and Reach; *D.A.R.*
C. viscosa (Pers.) Fr.—common (W. B. Grove); *J.H.*, *V.C.H.*: Aspley Wood; *Beds. N.H.S. Foray*, 1948: Clophill; *Beds. N.H.S. Foray*, 1950: common on coniferous tree stumps, Heath and Reach, Woburn, etc.; *D.A.R.*
Dacryomyces deliquescens (Bull.) Duby—common as *Tremella deliquescens*; *F.B.*: Clophill; *Beds. N.H.S. Foray*, 1949: Barton Hills; *Beds. N.H.S. Foray*, 1951: very common on old wooden fences, etc., Leighton Buzzard; *D.A.R.*
D. stillatus (Nees) Fr.—Bedford (W. B. Grove); *J.H.*, *V.C.H.*

TREMELLALES: TREMELLACEAE

Exidia glandulosa (Bull.) Fr., *Witches' butter*—common, as *Tremella arborea* and *S.* (? *T.*) *glandulosa*; *F.B.*: on hazel branches, Heath and Reach; *D.A.R.*
Tremella albida (Huds.) Fr.—common; *F.B.*
T. mesenterica (Retz.) Fr.—Ampthill; Warden; *F.B.*: Ampthill, 1889; *Mr. Ferraby in J.H.*, *V.C.H.*: common on fallen branches and twigs, Heath and Reach; *D.A.R.*

TREMELLALES: AURICULARIACEAE

Auricularia auricula-judae Schröet., *Jew's ear*—stump of a dead elm at the Friars, found by Mrs. Abbot; *Abbot's annotated F.B.* (The reference to elm makes one doubt the record): Heath and Reach; *Beds. N.H.S. Foray*, 1948: Rowney Warren; *Beds. N.H.S. Foray*, 1950: Barton Hills; *Beds. N.H.S. Foray*, 1951: very abundant on elder, Leighton Buzzard, etc.; *D.A.R.*: Cockayne Hatley; *T.L.*
A. mesenterica Fr.—common as *A. tremelloides*; *F.B.* (It is possible that Abbot's record of the above species also belongs here. He cites no plate, however, so one cannot be certain, but elm is quite a common host for this fungus): common (W. B. Grove); *J.H.*, *V.C.H.*: Barton Hills; *Beds. N.H.S. Foray*, 1951.
Helicobasidium purpureum (Tul.) Pat.—Cockayne Hatley; *T.L.*

AGARICALES: THELEPHORACEAE

Coniophora arida Fr. (*Corticium aridum* Fr.)—Ampthill, 1885 (W. G. Smith); *J.H.*, *V.C.H.*

C. membranacea (DC.) Massee—Bedford, as *Auricularia pulverulenta*; *Abbot's annotated F.B.*

C. puteana (Schum.) Karst.—Clophill; *Beds. N.H.S. Foray*, 1952: common on old logs, Leighton Buzzard, Heath and Reach; *D.A.R.*

Corticium caeruleum (Schrad.) Fr.—common, as *Byssus phosphorea*; *F.B.*

C. laeve (Pers.) Quél.—common on old twigs and logs, and often becoming distinctly pileate, Leighton Buzzard; *D.A.R.* 1952.

C. sambuci (Pers.) Fr.—common on elder, Leighton Buzzard, Heath and Reach, etc.; *D.A.R.* 1948.

Craterellus cornucopioides Fr.—Silsoe, as *Peziza cornucopioides*; *F.B.*: Aspley Wood; *Abbot's annotated F.B.*: Deadmansey Wood; *E.M.-R.*: Heath and Reach; *D.A.R.*

Cyphella lactea Bres.—on dead culms of *Glyceria maxima*, Leighton Buzzard, very rare; *D.A.R.*

C. villosa (Pers.) Karst.—common on decaying plant debris, especially old bramble stems, Leighton Buzzard; *D.A.R.* 1948.

Hymenochaete corrugata (Fr.) Lév. (*Corticium corrugatum* Fr.)—Sharnbrook, 1885 (W. G. Smith); *J.H.*, *V.C.H.*

H. rubiginosa (Dicks.) Lév. (*Stereum rubiginosum* (Schrad.) Fr.)—common; *F.B.*: common on oak stumps, Heath and Reach; *D.A.R.*

H. tabacina (Sow.) Lév. (*Stereum tabacinum* (Sow.) Fr.)—Clophill, Ampthill, as *Auricularia tabacinum*; *F.B.*

Peniophora aegerita von Hoehn. & Litsch.—Heath and Reach; *Beds. N.H.S. Foray*, 1948.

P. quercina (Pers.) Cooke (*Corticium quercinum* (Pers.) Fr.)—common, as *Auricularia corticalis*; *F.B.*

Phlebia aurantiaca (Sow.) Karst.—unlocalized, as *Auricularia aurantiaca*; *Abbot's annotated F.B.*

Stereum hirsutum (Willd.) Fr.—common, as *Auricularia reflexa*; *F.B.*: Northill, 1886 (W. G. Smith); *J.H.*, *V.C.H.*: Rowney Warren; *Beds. N.H.S. Foray*, 1950: Barton Hills; *Beds. N.H.S. Foray*, 1951: very common on fallen trunks, stumps and branches, Leighton Buzzard, etc.: *D.A.R.*: widespread, *T.L.*

S. purpureum (Pers.) Fr.—frequent, Bedford; Ampthill; *J.H.*, *V.C.H.*: Barton Hills; *Beds. N.H.S. Foray*, 1951: Clophill; *Beds. N.H.S. Foray*, 1952: not uncommon on birch, plum and poplar trees, Heath and Reach, Leighton Buzzard, etc.; *D.A.R.*: widespread; *T.L.*

S. rugosum (Pers.) Fr.—common on branches of deciduous trees, especially hazel, Heath and Reach; *D.A.R.* 1951.

S. spadiceum Fr.—Ampthill 1886 (W. G. Smith); *J.H.*, *V.C.H.*: common on oak, Heath and Reach; *D.A.R.*

Thelephora anthocephala (Bull.) Fr.—Sandy Warren, as *Clavaria anthocephala*; *F.B.*

T. coralloides Fr.—Warden, Beeston Leys, as *Clavaria coriacea*; *F.B.*

T. molissima (Pers.) Fr.—Ampthill, 1885 (W. G. Smith); *J.H.*, *V.C.H.*

T. terrestris (Ehrh.) Fr. (*T. laciniata* (Pers.) Fr.)—Ampthill fir plantations, as *Helvella caryophyllea*; *Plantae Bedford.*: Ampthill, Warden, as *Merulius caryophylleus*; *F.B.*: common (W. B. Grove); *J.H.*, *V.C.H.*: Clophill; *Beds. N.H.S. Foray*, 1949: Rowney Warren; *Beds. N.H.S. Foray*, 1950: common, Heath and Reach; *D.A.R.*

Tomentella mucidula Karst.—Rowney Warren; *Beds. N.H.S. Foray*, 1950.

T. phylacteris Bull.—Sheerhatch Wood, as *Auricularia phylacteris*; *F.B.* (W. B. Grove considered that this should be referred to *Thelephora biennis* Fr.).

AGARICALES: CLAVARIACEAE

Clavaria acuta (Sow.) Fr.—Clophill; *Beds. N.H.S. Foray*, 1952.

C. argillacea (Pers.) Fr.—Ampthill, 1887; *J.H.*, *V.C.H.*: Aspley Wood; *Beds. N.H.S. Foray*, 1948.

C. cinerea (Bull.) Fr.—Warden; Clophill; *F.B.*: Ampthill, 1887; *J.H.*, *V.C.H.*: Rowney Warren; *Beds. N.H.S. Foray*, 1950: common in mixed woodland, Heath and Reach; *D.A.R.*
C. coralloides (L.) Fr.—Ampthill, 1886; *J.H.*, *V.C.H.*
C. corniculata (Schaeff.) Fr. (*C. muscoides* Fr.)—common; *F.B.*: Ampthill, 1887 (W. B. Grove); *J.H.*, *V.C.H.*
Var. **pratensis** (Pers.) Cotton & Wakef.—Potton Woods, as *C. fastigiata*; *F.B.*
C. cristata (Holmsk.) Fr.—Clophill; *Beds. N.H.S. Foray*, 1952: common in mixed woods, Heath and Reach; *D.A.R.*
C. flaccida Fr.—Rowney Warren; *Beds. N.H.S. Foray*, 1950.
C. formosa (Pers.) Fr.—Ampthill, frequent (W. B. Grove); *J.H.*, *V.C.H.*
C. fusiformis (Sow.) Fr.—Ampthill, 1887 (W. B. Grove); *J.H.*, *V.C.H.*: Clophill; *Beds. N.H.S. Foray*, 1952: common in grassy places, woods and heaths, Heath and Reach, Woburn, etc.; *D.A.R.*
C. inaequalis (Müller) Quél.—Rowney Warren; *Beds. N.H.S. Foray*, 1950: common in woods and meadows, amongst grass and mosses, Heath and Reach; *D.A.R.*
C. luteo-alba Rea—Rowney Warren; *R. W. G. Dennis*, 1950.
C. rugosa (Bull.) Fr.—Ampthill, frequent (W. B. Grove); *J.H.*, *V.C.H.*
C. vermicularis Fr. (*C. fragilis* (Holmsk.) Fr.)—common, as *C. pistillaris*; *F.B.*
Sparassis crispa (Wulf.) Fr.—rare in coniferous woods, Heath and Reach; *D.A.R.* 1952.
Typhula erythropus (Bolt.) Fr.—Clophill; *Beds. N.H.S. Foray*, 1952.
T. gyrans (Batsch.) Fr.—Clapham between Woods, as *Clavaria gyrans*; *F.B.*

AGARICALES: HYDNACEAE

Acia fusco-atra (Fr.) Pat.—Clophill; *Beds. N.H.S. Foray*, 1952.
A. uda (Fr.) Bourd. and Galz.—Rowney Warren; *Beds. N.H.S. Foray*, 1950.
Hydnum auriscalpium (L.) Fr.—common; *F.B.*: Ampthill, 1885 (W. G. Smith); *J.H.*, *V.C.H.*: on buried cones of *Pinus sylvestris*, Heath and Reach; *D.A.R.*
H. imbricatum (L.) Fr.—fir plantations, Warden; *F.B.*
Odontia barba-jovis Fr.—'bases of birchen stumps', Southill Plantations, as *Hydnum barba-jovis*; *F.B.*
Radulum orbiculare Fr.—on fallen branches of wild cherry, Heath and Reach; *D.A.R.* 1951.

AGARICALES: POLYPORACEAE

Daedalea biennis Fr. (*Polyporus biennis* (Bull.) Fr., *Daedalea rufescens* Pers., *P. rufescens* (Pers.) Fr.)—Ampthill, as *Boletus biennis*; *Abbot's annotated F.B.*: Ampthill; *J.H.*, *V.C.H.*: Pavenham (W. B. Grove); *J.H.*, *V.C.H.*; on oak chips, Heath and Reach; *D.A.R.*
D. quercina (L.) Fr.—Everton Wood; Potton Wood; *Plantae Bedford.*: fallen elms, rare as *Agaricus quercinus*; *F.B.* (Abbot's statement that this fungus occurred on elm was due to confusion with *Lenzites betulina* (L.) Fr. This is borne out by Abbot citing two plates with his record, i.e., Bulliard, tab. 352, which is *Lenzites betulina*, and Sowerby, tab. 181, which is true *Daedalea quercina*. It seems probable that he saw both species but confused them under the same name): Ampthill, 1892; *E. M. Langley in J.H.*, *V.C.H.*: Heath and Reach; *Beds. N.H.S. Foray*, 1947: Aspley Wood; *Beds. N.H.S. Foray*, 1948: Clophill; *Beds. N.H.S. Foray*, 1949: Rowney Warren; *Beds. N.H.S. Foray*, 1950: confined to oak, and very common, Heath and Reach, etc.; *D.A.R.*
D. unicolor (Bull.) Fr.—common, as *Boletus unicolor*; *F.B.*: Ampthill, Kempston (W. B. Grove); *J.H.*, *V.C.H.*
Fistulina hepatica (Huds.) Fr., *Beefsteak fungus*—Haynes Wood, as *Boletus hepaticus*; *Plantae Bedford.*: common; *F.B.*: Ampthill Park, etc. (W. B. Grove); *J.H.*, *V.C.H.*: Heath and Reach; *Beds. N.H.S. Foray*, 1948: Clophill; *Beds. N.H.S. Foray*, 1949: Rowney Warren; *Beds. N.H.S. Foray*, 1950.
Fomes annosus (Fr.) Cooke (*Polyporus annosus* Fr.)—common (W. B. Grove); *J.H.*, *V.C.H.*: Clophill; *Beds. N.H.S. Foray*, 1949: abundant on *Pinus sylvestris* and occasionally on birch, Heath and Reach, Woburn, etc.; *D.A.R.*

F. conchatus (Pers.) Gillet—Renhold, by a drain near the wood, as *Boletus impuber*; *Abbot's annotated F.B.*

F. cystinus (Berk.) Gillet—Potton, Everton, as *Boletus suberosus*; *F.B.* (Berkeley (1836) states of *Polyporus cystinus*, 'This is certainly the same as *Boletus suberosus* Sow.' However, Abbot cited two plates. The first, Bolton tab. 162 (i.e., *Boletus tuberosus* Bolt.), is *Trametes suaveolens* (L.) Fr. He added later Sowerby, tab. 288, which is *Boletus suberosus*. It thus seems that Abbot confused *Trametes suaveolens*, which normally grows on willow, with *Fomes cystinus*.)

F. fomentarius (L.) Kickx. (*Polyporus fomentarius* (L.) Fr.)—Clapham, as *Boletus fomentarius*; *F.B.*:.unlocalized (W. B. Grove); *J.H., V.C.H.*: Both records probably refer to *Ganoderma applanatum* (Pers.) Pat., which until recently has been much confused with *F. fomentarius*.

F. fraxineus (Fr.) Cooke (*Polyporus fraxineus* Fr.)—Lidlington; *J.H., V.C.H.*

F. igniarius (L.) Gillet (*Polyporus igniarius* (L.) Fr.)—common, as *Boletus igniarius*; *F.B.*: Oakley; *E. M. Langley in J.H., V.C.H.* (Both of these records may refer to *F. pomaceus*): not uncommon, but confined to *Salix* spp., Leighton Buzzard; *D.A.R.*

F. nigricans (Fr.) Gillet—Heath and Reach; *Beds. N.H.S. Foray*, 1948.

F. pomaceus (Pers.) Bigeard & Guillermin (*Polyporus fulvus* (Bres.) Murrill)—Bromham; *J.H., V.C.H.*: Cockayne Hatley; *T.L.*: on *Prunus* spp., Eaton Bray, Totternhoe, Leighton Buzzard; *D.A.R.*

F. ulmarius (Sow.) Sacc. (*Polyporus ulmarius* (Sow.) Fr.)—common; *J.H., V.C.H.*

Ganoderma applanatum (Pers.) Pat. (*Polyporus applanatus* (Pers.) Wallr.)—Bedford, 1888; *J.H., V.C.H.*: exceedingly abundant on deciduous trees, especially beech, Heath and Reach, Woburn and Leighton Buzzard; *D.A.R.* (see also *Fomes fomentarius*).

G. lucidum (Leyss.) Karst. (*Polyporus lucidus* (Leyss.) Fr.)—Clapham Wood, as *Boletus rugosus*; *Plantae Bedford.*: Clapham Park Wood, as *B. obliquatus*; *F.B.*: Ampthill, 1887; *J.H., V.C.H.*

Lenzites betulina (L.) Fr. (*Lenzites flaccida* (Bull.) Fr.)—Clapham Park Wood, as *Agaricus coriaceus*; *F.B.*: Ampthill; *J.H., V.C.H.*: common on deciduous trees, Heath and Reach, Woburn, etc.; *D.A.R.* (see also *Daedalea quercina*).

Merulius corium Fr.—Sheerhatch Wood, as *Auricularia papyrina*; *F.B.*

M. himantioides Fr.—Rowney Warren; *Beds. N.H.S. Foray*, 1950.

M. lacrymans Fr., *Dry-rot fungus*—too common, as *Boletus lachrymans*; *F.B.*

M. tremellosus (Schrad.) Fr.—Aspley Wood; *Beds. N.H.S. Foray*, 1948.

Polyporus adustus (Willd.) Fr.—common, as *Boletus flabelliformis*; *F.B.*: Kempston, 1887; *J.H., V.C.H.*: very abundant on broad-leaved trees, Leighton Buzzard, Woburn, Heath and Reach; *D.A.R.*

P. amorphus Fr.—rare, on coniferous wood, Heath and Reach; *D.A.R.* 1950.

P. betulinus (Bull.) Fr.—Heath and Reach; *Beds. N.H.S. Foray*, 1947; Aspley Wood; *Beds. N.H.S. Foray*, 1948: Clophill; *Beds. N.H.S. Foray*, 1949: Rowney Warren; *Beds. N.H.S. Foray*, 1950: exceedingly abundant on birch, Leighton Buzzard, Heath and Reach, Woburn, etc.; *D.A.R.*

P. caesius (Schrad.) Fr.—'edges of pannels in old gates, Bedford', as *Boletus albidus*; *Abbot's annotated F.B.*: on conifers, Heath and Reach; *D.A.R.*

P. dryadeus (Pers.) Fr.—Carlton, 1887; *J.H., V.C.H.*: on oak, Heath and Reach; *D.A.R.*

P. fragilis Fr.—Barton Hills; *Beds. N.H.S. Foray*, 1951.

P. frondosus (Dicks.) Fr.—Maulden, Renhold, as *Boletus frondosus*; *F.B.* (Grove referred this to *P. intybaceus* Fr.): unlocalized; *J.H., V.C.H.*: on oak stumps, Heath and Reach; *D.A.R.*

P. fumosus (Pers.) Fr.—alder and birch stumps, Ampthill, as *Boletus pelloporus; Abbot's annotated F.B.*: frequent, 1887; *J.H., V.C.H.* (This fungus is not common now, and generally occurs on *Salix* spp.)

P. giganteus (Pers.) Fr.—Clophill; *Beds. N.H.S. Foray*, 1949: common on the ground, associated with the roots of beech and less commonly with oak, Leighton Buzzard, Heath and Reach, etc.; *D.A.R.* 1948.

P. hispidus (Bull.) Fr.—common, as *Boletus hispidus*; *F.B.*: frequent, 1887; *J.H.*, *V.C.H.*: Rowney Warren; *Beds. N.H.S. Foray*, 1950: common on ash and to a less extent on apple, Leighton Buzzard, Heath and Reach, Woburn; *D.A.R.*: Cockayne Hatley; *T.L.*

P. kymatodes Rost.—Rowney Warren; *Beds. N.H.S. Foray*, 1950.

P. lacteus Fr.—common on trunks of deciduous trees, especially beech, Heath and Reach; *D.A.R.* 1950.

P. schweinitzii Fr.—on stumps of conifers, Woburn; *D.A.R.* 1950.

P. semipileatus Peck (*Polyporus chioneus* Bres.)—frequent; *J.H.*, *V.C.H.*

P. squamosus (Huds.) Fr.—Haynes Wood, as *Boletus squamosus*; *Plantae Bedford.*: common; *F.B.*: common; *J.H.*, *V.C.H.*: Clophill; *Beds. N.H.S. Foray*, 1949: common on elm, Leighton Buzzard, Heath and Reach, etc.; *D.A.R.*: Cockayne Hatley; *T.L.*

P. stipticus (Pers.) Fr.—Barton Hills; *Beds. N.H.S. Foray*, 1951. (This fungus is normally one of coniferous wood, but it does occur on wood of deciduous trees.)

P. sulphureus (Bull.) Fr.—Hawnes, as *Boletus sulphureus*; *F.B.*: common, 1892; *J.H.*, *V.C.H.*: on oak and occasionally on plum, Heath and Reach; *D.A.R.*

P. varius Fr.—willows near Duck Mill, as *Boletus elegans*; Clapham Park Wood; as *B. nummularius*; *F.B.*: on willows, Leighton Buzzard, Heath and Reach, etc., *D.A.R.*

Polystictus abietinus (Dicks.) Fr. (*Polyporus abietinus* (Dicks.) Fr.)—Ampthill, 1887; *J.H.*, *V.C.H.*: Heath and Reach; *Beds. N.H.S. Foray*, 1948: Rowney Warren; *Beds. N.H.S. Foray*, 1950: on fallen coniferous trunks, Woburn; *D.A.R.*

P. perennis (L.) Fr. (*Polyporus perennis* (L.) Fr.)—Ampthill; *J.H.*, *V.C.H.*: Aspley Wood; *Beds. N.H.S. Foray*, 1948: Rowney Warren; *Beds. N.H.S. Foray*, 1950: not uncommon on burnt ground, especially in coniferous woods, Heath and Reach, Woburn; *D.A.R.*

P. versicolor (L.) Sacc. (*Polyporus versicolor* (L.) Fr.)—common, as *Boletus versicolor*; *F.B.*: common; *J.H.*, *V.C.H.*: abundant throughout the county and found on all forays of the *Beds. N.H.S.*

Poria versipora (Pers.) Baxter (*Poria mucida* Bres., *Irpex obliquus* (Schrad.) Fr.)—common on fallen wood; *D.A.R.* 1948.

Sistotrema confluens Pers. ex Fr.—Clapham Park Wood, as *Hydnum sublamellosum*: *F.B.*

Trametes confragosa (Bolt.) Jörstad (*T. bulliardii* Fr., *T. rubescens* (Alb. & Schw.) Fr.)—Fenlake, as *Boletus suaveolens*; Bromham, as *B. angustatus*; *F.B.*: Heath and Reach; *Beds. N.H.S. Foray*, 1948: Rowney Warren; *R. W. G. Dennis*: common on *Salix* and occasionally on *Sorbus*, Heath and Reach; *D.A.R.*

T. gibbosa (Pers.) Fr.—Ampthill, 1887; *Mr. Ferraby in J.H.*, *V.C.H.*: common, especially on beech; Heath and Reach; *D.A.R.*

T. suaveolens (L.) Fr.—common, as *Boletus salicinus*; *F.B.*: Bedford (W. B. Grove); *J.H.*, *V.C.H.*: rare on *Salix* spp., Leighton Buzzard; *D.A.R.*

AGARICALES: BOLETACEAE

Boletus badius Fr.—frequent (W. B. Grove); *J.H.*, *V.C.H.*: Clophill; *Beds. N.H.S Foray*, 1949: Rowney Warren; *Beds. N.H.S. Foray*, 1950: not uncommon in mixed woodland, Heath and Reach; *D.A.R.*

B. bovinus (L.) Fr.—Warden, as *B. gregarius*; *F.B.*: Ampthill, 1889; *J.H.*, *V.C.H.*: Aspley Wood; *Beds. N.H.S. Foray*, 1948: common in coniferous woods, Heath and Reach; *D.A.R.*

B. calopus Fr.—common, as *B. olivaceus*; *F.B.* (W. B. Grove considered that a further record of *B. aureus* from Warden was most likely also this species.)

B. chrysenteron (Bull.) Fr.—Northill, Ampthill (W. B. Grove); *J.H.*, *V.C.H.*: Aspley Wood; *Beds. N.H.S. Foray*, 1948: Clophill; *Beds. N.H.S. Foray*, 1952: a common fungus of beech woods, Heath and Reach; *D.A.R.*

B. edulis (Bull.) Fr.—common as *B. bovinus*; *F.B.*: not common (W. B. Grove); *J.H.*, *V.C.H.*: Aspley Wood; *Beds. N.H.S. Foray*, 1948: Clophill; *Beds. N.H.S. Foray*, 1949: common in mixed woodland, Heath and Reach; *D.A.R.*

B. elegans (Schum.) Fr.—Sandy, 1888; Ampthill, 1891; *J.H.*, *V.C.H.*: Rowney Warren; *R. W. G. Dennis*: a common fungus associated with larch, Heath and Reach; *D.A.R.*

B. erythropus Fr. non Pers.—common in mixed woods, and on heaths, Heath and Reach; *D.A.R.* 1948.

[**B.** ? **fragrans** Vitt.—near Bedford; *E. M. Langley in J.H., V.C.H.*]

B. granulatus (L.) Fr.—Ampthill; Sandy (W. B. Grove); *J.H., V.C.H.*: Aspley Wood; *Beds. N.H.S. Foray*, 1948: a common fungus of coniferous woods; Heath and Reach; Woburn; *D.A.R.*

B. impolitus Fr.—frequent, Ampthill; *J.H., V.C.H.* [a record of an uncommon species which needs confirmation].

B. luridus (Schaeff.) Fr.—Clophill; *Beds. N.H.S. Foray*, 1949: a not uncommon fungus of mixed woodland, especially on chalky soils; Heath and Reach; *D.A.R.*

B. luteus (L.) Fr.—common; *F.B.*: common at Ampthill; *J.H., V.C.H.*: Rowney Warren; *Beds. N.H.S. Foray*, 1950: a common species of coniferous woods, Heath and Reach; *D.A.R.*

B. piperatus (L.) Fr.—Warden; *F.B.*: Clophill; *Beds. N.H.S. Foray*, 1949: a common species, associated with birch, Heath and Reach, Woburn; *D.A.R.*

B. pulverulentus Opat.—a rare fungus of oak woods, Heath and Reach; *D.A.R.* 1952.

B. purpureus Fr.—Ampthill; *Mr. Ferraby, J.H., V.C.H.*

B. reticulatus (Schaeff.) Boud.—common in mixed woodland, Heath and Reach; *D.A.R.* 1949.

B. rubinus W. G. Sm.—under trees by the roadside, near Dunstable; *Cooke in Handbook of British Fungi*, 1871.

B. satanas Lenz—Oakley, 1892; *E. M. Langley in J.H., V.C.H.*: Woburn, 1897; *J.H., V.C.H.*

B. scaber (Bull.) Krombh.—Heath and Reach; *Beds. N.H.S. Foray*, 1948: Clophill; *Beds. N.H.S. Foray*, 1949: Rowney Warren; *Beds. N.H.S. Foray*, 1950: a common fungus associated with birch; Heath and Reach; Woburn; *D.A.R.*

B. subtomentosus (Schaeff.) Fr.—Southill, 1888 (W. B. Grove); *J.H., V.C.H.*: Aspley Wood; *Beds. N.H.S. Foray*, 1948: Rowney Warren; *Beds. N.H.S. Foray*, 1950: a common species of mixed woodland, Heath and Reach, Woburn; *D.A.R.*

B. variegatus (Swartz) Fr.—Barton Hills; *Beds. N.H.S. Foray*, 1951.

V. versipellis Fr.—Heath and Reach; *Beds. N.H.S. Foray*, 1947: Aspley Wood; *Beds. N.H.S. Foray*, 1948: a species associated with birch, Heath and Reach, Woburn; *D.A.R.*

B. viscidus (L.) Fr. (*B. laricinus* Berk.)—Ampthill, 1888; *J.H., V.C.H.*: Barton Hills; *Beds. N.H.S. Foray*, 1951; not uncommon in coniferous woods, Heath and Reach; *D.A.R.*

Gyroporus cyanescens (Bull.) Quél.—Aspley, Clapham between Woods, as *Boletus cyanescens*; *F.B.*

AGARICALES: AGARICACEAE: LEUCOSPORAE

Hamson recorded most toadstools under the genus *Agaricus*, and as his list is not readily accessible the full synonyms are not given here.

Amanita citrina (Schaeff.) Roques (*A. mappa* (Batsch) Fr.)—Aspley Wood; *Beds. N.H.S. Foray*, 1948: Clophill; *Beds. N.H.S. Foray*, 1949; Rowney Warren; *Beds. N.H.S. Foray*, 1950: common in mixed woodland and on heaths, Heath and Reach; *D.A.R.*

Var. **alba** Price—common, Heath and Reach; *D.A.R.* 1951.

A. echinocephala Vitt.—Barton Hills; *Beds. N.H.S. Foray*, 1951: a rare species of chalk pastures; *D.A.R.*

A. excelsa Fr. (*A. spissa* Fr.)—probably not uncommon in mixed woods, Heath and Reach; *D.A.R.* 1951.

A. muscaria (L.) Fr., *Fly Agaric*—common; *F.B.*: Ampthill Woods; *J.H., V.C.H.*: Aspley Wood; *Beds. N.H.S. Foray*, 1948: Clophill; *Beds. N.H.S. Foray*, 1949: Rowney Warren; *Beds. N.H.S. Foray*, 1950: common in association with birch, occasional in coniferous woods, very rare under beech; Heath and Reach, Woburn; *D.A.R.*

A. phalloides (Vaill.) Fr., *Death-cap fungus*—Silsoe, as *Agaricus bulbosus*; *F.B.*: Southill, Ampthill, Sharnbrook (W. B. Grove); *J.H.*, *V.C.H.*: common in oak and mixed woods, Heath and Reach; *D.A.R.*

A. rubescens (Pers.) Fr.—common, as *Agaricus verrucosus*; *F.B.* (W. B. Grove interprets this as the above, but Abbot cites Curtis *V* tab. 72, which shows three species, *Amanita rubescens*, *A. citrina* and *A.* sp.): common (W. B. Grove); *J.H.*, *V.C.H.*: Aspley Wood; *Beds. N.H.S. Foray*, 1948: Clophill; *Beds. N.H.S. Foray*, 1949: Rowney Warren; *Beds. N.H.S. Foray*, 1950; very common in mixed woodland, Heath and Reach; *D.A.R.*

A. solitaria (Bull.) Fr.—Barton Hills; *Beds. N.H.S. Foray*, 1951: a rare species confined to chalk pastures; *D.A.R.*

Amanitopsis fulva (Grev.) Rea—Elstow, as *Agaricus vaginatus*; *F.B.*: common in mixed woodland, Heath and Reach, Leighton Buzzard; *D.A.R.*

A. vaginata (Bull.) Roze—Clophill, Silsoe, as *Agaricus pulvinatus*; *Abbot's annotated F.B.*: common (W. B. Grove); *J.H.*, *V.C.H.*

Armillaria mellea (Vahl.) Fr., *Honey fungus*—common, as *Agaricus stipitis*; *F.B.*: common (W. B. Grove); *J.H.*, *V.C.H.*: Heath and Reach; *Beds. N.H.S. Foray*, 1948: Barton Hills; *Beds. N.H.S. Foray*, 1951: Clophill; *Beds. N.H.S. Foray*, 1952: a common species which is a serious parasite of woody shrubs and trees, Heath and Reach, Leighton Buzzard; *D.A.R.*: Cockayne Hatley; *T.L.*

Cantharellus cibarius Fr.—Ampthill; Woburn, as *Agaricus chantarellus*; *Plantae Bedford.*: common, as *Merulius chantarellus*; *F.B.*: Woburn, plentiful; *J.H.*, *V.C.H.*: common in mixed woodland, Heath and Reach; *D.A.R.*

C. cinereus (Pers.) Fr.—Aspley Wood; *Beds. N.H.S. Foray*, 1948.

C. tubaeformis (Bull.) Fr.—Aspley Wood; *Beds. N.H.S. Foray*, 1948; on decayed stump, Heath and Reach; *D.A.R.*

Clitocybe aurantiaca (Wulf.) Studer (*Cantharellus aurantiacus* (Wulf.) Fr.)—common and assuming various forms; *J.H.*, *V.C.H.*: Aspley Wood; *Beds. N.H.S. Foray*, 1948: Clophill; *Beds. N.H.S. Foray*, 1949: common in coniferous woods, Heath and Reach; *D.A.R.*

C. brumalis Fr.—common (W. B. Grove); *J.H.*, *V.C.H.*: common in mixed woodland, Heath and Reach; *D.A.R.*

C. candicans (Pers.) Fr.—Bedford, 1891 (W. B. Grove); *J.H.*, *V.C.H.*

C. cerussata Fr.—Clophill, as *Agaricus opacus*; *Abbot's annotated F.B.*: Clapham, 1891 (W. B. Grove); *J.H.*, *V.C.H.*: Rowney Warren; *Beds. N.H.S. Foray*, 1950: Barton Hills; *Beds. N.H.S. Foray*, 1951: common in mixed woods and shrubberies, Leighton Buzzard; *D.A.R.*

C. clavipes (Pers.) Fr.—common (W. B. Grove); *J.H.*, *V.C.H.*: Rowney Warren; *Beds. N.H.S. Foray*, 1950: Clophill; *Beds. N.H.S. Foray*, 1952.

C. cyathiformis (Bull.) Fr.—Warden, Ampthill, as *Agaricus sordidus*; *F.B.*

C. dealbata (Sow.) Fr.—Barton Hills; *Beds. N.H.S. Foray*, 1951.

[C. elixa (Sow.) Berk.—common, as *Agaricus elixus*; *F.B.*: Ampthill, 1887; *J.H.*, *V.C.H.* (According to Pearson and Dennis (1948) this species is either *C. inornata* or *C. alexandrii*)].

C. flaccida (Sow.) Fr. (*C. inversa* (Scop.) Fr.)—common, as *Agaricus infundibuliformis*; *F.B.* (I agree with W. B. Grove that judging from Abbot's comments this is most likely to be *C. flaccida*; but the plate he cites, Bulliard tab. 286, is of *C. infundibuliformis*): Houghton Conquest, as *Agaricus lobatus*; *Abbot's annotated F.B.*: Rowney Warren; *Beds. N.H.S. Foray*, 1950.

C. fragrans (Sow.) Fr.—Silsoe, as *Agaricus fragrans*; *F.B.*: common (W. B. Grove); *J.H.*, *V.C.H.*: not uncommon in mixed woods; Heath and Reach; *D.A.R.*

C. geotropa (Bull.) Fr.—frequent at Ampthill (W. B. Grove); *J.H.*, *V.C.H.*; Rowney Warren; *Beds. N.H.S. Foray*, 1948: not uncommon in woods and pastures, either singly or in large rings, Woburn, Heath and Reach; *D.A.R.*

C. gigantea (Sow.) Fr. (*Paxillus giganteus* (Sow.) Fr.)—common, 1892 (W. B. Grove); *J.H.*, *V.C.H.*: a fungus of grassy places—gardens, pastures, etc., Heath and Reach; *D.A.R.*

C. hirneola Fr. var. **ovispora** Lange—Clophill; *Beds. N.H.S. Foray*, 1952.

C. infundibuliformis (Schaeff.) Fr.—Ampthill, Maulden, as *Agaricus cyathiformis*; *F.B.* (W. B. Grove thought that this could be referred to the above or to *C. brumalis*): typical, near Bedford, 1891; *E. M. Langley in J.H., V.C.H.*: Rowney Warren; *Beds. N.H.S. Foray*, 1950: Barton Hills; *Beds. N.H.S. Foray*, 1951: common in mixed woodland, Heath and Reach; *D.A.R.*

C. inornata (Sow.) Fr.—Barton Hills, an uncommon species of mixed woods and having an unpleasant smell; *Beds. N.H.S. Foray*, 1951 (see also *C. elixus*).

C. metachroa Fr.—Ampthill, 1887; *J.H., V.C.H.*

C. monstrosa (Sow.) Gillet—Warden, Henlow, as *Agaricus monstrosa*; *Abbot's annotated F.B.*

C. nebularis (Batsch) Fr.—common, as *Agaricus caseus*; *F.B.*: Milton Ernest, 1891; *E. M. Langley in J.H., V.C.H.*: Rowney Warren; *Beds. N.H.S. Foray*, 1950: Clophill; *Beds. N.H.S. Foray*, 1952: common in woods, sometimes forming large rings; Woburn, etc.; *D.A.R.*

C. obsoleta (Batsch) Fr.—Aspley Wood; *Beds. N.H.S. Foray*, 1948: Woburn, etc.; *D.A.R.*

C. odora (Bull.) Fr. (*C. viridis* (With.) Fr.)—Clophill, as *Agaricus viridis*; Warden, as *A. odoras*; *F.B.*: common (W. B. Grove); *J.H., V.C.H.*: Clophill; *Beds. N.H.S. Foray*, 1949: a fungus of mixed woods, smelling of anise, Heath and Reach; *D.A.R.*

C. phyllophila Fr.—Bedford, 1891; *E. M. Langley in J.H., V.C.H.*

C. pithyophila Fr.—common, as *Agaricus dealbatus*; *F.B.*

C. subinvoluta W. G. Sm. sensu Lange—Heath and Reach; *D.A.R.*

C. vibecina Fr.—Rowney Warren; *Beds. N.H.S. Foray*, 1950: Clophill; *Beds. N.H.S. Foray*, 1952: a common fungus of woods and heaths, smelling of new meal, Woburn, Heath and Reach, Leighton Buzzard; *D.A.R.* 1948.

Collybia ambusta Fr.—Clophill; *Beds. N.H.S. Foray*, 1952.

C. butyracea (Bull.) Fr.—(Grove interprets Abbot's record from Eversholt of *Agaricus aquosus* as being this species; but I think this is doubtful): common (W. B. Grove); *J.H., V.C.H.*: Rowney Warren; *Beds. N.H.S. Foray*, 1950: Clophill; *Beds. N.H.S. Foray*, 1952: common in mixed and coniferous wood, Woburn, Heath and Reach; *D.A.R.*

C. distorta Fr.—Milton Ernest, 1892; *J.H., V.C.H.*

C. clusillis Fr.—Southill, Aspley, as *Agaricus umbilicatus*; *F.B.*

C. fusipes (Bull.) Berk.—common, as *Agaricus crassipes*; *F.B.*: Kempston, etc., 1889 (W. B. Grove); *J.H., V.C.H.*: common on old stumps, often in large clusters, Heath and Reach, Woburn; *D.A.R.*

C. longipes (Bull.) Berk.—Southill (W. B. Grove); *J.H., V.C.H.*

C. maculata (Alb. & Schw.) Fr.—common, as *Agaricus carnosus*; *F.B.*: Ampthill; Sandy (W. B. Grove); *J.H., V.C.H.*: found on all *Beds. N.H.S. Forays* except 1951: a common pine-wood species, Woburn; *D.A.R.*

C. platyphylla (Pers.) Fr.—not uncommon in oak woods at Heath and Reach; *D.A.R.* 1948.

C. racemosa (Pers.) Fr.—Melchbourne Wood, as *Agaricus racemosa*; *Abbot's annotated F.B.*

C. radicata (Rehl) Berk.—Silsoe New Wood, as *Agaricus radiatus*; *Plantae Bedford.*: Bromham, as *A. umbraculum*; *F.B.*: common (W. B. Grove); *J.H., V.C.H.*: Heath and Reach; *Beds. N.H.S. Foray*, 1947: Aspley Wood; *Beds. N.H.S. Foray*, 1948: Barton Hills; *Beds. N.H.S. Foray*, 1951: a species attached to buried wood by a long tapering 'root' which may reach 18 ins. in length, Woburn; *D.A.R.*

C. tuberosa (Bull.) Fr.—common, as *Agaricus albus*; *F.B.*: Ampthill, 1887 (W. B. Grove); *J.H., V.C.H.*

C. velutipes (Curt.) Fr.—common, as *Agaricus velutipes*; *F.B.*: common (W. B. Grove); *J.H., V.C.H.*: common on dead and living trees of elm, maple, elder, etc.; *D.A.R.*

Hygrophorus calyptraeformis Berk.—Carlton, 1887 (W. B. Grove); *J.H., V.C.H.*

H. ceraceus (Wulf.) Fr.—Thurleigh, Aspley, as *Agaricus ceraceus*; *F.B.*: Ampthill (W. B. Grove); *J.H., V.C.H.*: Deadmansey Wood; *E.M.-R.*

H. chlorophanus Fr.—Ampthill (W. B. Grove); *J.H., V.C.H.*: a species of grassy places, Heath and Reach; *D.A.R.*

H. coccineus (Schaeff.) Fr.—frequent (W. B. Grove); *J.H., V.C.H.*: Deadmansey Wood; *E.M.-R.*: Rowney Warren; *Beds. N.H.S. Foray*, 1950.

H. conicus Fr.—common, as *Agaricus aurantius*; *F.B.*: common (W. B. Grove); *J.H., V.C.H.*: Deadmansey Wood; *E.M.-R.*: Rowney Warren; *R. W. G. Dennis*: in short grass, Heath and Reach; *D.A.R.*

[**H. cossus** (Sow.) Fr.—(W. B. Grove refers Abbot's record of *Agaricus cossus*, common, to either *H. cossus* or *H. eburneus*)].

H. eburneus (Bull.) Fr.—not uncommon in mixed woods, Heath and Reach; *D.A.R.* 1950.

H. hypothejus Fr.—common, Ampthill (W. B. Grove); *J.H., V.C.H.*: Rowney Warren; *R. W. G. Dennis*: common in coniferous woods, Heath and Reach; *D.A.R.*

H. miniatus Fr.—Rowney Warren; *Beds. N.H.S. Foray*, 1950: [a species of heaths and probably more common than this single record indicates; *D.A.R.*].

H. nitratus (Pers.) Fr.—Bromham; *J.H., V.C.H.*

H. niveus (Scop.) Fr.—common as *Agaricus niveus*; *F.B.*: Rowney Warren; *Beds. N.H.S. Foray*, 1950: Barton Hills; *Beds. N.H.S. Foray*, 1951: common in short grass on lawns and heaths, Leighton Buzzard; *D.A.R.*

H. ovinus (Bull.) Fr.—Carlton, 1891; *J.H., V.C.H.*

H. pratensis (Pers.) Fr.—Ampthill, Clapham Springs, as *Agaricus miniatus*; *Plantae Bedford.*: common (W. B. Grove); *J.H., V.C.H.*: Rowney Warren; *Beds. N.H.S. Foray*, 1950.

H. psittacinus (Schaeff.) Fr.—Ampthill Park, as *Agaricus psittacinus*; *F.B.*: common (W. B. Grove); *J.H., V.C.H.*: Rowney Warren; *R. W. G. Dennis*: common in grassy places, especially on lawns, Leighton Buzzard; *D.A.R.*

H. puniceus Fr.—amongst moss, Bedfordshire; *Cooke in Illustrations of British Fungi*: frequent (W. B. Grove); *J.H., V.C.H.*

H. virgineus (Wulf.) Fr.—common, as *Agaricus ericeus* and *A. eburneus*; *F.B.*: common (W. B. Grove); *J.H., V.C.H.*: common in lawns and on heaths, Leighton Buzzard; *D.A.R.*

Laccaria amethystina (Vaill.) Cooke—Clapham; Silsoe Wood, as *Agaricus amethystinus*; *Plantae Bedford.*: common (W. B. Grove); *J.H., V.C.H.*: Deadmansey Wood; *E.M.-R.*: Clophill; *Beds. N.H.S. Foray*, 1952: not uncommon in mixed woodland, Heath and Reach; *D.A.R.*

L. laccata (Scop.) Cooke—common, as *Agaricus farinaceus*; *F.B.*: common (W. B. Grove); *J.H., V.C.H.*: Deadmansey Wood; *E.M.-R.*: Rowney Warren; *Beds. N.H.S. Foray*, 1950: Clophill; *Beds. N.H.S. Foray*, 1952.

L. proxima Boud.—Rowney Warren; *Beds. N.H.S. Foray*, 1950.

Lactarius blennius Fr.—Aspley Wood; *Beds. N.H.S. Foray*, 1948: Rowney Warren; *Beds. N.H.S. Foray*, 1950: Barton Hills; *Beds. N.H.S. Foray*, 1951: common in mixed woods, Heath and Reach; *D.A.R.*

L. camphoratus (Bull.) Fr.—uncommon, under pines, Heath and Reach; *D.A.R.* 1949.

L. chrysorheus Fr.—uncommon, under oaks, Heath and Reach; *D.A.R.* 1951.

L. deliciosus (L.) Fr.—common, as *Agaricus deliciosus*; *Abbot's annotated F.B.*: Sandy, 1887; Ampthill; Southill, etc. (W. B. Grove); *J.H., V.C.H.*

L. glycyosmus Fr.—Deadmansey Wood; *E.M.-R.* 1948: Aspley Wood; *Beds N.H.S Foray*, 1948: Rowney Warren; *R. W. G. Dennis*: Clophill; *Beds. N.H.S. Foray*, 1952: common in birch woods and with a strong smell of coconut, Heath and Reach; *D.A.R.*

L. hepaticus Plowr.—Rowney Warren; *R. W. G. Dennis*, 1950.

L. insulsus Fr.—frequent (W. B. Grove); *J.H., V.C.H.*: uncommon under hazel, Heath and Reach; *D.A.R.*

L. mitissimus Fr.—unlocalized, 1892; *E. M. Langley in J.H., V.C.H.*: Rowney Warren; *Beds. N.H.S. Foray*, 1950.

L. piperatus (Scop.) Fr.—common, Heath and Reach; *D.A.R.* 1950 (see also under *L. vellereus* Fr.).

L. plumbeus Fr. (*L. turpis* Fr.)—Aspley Wood; *Beds. N.H.S. Foray*, 1948: Clophill, as *L. turpis*; *Beds. N.H.S. Foray*, 1949: Rowney Warren; *Beds. N.H.S. Foray*, 1950: Clophill; *Beds. N.H.S. Foray*, 1952: very common in birch woods, Heath and Reach, Woburn; *D.A.R.*

L. pyrogalus (Bull.) Fr.—Sandy, 1887; Great Warden, 1892 (W. B. Grove); *J.H., V.C.H.*: Deadmansey Wood; *E.M.-R.*

L. quietus Fr.—common (W. B. Grove); *J.H., V.C.H.*: Aspley Wood; *Beds. N.H.S. Foray*, 1948: common under oaks, Heath and Reach; *D.A.R.*

L. rufus (Scop.) Fr.—common (W. B. Grove); *J.H., V.C.H.*: Aspley Wood; *Beds. N.H.S. Foray*, 1948: Clophill; *Beds. N.H.S. Foray*, 1949: Rowney Warren; *Beds. N.H.S. Foray*, 1950: common in pine woods, Heath and Reach, Woburn; *D.A.R.*

L. serifluus (DC.) Fr.—not uncommon amongst grass in woodland glades, Heath and Reach; *D.A.R.* 1949.

L. subdulcis (Pers.) Fr.—common, as *Agaricus lactifluus*; *F.B.*: frequent (W. B. Grove); *J.H., V.C.H.*: Deadmansey Wood; *E.M.-R.*: Rowney Warren; *Beds. N.H.S. Foray*, 1950: common in mixed woodland, Heath and Reach; *D.A.R.*

L. torminosus Fr.—Silsoe, as *Agaricus piperatus*; *F.B.*: frequent (W. G. Smith and W. B. Grove); *J.H., V.C.H.*: Clophill; *Beds. N.H.S. Foray*, 1952: common in birch woods, Heath and Reach, Woburn; *D.A.R.*

L. trivialis Fr.—frequent, 1892; *J.H., V.C.H.*

L. uvidus Fr.—common, as *Agaricus livido-rubescens*; *F.B.*

L. vellereus Fr.—Clapham Park Wood; Silsoe, as *Agaricus listeri*; *F.B.* (W. B. Grove lists this as *L. piperatus*, but on account of the thick, yellowish, distant gills, I think it is more probably *L. vellereus*): Oakley, 1892 (W. B. Grove), *E. M. Langley in J.H., V.C.H.*: Aspley Wood; *Beds. N.H.S. Foray*, 1948: very common in mixed woods, Heath and Reach; *D.A.R.*

L. vietus Fr.—Rowney Warren; *Beds. N.H.S. Foray*, 1950: common under birches, Heath and Reach, Woburn; *D.A.R.*

L. volemus Fr.—Clapham, 1886; *J.H., V.C.H.*

Lentinus cochleatus (Pers.) Fr.—Clapham Park Wood, as *Agaricus confluens*; *F.B.*

L. lepideus Fr.—common on railway bridges, etc. (W. B. Grove); *J.H., V.C.H.* [a fungus often found on railway sleepers; *D.A.R.*].

L. tigrinus (Bull.) Fr.—banks of the Ouse (W. B. Grove); *J.H., V.C.H.*

Lepiota acutesquamosa (Weinm.) Fr.—Adelaide Square, Bedford, 1896; *E. M. Langley in J.H., V.C.H.*

L. amianthina (Scop.) Fr.—Ampthill fir plantations, as *Agaricus croceus*; *Plantae Bedford.*: Ampthill; Warden; *F.B.*: Aspley Wood; *Beds. N.H.S. Foray*, 1948: Rowney Warren; *Beds. N.H.S. Foray*, 1950: Clophill; *Beds. N.H.S. Foray*, 1952: common in sandy places among mosses, Heath and Reach, Leighton Buzzard; *D.A.R.*

L. badhamii Berk. & Broome—frequent, 1892 (W. B. Grove); *E. M. Langley in J.H., V.C.H.* (This species is certainly not common now).

L. clypeolaria (Bull.) Fr.—Husborne Crawley, as *Agaricus clypeolarius*; *F.B.*: not common, Woburn; *D.A.R.*

L. cristata (Alb. & Schw.) Fr.—common (W. B. Grove); *J.H., V.C.H.*: Rowney Warren; *R. W. G. Dennis*: Barton Hills; *Beds. N.H.S. Foray*, 1951.

L. excoriata (Schaeff.) Fr.—Ampthill (W. B. Grove); Willington; *E. M. Langley in J.H., V.C.H.*

L. gracilenta (Krombh.) Fr.—frequent, 1892, *E. M. Langley in J.H., V.C.H.*

L. granulosa (Batsch) Fr.—common (W. B. Grove); *J.H., V.C.H.*

L. lilacea Bres.—Heath and Reach; *Beds. N.H.S. Foray*, 1948.

L. procera (Scop.) Fr., *Parasol fungus*—Ampthill fir plantations, as *Agaricus procerus*; *Plantae Bedford.*: common; *F.B.*: common (W. B. Grove); *J.H., V.C.H.*: Sandy; *T.L.*: not uncommon in pastures and grassy woodland glades, Woburn, Leighton Buzzard; *D.A.R.*

L. rachodes (Vitt.) Fr.—frequent, especially near Bedford (W. B. Grove); *E. M. Langley in J.H., V.C.H.*: Clophill; *Beds. N.H.S. Foray*, 1949: Rowney Warren; *Beds. N.H.S. Foray*, 1950: common in mixed woods, Heath and Reach, Woburn; *D.A.R.*

Leptotus muscigenus (Bull.) Maire—on *Hypnum sericeum*, on Hassett's walls, as *Merulius muscigenus*; *Abbot's annotated F.B.*

L. retirugis (Bull.) Karst.—Ampthill Park; Hassocks Walls, as *Merulius membranaceus*; *F.B.*

Marasmius acervatus (Fr.) Pearson and Dennis (*Marasmius erythropus* (Pers.) Fr.)—Aspley Wood; *Beds. N.H.S. Foray*, 1948: Clophill; *Beds. N.H.S. Foray*, 1949: Clophill; *Beds. N.H.S. Foray*, 1952: not uncommon, Heath and Reach; *D.A.R.*

M. androsaceus (L.) Fr.—common, as *Agaricus androsaceus*; *F.B.*: Sandy, 1887 (W. B. Grove); *J.H., V.C.H.*: common on fallen twigs and also on heather, Heath and Reach; *D.A.R.*

M. confluens (Pers.) Karst. (*Collybia confluens* Fr.)—Rowney Warren; *Beds. N.H.S. Foray*, 1950: common in mixed woods, Heath and Reach; *D.A.R.*

M. conigenus (Pers.) Karst.—Rowney Warren; *Beds. N.H.S. Foray*, 1950: common, attached to buried fir cones by a long 'root' up to 12 ins. in length, Heath and Reach; *D.A.R.*

M. dryophilus (Bull.) Karst. (*Collybia dryophila* (Bull.) Fr.)—common, as *Agaricus dryophilus*; *F.B.*: Clapham, 1892; *E. M. Langley in J.H., V.C.H.*: Heath and Reach; *Beds. N.H.S. Foray*, 1948: Clophill; *Beds. N.H.S. Foray*, 1949: common in mixed woods, Heath and Reach, Woburn; *D.A.R.*

M. epiphyllus (Pers.) Fr.—on poplar leaves, Clapham Park Wood; Renhold Wood, as *Merulius squamula*; *F.B.*

M. hudsonii (Pers.) Fr.—Warden, Barton, as *Agaricus pilosus*; *F.B.*

M. oreades (Bolt.) Fr., *Fairy-ring fungus*—common, as *Agaricus oreades*; *F.B.*: common, but rare in 1887 (W. B. Grove); *J.H., V.C.H.*: Clophill; *Beds. N.H.S. Foray*, 1952: very abundant in pastures and on lawns, forming large rings, Leighton Buzzard, Heath and Reach, Woburn, etc.; *D.A.R.*

M. peronatus (Bolt.) Fr.—Clophill; Maulden, as *Agaricus peronatus*; *F.B.*: common (W. B. Grove); *J.H., V.C.H.*: Clophill; *Beds. N.H.S. Foray*, 1949: Rowney Warren; *Beds. N.H.S. Foray*, 1950: Barton Hills; *Beds. N.H.S. Foray*, 1951: common in mixed woods, Heath and Reach; *D.A.R.*

M. porreus (Pers.) Fr.—Silsoe, as *Agaricus alliaceus*; *F.B.*

M. ramealis (Bull.) Fr.—common, as *Agaricus ramealis*; *F.B.*: frequent (W. B. Grove); *J.H., V.C.H.*

M. rotula (Scop.) Fr.—common, as *Agaricus rotula*; *F.B.*: Kempston, 1887 (W. B. Grove); *J.H., V.C.H.*: not uncommon on fallen twigs, Leighton Buzzard, etc.; *D.A.R.*

M. undatus Berk.—Rowney Warren; *Beds. N.H.S. Foray*, 1950.

Mycena acicula (Schaeff.) Fr.—common, as *Agaricus clavus*; *F.B.* (W. B. Grove considered this to be synonymous with *M. adonis* (Bull.) Fr. Abbot, however, cited Bulliard tab. 148, which shows a yellow-stemmed species. *M. adonis* is white-stemmed and differs from *M. acicula*, a yellow-stemmed species, in this respect): common, as *Agaricus acicularis*; *Abbot's annotated F.B.*

M. aetites Fr.—under bushes and shrubs in sandpits, Leighton Buzzard; *D.A.R.* 1950.

M. alcalina Fr.—Carlton, 1887 (W. B. Grove); *J.H., V.C.H.*: Rowney Warren; *Beds. N.H.S. Foray*, 1950: common on old stumps of conifers and other trees, Leighton Buzzard; *D.A.R.*

M. ammoniaca Fr.—Sandy, 1887 (W. B. Grove); *J.H., V.C.H.*: Clophill; *Beds. N.H.S. Foray*, 1952: common in lawns, Leighton Buzzard; *D.A.R.*

M. avenacea Fr.—common in short grass on lawns, commons, fields, etc., Leighton Buzzard; *D.A.R.* 1951.

M. cinerella Karst.—exceedingly abundant in short grass, and with *Polytrichum* in sandpits, Leighton Buzzard; *D.A.R.* 1951.

M. epipterygia (Scop.) Fr.—Southill, as *Agaricus nutans*; *Abbot's annotated F.B.*: Ampthill, 1887; *J.H., V.C.H.*: Clophill; *Beds. N.H.S. Foray*, 1949: Rowney Warren; *Beds. N.H.S. Foray*, 1950: common on decaying stumps and amongst grass, Leighton Buzzard; *D.A.R.*

M. flavo-alba Fr.—among grass in sandpits, Leighton Buzzard; *D.A.R.* 1951.

M. galericulata (Scop.) Fr. (*M. rugosa* Fr.)—common, as *Agaricus macer* and as *A. varius*; *F.B.*: Ampthill (W. B. Grove); *J.H., V.C.H.*: found on all forays of the *Beds. N.H.S.* except 1951: exceedingly abundant throughout the county on old stumps, and dead wood, Leighton Buzzard, etc.; *D.A.R.*

M. galopus (Pers.) Fr.—Deadmansey Wood; *E.M.-R.* 1948: Clophill; *Beds. N.H.S. Foray*, 1949: Rowney Warren; *R. W. G. Dennis*: Clophill; *Beds. N.H.S. Foray*, 1952: abundant in mixed and coniferous woods, Heath and Reach, Woburn, etc.; *D.A.R.*
Var. **alba** Hornem.—not uncommon in coniferous woods, Heath and Reach; *D.A.R.* 1951.
Var. **nigra** Hornem.—Rowney Warren; *R. W. G. Dennis*, 1950: not uncommon in coniferous woods, Heath and Reach; *D.A.R.*
M. gypsea Fr. (*M. nivea* Quél.)—Clophill; *Beds. N.H.S. Foray*, 1949.
M. hiemalis (Osbeck) Fr.—Barton Hills; *Beds. N.H.S. Foray*, 1951.
M. inclinata Fr.—Clophill; *Beds. N.H.S. Foray*, 1952: on oak stumps, smelling of iodoform, Heath and Reach; *D.A.R.*
M. lasiosperma Bres.—Clophill; *Beds. N.H.S. Foray*, 1952.
M. lineata (Bull.) Fr.—Rowney Warren; *R. W. G. Dennis*, 1950.
M. metata Fr.—Rowney Warren; *Beds. N.H.S. Foray*, 1950: Clophill; *Beds. N.H.S. Foray*, 1952.
M. polygramma (Bull.) Fr.—common, as *Agaricus polygrammus*; *Abbot's annotated F.B.*: Barton Hills; *Beds. N.H.S. Foray*, 1951: Clophill; *Beds. N.H.S. Foray*, 1952.
M. pura (Pers.) Fr.—Ampthill (W. B. Grove); *J.H., V.C.H.*: Deadmansey Wood; *E.M.-R.*: Rowney Warren; *Beds. N.H.S. Foray*, 1950.
M. sanguinolenta (Alb. & Schw.) Fr.—Heath and Reach; *Beds. N.H.S. Foray*, 1948: Clophill; *Beds. N.H.S. Foray*, 1949: Rowney Warren; *R. W. G. Dennis*.
Nyctalis asterophora Fr.—uncommon, on *Russula nigricans*, Heath and Reach; *D.A.R.* 1951 (see also *N. parasitica*).
N. parasitica (Bull.) Fr.—Silsoe, as *Agaricus lycoperdonoides*; *F.B.* (W. B. Grove listed this as *N. asterophora*; but Abbot cited Bulliard tab. 166, which is undoubtedly *N. parasitica*, and Abbot's reference to gills also indicates this species).
Omphalia fibula (Bull.) Fr.—common; *F.B.*: Sandy, 1887 (W. B. Grove); *J.H., V.C.H.*: Clophill; *Beds. N.H.S. Foray*, 1949: Rowney Warren; *R. W. G. Dennis*: very common, on heaths among short grass and moss, Heath and Reach, Leighton Buzzard, etc.; *D.A.R.*
Var. **swartzii** Fr.—Ampthill, 1887 (W. B. Grove); *J.H., V.C.H.*: Clophill; *Beds. N.H.S. Foray*, 1949: Heath and Reach; *D.A.R.*
O. gracillima Weinm.—Rowney Warren; *Beds. N.H.S. Foray*, 1950.
O. griseo-pallida Desm.—Rowney Warren; *Beds. N.H.S. Foray*, 1950.
O. hydrogramma (Bull.) Fr.—Barton Hills; *Beds. N.H.S. Foray*, 1951.
O. umbellifera (L.) Fr.—common, as *Agaricus umbellifera*; *F.B.*: very common on heaths among short grass and moss, Heath and Reach; *D.A.R.*
O. viridis (Hornem.) Lange—Rowney Warren; *Beds. N.H.S. Foray*, 1950.
Panus stipticus Karst.—Clophill, as *Agaricus stypticus*; *F.B.*: common on tree trunks of oak, beech, etc., Leighton Buzzard, Heath and Reach; *D.A.R.*
P. torulosus (Pers.) Fr.—Thurleigh, as *Agaricus conchatus*; *F.B.*
Pleurotus applicatus (Batsch.) Fr.—Clophill; Silsoe, as *Agaricus applicatus*; *F.B.*
P. atrocaeruleus Fr.—on mountain-ash and oak, rare, Heath and Reach; *D.A.R.* 1952.
P. dryinus (Pers.) Fr. (*P. corticatus* Fr.)—unlocalized, 1892; *E. M. Langley in J.H., V.C.H.*
P. ostreatus (Jacq.) Fr.—Kempston; Clapham, as *Agaricus ostreatus*; *F.B.*: common (W. B. Grove); *J.H., V.C.H.*
Var. **euosmus** (Berk.) Cooke—Bedford, 1889 (W. G. Smith); *J.H., V.C.H.*
P. palmatus (Bull.) Fr.—Barton Hills; *Beds. N.H.S. Foray*. 1951.
P. petaloides (Bull.) Fr.—Clapham Park Wood, as *Agaricus petaloides*; *F.B.* (W. B. Grove referred this record to *P. serotinus* (Schrad.) Fr.)
P. septicus Fr.—Ampthill, 1887 (W. B. Grove); *J.H., V.C.H.*
P. tremulus (Schaeff.) Fr.—*Eng. Fungi* tab. 242, Jan. 1800, was drawn from material sent by Abbot.
P. ulmarius (Bull.) Fr.—sides of elms, Ravensden, as *Agaricus ulmarius*; *F.B.*: common (W. B. Grove); *J.H., V.C.H.*
Russula adusta Fr.—Southill, 1887 (W. B. Grove); *J.H., V.C.H.*: Aspley Wood; *Beds. N.H.S. Foray*, 1948: uncommon, in mixed woods, Heath and Reach; *D.A.R.*

R. aeruginea Lindb.—not uncommon in birch woods, Heath and Reach; *D.A.R.* 1951.

R. alutacea (Pers.) Fr.—Rowney Warren; *R. W. G. Dennis* 1950.

R. atropurpurea Kromb.—very common in mixed woodland, Heath and Reach; *D.A.R.* 1948.

R. azurea Bres.—Clophill; *Beds. N.H.S. Foray*, 1952.

R. brunneo-violacea Crawshay—Rowney Warren; *Beds. N.H.S. Foray*, 1950.

R. claroflava Grove—uncommon, in birch woods, Heath and Reach; *D.A.R.* 1952.

R. cyanoxantha (Schaeff.) Fr.—the commonest species (W. B. Grove); *J.H., V.C.H.*: Clophill; *Beds. N.H.S. Foray*, 1949: Rowney Warren; *Beds. N.H.S. Foray*, 1950: very common in mixed woods and very variable; Heath and Reach, Woburn, Leighton Buzzard; *D.A.R.*

R. delica Fr.—frequent, 1892; *E. M. Langley in J.H., V.C.H.*: Aspley Wood; *Beds. N.H.S. Foray*, 1948: Heath and Reach; *D.A.R.*

R. drimeia Cooke—Clophill; *Beds. N.H.S. Foray*, 1949.

R. emetica (Schaeff.) Fr.—Aspley Wood; *Beds. N.H.S. Foray*, 1948: Barton Hills; *Beds. N.H.S. Foray*, 1951: a common fungus of mixed woods, Heath and Reach; *D.A.R.*

R. fallax (Fr.) Cooke—Clophill; *Beds. N.H.S. Foray*, 1952: common in birch woods, Heath and Reach; *D.A.R.*

R. fellea Fr.—Aspley Wood; *Beds. N.H.S. Foray*, 1948: Rowney Warren; *Beds. N.H.S. Foray*, 1950: common in mixed woods; *D.A.R.*

R. foetens Fr.—frequent (W. B. Grove); *J.H., V.C.H.*: common in mixed woods, Heath and Reach; *D.A.R.*

R. fragilis Fr.—Rowney Warren; *Beds. N.H.S. Foray*, 1950: common in mixed woods, especially on birch, Heath and Reach; *D.A.R.*
Var. **nivea** (Pers.) Cooke—common in mixed woods, especially on birch; Heath and Reach; *D.A.R.*

R. heterophylla Fr.—Southill, 1890 (W. B. Grove); *J.H., V.C.H.*

R. integra (L.) Fr.—Oakley, 1891; *E. M. Langley in J.H., V.C.H.*

R. laurocerasi Melz.—uncommon in mixed woods, Heath and Reach; *D.A.R.* 1951.

R. lepida Fr.—common in mixed woods, Heath and Reach; *D.A.R.* 1949.

R. lutea Fr. forma **luteorosella** Britz.—Heath and Reach; *D.A.R.* 1950.

R. nigricans Fr.—common, as *Agaricus elephantinus*; *F.B.*: Ampthill; Southill (W. B. Grove); *E. M. Langley in J.H., V.C.H.*: Aspley Wood; *Beds. N.H.S. Foray*, 1948: very common in mixed woods; *D.A.R.*

R. ochroleuca Fr.—Ampthill, 1886 (W. B. Grove); Aspley Wood; *Beds. N.H.S. Foray*, 1948: Rowney Warren; *Beds. N.H.S. Foray*, 1950: Clophill; *Beds. N.H.S. Foray*, 1952: very common in mixed woodland; *D.A.R.*

R. pectinata (Bull.) Fr.—common under oaks, Heath and Reach; *D.A.R.* 1949.

R. puellaris Fr.—uncommon, Heath and Reach; *D.A.R.* 1952

R. queletii Fr.—Ampthill (W. B. Grove); *J.H., V.C.H.*

R. sororia Fr.—Rowney Warren; *R. W. G. Dennis*: Clophill; *Beds. N.H.S. Foray*, 1952: common under oaks, Heath and Reach; *D.A.R.* 1948.

R. venosa Vel.—Aspley Wood; *Beds. N.H.S. Foray*, 1948: Rowney Warren; *R. W. G. Dennis*: Clophill; *Beds. N.H.S. Foray*, 1952: common, especially in damp birch woods, Heath and Reach; *D.A.R.*
Var. **pallida** Lange—Rowney Warren, 1950; *R. W. G. Dennis*.

Schizophyllum commune Fr.—common on fallen tree trunks, Leighton Buzzard, Heath and Reach; *D.A.R.* 1948.

Tricholoma atrosquamosum (Chév.) Sacc.—Ampthill, 1887 (W. B. Grove); *E. M. Langley in J.H., V.C.H.*

T. brevipes (Bull.) Fr.—Kempston; Bedford (W. B. Grove); *J.H., V.C.H.*

T. carneum (Bull.) Fr.—Barton Hills; *Beds. N.H.S. Foray*, 1951.

T. cinerascens (Bull.) Fr.—Clophill; *Beds. N.H.S. Foray*, 1952.

T. cingulatum (Fr.) Lange—common under *Salix* sp. in sandpit, Grovebury Road, Leighton Buzzard; *D.A.R.* 1951. [The first British record: see *Trans. Beds. Nat. Hist. Soc.* 1951, 35.]

T. gambosum Fr.—Bedford, 1892; *E. M. Langley in J.H., V.C.H.*

T. loricatum Fr.—Ampthill, 1892; *J.H.*, *V.C.H.*

T. melaleucum (Pers.) Fr. (*T. humile* (Pers.) Fr., T. *subpulverulentum* (Pers.) Fr.)—Kempston (W. G. Smith); Bedford; Sandy (W. B. Grove); *J.H.*, *V.C.H.*: Barton Hills; *Beds. N.H.S. Foray*, 1951.

T. nudum Fr., *Wood Blewit*—Clapham Park, Warden, as *Agaricus nudus*; *F.B.*: common (W. B. Grove); *J.H.*, *V.C.H.*: Aspley Wood; *Beds. N.H.S. Foray*, 1948: Rowney Warren; *Beds. N.H.S. Foray*, 1950: Barton Hills; *Beds. N.H.S. Foray*, 1951: Clophill; *Beds. N.H.S. Foray*, 1952: common in mixed woods late in the season, and often persisting until the end of the year; *D.A.R.* [See also under *Cortinarius violaceus* Fr.]

T. personatum Fr., *Blewit*—Ampthill (W. B. Grove); *J.H.*, *V.C.H.*: not common, but sometimes forming large rings in pastures, Leighton Buzzard; *D.A.R.*

T. portentosum Fr.—near Bedford, 1892; *R. M. Langley in J.H.*, *V.C.H.*: Aspley Wood; *Beds. N.H.S. Foray*, 1948.

T. psammopus (Kalchbr.) Fr.—Rowney Warren, 1951; *R. W. G. Dennis*: uncommon in pine woods, Heath and Reach; *D.A.R.*

T. robustum (Alb. & Schw.) Fr. (*Armillaria focalis* Fr.)—Ampthill Woods, 1887 (W. B. Grove); *J.H.*, *V.C.H.*

T. rutilans (Schaeff.) Fr.—Warden, Ampthill, as *Agaricus xerampelinus*; *F.B.*: common in 1888 (W. B. Grove); *J.H.*, *V.C.H.*: Aspley Wood; *Beds. N.H.S. Foray*, 1948: Clophill; *Beds. N.H.S. Foray*, 1952: a common species in pine woods, growing in clusters on old stumps and buried wood, Woburn, Heath and Reach; *D.A.R.*

T. saponaceum Fr. (*T. luridum* (Schaeff.) Fr.)—Clapham, 1891 (W. B. Grove); *E. M. Langley in J.H.*, *V.C.H.*: a species with a sweet soapy smell, found in mixed woods, Heath and Reach; *D.A.R.*

T. scalpturatum Fr. (*T. chrysites* (Jungh.) Gillet)—Barton Hills; *Beds. N.H.S. Foray*, 1951.

T. sordidum Fr.—Rowney Warren; *Beds. N.H.S. Foray*, 1950.

T. sulphureum (Bull.) Fr.—Silsoe, Warden, as *Agaricus sulphureus*; *F.B.*

T. terreum Fr.—common, as *Agaricus terreus*; *F.B.*: common (W. B. Grove); *E. M. Langley in J.H.*, *V.C.H.*: Barton Hills; *Beds. N.H.S. Foray*, 1951.

AGARICALES: AGARICACEAE: RHODOSPORAE

Clitopilus cretatus Berk. & Broome—common (W. B. Grove); *J.H.*, *V.C.H.*

C. prunulus (Scop.) Fr. (*C. orcella* (Bull.) Fr.)—frequent (W. G. Smith and W. B. Grove); *J.H.*, *V.C.H.*: Deadmansey Wood; *E.M.-R.*: smelling of new meal and common in mixed woods, Heath and Reach; *D.A.R.*

Eccilia undata (Fr.) Bigeard & Guillermin—Rowney Warren; *Beds. N.H.S. Foray*, 1950.

Entoloma clypeatum (L.) Fr.—uncommon, under hawthorn and in shrubberies, Leighton Buzzard; *D.A.R.* 1953.

E. porphyrophaeum Fr.—Rowney Warren; *Beds. N.H.S. Foray*, 1950.

E. sericeum (Bull.) Fr.—Ampthill, etc., 1887 (W. B. Grove); *J.H.*, *V.C.H.*: Rowney Warren; *Beds. N.H.S. Foray*, 1950: Clophill; *Beds. N.H.S.* Foray, 1952: common in grassy places on heaths, Leighton Buzzard; Heath and Reach; *D.A.R.*

Leptonia chalybaea (Pers.) Fr.—common, as *Agaricus columbarius*; *F.B.*

L. euchroa (Pers.) Fr.—Barton Hills; *Beds. N.H.S. Foray*, 1951.

L. lampropus Fr.—Deadmansey Wood; *E.M.-R.*

Nolanea cetrata (Fr.) Schroet.—common (W. B. Grove); *J.H.*, *V.C.H.*: Rowney Warren; *R. W. G. Dennis*: Heath and Reach; *D.A.R.*

N. mammosa (L.) Fr.—Rowney Warren; *Beds. N.H.S. Foray*, 1950.

N. staurospora Bres.—common, as *Agaricus rufus*; *F.B.* [W. B. Grove thought that this might be *A. pascuus* Pers.—a name applied in the past to *N. staurospora*—see Peason and Dennis, 1948]: common, as *A. pascuus* (W. B. Grove); *J.H.*, *V.C.H.*: Rowney Warren, 1950; *R. W. G. Dennis*: common in grassy places, lawns and woods, Heath and Reach, Leighton Buzzard; etc. *D.A.R.*

Pluteus cervinus (Schaeff.) Fr.—Pavenham, 1887 (W. B. Grove); *J.H.*, *V.C.H.*:
Rowney Warren; *Beds. N.H.S. Foray*, 1950: Barton Hills; *Beds. N.H.S. Foray*,
1951: Clophill; *Beds. N.H.S. Foray*, 1952: very common on old stumps and very
variable, Heath and Reach; *D.A.R.*
P. chrysophaeus (Schaeff.) Fr.—Kempston, 1887 (W. B. Grove); *J.H.*, *V.C.H.*
P. cinereo-fuscus Lange—Barton Hills; *Beds. N.H.S. Foray*, 1951.
P. nanus (Pers.) Fr.—Deadmansey Wood; *E.M.-R.* 1948.
P. phlebophorus (Ditmar) Fr.—Barton Hills; *Beds. N.H.S. Foray*, 1951.
P. salicinus (Pers.) Fr.—not uncommon on fallen branches, Heath and Reach;
D.A.R. 1952.
Volvaria pusilla (Pers.) Fr. (*V. parvula* (Weinm.) Fr.)—Biddenham, 1887
(W. B. Grove); *J.H.*, *V.C.H.*
V. speciosa Fr.—Ampthill, abundant 1892 (W. B. Grove); Bedford, 'the thin
grey variety', 1898; *J.H.*, *V.C.H.*: rare in pastures, Heath and Reach; *D.A.R.*

AGARICALES: AGARICACEAE: OCHROSPORAE

Bolbitius titubans (Bull.) Fr.—with no details, as *Agaricus titubans*; *F.B.*
[Peason and Dennis think this may be synonymous with *B. vitellinus*].
B. vitellinus (Pers.) Fr.—Barton Hills; *Beds. N.H.S. Foray*, 1951: not uncommon
in richly manured pastures, sawdust heaps, etc., Leighton Buzzard; *D.A.R.*
Cortinarius anomalus Fr.—Deadmansey Wood; *E.M.-R.*, 1948.
C. biformis Fr.—in mixed wood, Heath and Reach; *D.A.R.*, 1951.
C. callisteus Fr.—Great Warden, 1892; *J.H.*, *V.C.H.*
C. cinnamomeus Fr.—Ampthill, as *Agaricus cinnamomeus*; *F.B.*: Rowney Warren,
1950; *R. W. G. Dennis*: Barton Hills; *Beds. N.H.S. Foray*, 1951.
[Abbot lists *Cortinarius collinitus* (Pers.) Fr. (*Agaricus collinitus*), with no details in
his annotated copy of *F.B.*]
C. elatior Fr.—Warden; Ampthill, as *Agaricus collinitus*; *F.B.* [W. B. Grove thought
this might be referred to this species]: common in oak woods, Heath and Reach;
D.A.R.
C. glandicolor Fr.—not uncommon in mixed woods, Heath and Reach; *D.A.R.*
1952.
C. glaucopus (Schaeff.) Fr.—common, as *Agaricus glaucopus*; *F.B.* [W. B. Grove
thought that this might be referred to *C. cyanopus* (Secr.) Fr.].
C. hemitrichus (Pers.) Fr.—common in birch woods, Heath and Reach;
Woburn; *D.A.R.* 1951.
C. hinnuleus Fr.—Warden, as *Agaricus hinnuleus*; *F.B.*: Bedford, 1887 (W. B.
Grove); *J.H.*, *V.C.H.*
C. malachius (Fr.) Pearson—Aspley Wood; *Beds. N.H.S. Foray*, 1948.
C. orichalceus (Batsch) Fr.—Ampthill, as *Agaricus orichalceus*; *F.B.*
C. paleaceus Fr.—under birch trees, Leighton Buzzard; *D.A.R.* 1952.
C. punctatus (Pers.) Fr.—Barton Hills; *Beds. N.H.S. Foray*, 1951.
C. purpurascens Fr. var. **subpurpurascens** Fr.—common, as *Agaricus subpur-
purascens*; *F.B.*: Deadmansey Wood; *E.M.-R.*
C. semi-sanguineus Gillet—Rowney Warren; *Beds. N.H.S. Foray*, 1950.
C. tabularis (Bull.) Fr.—Ampthill, 1887 (W. B. Grove); *J.H.*, *V.C.H.*
C. triumphans Fr.—an uncommon species of birch woods, Heath and Reach;
D.A.R.
C. urbicus Fr.—common in dense clusters under *Salix* sp. in sandpits, Grovebury
Road, Leighton Buzzard; *D.A.R.*, 1951. [The first certain record of the species
for Britain: see *Trans. Beds. Nat. Hist. Soc.* 1951, 35.]
C. violaceus Fr.—common, as *Agaricus violaceus*; *F.B.* [Abbot cites Bulliard tab.
585, which is true *C. violaceus*, and Sowerby tab. 209, which is *Tricholoma nudum*].
Crepidotus mollis (Schaeff.) Fr.—common, as *Agaricus mollis*; *F.B.*
C. pubescens Bres.—common on fallen twigs and herbaceous stems, Leighton
Buzzard; Heath and Reach; *D.A.R.* 1952.
C. variabilis (Pers.) Fr.—common, as *Agaricus sessilis*; *F.B.* [W. B. Grove inter-
prets Abbot's record as this, but it is possible that he included several species
under the one name]: Clophill; *Beds. N.H.S. Foray*, 1952: common on plant
debris, Leighton Buzzard, Heath and Reach; *D.A.R.*

Flammula agardhii (Lund) Fr.—under *Salix* sp., sandpits, Grovebury Road, Leighton Buzzard; *D.A.R.* 1951 [the first British record: see *Trans. Beds. Nat. Hist. Soc.* 1951, 35].

F. alnicola Fr.—Aspley Wood; *Beds. N.H.S. Foray,* 1948.

F. carbonaria Fr.—Aspley Wood; *Beds. N.H.S. Foray,* 1948: Clophill; *Beds. N.H.S. Foray,* 1949.

F. flavida (Schaeff.) Fr.—Ampthill, Bedford (W. B. Grove); *J.H., V.C.H.*

F. gummosa (Lasch) Quél.—Clophill; *Beds. N.H.S. Foray,* 1949.

F. lenta (Pers.) Fr. (*Hebeloma glutinosum* (Lindgr.) Fr.)—near Bedford (W. B. Grove); *E. M. Langley in J.H., V.C.H.*

F. sapinea Fr.—Heath and Reach; *Beds. N.H.S. Foray,* 1947: Aspley Wood; *Beds. N.H.S. Foray,* 1948: Rowney Warren; *Beds. N.H.S. Foray,* 1950: Clophill; *Beds. N.H.S. Foray,* 1952: common, attached to buried wood, in pine woods, Woburn; *D.A.R.*

Galera clavata Vel.—Rowney Warren; *R. W. G. Dennis,* 1950.

G. hypnorum (Batsch) Fr.—common, as *Agaricus hypni*; *F.B.*: Aspley Wood; *Beds. N.H.S. Foray,* 1947: Rowney Warren; *Beds. N.H.S. Foray,* 1950: Clophill; *Beds. N.H.S. Foray,* 1952: common with moss on heaths, Leighton Buzzard, Heath and Reach; *D.A.R.*

G. mycenoides (Fr.) Quél. sensu Kühner—Heath and Reach; *Beds. N.H.S. Foray,* 1948.

G. mycenopsis Fr. sensu Ricken—Rowney Warren; *R. W. G. Dennis,* 1950.

G. pilosella (Pers.) Fr. sensu Kühner—Barton Hills; *Beds. N.H.S. Foray,* 1951.

G. pygmaeo-affinis Fr.—Kempston, as *Agaricus striaepes* (W. B. Grove); *J.H., V.C.H.*

G. spicula (Lasch) Fr. forma **macrospora** Kühner—Clophill; *Beds. N.H.S. Foray,* 1952.

G. tenera (Schaeff.) Fr.—common, as *Agaricus tener*; *F.B.*: common (W. B. Grove); *J.H., V.C.H.*: common in grassy places, Heath and Reach, Leighton Buzzard: *D.A.R.*

Var. **minor** Lange—Heath and Reach; *Beds. N.H.S. Foray,* 1948.

Hebeloma crustuliniforme (Bull.) Fr.—Putnoe, as *Agaricus crustuliniformis*; *F.B.*: Ampthill, 1892; *J.H., V.C.H.*: Clophill; *Beds. N.H.S. Foray,* 1949: Rowney Warren; *Beds. N.H.S. Foray,* 1950: Barton Hills; *Beds. N.H.S. Foray,* 1951: common, Leighton Buzzard; *D.A.R.*

H. fastibile (Pers.) Fr.—near Bedford, Kempston (W. G. Smith); *E. M. Langley in J.H., V.C.H.*

H. mesophaeum (Pers.) Fr.—Barton Hills; *Beds. N.H.S. Foray,* 1951: Clophill; *Beds. N.H.S. Foray,* 1952: common, especially under willows, Leighton Buzzard; *D.A.R.* 1949.

H. nudipes Fr.—near Bedford, 1892; *E. M. Langley in J.H., V.C.H.* [Pearson and Dennis think this species is most likely to be *H. longicaudum* (Pers.) Fr.]

H. sacchariolens Quél.—Deadmansey Wood; *E.M.-R.* 1948.

H. sinapizans (Paul.) Fr.—Clophill; *Beds. N.H.S. Foray,* 1952.

Inocybe asterospora Quél.—Bedford, 1887 (W. B. Grove); *J.H., V.C.H.*: Clophill; *Beds. N.H.S. Foray,* 1952: a common species of heaths and grassy places in woods, Heath and Reach; *D.A.R.* [See also *I. rimosa.*]

I. bongardii (Weinm.) Quél.—Southill Park, 1887 (W. B. Grove); *J.H., V.C.H.*

I. cincinnata Fr.—Clophill; *Beds. N.H.S. Foray,* 1952.

I. cookei Bres.—rare in mixed woods, Heath and Reach; *D.A.R.* 1952.

I. corydalina Quél.—Barton Hills; *Beds. N.H.S. Foray,* 1951.

I. deglubens Fr.—Rowney Warren; *R. W. G. Dennis,* 1950.

I. fastigiata (Schaeff.) Fr.—common, as *Agaricus rimosus*; *F.B.*: Barton Hills; *Beds. N.H.S. Foray,* 1951: common in mixed woods, Heath and Reach; *D.A.R.*

I. geophylla (Sow.) Fr.—common, as *Agaricus geophyllus*; Clapham Park Wood, as *A. auricomus*; *F.B.* [W. B. Grove had some doubt as to the identity of the last named]: Ampthill, 1887 (W. B. Grove); *J.H., V.C.H.*: Deadmansey Wood; *E.M.-R.* 1948: Clophill; *Beds. N.H.S. Foray,* 1952: common in mixed woodland, Heath and Reach; *D.A.R.*

Var. **lilacina** Fr.—Clophill; *Beds. N.H.S. Foray,* 1952: common in mixed woods, Heath and Reach; *D.A.R.* 1950.

I. griseo-lilacina Lange—Leighton Buzzard; *D.A.R.* 1950.

I. lacera Fr.—common in sandy places, Heath and Reach, Woburn, Leighton Buzzard; *D.A.R.* 1950.

I. lanuginosa (Bull.) Fr.—Rowney Warren; *Beds. N.H.S. Foray,* 1950.

I. maculata Boud.—common in mixed woods, Heath and Reach; *D.A.R.* 1952.

I. petiginosa (Fr.) Gillet forma **rufo-alba** (Pat. & Doass.) Heim—Rowney Warren; *Beds. N.H.S. Foray,* 1950.

I. praetervisa Quél.—Heath and Reach; *D.A.R.* 1952.

I. rimosa (Bull.) Fr.—common (W. B. Grove); *J.H., V.C.H.* [Pearson and Dennis state that this name has been applied to *I. asterospora* by earlier botanists].

I. umboninata Peck—Rowney Warren; *R. W. G. Dennis,* 1950: Woburn, Heath and Reach; *D.A.R.*

Naucoria jennyae Karst.—Rowney Warren; *Beds. N.H.S. Foray,* 1950. [The first record for Britain.]

N. melinoides Fr.—Bedford 1887, as *Galera mniophila* (Lasch) Fr.; *J.H., V.C.H.*

N. semi-orbicularis (Bull.) Fr.—not uncommon in short grass, Heath and Reach; *D.A.R.* 1952.

N. sideroides (Bull.) Fr.—Rowney Warren; *Beds. N.H.S. Foray,* 1950.

N. pediades Fr.—Pavenham, 1887; *J.H., V.C.H.*

Paxillus atrotomentosus (Batsch) Fr.—Ampthill, Sandy (W. B. Grove); *J.H., V.C.H.*: Clophill; *Beds. N.H.S. Foray,* 1949: Rowney Warren; *Beds. N.H.S. Foray,* 1950: on old coniferous stumps, Heath and Reach; *D.A.R.*

P. involutus (Batsch) Fr.—common, as *Agaricus contiguus*; *F.B.*: common (W. B. Grove); *J.H., V.C.H.*: Aspley Wood; *Beds. N.H.S. Foray,* 1948: Clophill; *Beds. N.H.S. Foray,* 1949: Rowney Warren; *Beds. N.H.S. Foray,* 1952: common, especially in birch woods, Heath and Reach, Leighton Buzzard, Woburn; *D.A.R.*

P. lepista Fr.—unlocalized, 1901; *E. M. Langley in J.H., V.C.H.* [Pearson and Dennis say that this name has been applied to *Clitocybe mundula* (Lasch) Pearson and Dennis and *C. amara* Fr.]

Pholiota aegerita Brig. (P. pudica (Bull.) Fr.)—on elder, Bedford, 1887; *J.H., V.C.H.*

P. aurivella (Batsch) Fr.—Great Barford, 1896; *J.H., V.C.H.*

P. blattaria Fr. non Ricken—Barton Hills; *Beds. N.H.S. Foray,* 1951.

P. dura (Bolt.) Fr.—common at Bedford (W. B. Grove); *J.H., V.C.H.*: common in grass, Leighton Buzzard; *D.A.R.*

P. erebia Fr.—in mixed woods, Woburn; *D.A.R.* 1952.

P. marginata (Batsch) Fr.—common, as *Agaricus marginatus*; *F.B.*: Barton Hills; *Beds. N.H.S. Foray,* 1951.

P. mutabilis (Schaeff.) Fr.—Aspley Wood; *Beds. N.H.S. Foray,* 1948: on old stumps, uncommon, Heath and Reach; *D.A.R.* 1952.

P. radicosa (Bull.) Fr.—Silsoe, Clapham Park Wood, as *Agaricus radicosus*; *F.B.*

P. spectabilis Fr.—Clapham Park Wood, as *Agaricus aureus*; *F.B.*: Clophill; *Beds. N.H.S. Foray,* 1949.

P. squarrosa (Mull.) Fr.—between Bedford and Elstow, as *Agaricus floccosus*; *F.B.*: common (W. B. Grove); *J.H., V.C.H.*

P. togularis (Bull.) Fr. non Ricken sensu Kühner—Barton Hills; *Beds. N.H.S. Foray,* 1951.

Ripartites tricholoma (Alb. & Schw.) Karst.—not uncommon, Heath and Reach; *D.A.R.* 1952.

Tubaria furfuracea (Pers.) Gillet—Clapham between Woods, as *Agaricus zylophilus*; *F.B.*: common (W. B. Grove); *J.H., V.C.H.*: common on fallen twigs and bare soil, throughout the winter, Heath and Reach, Leighton Buzzard, Woburn; *D.A.R.*

T. inquilina (Fr.) Gillet—Bedford, 1887 (W. B. Grove); *J.H., V.C.H.*: Rowney Warren; *Beds. N.H.S. Foray,* 1950.

T. pellucida (Bull. ex Fr.) Gillet—Heath and Reach; *Beds. N.H.S. Foray,* 1948.

AGARICALES: AGARICACEAE: PORPHYROSPORAE

Hypholoma candolleanum Fr.—Kempston (W. G. Smith and W. B. Grove);
common, as *H. appendiculatum* Fr. (probably this species); *J.H.*, *V.C.H.*: common
round bases of old stumps and fences, etc., Heath and Reach, Leighton Buzzard;
D.A.R.

H. fasciculare (Huds.) Fr.—common, as *Agaricus fascicularis*; *F.B.*: common
(W. B. Grove); *J.H.*, *V.C.H.*: Aspley Wood; *Beds. N.H.S. Foray*, 1948: Clophill; *Beds. N.H.S. Foray*, 1949: Rowney Warren; *Beds. N.H.S. Foray*, 1950:
common on stumps and often forming large clusters, Heath and Reach, etc.;
D.A.R.

H. hydrophilum (Bull.) Fr. (*Bolbitius hydrophilus* (Bull.) Fr.)—Ampthill, 1887
(W. B. Grove); *J.H.*, *V.C.H.*: Clophill; *Beds. N.H.S. Foray*, 1949: common on
old stumps, etc.; *D.A.R.*

H. lacrymabundum Fr. non Bull.—Kempston, 1885 (W. G. Smith); *J.H.*, *V.C.H.*

H. sublateritium Fr.—frequent (W. B. Grove); *J.H.*, *V.C.H.*: Heath and Reach;
Beds. N.H.S. Foray, 1948: Clophill; *Beds. N.H.S. Foray*, 1949: not uncommon
later in the season, Heath and Reach; *D.A.R.*

H. velutinum (Pers.) Fr.—common (W. B. Grove); *J.H.*, *V.C.H.*

Psalliota arvensis (Schaeff.) Fr.—common (W. B. Grove); *J.H.*, *V.C.H.*:
common in pastures, Heath and Reach; Leighton Buzzard; *D.A.R.*

P. campestris (L.) Fr., *Mushroom.*—Clophill; *Beds. N.H.S. Foray*, 1949: common
in pastures, Heath and Reach, Leighton Buzzard, etc.; *D.A.R.*: Cockayne
Hatley; *T.L.*

   Var. **praticola** (Vitt.) Fr.—common; *J.H.*, *V.C.H.*

   Var. **rufescens** Berk.—unlocalized, 1901; E. M. Langley in *J.H.*, *V.C.H.*

P. elvensis Berk. & Broome—near Bedford (W. B. Grove); E. M. Langley in
J.H., *V.C.H.*

P. silvatica (Schaeff.) Fr.—near Bedford; *E. M. Langley in J.H.*, *V.C.H.*: Rowney
Warren; *Beds. N.H.S. Foray*, 1950.

P. silvicola (Vitt.) Sacc.—Rowney Warren; *Beds. N.H.S. Foray*, 1950: common
in mixed woods, Heath and Reach, Leighton Buzzard; *D.A.R.*

Psilocybe areolata (Kl.) Berk.—Kempston (W. G. Smith); *J.H.*, *V.C.H.*
[Pearson and Dennis write that this name has uncertain interpretation].

P. atrorufa (Schaeff.) Fr.—Rowney Warren; *Beds. N.H.S. Foray*, 1950.

P. bullacea (Bull.) Fr.—common in sandpits, Grovebury Road, Leighton
Buzzard—persisting until February; *D.A.R.* 1951.

P. elongata Fr.—Aspley Wood; *Beds. N.H.S. Foray*, 1948. [This species is uncommon as it grows only in sphagnum bogs; D.A.R.].

P. foenisecii (Pers.) Fr.—frequent (W. B. Grove); *J.H.*, *V.C.H.*: common on
lawns, Leighton Buzzard; *D.A.R.*

P. semi-lanceata Fr.—Rowney Warren; *Beds. N.H.S. Foray*, 1950: common
on lawns and grassy places in woods, Heath and Reach, Leighton Buzzard;
D.A.R.

P. sub-ericaea Fr.—Heath and Reach; *Beds. N.H.S. Foray*, 1948.

Stropharia aeruginosa (Curt.) Fr.—Ampthill, Silsoe; as *Agaricus aeruginosus*;
F.B.: common (W. B. Grove); *J.H.*, *V.C.H.*: Rowney Warren; *R. W. G.
Dennis* 1950: Clophill; *Beds. N.H.S. Foray*, 1952: common, Heath and Reach;
D.A.R.

S. coronilla (Bull.) Fr.—Kempston, 1887 (W. B. Grove); *J.H.*, *V.C.H.*: common
in pastures, Heath and Reach; *D.A.R.*

S. inuncta Fr.—Barton Hills; *Beds. N.H.S. Foray*, 1951.

S. semiglobata (Batsch) Fr.—common, as *Agaricus semiglobatus*; *F.B.*: common
(W. B. Grove); *J.H.*, *V.C.H.*: common in heaths and pastures and on manure,
Leighton Buzzard, Heath and Reach; *D.A.R.*

AGARICALES: AGARICACEAE: MELANOSPORAE

Annellaria fimiputris (Fr.) Karst.—Clophill; Warden, as *Agaricus clypeatus*;
common, as *A. fimiputris*; *F.B.*: common; *J.H.*, *V.C.H.*

A. semi-ovata (Sow. ex Fr.) Pearson & Dennis (*A. separata* (L.) Karst.)—
common, as *Agaricus semi-ovatus*; *F.B.*: common in 1890; *J.H.*, *V.C.H.*: Barton
Hills; *Beds. N.H.S. Foray*, 1951.

Coprinus atramentarius (Bull.) Fr.—Ampthill, as *Agaricus luridus*; common as *A. ovatus*; *F.B.*: Heath and Reach; *Beds. N.H.S. Foray*, 1948: Barton Hills; *Beds. N.H.S. Foray*, 1951: common at base of old stumps and on buried wood, Leighton Buzzard, etc.; *D.A.R.*

C. brasicae Peck—on leaves of *Glyceria maxima*; *D.A.R.*

C. comatus (Fl. Dan.) Fr., *Shaggy Ink-cap*—common, as *Agaricus cylindricus*; *F.B.*: common (W. B. Grove); *J.H., V.C.H.*: Clophill; *Beds. N.H.S. Foray*, 1949: common in pastures and gardens and by roadsides, Leighton Buzzard, Heath and Reach; *D.A.R.*

C. domesticus (Pers.) Fr.—Ampthill, 1887; *J.H., V.C.H.*

C. friesii Quél.—uncommon on grass stems, Stanbridge; *D.A.R.* 1948.

C. funarianum Métrod—Clophill; *Beds. N.H.S. Foray*, 1952 [The first record for Britain].

C. macrorhizus (Pers.) Rea—on manure heaps, early in year, Leighton Buzzard; *D.A.R.* 1951.

C. micaceus (Bull.) Fr.—common as *Agaricus micaceus*; *F.B.*: common (W. B. Grove); *J.H., V.C.H.*: Barton Hills; *Beds. N.H.S. Foray*, 1951: common on stumps or attached to buried wood, Heath and Reach, Leighton Buzzard; *D.A.R.*

C. niveus (Pers.) Fr.—common, as *Agaricus momentaneus*; *F.B.*: common; *J.H., V.C.H.*: Barton Hills; *Beds. N.H.S. Foray*, 1951.

C. plicatilis (Curt.) Fr.—common, as *Agaricus plicatilis*; *F.B.*: common (W. B. Grove); *J.H., V.C.H.*: Barton Hills; *Beds. N.H.S. Foray*, 1951: common in grassy places, especially woodland rides; Heath and Reach; Leighton Buzzard; *D.A.R.*

C. picaceus (Bull.) Fr.—in grass, Stanbridge; *D.A.R.* 1944.

C. radiatus (Bolt.) Fr.—common on horse-dung, Leighton Buzzard; *D.A.R.* 1951.

Gomphidius glutinosus (Schaeff.) Fr.—Barton Hills; *Beds. N.H.S. Foray*, 1951.

G. rutilus (Schaeff.) Fr. (G. viscidus (L.) Fr.)—frequent (W. B. Grove); *J.H., V.C.H.*: Rowney Warren; *Beds. N.H.S. Foray*, 1950: common in pine woods, Heath and Reach; *D.A.R.*

Panaeolus campanulatus (L.) Fr.—common; *J.H., V.C.H.*: Rowney Warren; *Beds. N.H.S. Foray*, 1950: Barton Hills; *Beds. N.H.S. Foray*, 1951: common on lawns, in pastures and in richly manured ground, Leighton Buzzard; *D.A.R.* var. **sphinctrinus** (Fr.) Bres. (*P. sphinctrinus* Fr.)—Barton Hills; *Beds. N.H.S. Foray*, 1951.

P. papilionaceus (Bull.) Fr.—common; *J.H., V.C.H.*: on lawns and in pastures, Leighton Buzzard; *D.A.R.*

Psathyrella atomata Fr.—Bromham; *J.H., V.C.H.*

P. conopilea Fr.—Rowney Warren; *Beds. N.H.S. Foray*, 1950: common in woods, Heath and Reach; *D.A.R.*

P. disseminata (Pers.) Fr.—common, as *Agaricus striatus* and *A. minutulus*; *F.B.*: Barton Hills; *Beds. N.H.S. Foray*, 1951: on old stumps, often abundant, Heath and Reach; *D.A.R.*

P. gossypina (Bull. ex Fr.) Pearson & Dennis—Heath and Reach; *D.A.R.* 1951.

P. gracilis Fr.—Bedford (W. B. Grove); *J.H., V.C.H.*: Barton Hills; *Beds. N.H.S. Foray*, 1951: common in grass and woodland rides, by roadsides, etc., Heath and Reach, Leighton Buzzard; *D.A.R.* var. **corrugis** (Pers.) Lange—frequent (W. B. Grove); *J.H., V.C.H.*

P. pennata (Fr.) Pearson & Dennis—Aspley Wood; *Beds. N.H.S. Foray*, 1948.

P. subatomata Lange—Barton Hills; *Beds. N.H.S. Foray*, 1951.

P. spadicea-grisea (Schaeff. ex. Fr.) A. H. Smith (*Psathyra spadiceo-grisea* (Schaeff.) Fr.)—Barton Hills; *Beds. N.H.S. Foray*, 1951.

GASTEROMYCETES

PHALLACEAE

Mutinus caninus (Huds.) Fr. (*Cynophallus caninus* (Huds.) Fr.)—Silsoe New Wood, as *Phallus caninus*; *Plantae Bedford*: Luton Hoo Park (W. B. Grove); *J.S. in J.H., V.C.H.*

Phallus impudicus (L.) Pers. *Stink-horn*—Silsoe New Wood; Mr. Sibley's garden, Market Street; *Plantae Bedford*: common at Ampthill; *J.H.*, *V.C.H.*: Aspley Wood; *Beds. N.H.S. Foray*, 1948: Clophill; *Beds. N.H.S. Foray*, 1949: Rowney Warren; *Beds. N.H.S. Foray*, 1950: common in mixed, and coniferous woods, Heath and Reach; Woburn; *D.A.R.*

LYCOPERDACEAE   *Puff Balls and Earth Stars*

Bovista plumbea Fr.—Ford End, as *Lycoperdon ardosiaceum*; *F.B.*: common; *J.H.*, *V.C.H.*

Calvatia caelata (Bull.) Morg. (*Lycoperdon caelatum* (Bull.) Fr.)—Ampthill, 1887; *J.H.*, *V.C.H.*

C. saccatum (Vahl.) Morg. (*Lycoperdon saccatum* (Vahl.) Fr.)—Deadmansey Wood; *E.M.-R.* 1948.

Geaster fimbriatus Fr.—Ampthill Park, as *Lycoperdon stellatum*; *Plantae Bedford* [Abbot gives this station for *L. recolligens* (*G. fimbriatus*) in *F.B.* and Ford End for *L. stellatum*]: Barton Hills; *Beds. N.H.S. Foray*, 1951.

G. limbatus Fr.—Ford End, as *Lycoperdon stellatum*; *F.B.*: Barton Hills; *Beds. N.H.S. Foray*, 1951.

G. triplex Jungh.—Barton Hills; *Beds. N.H.S. Foray*, 1951.

Lycoperdon depressum Bon. (*L. hyemale* (Pers.) Vitt.)—Heath and Reach; *Beds. N.H.S. Foray*, 1948: common in sandy places in short grass and moss, Leighton Buzzard, Heath and Reach; *D.A.R.*

L. echinatum Pers.—Barton Hills; *Beds. N.H.S. Foray*, 1951: Clophill; *Beds. N.H.S. Foray*, 1952.

L. giganteum (Batsch) Pers. (*Calvatia gigantea* (Pers.) Lloyd—common, as *L. proteus*; *F.B.*: common (W. B. Grove); *J.H.*, *V.C.H.*

L. nigrescens Pers.—Clophill; *Beds. N.H.S. Foray*, 1952: uncommon on heaths and in dry sandy places, Heath and Reach; *D.A.R.*

L. perlatum Pers. (L. germatum Batsch.)—common (W. B. Grove); *J.H.*, *V.C.H.*: Clophill; *Beds. N.H.S. Foray*, 1949: Rowney Warren; *Beds. N.H.S. Foray*, 1950: Barton Hills; *Beds. N.H.S. Foray*, 1951: abundant in mixed woods, Heath and Reach, Woburn, etc.; *D.A.R.*

L. pyriforme (Schaeff.) Pers.—Kempston, etc. 1887 (W. B. Grove); *J.H.*, *V.C.H.*: Clophill; *Beds. N.H.S. Foray*, 1949: abundant on old rotting stumps of trees, Heath and Reach, Woburn, Leighton Buzzard; *D.A.R.*

Tulostoma mammosum (Mich.) Fr.—walls above Bedford, as *Lycoperdon pedunculatum*; *Plantae Bedford*.: common; *F.B.*

NIDULARIACEAE   *Bird's-nest Fungi*

Crucibulum vulgare Tul.—Warden, as *Peziza laevis*; *F.B.*

Cyathus striatus (Huds.) Pers.—Warden, as *Peziza striata*; *F.B.*: Ampthill, 1887 (W. B. Grove); *J.H.*, *V.C.H.*

C. vernicosus (Bull.) DC. (*C. olla* Pers.)—common, as *Peziza campanulata*; *F.B.*: Bedford—1887 (W. B. Grove); *J.H.*, *V.C.H.*: Clophill; *Beds. N.H.S. Foray*, 1949.

SCLERODERMATACEAE

Scleroderma verrucosum (Vaill.) Pers.—Ampthill, as *Lycoperdon verrucosum*; *F.B.*: Southill, etc. (W. B. Grove); *J.H.*, *V.C.H.*: Aspley Wood; *Beds. N.H.S. Foray*, 1948: Rowney Warren; *Beds. N.H.S. Foray*, 1950: Barton Hills; *Beds. N.H.S. Foray*, 1951: common on heaths, Heath and Reach; *D.A.R.*

S. vulgare (Horn.) Fr. (*S. aurantium* Pers.)—Bedford, etc., 1887 (W. B. Grove); *J.H.*, *V.C.H.*: Heath and Reach; *Beds. N.H.S. Foray*, 1947: Aspley Wood; *Beds. N.H.S. Foray*, 1948: Clophill; *Beds. N.H.S. Foray*, 1949: Rowney Warren; *Beds. N.H.S. Foray*, 1950: abundant on heaths, etc., Heath and Reach; *D.A.R.*

SPHAEROBOLACEAE

Sphaerobolus stellatus Pers.—Ampthill Park, as *Lycoperdon carpobolus*; *Plantae Bedford*.

FUNGI IMPERFECTI

SPHAEROPSIDALES

Ascochyta imperfecta Peck—unlocalized, on lucerne; *Min. of Agric. and Fish. Bulletin*, 126, 38.
A. pinodella L. K. Jones—Cockayne Hatley; *T.L.*
Ceuthospora phacidioides Grev.—White Wood, Everton, ? Beds., as *Sphaeria abifrons*; *Abbot's annotated F.B.*
Cytospora leucospora Pers.—Renhold Wood, as *Sphaeria cirrhata*; *F.B.*
Phoma eupyrena Sacc.—Ampthill; *F. H. Warcup*, 1952.
Septoria apii Chester—Cockayne Hatley; *T.L.*

MELANCONIALES

Actinonema rosae (Lib.) Fr.—common on roses, forming 'black spot', Leighton Buzzard, Totternhe; *D.A.R.* 1952: Cockayne Hatley; *T.L.*
Colletotrichum atramentarium (Berk & Broome) Taubenh.—Cockayne Hatley; *T.L.*

MONILIALES

Aegerita candida Pers. ex Fr. (the imperfect stage of *Peniophora candida* Lyman) —Heath and Reach; *Beds. N.H.S. Foray*, 1948.
Aspergillus glaucus Link ex Fr.—common, as *Mucor glaucus*; *F.B.*
A. niger van Tiegh.—on *Dracunculus vulgare* Schott., Leighton Buzzard; *D.A.R.* 1952.
Bactridiun flavum Kunze ex Fr.—Heath and Reach; *Beds. N.H.S. Foray*, 1948.
Botrytis cinerea Pers. ex Fr.—abundant on decaying vegetation, Leighton Buzzard; *D.A.R.* 1948: widespread; *T.L.*
B. tulipae (Lib.) Lind.—Cockayne Hatley; *T.L.*
B. vera Fr.—'on caps of old Boleti', Clapham Lane, as *Mucor botrytis*; *F.B.* [Grove referred this to *Polyactis cana* Kuntze & Schm. but Wakefield and Bisby state that *M. botrytis* is probably *B. cinerea*].
Cercosporella herpotrichoides Fron.—Cockayne Hatley; *T.L.*
Cladosporium fulvum Cooke—Cockayne Hatley; *T.L.*
C. herbarum Link ex Fr.—Heath and Beach; *Beds. N.H.S. Foray*, 1948.
Cylindrocarpon radicicola Wollenw.—Ampthill; *F. H. Warcup*, 1952.
Cylindrocolla urticae (Pers. ex Fr.) Bon.—on nettle stems, Leighton Buzzard; *D.A.R.* 1952.
Dendryphion comosum Wallr.—frequent on nettle stems, Leighton Buzzard; *D.A.R.* 1952.
Fusarium caeruleum (Lib.) Sacc.—on potato, unlocalized, *Min. of Agric. and Fish. Bulletin*, 126, 15: Cockayne Hatley; *T.L.*
F. culmorum (W. G. Sm.) Sacc.—Ampthill; *F. H. Warcup*, 1952.
F. equiseti (Corda) Sacc.—Ampthill; *F. H. Warcup*, 1952.
F. oxysporum sensu Snyder & Hansen—Ampthill; *F. H. Warcup*, 1952.
F. sambucinum Fuckel—Ampthill; *F. H. Warcup*, 1952.
F. solani (Mart.) Sacc. var **martii** Appel & Wollenw.—on runner bean, unlocalized; *Min. of Agric. and Fish. Bulletin*, 126, 34.
F. vasinfectum Atk var. **lutulatum** (Sherb. ut. sp.) Wollenw.—on runner bean, unlocalized; *Min. of Agric. and Fish. Bulletin*, 126, 34.
Fusicladium dendriticum (Wallr.) Fuckel (the imperfect stage of *Venturia inaequalis* (Cooke) Wint.)—abundant on apple leaves, Leighton Buzzard; *D.A.R.* 1948.
F. pirinum (Lib.) Fuckel (the imperfect stage of *Venturia pirina* Aderh.)— abundant on leaves of pear trees; *D.A.R.* 1948.
Monilia fructigena Pers. ex Westend. (the imperfect stage of *Sclerotinia fructigena* Aderh. & Ruhl.)—abundant on apples, Leighton Buzzard; *D.A.R.*
M. racemosa Pers. ex Fr.—unlocalized, rare, as *Mucor caespitosus*; *F.B.* (this was listed as *Penicillium glaucum* Link by W. B. Grove).
Oidium chrysanthemi Rabenh.—Cockayne Hatley; *T.L.*

Ovularia obliqua (Cooke) Oudem.—abundant on *Rumex* spp., Heath and Reach, Leighton Buzzard, etc.; *D.A.R.* 1951.

O. primulana Karst.—common on *Primula vulgaris*, Leighton Buzzard; *D.A.R.* 1952.

Ozonium auricomum Link ex Wallr. (the mycelial condition of *Coprinus* spp.)—common on old stumps and in log piles, Leighton Buzzard; *D.A.R.* 1952.

Ptychogaster albus Corda—common in pine woods (W. B. Grove); *J.H., V.C.H.*: Rowney Warren; *Beds. N.H.S. Foray*, 1950.

Scolecotrichum clavariarum (Desm.) Sacc.—Rowney Warren; *Beds. N.H.S. Foray*, 1950.

Sepedonium chrysospermum Fr.—'on caps of old Boleti', Clapham Lane, as *Mucor chrysospermum*; *F.B.*: Aspley Wood; *Beds. N.H.S. Foray*, 1948: common on caps of old Boleti, Heath and Reach, Woburn, etc.; *D.A.R.*

Sporendonema casei Desm. ex Fr.—'cheese in damp cellars', common, as *Mucor caseus* (W. B. Grove listed this as *Torula sporendonema* Berk. & Broome (*Oospora crustacea* (Bull.) (Sacc.)) ; Potton Wood, as *Mucor crustacea*; *F.B.*

Sporotrichum aurantiacum Fr.—Goldington, as *Mucor aurantiacus*; *F.B.*

Stemphylium botryosum Wallr.—on onion leaves, Leighton Buzzard; *D.A.R.* 1952.

Stilbum fimetarium (Pers. ex Fr.) Berk. & Broome—Heath and Reach; *Beds. N.H.S. Foray*, 1948.

Thielaviopsis basicola (Berk. & Broome) Ferraris—Cockayne Hatley; *T.L.*

Trichoderma viride Pers. ex Fr.—common, as *Mucor lignifractus*; *F.B.*: Heath and Reach; *Beds. N.H.S. Foray*, 1948.

Trichothecium roseum Link ex Fr.—Heath and Reach; *Beds. N.H.S. Foray*, 1948.

Torula herbarum Link ex Fr.—common on herbaceous stems, Leighton Buzzard; *D.A.R.* 1952.

Tubercularia vulgaris Tode ex Fr. (*T. nigricans* (Bull.) Gmel.)—common, as *Sphaeria tremelloides*; Southill, Clophill, as *T. nigricans*; *F.B.*: common on currant twigs (W. B. Grove); *J.H., V.C.H.*

Verticillium agaricinum Corda—on *Agaricus ostreatus*, Ampthill (W. B. Grove); *Mr. Ferraby in J.H., V.C.H.*

V. albo-atrum Reinke & Berth.—Cockayne Hatley; *T.L.*

MYCELIA STERILIA

Sclerotium cepivorum Berk.—on onions, unlocalized; *Min. of Agric. and Fish. Bulletin*, 126, 49: Cockayne Hatley; *T.L.*

S. tuliparum Kleb.—on tulips, unlocalized; *Min. of Agric. and Fish. Bulletin*, 126. 83.

INDEX OF BOTANISTS

[1] Saunders also credited some records to R. A. Chambers.

[1] Ellen Crouch (1846–1914), sister of Charles.

GEOGRAPHICAL INDEX

This contains place names used in the Flora. Names of parishes are shown in capitals and places other than parishes in lower case. The botanical districts are shown in italics. Names including the name of the parish in which they are situated are not listed. When more than one botanical district is given it must be understood that the place or parish is in more than one botanical district. Stations have sometimes been given with reference to the nearest name on the Ordnance Survey map, which may, in some cases, be in a neighbouring county.

Aley Green, CADDINGTON, *Colne*, 52/065185
AMPTHILL, *Ouse, Ivel*, 52/035380
Apesfield Spring, WOBURN, *Ouzel*, 42/930320
Appley End, HAYNES, *Ivel*, 52/106410
Arden Dell Wood, HYDE, *Lea*, 52/123190
ARLESEY, *Ivel*, 52/190356
ASPLEY GUISE, *Ouzel*, 42/940360
ASPLEY HEATH, *Ouzel*, 42/930350
Astey Wood, STAGSDEN, *Ouse*, 42/990485
ASTWICK, *Ivel*, 52/215385
Austin Canons, BEDFORD, *Ouse*, 52/040487
Badger Dell Wood, LUTON, *Colne*, 52/063210
Baker's Wood, HEATH AND REACH, *Ouzel*, 42/920290
BARTON-IN-THE-CLAY, *Ivel, Lea*, 52/080310
Basmead (see Bushmead)
BATTLESDEN, *Ouzel*, 42/970290
Beadlow, CLOPHILL, *Ivel*, 52/106385
Beckeringspark, MILLBROOK, *Ivel*, 52/005366
BEDFORD, *Ouse*, 52/050500
Beeston, SANDY, *Ivel*, 52/170480
BIDDENHAM, *Ouse*, 52/015500
Bidwell, HOUGHTON REGIS, *Ouzel*, 52/012248
BIGGLESWADE, *Ivel*, 52/190450
BILLINGTON, *Ouzel*, 42/940230
Birchmore, WOBURN, *Ouzel*, 42/945337
Biscot, LUTON, *Lea*, 52/075237
Blackgrove Wood, TILSWORTH, *Ouzel*, 42/980235
BLETSOE, *Ouse, Kym*, 52/022584
Blow's Downs, DUNSTABLE, *Lea, Colne*, 52/035215
BLUNHAM, *Ouse*, 52/150510

BOLNHURST, *Ouse*, 52/088598
Box End, KEMPSTON, *Ouse*, 52/010490
Bradger's Hill, LUTON, *Lea*, 52/095237
Bramingham, STREATLEY, *Lea*, 52/070260
Bramingham Shott (see Wardown)
Briar Stockings, EVERSHOLT, *Ivel*, 42/990340
Brogborough, RIDGMONT, *Ouse*, 42/960390
BROMHAM, *Ouse*, 52/010510
Broom, SOUTHILL, *Ivel*, 52/170430
Buckle Grove, SILSOE, *Ivel*, 52/086344
Bunker's Hill, SANDY, *Ivel*, 52/190480
Bushmead, EATON SOCON, *Ouse*, 52/115605
Byslip Wood, STUDHAM, *Colne*, 52/032164
CADDINGTON, *Colne*, 52/065197
Cadwell, HOLWELL (Herts.) *Ivel*, 52/189325
Cainhoe, CLOPHILL, *Ivel*, 52/100368
Caldecote, NORTHILL, *Ivel*, 52/165460
CAMPTON, *Ivel*, 52/130380
Candle Ford End (one of Abbot's stations, probably same as Ford End)
CARDINGTON, *Ouse*, 52/090480
CARLTON,[1] *Ouse*, 42/955552
Castle Mills, RENHOLD, *Ouse*, 52/090510
Cauldwell, BEDFORD, *Ouse*, 52/050494
CHALGRAVE, *Ouzel, Ivel*, 52/008274
Chalton, TODDINGTON, *Ivel*, 52/032265
Chapel End, CARDINGTON, *Ouse*, 52/095480
Chaul End, CADDINGTON, *Lea, Colne*, 52/055215
Chawston, EATON SOCON, *Ouse*, 52/160560
CHELLINGTON,[1] *Ouse*, 42/956558

[1] CARLTON and CHELLINGTON are joined for administrative purposes.

[1] DEAN and SHELTON are joined for administrative purposes.

High Down, PIRTON (Herts.), *Ivel*,
52/145303
HIGHAM GOBION, *Ivel*, 52/105330
Hill Plantation (one of McLaren's
stations which I am unable to
identify)
Hillfoot Farm, CARDINGTON, *Ouse*,
52/108449
Hinwick, PODINGTON, *Nene*, 42/935620
HOCKLIFFE, *Ouzel*, 42/970270
HOLCOT,[1] *Ouse*, *Ouzel*, 42/945390
Hollington, FLITTON, *Ivel*, 52/065366
Hollwell Bury Field, UPPER STONDON,
Ivel, 52/166345
HOLWELL (Herts.), *Ivel*, 52/165332
Honey Hills, BEDFORD, *Ouse*, 52/025492
Honeydon, EATON SOCON, *Ouse*,
52/125590
Horsemoor Farm, WOBURN, *Ouzel*,
42/937335
Horsley's (Hostler's) Wood, HYDE, *Lea*,
52/130190
HOUGHTON CONQUEST, *Ouse*, 52/044415
HOUGHTON REGIS, *Ouzel*, *Lea*, 52/020240
Howbury, RENHOLD, *Ouse*, 52/095515
Hunter's Nursery (a station given by
A. B. Sampson which I cannot
identify)
HUSBORNE CRAWLEY, *Ouzel*, 42/955360
HYDE, *Lea*, *Colne*, 52/128174
Ickwell, NORTHILL, *Ivel*, 52/150455
Ireland, SOUTHILL, *Ivel*, 52/134415
Isle of Wight Farm, KENSWORTH, *Colne*,
52/018188
Jackdaw Hill, LIDLINGDON, *Ouse*,
42/995383
Judge's Spinney, OAKLEY, *Ouse*,
52/018542
Keeper's Warren, SOUTHILL, *Ivel*,
52/130420
KEMPSTON, *Ouse*, 52/035480
KENSWORTH, *Colne*, 52/030190
KEYSOE, *Kym*, 52/077630
Kidney Wood, HYDE, *Lea*, *Colne*,
52/092192
Kimbolton Station, TILBROOK (Hunts.),
Kym, 52/088710
King's Mead Marsh, BEDFORD, *Ouse*,
52/078492
King's Wood, HEATH AND REACH, *Ouzel*,
42/930300
King's Wood, HOUGHTON CONQUEST,
Ouse, 52/045404
Kinsbourne Green, HARPENDEN
(Herts.), *Colne*, 52/110160
Kitchen End, PULLOXHILL, *Ivel*,
52/074332
Knapwell Hill, PODINGTON, *Nene*,

Knocking Hoe, SHILLINGTON, *Ivel*,
52/132312
KNOTTING,[1] *Kym*, 52/003635
LANGFORD, *Ivel*, 52/187410
Leagrave, LUTON, *Lea*, 52/065240
Leete Wood, BARTON, *Ivel*, 52/085295
LEIGHTON BUZZARD, *Ouzel*, 42/920250
LIDLINGTON, *Ouse*, 42/990390
Lilley Hoo, LILLEY (Herts.), *Lea*,
52/123280
Limbersey Lane, MAULDEN, *Ivel*,
52/060395
Litany, TOTTERNHOE, *Ouzel*, 42/980232
LITTLE BARFORD, *Ouse*, 52/180570
Little Billington (see BILLINGTON)
LITTLE STAUGHTON, *Ouse*, *Kym*,
52/100625
Longholme, BEDFORD, *Ouse*, 52/058495
Long Lane, TINGRITH, TODDINGTON,
Ivel, 52/010294
Long Wood, STUDHAM, *Colne*, 52/023147
Lord's Hill, CHALGRAVE, HOUGHTON
REGIS, *Ouzel*, 52/010260
Lowe's Wood, WOBURN, *Ouzel*,
42/927325
LUTON, *Lea*, *Colne*, 52/090210
Luton Hoo, HYDE, *Lea*, *Colne*,
52/105185
Lynches, LUTON, *Lea*, 52/055219
Maiden Bower, HOUGHTON REGIS, *Ouzel*,
42/997225
Maiden Common, LUTON, *Lea*,
52/074226
Market Street (see MARKYATE)
Markham Hills, STREATLEY, *Ivel*,
52/064292
MARKYATE (Herts.), *Colne*, 52/060160
Marslets, LUTON, *Lea*, 52/103205
MARSTON MORTAINE,[2] *Ouse*, 42/995415
MAULDEN, *Ivel*, 52/050380
Maulden Firs, STREATLEY, *Lea*,
52/093275
Mead Hook, PULLOXHILL, *Ivel*,
52/065330
Medbury, ELSTOW, *Ouse*, 52/060465
MELCHBOURNE,[3] *Kym*, 52/030660
MEPPERSHALL, *Ivel*, 52/137363
Mermaid's Pond, ASPLEY GUISE, *Ouzel*,
42/938347
MILLBROOK, *Ouse*, *Ivel*, 52/014385
Millow, DUNTON, *Cam*, 52/230435
MILTON BRYAN, *Ouzel*, 42/970310
MILTON ERNEST, *Ouse*, 52/020560
Mob's Hole, EYEWORTH, *Cam*,
52/262438
MOGGERHANGER, *Ouse*, 52/140490

[1] HOLCOT and SALFORD are joined for administrative purposes.

[1] KNOTTING and SOULDROP are joined for
administrative purposes.
[2] Spelt variously, Marston Moreteyne,
Marston Moretaine, etc.
[3] MELCHBOURNE and YELDEN are joined for
administrative purposes.

---

[1] HOLCOT and SALFORD are joined for administrative purposes.

[2] DEAN and SHELTON are joined for administrative purposes.

[3] KNOTTING and SOULDROP are joined for administrative purposes.

[1] MELCHBOURNE and YELDEN are joined for administrative purposes.

INDEX OF PLANT NAMES

Names of groups higher than genera are printed in LARGE CAPITALS and of the principal genera in SMALL CAPITALS. Specific epithets of Bedfordshire plants in the principal genera, other Bedfordshire species and English names of plants are printed in ordinary roman type. Other species mentioned in the text and synonyms are printed in *italics*. The species of Fungi are not indexed as they are listed alphabetically in their genera in the text. When more than one reference is given for a genus of Fungi the second and subsequent references are to synonyms.

KK

LIST OF SUBSCRIBERS

Acton Technical College Library.
Frank W. Adams, Esq., Sheffield.
George W. Adams, Esq., Dunstable.
Laurence G. Adams, Esq., Dunstable.
L. J. Adams, Esq., Luton.
Messrs. Edward G. Allen & Son, Ltd., London, W.C.2.
H. G. Allen, Esq., Wootton, Northampton.
Miss Jean Allison, M.A., Sandy.
Claude E. A. Andrews, Esq., B.Sc., F.L.S., Moseley, Birmingham.
G. M. Ash, Esq., F.L.S., Haslemere, Surrey.
Mrs. M. R. Ashton, J.P., Dunstable.
K. T. Atkins, Esq., Bedford.
N. T. Bagshawe, Esq., M.A., Aspley Guise.
T. W. Bagshawe, Esq., F.S.A., F.R.Hist.S., F.M.A., Aspley Guise.
Messrs. Bailey Bros. & Swinfen, Ltd., London, W.C.2.
Messrs. James Bain, Ltd., London, W.C.2.
H. E. Bannister, Esq., Potten End, Herts.
H. F. Barnes, Esq., M.A., Ph.D., Bedford.
Miss F. M. Barton, Bath.
Miss I. C. Bates, Luton.
Col. G. A. Battcock, Carlton.
Col. G. A. Battcock, Bedford.
Mrs. Marjorie E. Bawden, Harpenden.
Bedford College for Women (University of London).
Bedford Corporation.
Bedford Corporation, Cecil Higgins Museum and Art Gallery.
Bedford Modern School Library.
Bedford Modern School Museum.
Bedford Public Library (T. Cooper, Esq., F.L.A.).
Bedford School.
Bedfordshire County Library.
Bedfordshire County Record Office.
Bedfordshire Education Committee.
Bedfordshire Education Committee, Bedford College of Physical Education.
Bedfordshire Education Committee, Sandy Secondary School.
Bedfordshire Natural History and Archaeological Society.

Bedfordshire Natural History Society and Field Club.
Bedfordshire Times (The Editor).
Bedfordshire Times Publishing Co., Ltd.
Miss D. Bexon, M.Sc., Nottingham University.
Harold Beynon, Esq., Sandy.
F. J. Bingley, Esq., M.A., Flatford Mill Field Centre.
Birmingham Public Libraries.
Dr. Kathleen B. Blackburn, King's College, Newcastle-upon-Tyne.
J. E. H. Blackie, Esq., M.A., F.R.E.S., Alconbury, Huntingdon.
Messrs. B. H. Blackwell, Ltd., Oxford.
E. T. Blundell, Esq., Houghton Regis.
Messrs. G. Blunt & Sons, Ltd., London, N.W.10.
Hylton Blythe, Esq., Flitwick.
R. A. Boniface, Esq., Chiswick, W.4.
Messrs. Bookland & Co., Bangor, North Wales.
The Bookshop, Luton.
Boots the Chemists, Luton.
N. L. Bor, Esq., C.I.E., M.A., D.Sc., Royal Botanic Gardens, Kew.
The Botanical Institute, Pavia, Italy.
Aubrey Boutwood, Esq., Toddington.
S. R. Bowden, Esq., Letchworth.
J. P. M. Brenan, Esq., M.A., Kew.
P. W. Briar, Esq., Ampthill.
Sister Brigid, B.Sc., Convent of the Holy Ghost, Bedford.
Fred. A. Brokenshire, Esq., Barnstaple.
W. J. Brook, Esq., Dunstable.
Miss Margaret Brown, Ilminster, Somerset.
Buckinghamshire County Library.
Miss Anne M. Buck, B.A., F.M.A., Prestbury, Cheshire.
Oliver Buckle, Esq., Worthing.
Miss Winifred F. Buckle, Watford.
Kenneth E. Bull, Esq., Tunbridge Wells.
Wilfred A. C. Bullock, Esq., B.Sc., Repton, Derby.
H. E. Bunker, Esq., Ashton Preston, Lancs.
Dr. R. C. L. Burges, M.A., F.L.S., Birmingham.
Messrs. Burgess' (Stationers) Ltd., Hitchin.

Alderman John Burgoyne, O.B.E., J.P., Luton.

J. H. Burgoyne, Esq., D.Sc., Luton.

John Burrell, Esq., Cranfield.

Lt.-Col. J. H. Busby, M.B.E., British Commonwealth Forces H.Q., Korea.

William E. Butcher, Esq., Sandy.

E. F. M. Butler, Esq., Hitchin.

Philip E. N. Butt, Esq., Caddington County School.

Professor P. A. Buxton, F.R.S., London School of Hygiene and Tropical Medicine.

Miss Dorothy A. Cadbury, Edgbaston, Birmingham.

Cambridge City Library.

Miss M. S. Campbell, F.L.S., Aberfeldy, Perthshire.

N. R. Campbell, Esq., Ph.D., Hertford.

Major J. W. Cardew, Bedford.

Miss Ruth Carey, Buxton.

Carlisle Public Library (Kenneth Smith, Esq., F.R.E.S., F.L.A., City Librarian).

Mrs. M. L. Carter, Hemel Hempstead.

C. Cartwright, Esq., Walsall.

V. H. Chambers, Esq., Ph.D., A.R.C.S., Luton.

S. Charlesworth, Esq., Ph.D., M.A., Luton.

John W. P. Chidell, Esq., Bembridge, Isle of Wight.

W. Church, Esq., B.Sc., Luton.

Edgar Clarke, Esq., Luton.

W. D. Coales, Esq., B.Sc., Dunstable.

H. Cole, Esq., Luton.

T. G. Collett, Esq., Ealing.

Miss Ann Conolly, M.A., University College of Leicester.

Conservatoire et Jardin botaniques, Geneva, Switzerland.

H. H. Cooper, Esq., Henlow.

Mrs. I. E. Cooper, Lower Stondon.

Henry Cork, Esq., London, S.E.23.

Miss Margaret Cornish, M.Sc., Croydon.

Countryside Libraries, Ltd., Hitchin.

Miss Peggy Cox, Luton.

Dr. William J. Cox, Watford.

Rev. J. C. Culshaw, M.A., Bath.

Miss J. Curle, Bletchley Park Training College.

P. W. Curnow, Esq., D.F.M., B.Sc., F.G.S., Glenbuchat, Aberdeenshire.

Ivo L. Currall, Esq., Luton.

A. A. Dallman, Esq., Doncaster.

Dame Alice Harpur School, Bedford.

D. Daniels, Esq., Harrow Weald, Middx.

R. W. David, Esq., Cambridge.

Miss G. H. Day, Harrold.

Mrs. Vera Day, Harpenden.

R. H. Deakin, Esq., Canford Magna, Dorset.

Messrs. Deighton Bell & Co., Ltd., Cambridge.

R. W. G. Dennis, Esq., Ph.D., Royal Botanic Gardens, Kew.

Miss E. Disney, Bedford.

C. S. Downer, Esq., Nottingham.

Miss Ursula K. Duncan, F.L.S., Arbroath, Angus.

J. S. Dunn, Esq., B.Sc., Northill.

Wilfred Durant, Esq., Bedford.

Miss O. N. M. Dyke, Luton.

Mrs. M. E. Dymond, Bedford.

E. S. Edees, Esq., M.A., Newcastle, Staffs.

Miss B. E. J. Edmunds, Hunton Bridge, Herts.

D. W. Elliott, Esq., Stagsden.

E. A. Ellis, Esq., Castle Museum, Norwich.

John F. Ferns, Esq., M.Sc., A.R.I.C., Kew.

F. Fincher, Esq., Bromsgrove, Worcs.

Dr. G. W. T. H. Fleming, F.L.S., Gloucester.

Herbert Fordham, Esq., Baldock.

Lt.-Col. C. J. Fox, T.D., Clacton-on-Sea.

C. E. Freeman, Esq., F.M.A., and Mrs. Freeman, Barton-in-the-Clay.

Miss L. Winifred Frost, B.Sc., Salford, 6.

J. C. Gardiner, Esq., London, S.W.5.

Rev. P. M. Garnett, Southampton.

Mrs. B. E. M. Garratt, Battle, Sussex.

F. Garrett, Esq., Luton.

W. P. Gatward, Esq., Harpenden.

Mrs. P. J. Gent, Wellingborough.

Miss M. E. Gibbs, M.A., F.Z.S., St. Albans.

Mrs. A. N. Gibby, B.Sc., A.R.I.C., Durham.

Andrew G. C. Gibson, Esq., King's Stanley, Glos.

John L. Gilbert, Esq., Wansford, Peterborough.

C. H. Gimingham, Esq., Ph.D., University of Aberdeen.

Isaac Godber, Esq., Willington.

Miss Joyce Godber, M.A., F.S.A., Willington.

Rev. R. H. Goode, M.A., Marston Mortaine.

Miss C. M. Goodman, Birmingham, 17.

J. Henry Gough, Esq., Buckingham.

R. A. Graham, Esq., F.L.S., Northwood, Middlesex.

Grassland Research Station, Stratford-on-Avon.

Universität Graz, Institut für systematische Botanik, Graz, Austria.
P. S. Green, Esq., B.Sc., Royal Botanic Garden, Edinburgh.
P. H. Gregory, Esq., Ph.D., D.Sc., Harpenden.
Frank C. Gribble, Esq., Bedford.
Donald Grose, Esq., Liddington, Wilts.
Mrs. V. M. Craigie Halkett, Bedford.
P. C. Hall, Esq., Erith, Kent.
F. D. Hanson, Esq., Birmingham, 27.
W. H. Hardaker, Esq., Birmingham, 17.
Raymond Harley, Esq., Broadwell, Glos.
Henry Ralph Harman, Esq., Luton.
Harvard University, Library Arnold Arboretum, Jamaica Plain, Mass., U.S.A.
Harvard University, Gray Herbarium, Cambridge, Mass., U.S.A.
Haslemere Educational Museum, Surrey.
Richard A. Haynes, Esq., B.A., Chorlton-cum-Hardy, Manchester.
J. H. Hawkins, Esq., M.Sc., Arnside, Westmorland.
Hawnes School, Haynes Park, Bedford.
Messrs. W. Heffer & Sons, Ltd., Cambridge.
Hendon Central Library, London, N.W.4.
W. N. Henman, Esq., Bedford.
F. N. Hepper, Esq., B.Sc., Royal Botanic Gardens, Kew.
Hereford City Library, Museum and Art Gallery.
H.M. Stationery Office, Edinburgh, 11.
Hertfordshire County Library.
Hertfordshire Natural History Society and Field Club.
Mrs. G. W. Higgs, Dunstable.
Hitchin Museum.
Mrs. R. B. Hobourn, Woburn.
Messrs. F. R. Hockliffe, Ltd., Bedford.
Miss Margaret Holden, B.Sc., Harpenden.
Sir Jim S. Holland, Bart., M.A., T.D., London, E.C.2.
County Borough of Huddersfield, Tolson Memorial Museum.
Ernest R. Hunt, Esq., Luton.
Huntingdon Grammar School.
Huntingdonshire County Library.
Cyril A. Hutson, Esq., Luton.
Imperial Chemical Industries, Ltd., Welwyn.
Miss E. M. C. Isherwood, Redhill, Surrey.
R. B. Ivimey-Cook, Esq., Dinas Powis, Glam.
E. Jackson, Esq., Toddington.

Major J. P. A. Jackson, London, S.W.1.
Professor Frank W. Jane, Ph.D., D.Sc., Royal Holloway College.
C. MacKechnie Jarvis, Esq., F.L.S., London, S.W.1.
Dr. D. M. Jeffreys, Bedford.
A. Clive Jermy, Esq., Brundall, Norfolk.
Miss Nora H. Johnson, B.Sc., Harpenden.
E. W. Jones, Esq., Ph.D., Imperial Forestry Institute, Oxford.
J. L. Jones, Esq., St. Albans.
A. C. Jordan, Esq., and Mrs. Jordan, Luton.
E. G. Kellett, Esq., M.A., D.Phil., Welwyn Garden City.
Kent County Library, Maidstone.
Douglas H. Kent, Esq., London, W.13.
Kettering Public Library.
H. A. S. Key, Esq., M.P.S., Bedford.
King's College (The Librarian), London.
E. Knott, Esq., Binbrook, Lincs.
F. W. Kuhlicke, Esq., B.A., F.S.A., A.M.A., Bedford.
Keith A. Larking, Esq., Dunstable.
B. R. Laurence, Esq., B.Sc., London, S.E.23.
Miss Joyce R. Lawrie, Sandy.
Sir Eric Leadbitter, C.V.O., Croydon.
The Leagrave Press, Ltd., Luton.
Leeds Public Libraries (F. G. B. Hutchings, Esq., F.L.A., City Librarian).
University College, Leicester (The Librarian).
Letchworth Museum and Art Gallery.
Messrs. H. K. Lewis & Co., Ltd., London, W.C.1.
R. Lewis, Esq., Kendall, Westmorland.
R. P. Libbey, Esq., B.Sc., King's Lynn.
The Linnean Society of London.
Miss Katherine D. Little, Hitchin.
Liverpool Botanical Society.
Liverpool Public Libraries.
Miss Cynthia Longfield, London, W.8.
J. E. Lousley, Esq., London, S.W.16.
J. F. Lovatt, Esq., Pathological Laboratory, Harpenden.
Lund University Library, Sweden.
Luton Committee for Education:
Beech Hill Secondary Boys' School (S. H. Porter, Esq.).
Beech Hill Secondary Girls' School (Miss M. Leigh Pirie).
Challney Secondary Boys' School (S. Lindley, Esq.).
Challney Secondary Girls' School (Miss D. M. Thomson).
Denbigh Road Secondary Boys' School (G. Tripp, Esq.).

Langley Street Secondary Girls' School (Mrs. W. M. Froud).

Luton Secondary Technical School (S. Charlesworth, Esq., Ph.D., M.A.).

Luton Grammar School (K. B. Webb, Esq., M.A., B.Sc.).

Old Bedford Road Secondary Boys' School (C. H. Warman, Esq., B.Sc.).

Stopsley Secondary Girls' School (Miss M. Millen).

Surrey Street Secondary Boys' School (C. Page, Esq.).

Luton Corporation Parks Department (R. J. English, Esq., F.Inst.P.A.).

Luton Public Libraries (F. M. Gardner, Esq., F.L.A.).

Luton Industrial Co-operative Society, Ltd., Bookshop.

The Mayor of Luton (Alderman H. C. Janes, J.P.).

Simon Luxemburg, Esq., Markyate, Herts.

David McClintock, Esq., M.A., Platt, Kent.

Dr. J. M. McCurdy, Ashton-in-Makerfield, Lancs.

Miss Irene MacKerness, Bedford.

Manchester Museum (R. U. Sayce, Esq., M.A., M.Sc., F.M.A.).

Sir Frederick Mander, M.A., B.Sc., F.E.I.S., J.P., Toddington.

The Albert R. Mann Library, Ithaca, N.Y., U.S.A.

Dr. Harold H. Mann, Woburn Experimental Station, Husborne Crawley.

F. J. Manning, Esq., and Mrs. Manning, Streatley.

Miss Mildred Marriott, Oxford.

Arthur E. Meeks, Esq., Luton.

R. Melville, Esq., Ph.D., Royal Botanic Gardens, Kew.

Harry Meyer, Esq., and Miss Doris Meyer, Letchworth.

E. A. Middle, Esq., Newport, Mon.

Miss G. E. Middleton, Bedford.

Dr. J. N. Mills, Didsbury, Manchester.

Dr. W. H. Mills, F.R.S., Jesus College, Cambridge.

Alderman Mrs. K. M. Milner, M.B.E., J.P., Luton.

E. Milne-Redhead, Esq., M.A., F.L.S., Petersham, Surrey.

Dr. H. Milne-Redhead, Mainsriddle, Dumfries.

Miss M. E. Milward, The King's School, Canterbury.

The Modern Book Co., London, W.2.

W. E. Montgomery, Esq., Forensic Science Laboratory, Nottingham.

John Moor, Esq., Teynham, Kent.

Dr. F. Joan Moore, Harpenden.

Miss B. M. C. Morgan, Horley, Surrey.

Mrs. R. H. Mortis, Hertford.

National Institute of Agricultural Engineering, Silsoe.

National Museum of Wales, Cardiff.

The Nature Conservancy, London, S.W.1.

Miss C. Neave, Bedford.

T. G. Newcomen, Esq., Ampthill.

H. W. Newman, Esq., Bromham.

The New York Botanical Garden, U.S.A.

Miss J. M. Nicholls, Bedford.

Herbert Nickols, Esq., Pelynt, Cornwall.

Mrs. C. J. Norman, Great Barford.

Northamptonshire Natural History Society and Field Club.

Northampton Public Libraries.

North Bedfordshire College of Further Education, Bedford.

North Wales University College, Bangor.

Miss M. Norton, Easebourne, Sussex.

Ivan J. O'Dell, Esq., Hull.

Offley Training College, Herts.

Oldham Public Libraries, Art Gallery and Museum (Miss T. Simpson, A.M.A., Director).

Onesters (1st Bedfordshire) Rover Crew (K. T. Madex, Esq.).

John Ounsted, Esq., M.A., Leighton Park School, Reading.

Oxford University: Department of Botany (The Librarian).

Ray Palmer, Esq., Rosehill, Cornwall.

W. E. Palmer, Esq., M.A., B.Sc., Yeovil.

Brian Patterson, Esq., Luton.

W. E. Peacock, Esq., Hitchin.

F. Pearson, Esq., Luton.

Mrs. A. T. Peppercorn, Amersham Common, Bucks.

Franklyn Perring, Esq., B.A., Cambridge.

Peterborough Public Library.

J. H. G. Peterken, Esq., F.L.S., Leytonstone, E.11.

John H. Peters, Esq., B.Sc., Bedford.

D. H. Phillips, Esq., M.Sc., M.I.Biol., States Experimental Station, Trinity, Jersey, C.I.

W. E. K. Piercy, Esq., B.Sc., Clifton.

Messrs. H. Pipler, Ltd., Bedford.

Mrs. E. M. Pitcher, Luton.

Mrs. W. E. Pollard, Luton.

D. L. H. Porter, Esq., London, N.W.1.

T. G. Y. Porter, Esq., Bedford.

C. T. Prime, Esq., M.A., F.L.S., Sanderstead, Surrey.
O. G. Prudden, Esq., Bedford.
Miss M. R. Pugh, Bromham.
G. D. W. Randall, Esq., Luton.
G. H. Rawlins, Esq., Kempston.
Robert C. Readett, Esq., Birmingham, 11.
Reading University Library.
Reading Public Libraries.
Basil W. Ribbons, Esq., B.Sc., Glasgow University.
Miss C. M. Rob, F.L.S., Thirsk, Yorks.
F. W. Roberts, Esq., F.R.I.C., Luton.
Mrs. G. H. Robinson, Stevington.
David C. Rochester, Esq., Luton.
Ian Henderson Rorison, Esq., Department of Botany, Oxford.
Dr. Effie M. Rosser, Manchester Museum.
Rothamsted Experimental Station, Harpenden.
Dr. S. P. Rowlands, Doncaster.
Royal Botanic Gardens, Edinburgh.
Royal Holloway College (The Librarian), University of London.
Mrs. B. H. S. Russell, London, S.W.1.
St. Andrews University, United College.
Sir Edward Salisbury, C.B.E., F.R.S., D.Sc., Royal Botanic Gardens, Kew.
R. E. Sandell, Esq., Devizes.
N. Y. Sandwith, Esq., M.A., F.L.S., Royal Botanic Gardens, Kew.
J. T. Sansom, Esq., Luton.
H. B. Sargent, Esq., F.R.E.S., Porthleven, Cornwall.
Mrs. Patrick Saunders, Gretton, Glos.
Dr. E. Scott, Ashford, Kent.
S. H. Scott, Esq., Luton.
Lady Severn, Wallingford, Berks.
R. Sharp, Esq., Flitton.
Mrs. L. Sharpe, Greenfield.
Peter J. Shipton, Esq., Bedford.
N. Douglas Simpson, Esq., M.A., F.L.S., Bournemouth.
Mrs. R. Skinner, Bromham.
Dr. W. J. L. Sladen, Edward Grey Institute, Botanical Gardens, Oxford.
W. A. Sledge, Esq., Ph.D., B.Sc., Leeds, 6.
W. G. Small, Esq., London, W.5.
Messrs. John Smith & Son (Glasgow) Ltd., Glasgow, C.2.
Messrs. W. H. Smith & Son, Ltd., Bookstall, Harpenden.
Messrs. W. H. Smith & Son, Ltd., Harpenden.
Messrs. W. H. Smith & Son, Ltd., Hemel Hempstead.
Messrs. W. H. Smith & Son, Ltd., London, W.C.2.

Messrs. W. H. Smith & Son, Ltd., Luton.
Messrs. W. H. Smith & Son, Ltd., St. Albans.
F. G. R. Soper, Esq., Bedford.
Mrs. G. M. Souster, Luton.
H. B. Souster, Esq., Luton.
Southampton Central Library (R. W. Lynn, Esq., A.L.A., Chief Librarian).
Southampton University: The Library (Miss M. I. Henderson, M.A.).
South London Botanical Institute, S.E.24.
F. A. Sowter, Esq., Leicester.
L. A. Speed, Esq., Bedford.
Messrs. J. Staddon & Son (Luton) Ltd., Luton.
Messrs. Stalker's Bookshop, Luton.
R. C. Starke, Esq., Luton
V. Edward Steers, Esq., Bedford.
Messrs. B. F. Stevens & Brown, Ltd., London, W.1.
Students' Bookshops, Ltd., Cambridge.
J. Sutherland-Smith, Esq., Bedford.
Eric L. Swann, Esq., King's Lynn.
T. W. Sweby, Esq., Luton.
Douglas Swinscow, Esq., Knebworth.
Sir Arthur Tansley, M.A., F.R.S., F.L.S., Grantchester, Cambridge.
F. J. Taylor, Esq., B.Sc., F.L.S., Queen Mary College.
G. Taylor, Esq., D.Sc., British Museum (Natural History).
Mrs. S. E. Taylor, Bedford.
Miss E. R. Tetley, Bedford.
Rev. Clifford W. Thomson, Marsworth, Herts.
Dr. Agnes E. Towers, Bedford.
Miss G. E. Toyer, Luton.
A. F. Twist, Esq., Chettle, Dorset.
Prof. D. H. Valentine, Ph.D., Durham University.
Bernard Verdcourt, Esq., B.Sc., F.L.S., Nairobi, Kenya.
Dr. G. M. Vevers, Whipsnade Park.
E. C. Wallace, Esq., Sutton, Surrey.
B. T. Ward, Esq., Chingford, E.4.
Watford Public Libraries, Watford.
L. C. H. Watkins, Esq., Broomedge, Cheshire.
E. V. Watson, Esq., Ph.D., University of Reading.
Mrs. W. Boyd Watt, F.Z.S., M.B.O.U., Bournemouth.
Miss Mary McCullum Webster, Macduff, Banff.
Mrs. B. Welch, B.Sc., Richmond, Surrey.
Wellingborough Public Library.
Sir Richard Wells, Bart., Felmersham.
Lady Wells, Felmersham.

Bernard West, Esq., Bedford.
Dr. Cyril West, O.B.E., Aylesford, Kent.
Westminster Public Library, W.C.2.
Messrs. Wheldon & Wesley, Ltd., London, W.1.
Major Simon Whitbread, D.L., J.P., Southill.
Mrs. A. White, Letchworth.
Harold White, Esq., J.P., Luton.
H. L. K. Whitehouse, Esq., M.A., Ph.D., Botany School, Cambridge.

Miss M. Whitmore, Bedford.
J. E. Willé, Esq., St. Albans.
Dr. W. B. Williams, Portslade, Sussex.
J. E. Woodhead, Esq., B.Sc., London, S.E.21.
G. H. Wyatt, Esq., Chorleywood, Herts.
Messrs. Wyman & Sons, Ltd., London, W.5.
D. P. Young, Esq., Ph.D., Sanderstead, Surrey.
Messrs. Henry Young & Sons, Ltd., Liverpool, 2.